Springer Proceedings in Materials

Volume 7

Springer Proceedings in Materials publishes the latest research in Materials Science and Engineering presented at high standard academic conferences and scientific meetings. It provides a platform for researchers, professionals and students to present their scientific findings and stay up-to-date with the development in Materials Science and Engineering. The scope is multidisciplinary and ranges from fundamental to applied research, including, but not limited to:

- Structural Materials
- Metallic Materials
- Magnetic, Optical and Electronic Materials
- Ceramics, Glass, Composites, Natural Materials
- Biomaterials
- Nanotechnology
- Characterization and Evaluation of Materials
- Energy Materials
- Materials Processing

To submit a proposal or request further information, please contact one of our Springer Publishing Editors according to your affiliation:

European countries: **Mayra Castro** (mayra.castro@springer.com)
India, South Asia and Middle East: **Priya Vyas** (priya.vyas@springer.com)
South Korea: **Smith Chae** (smith.chae@springer.com)
Southeast Asia, Australia and New Zealand: **Ramesh Nath Premnat** (ramesh.premnath@springernature.com)
The Americas: **Michael Luby** (michael.luby@springer.com)
China and all the other countries or regions: **Mengchu Huang** (mengchu.huang@springer.com)

More information about this series at http://www.springer.com/series/16157

T. Rajmohan · K. Palanikumar · J. Paulo Davim
Editors

Advances in Materials and Manufacturing Engineering

Select Proceedings of ICMME 2019

Springer

Editors
T. Rajmohan
Department of Mechanical Engineering
SCSVMV
Kanchipuram, Tamil Nadu, India

K. Palanikumar
Department of Mechanical Engineering
Sri Sairam Institute of Technology
Chennai, Tamil Nadu, India

J. Paulo Davim
Department of Mechanical Engineering
University of Aveiro
Aveiro, Portugal

ISSN 2662-3161 ISSN 2662-317X (electronic)
Springer Proceedings in Materials
ISBN 978-981-15-6266-2 ISBN 978-981-15-6267-9 (eBook)
https://doi.org/10.1007/978-981-15-6267-9

This Springer imprint is published by the registered company Springer Nature Singapore Pte Ltd.
The registered company address is: 152 Beach Road, #21-01/04 Gateway East, Singapore 189721, Singapore

Committee Members

Chief Patron

Pujyasri Sankara Vijayendra Saraswathi Swamigal

Patrons

Prof. Dr. S. Jayarama Reddy, Chancellor, SCSVMV
Prof. Dr. Vishnu Potty V. S., Vice-Chancellor, SCSVMV
Prof. Dr. G. Srinivasu, Registrar, SCSVMV

Chairman

Prof. Dr. G. Sriram, Dean, Engineering and Technology, SCSVMV

Convenor

Dr. T. Rajmohan, Head, Department of Mechanical Engineering, SCSVMV

Organizing Secretaries

Dr. R. Vinayagamoorthy, Assistant Professor, Department of Mechanical Engineering, SCSVMV
Dr. S. Vijaya Bhaskar, Assistant Professor, Department of Mechanical Engineering, SCSVMV

Editorial and Technical Committee

Preface

Advances in Materials and Manufacturing Engineering comprise selected papers from the fourth International Conference on Materials and Manufacturing Engineering (ICMME 2019) at Sri Chandrasekharendra Saraswathi Viswa Mahavidyalaya, Kanchipuram, India. This book covers a wide range of topics within the materials and manufacturing disciplines. The content includes technical papers and review articles mainly in two different areas, namely synthesis, development and characterization of new materials and machining-associated studies on different materials. In addition, the book also covers some recent advancements in optimization techniques, computational fluid dynamics, tribology, alternate fuels, turbomachinery, renewable energy, thermal engineering and metal forming. The editorial team's expertise covers a variety of aspects of mechanical engineering. This collective expertise enabled the team to manage the editorial process efficiently in particular by providing a high-quality review process. This book will be a technological and scientific platform for researchers, scientists, engineers and academicians seeking advancements in the area of materials and manufacturing engineering.

Kanchipuram, India
Chennai, India
Aveiro, Portugal

T. Rajmohan
K. Palanikumar
J. Paulo Davim

Acknowledgements We express our sincere gratitude to the management of Sri Chandrasekharendra Saraswathi Viswa Mahavidyalaya, for giving us the opportunity to organize this international conference and also thank all the delegates for making this event a grand success. We thank the keynote speakers, conference chairs, advisory, program, and technical committees, reviewers, colleagues, college administration, our friends, students and well-wishers for making this event a memorable one. We thank Springer publications for having consented to publish our papers through a book. We thank one and all.

About the Conference—ICMME 2019

The fourth International Conference on Materials and Manufacturing Engineering (ICMM-2019) is organized by the Department of Mechanical Engineering, Sri Chandrasekharendra Saraswathi Viswa Mahavidyalaya, Kanchipuram, India, during March 21 and 22, 2019. The conference is aimed at providing a common platform for researchers, industry personnel, academicians, students and participating professionals to interact and discuss the advances made in various areas of Materials and Manufacturing Engineering. Special invited lectures by scientists and experts from foreign universities, leading institutions, research organizations and industries have been planned in addition to paper presentation. The conference will focus on today's technical challenges, research updates and breakthrough innovations that are shaping the future of Materials and Manufacturing Engineering. The conference convenes engineers, scientists and technologists for the purposes of exploring solutions to global challenges and for the advancement of Materials and Manufacturing Engineering worldwide.

SRI CHANDRASEKHARENDRA SARASWATHI VISWA MAHAVIDYALAYA
(Deemed to be University U/S 3 of Act 1956)
(Accredited with "A" Grade by NAAC) Approved by AICTE, New Delhi
Enathur, Kanchipuram - 631561. Tamilnadu, India.
FOURTH INTERNATIONAL CONFERENCE
ON
MATERIALS AND MANUFACTURING
ENGINEERING

FOURTH INTERNATIONAL CONFERENCE
MATERIALS AND MANUFACTURING
ENGINEERING
ICMME 2019
SCSVMV

Contents

About the Editors

Dr. T. Rajmohan is currently working as an Associate Professor and Head for the Department of Mechanical Engineering, SCSVMV Deemed to be University, Kancheepuram. He obtained his Bachelors in Mechanical Engineering from the University of Madras and Masters in Production Engineering from Annamalai University, Chidambaram. He did his research in the area of Machining of composites and received his Doctoral degree from SCSVMV Deemed to be University. His major areas of research includes material processing, machining and tribological behaviour of composite materials etc. He has published more than 75 research articles in refereed international journals and conferences and he is also serving as an editorial member for several journals in Elsevier, Sage, Taylor & Francis and many other reputed publications. He has more than 20 years of teaching experience and has more than 8 research scholars under guidance.

Dr. K. Palanikumar is currently working as Professor and Principal in Sri Sairam Institute of Technology, Chennai. He obtained his A.M.I.E in Mechanical Engineering from the Institution of Engineers, India and Masters in Production Engineering from Annamalai University, Chidambaram. He received his Doctoral degree from Anna University, Chennai and did his Post-Doctoral in the University of Averio, Portugal. His major areas of research include material processing and applications of composite materials, natural fibers, total quality management and optimization. He has published more than 260 research articles in refereed international journals and conferences and he is also serving as an editorial member for several journals in Elsevier, Sage, Taylor & Francis and many other reputed publications. He has more than 28 years of teaching experience and he has successfully guided more than 14 research scholars. He has successfully completed several funded research projects from ISTE, AICTE and DST and has more than 10 patents to his credits. He is being honoured with several awards during his carrier like Best Principal Award, Publons Peer review Award, Certified Sentinel of Science Award, Outstanding Reviewer Award, Best Teacher Award, Best Academic Researcher Award etc.

Dr. J. Paulo Davim, received his Ph.D. degree in Mechanical Engineering in 1997, M.Sc. degree in Mechanical Engineering (materials and manufacturing processes) in 1991, Mechanical Engineering degree (5 years) in 1986, from the University of Porto (FEUP), the Aggregate title (Full Habilitation) from the University of Coimbra in 2005 and the D.Sc. from London Metropolitan University in 2013. He is Eur Ing by FEANI-Brussels and Senior Chartered Engineer by the Portuguese Institution of Engineers with an MBA and Specialist title in Engineering and Industrial Management. Currently, he is Professor at the Department of Mechanical Engineering of the University of Aveiro, Portugal. He has more than 30 years of teaching and research experience in Manufacturing, Materials, Mechanical and Industrial Engineering with special emphasis in Machining & Tribology. He has also interest in Management, Engineering Education and Higher Education for Sustainability. He has guided large numbers of postdoc, Ph.D. and master's students as well as coordinated & participated in several financed research projects. He has received several scientific awards. He has worked as evaluator of projects for international research agencies as well as examiner of Ph.D. thesis for many universities.

He is the Editor in Chief of several international journals, Guest Editor of journals, books Editor, book Series Editor and Scientific Advisory for many international journals and conferences. Presently, he is an Editorial Board member of 25 international journals and acts as reviewer for more than 80 prestigious Web of Science core collection journals. In addition, he has also published as editor (and co-editor) more than 100 books and as author (and co-author) more than 10 books, 80 book chapters and 400 articles in journals and conferences (more than 200 articles in journals indexed in Web of Science core collection/h-index 45+/6500+ citations and SCOPUS/h-index 53+/8500+ citations).

Impact Model for Grinding Process in the Framework of Sustainable Manufacturing

Jigneshkumar M. Parmar, Chetankumar M. Patel, and Ajitkumar N. Shukla

Abstract Exposure to metalworking fluids (MWF) and grinding process is common in machining processes across the globe and leads to skin and respiratory disease to the operator. The aim of the study is to propose an impact model resulting due to the exposure and adverse effect of the grinding process on the health of the operator as studied in Rajkot, India. The impact model generated gives the guideline to perform grinding process safely and make it more sustainable. Pulmonary function test (PFT) and peripheral capillary oxygen saturation (S_PO_2) were taken for building this model, performed on 100 grinding operators and 100 control group subjects to compare the results. Various parameters considered for the study are forced expiratory volume in 1 s (FEV_1), forced vital capacity (FVC), ratio of forced expiratory volume in 1 s to forced vital capacity (FEV_1/FVC), peak expiratory flow rate (PEFR) and peripheral capillary oxygen saturation (S_PO_2). It successfully shows that 80 grinding operators and 12 control group subjects have low lung functions. This is the first investigation describing the exposure to MWF and grinding process and its adverse effect on the health of the grinding operators in Rajkot, India. In general working condition in the companies were acceptable, but the negligence of the safety and health precautions was a major reason behind the adverse effect.

Keywords Metalworking fluid · Grinding process · Occupational hazards · Pulmonary function test · Peripheral capillary oxygen saturation

1 Introduction

Grinding process is the primary finishing process, which performed on all the machined parts to get the better surface finish. To get better results, metalworking fluid (MWF) is used in the grinding process. During the grinding process, material

J. M. Parmar (✉) · C. M. Patel
RK University, Rajkot 360020, India
e-mail: parmarjignesh.m@gmail.com

A. N. Shukla
Pranveer Singh Institute of Technology, Kanpur 209305, India

T. Rajmohan et al. (eds.), *Advances in Materials and Manufacturing Engineering*, Springer Proceedings in Materials 7, https://doi.org/10.1007/978-981-15-6267-9_1

removed in the form of powder and due to the use of MWF mist gets generated, which may enter into the respiratory system of the operator. This is a prime hazard for the grinding operators, as it led to respiratory disease. The accuracy obtained from the grinding process and the surface finish is far better than the turning and milling operation [1].

The MWF limits heat generation by reducing the amount of friction in the grinding area as a result of its lubricating properties. It reduces the heat by putting some energy into the liquid instead of the workpiece. Thus, cold fluid is more effective for heat transfer [2]. In addition, the MWF is used to remove the chip from the grinding operation [3, 4]. The grinding process is either dry or wet. The material removed through the dry process is in the form of a powder which generally enters into the respiratory of the operator. The wet grinding process laden with MWF is even more dangerous as it produces mist during the operation of the process but gives the better surface and extended tool life. During the grinding process, such a situation leads to hazards like skin irritation and respiratory issues to the operators. Such symptoms are common in the grinding operators after long exposure to a similar working environment. The study on agate grinders shows the risk of digestive and respiratory during the grinding [5, 6].

2 Material

The present study is performed to know the adverse effect of the grinding process on the grinding operators of a diverse age group and build a model. Hence, two different groups of subjects are examined, (i) subjects employed in the grinding shop of various industries of diverse age group and (ii) subjects having no exposure of the grinding process.

To perform this study, an ethical approval (RKU/SPT/2016/08/27, Dated 05/08/2016) was taken prior to the performance from the ethics committee of School of Physiotherapy, RK University (ECR/259/Indt/GJ/2016).

This study is performed in accord with the consent of the organization to retain confidentiality and the consent taken from each subject to participate in this study. The study begins with the identification of the subjects on the random sampling basis, for which the whole process in depth was explained to the subject. They were made aware of the consequence as well as the purpose of the study.

All the subjects were examined using tools as under:

- Pulmonary Function Test (PFT): This test is a practical examination of lung function. It determines forced expiratory volume in 1 s (FEV_1), forced vital capacity (FVC), forced expiratory volume in 1 s to forced vital capacity (FEV_1/FVC), peak expiratory flow rate (PEFR).
- Peripheral Capillary Oxygen Saturation (S_PO_2): This test gives the oxygen saturation rate in the blood of the subject.
- Weighing Machine
- Measure Tape

3 Method

In this study, spirometric data of 200 persons were employed, as 100 grinding operators, working for 8–10 h a day on grinding machine and 100 subjects having no grinding exposure of the grinding process formed the control group. Subjects were called randomly, and a demo performance was given. The hygienic mouthpiece was employed to train the subject for the exhalation and inhalation under normal condition. Later, they were requested to inhale through the mouthpiece connected with the apparatus to get the result of the test in a minimum of three trials. Afterward, they were asked to make inhalation under slow maneuver condition. The value of FEV_1, FVC, FEV_1/FVC, and PEFR was recorded and examined. Statistical analysis was performed using Office 365. Mean values and SD of age, height, weight, and exposure to the grinding environment (EGE) were computed for both groups. Statistical significance was evaluated using analysis of variance (ANOVA).

4 Result and Discussion

Table 1 shows the demographic characteristics of the grinding operators and control group. There was no substantial change in age, height, and weight apart from Exposure to Grinding Environment (EGE).

Table 2 shows the spirometric statistics of the grinding operators and control group. The predicted values calculated using spirometer reading and measured values are expressed in the absolute units and as a percentage of the predicted values. The values of the grinding operators and control group were compared.

Table 3 shows the status of the pulmonary functions in the grinding operators and control group. The results of the pulmonary functions propose that 80 grinding operators suffered from some pulmonary disorder.

Table 4 shows the statistics of the S_PO_2 in both groups. The result suggests that there is no impact of the EGE on oxygen saturation rate of any subjects of both groups.

Table 1 Details of grinding operators and control group

	Group	
	Grinding operators (n = 100)	Control group (n = 100)
Age[a], y	30 ± 10	21 ± 3
Height[a], cm	165 ± 7	170 ± 7
Weight[a], kg	60 ± 11.5	60 ± 13
EGE[a], y	8 ± 8	0

Note [a]Indicates M ± SD

Table 2 Measured and predicted values of pulmonary functions in the grinding operators and control group

		FEV_1, L	FVC, L	FEV_1/FVC, %	PEFR, L/min
Grinding operator	Predictable[a]	2.9 ± 0.4	3.9 ± 0.4	85 ± 3	589 ± 1
	Measured[a]	3.0 ± 0.5	3.3 ± 0.5	92 ± 7	394 ± 111
	Measured predictable[a]	104 ± 16	97 ± 16	108 ± 8	74 ± 23
Control group	Predictable[a]	3 ± 0.3	4 ± 0.7	84 ± 5	467 ± 8
	Measured[a]	4 ± 0.3	5 ± 0.3	81 ± 6	490 ± 66
	Measured predictable[a]	86 ± 8	115 ± 10	76 ± 2	105 ± 33

Note [a]Indicates M ± SD

Table 3 Status of the pulmonary functions in the grinding operators and control group

	Status of pulmonary functions		
	Normal	Obstructive	Restrictive
Grinding operators (%)	20 (20)	1 (1)	79 (79)
Control group (%)	88 (88)	2 (2)	10 (10)

Table 4 S_PO_2 indices in the grinding operators and control group

	Group	
	Grinding operators	Control group
S_PO_2, %	97.8 ± 1.5	97.8 ± 1.1

Table 5 shows that operators considered for the study from Rajkot having average EGE are around 8 years. So, all the indicators are benchmarked with this average value of the exposure.

Table 6 shows the comparison of *F* and *P* value between two groups, which suggests the significant difference between the two groups.

Total 200 subjects, 100 grinding operators and 100 control group subjects satisfy all the inclusion and exclusion criteria and participated in PFT and S_PO_2.

In the restrictive lung disease, values of FVC, FEV_1, PEFR decreases, and FEV_1/FVC increased compared to the normal value. In the obstructive lung disease, FVC, FEV_1, FEV1/FVC, and PEFR decrease compared to the normal value. Hence,

Table 5 Details of EGE of the grinding operators

	EGE, y
Range	0.1 ± 35
Mean ± SD	8 ± 8

Table 6 Statistical analysis of pulmonary functions' indices and S_PO_2 in the grinding operators and control group

Index	*F*	p^a- values	*F* crit
FEV_1, L	165.97	5.56E–28	3.89
FVC, L	464.06	8.5E–54	3.89
FEV_1/FVC, %	139.4	1.04E–24	3.89
PEFR, L/min	54.45	4.3E–12	3.89
S_PO_2, %	0.01	0.92	3.89

Note [a]Indicates a significant difference between the grinding operators and control group. Statistical significance assessed by ANOVA

FEV_1/FVC is the main parameter to determine the lung disease, whether it is obstructive or restrictive [7].

The comparison of the FEV_1/FVC of both groups clearly suggests that the value of FEV_1/FVC is high in the majority of the grinding operators compared to control group subjects. It indicates that the majority (79%) of the grinding operators are suffering from restrictive lung disease compared to the control group (10%). The obstructive disease left out as only one in grinding operators and two subjects in the control group having a low ratio than the normal value. In the current study, the important variances were in the spirometric indices specified as a percentage of predicted values in grinding operators ($p < 0.001$).

5 Impact Model

Lack of fluid management program has contributed to potentially increasing the health risk associated with MWF oil mist. This reflects the need for an effective MWF program. The effective MWF program will include a comprehensive and well written Standard Operating Process (SOP) for all aspects of the program. The program should include employee's proper training, implementation guidelines, preventive maintenance procedures, and follow-up action. This is an active approach rather than a reactive. Adoption of SOP will solve many common problems [8].

Therefore, the strategy should be.

- Provide the manual of safety guidance that represents good practice.
- Provide various safety tools/gadgets that could be actively used in the workplace like, laminated task sheets that establish detail safety procedure for common tasks, posters containing dos and don'ts of the process, process monitoring strategies, control sheets detailing the key issues.
- Demonstrate that the effective management of MWF is good for employees' health as well as for the organization [9].

Recommended preventive action for the safe grinding process is:

- Ensure that the workers have been properly trained to use the machine safely and the grinder must be operated on the given standard operating range given by its manufacturer.
- Always wear proper eye and face protection, such as a full-face shield while using a grinder. Also wear protective shoes, adequate body cover clothes and hearing protectives.
- Regularly inspect the properly guarded guard and handle.

Recommended PPEs for grinding process are:

- Eye & face protection: safety spectacles/goggles, face shield (if needed)
- Foot & leg protection: leggings (in case of extreme heat generation), combination foot and shin guards, safety shoes
- Hand & arm protection: sleeves and gloves (leather gloves, aramid fiber gloves to be chosen as per material to grind with MWF)
- Body protection: apron
- Hearing protection: single-use earplugs/earmuff [10]
- Respiratory protection: particulate respirator/Dust mask [10, 11].

6 Conclusion

This study documented that 80 grinding operators having lower lung functions (79 restrictive and one obstructive type disorder) compare to 12 control group subjects (10 restrictive and two obstructive type disorder). In the control group subjects, this disorder may due to hierarchy or due to some other reason. This study documents the occupational changes in the pulmonary functions of the grinding operators compared to control group. To diagnose the precise hostile consequence of the grinding process on the operators' health, it is essential to study individually by the process, MWF, and material to be a grind. On the other hand, grinding does not produce only mist and powder in the form of a chip, and also, it generates the hazardous gases during the grinding process, depending upon the material, which very hazardous for the grinding operators. So, the further study required on the type of process, material to grind, MWF used and EGE on pulmonary functions. This study suggests the impact model including the guideline for the safe practice in the grinding shop. By implementing this model, one may have an improved working environment and the safe practice, which may save the grinding operators from the adverse effect of the grinding process on the health of the operators. This model may not eliminate all the hazards from the working area, though may provide a better and improved working environment, which may safer for the grinding operators and more sustainable toward the environment.

References

1. Irani RA, Bauer RJ, Warkentin A (2005) A review of cutting fluid application in the grinding process. Int J Mach Tools Manuf 45(15):1696–1705. https://doi.org/10.1016/j.ijmachtools.2005.03.006
2. Rowe WB (1990) Grinding technology-theory and applications of machining with abrasives. Tribol Int 23(6):443. https://doi.org/10.1016/0301-679x(90)90061-s
3. Ebbrell S, Woolley NH, Tridimas YD, Allanson DR, Rowe WB (2000) Effects of cutting fluid application methods on the grinding process. Int J Mach Tools Manuf. https://doi.org/10.1016/S0890-6955(99)00060-7
4. Webster JA, Cui C, Mindek RB, Lindsay R (1995) Grinding fluid application system design. CIRP Ann Manuf Technol. https://doi.org/10.1016/S0007-8506(07)62337-3
5. Rafeemanesh E, Majdi MR, Ehteshamfar SM, Fahoul MJ, Sadeghian Z (2014) Respiratory diseases in agate grinding workers in Iran. Int J Occup Environ Med
6. Sharma DB Patel TA Varshney AM (2011) A study on health aspects of agate workers in Shakarpura-Khambat Heal. J Indian Assoc Prev Soc Med
7. Hyatt RE, Scanlon PD, Nakamura M (2009) Interpretation of pulmonary function tests—a practical guide, 3rd edn. Wolters Kluwer (India) Pvt. Ltd., New Delhi
8. Gauthier SL (2003) Metalworking Fluids: Oil Mist and Beyond. Appl Occup Environ Hyg 18:818–824. https://doi.org/10.1080/10473220390237313
9. UKLA (2018) Good practice guide for safe handling and disposal of metalworking fluids. 45. https://doi.org/10.1016/j.cam.2009.02.104
10. Creek C, Drive N (2009) Personal protective
11. OSHA Bulletin: general respiratory protection guidance for employers and workers—occupational safety and health administration, https://www.osha.gov/dts/shib/respiratory_protection_bulletin_2011.html. Last Accessed 13 Feb 2019

Execution Analysis of Vapor-Compression Refrigeration System Using R12, R134a, R290 and R600a as Working Medium

M. Sivakumar and S. Mahalingam

Abstract In this paper, the execution of vapor pressure refrigeration framework is broke down with R12, R134a, R290 and R600a as working liquid. At the point when R12 refrigerant is utilized as working liquid, its outcome is high ozone exhausting potential and an unnatural weather change potential. Different exploratory examinations were completed for long haul substitution of R134a rather than R12 on account of zero ozone exhausting potential, non-combustibility, security and comparable vapor weight as that of R12. R134a has zero ozone consumption layer however there is a high a dangerous atmospheric deviation. But R600a has zero ozone exhaustion layer and negligible global warming. Unadulterated R290 or R600a is not reasonable for direct swap for R134a in light of their immersion properties. Any blends of R134a/R290/R600a at any mole part give better execution and COP level when contrasted with R12.

Keywords Vapor-compression refrigeration system · Refrigerants · COP · And compressor

1 Introduction

Refrigeration is the study of delivering and keeping up temperature beneath that of the encompassing environment. This implies the expelling of warmth from a substance to be cooled. Warmth constantly passes downhill, from a warm body to a cooler one, until the point when the two bodies are at a similar temperature. Keeping up perishables at their required temperature is finished by refrigeration. Perishables as well as today numerous human work spaces in workplaces and production line structures are air contingent also; a refrigeration unit is the core of the framework.

M. Sivakumar (✉)
Department of Mechanical Engineering, Annai Mathammal Sheela Engineering College, Namakkal 637013, India
e-mail: rvssiva@gmail.com

S. Mahalingam
Department of Mechanical Engineering, Sona College of Technology, Salem 636005, India

T. Rajmohan et al. (eds.), *Advances in Materials and Manufacturing Engineering*, Springer Proceedings in Materials 7, https://doi.org/10.1007/978-981-15-6267-9_2

In basic, refrigeration implies the cooling of or expulsion of warmth from a framework. The gear utilized to keep up the framework at a low temperature is named as refrigerating framework, and the framework which is kept at lower temperature is called refrigerated framework. Refrigeration is for the most part delivered in one of the accompanying three different ways: (i) by liquefying of a strong, (ii) by sublimation of a strong, (iii) by dissipation of a fluid. A large portion of the business refrigeration is delivered by the dissipation of a fluid called refrigerant. It is utilized for conservation of sustenance, assembling of ice, strong carbon dioxide and control of air-molding framework. The refrigerants' thermodynamic properties broke down their exhibition in salt water-to-air and brackish water-to-water GSHPs, and talked about ongoing advancement in their utilization in GSHPs. Investigations of CO_2 were the most widely recognized because of its good properties, covering propelled cycles, direct extension, optional liquid and crossbreed GSHPs. In spite of the fact that with poisonousness concerns, NH_3 was the second generally examined, including fume pressure GSHPs for warming, ingestion-type GSHPs to take out ground irregularity and half-breed pressure assimilation GSHPs to augment the working temperature go. A couple of studies assessed water as a refrigerant for ingestion-type GSHPs, including applications for sun-based cooling, ground awkwardness and region warming. Propane was the main hydrocarbon considered for GSHPs, remembering examinations for refrigerant charge, execution investigation and propane as an auxiliaryliquid. We compared there frigerants' thermodynamic properties, analyzed their performance in brine-to-air and brine-to water GSHPs and discussed recent progress in their use in GSHPs. Though with toxicity concerns, NH_3 was the second most studied, including vapor-compression GSHPs for heating, absorption-type GSHPs to eliminate ground imbalance and hybrid compression-absorption GSHPs to widen the operating temperature range [1]. However, due to flammability and submission to severe regulations, the use of hydrocarbons requires improvements in design and optimization of components to reduce the refrigerant charge [2]. A theoretical performance study on a traditional vapor-compression refrigeration system with refrigerant mixtures based on HFC134a, HFC152a, HFC32, HC290, HC1270, HC600 and HC600a was done for various ratios, and their results are compared with CFC12, CFC22 and HFC134a as possible alternative replacements. In spite of the HC refrigerants' highly flammable characteristics, they are used in many applications, with attention being paid to the safety of the leakage from the system, as other refrigerants in recent years are not related with any effect on the depletion of the ozone layer and increase in global warming [3]. A performance analysis on a vapor-compression refrigeration system with various eco-friendly refrigerants of HFC152a, HFC32, HC290, HC1270, HC600a and RE170 was done, and their results were compared with R134a as possible alternative replacement. The results showed that the alternative refrigerants investigated in the analysis RE170, R152a and R600a have a slightly higher performance coefficient (COP) than R134a for the condensation temperature of 50 °C and evaporating temperatures ranging between −30 and 10 °C [4]. A performance analysis on a vapor-compression refrigeration system with various refrigerant mixtures of R152a, RE170, R600a and R290 which were done for various mixture ratio, and their results were compared with R134a as possible alternative replacement. The results showed that all of the alternative refrigerants investigated in the analysis except R431A, [R152a (29%), R290 (71%)] have a slightly higher performance coefficient (COP) than R134a for the condensation temperature of 50 °C and evaporating temperatures ranging between −30 and 10 °C. Refrigerant blend of R152a/RE170 (20/80 by wt%) instead of R134a was found to be a replacement refrigerant among other alternatives [5]. The parametric investigations of actual vapor-compression refrigeration (VCR) cycle in terms of coefficient of performance (COP), exergy destruction (ED) and exergetic efficiency of refrigerants R134a, R152a and RE170 have been predicted with the help of vapor-compression cycle design software (CYCLE_D4.0). The present investigation has been carried out for evaporator and condenser temperatures in the range of −30 °C to 10 °C and 25 °C to 50 °C, respectively. The overall efficiency defect in the refrigeration cycle working with RE170 is consistently better than those of R134a and R152a. It has also been found that at higher evaporating temperatures, the exergy losses are minimal for the refrigerants in the four components. The results indicate that COP and exergetic efficiency for RE170 are higher in comparison with R134a and R152a [6]. The experimental analysis allowed the determination of cooling capacity, the electrical power absorbed, the COP and other variables characterizing the working of the plant. The results demonstrated that the cooling capacity for R422D was lower than for R22, while the electrical power absorbed with R422D was higher than that with R22. As a consequence, the COP of R422D was lower than that of R22 [7]. Long-period food storage vertical freezers are small cooling capacity appliances usually working with HC. The particular system architecture and the small thermal load allow the use of natural draft wires and tube heat exchangers with large heat transfer areas. Despite of the good performance achieved, high refrigerant charges with respect to the cooling capacity of the system are required, due to the high volumes of the heat exchangers. In this work, an experimental analysis of an apparatus with a typical arrangement of the heat exchangers and capillary tube is presented. The system was first designed according to the current standards to reach high performances. Then, the working conditions of the system were completely characterized experimentally in the whole range of refrigerant charges compatible with steady operation [8].

2 Refrigerants

Refrigerant is characterized as any substance that ingests warm through extension or vaporization and loses it through buildup in a refrigeration framework. The term refrigerant in the broadest sense is additionally connected to such optional cooling mediums as virus water or saline solution, arrangements. These substances retain warm at one place at low temperature level and reject the equivalent at some other place having higher temperature and weight. The dismissal of warmth happens at the expense of some mechanical work. Consequently, circling cold mediums and cooling mediums are not essential refrigerants. In the early day's just four refrigerants, Air, Ammonia (NH3), Carbon dioxide (CO_2), Sulfur dioxide (SO_2), having substance, physical and thermodynamics properties allowing their proficient application and administration in the useful plan of refrigeration hardware were utilized. All the refrigerants change from fluid state to fume state during the procedure.

A. *R12 (Chlorofluorocarbon)*

The refrigerants chlorofluorocarbon (CFCs) and hydro-chlorofluorocarbon (HCFCs) both have high ozone draining potential (ODP) and an Earth-wide temperature boost potential (GWP) and contribute to ozone layer consumption and an Earth-wide temperature boost.

B. *R134a (Tetrafluoroethane)*

The ODP of R134a is zero, yet it has a moderately high an unnatural weather change potential. Numerous examinations are being completed which are focusing on the utilization of condition count agreeable refrigerants in refrigeration frameworks.

C *R290 (Propane)*

R290 is the normal name for high immaculateness propane (C3H8) reasonable for use in the refrigeration and cooling industry. The item ordinarily is in any event 99.5% unadulterated with negligible dimensions of basic polluting influences including dampness and unsaturated hydrocarbons.

D. *R600a (Isobutane)*

This work exhibits an investigation on the utilization of HFC134a to supplant R600a in a residential cooler. A fridge intended to work with R600a with a gross limit experiment. R600a low weight level is associated with a generally high basic temperature, great performance, increased effectiveness, zero ozone exhaustion and irrelevant worldwide cautioning potential (Table 1).

3 Experimental Setup of Vapor-Compression Refrigeration System

The framework under investigation was made out of five fundamental parts, i.e., a blower, an evaporator, a condenser, fine cylinders and a fluid line filter–drier. In this examination, the following were utilized arrangement of a three-stage, 220 V, responding blower. The information intensity of the blower inside the framework differed somewhere in the range of 230 and 300 W.

Globally, the production of cold housing is seen as a major energy challenge of this new century. The economic development of developing countries, submitted their majority in hot climates, will lead to a growing demand chilling requirements. However as of now, the creation of cold arrangements is essentially founded on refrigeration frameworks significant customers of electrical vitality. For some years

Table 1 Properties of R12, R134A, R290 and R600a

Sl. No.	Refrigerant	R12	R134A	R290	R600a
1	Sub-atomic mass (g)	100	100	100	100
2	Breaking point at barometrical weight 14.7 Psia, 1 bar abs (°F)	−21.8	−26.1	−44	−31.2
3	The point of solidification at barometrical weight 14.7 Psia, 1 bar abs (°F)	−252	−280.5	−309.8	−217
4	*Basic Point of R12, R134a, R290 and R600a*				
	Temperature (°F)	234	220	206	306
	Weight (Psia)	597	607	617	551
	Explicitvolume (CU.Ft/lb)	0.0287	0.0507	0.0728	0.0702
5	Inactive warmth of KJ/Kg	165.24	216.87	423.33	364.25
6	Touchy limits in air, % by volume	Non-flammable	Non-flammable	2.3–7.3	1.8–8.4
7	ODP	0.82	0	0	0
8	GWP	8100	1300	20	20

now, because of their impact on the environment, the use of halogenated refrigerants has been progressively subject to quotas. In this context, the use of “natural” refrigerants becomes a possible solution. The solution to the environmental impacts of refrigerant gases would therefore pass by a gas which contains no chlorine, no fluorine and does not reject any CO_2 emissions in the atmosphere, in brief a green gas. The aim of our project is to contribute to the protection of our environment [9].

The other evaluating and controlling fragments were used in the structure, that were, an electrical switch, an imperativeness meter, a voltmeter, an ampere meter, a modernized indoor controller for controlling the evaporator temperature, bourdon tube type low weight measure and high weight check, thermocouples and pointer what’s more, gas stream control valves (Fig. 1).

4 Experimental Procedure

The temperatures and weights of the refrigerant and beginning water temperatures were estimated at different areas in the exploratory setup. The blower vitality utilization is estimated utilizing a vitality meter. Three manual sort development gadgets are utilized to manage the mass stream rate of refrigerant and to set weight contrast. The refrigerant is charged after the framework had been emptied. The refrigerant

Fig. 1 Experimental setup of vapor-compression refrigeration system

R290/R600a is azeotropic mix, which is charged in the fluid stage because of its creation move and temperature float. Drop-in tests are completed with no alterations to the test mechanical assembly. The investigation is begun with R134a to set up the base reference for further examinations with the other two refrigerants. The warm heap of the framework is changed with an outer electrically warmed unit.

A trial concentrate on the substitution of R12 and R134a by the new R290/R600a refrigerant blend as drop-in supplanting refrigerant with and without the impact of attractive field. With no change to the framework segments, drop-in exploratory tests were performed on a fume pressure refrigeration framework with a responding blower, which was initially intended to work with R12. Without any modification to the system components, drop-in experimental tests were performed on a vapor-compression refrigeration system with a reciprocating compressor, which was originally designed to operate with CFC12. The test results with no magnets showed that the refrigerant R290/R600a had 19.9–50.1% higher refrigerating capacity than R12 and 28.6–87.2% than R134a. The mixture R290/R600a consumed 6.8–17.4% more energy than R12. The coefficient of performance of R290/R600a mixture increases from 3.9–25.1% than R12 at lower evaporating temperatures and 11.8–17.6% at higher evaporating temperatures [10].

5 Result and Discussion

The trial results got from the execution examination of R134a, R290 and R600a are talked about as for COP. The exploratory aftereffects of the parameters influence the COP of framework. A greater part of refrigeration frameworks in the India is utilizing R134a as their refrigerant. In a refrigeration framework, the most costly segment is the blower. Along these lines if a surrogate to R134a is accomplished which could be utilized without the substitution of the blower, the substitute would be exceptionally practical. Therefore, the most critical execution parameter that is considered for choosing a particular structure from countless was the coordinating of the immersion properties. The immersion properties of the HC blend (50%R600a/50%R290) coordinate intently the immersion properties of R134a. Accordingly, the properties of blend included at the amount of R134a (100 g), R600a (100 g) and R290 (100 g). Consequently, any blend of R134a/R600a/R290 at any mole parts can have immersion properties near R134a. The proposed elective ternary blend can be considered as drop-in substitution for R134a iceboxes. Enable the framework to keep running for couple of minutes. The esteem position is chosen to development valve condition. The device is permitted to achieve the relentless state condition by seeing the temperature readings in the advanced temperature markers for around 5 min roughly (Table 2).

A test execution concentrates on a fume pressure refrigeration framework with the new R290/R600a refrigerant blend as drop-in supplanting was led and contrasted and CFC12 and HFC134a. The fume pressure refrigeration framework was at first intended to work with R12. Trial results indicated that the refrigerant R290/R600a had 19.9% to 50.1% higher refrigerating limit than R12 and 28.6% to 87.2% than R134a.The refrigerant R134a demonstrated somewhat lower refrigerating limit than R12. The blend R290/R600a devoured 6.8% to 17.4% more vitality than R12 [11].

At the point when the framework achieves unfaltering state, the temperature readings $T1$, $T2$ values are taken. A similar methodology is rehashed as taking another a few readings and the readings are organized. At starting refrigerant temperature of ($T1 = 20.5$ °C and $T2 = 19.5$ °C) in the framework, Fig. 5.1 demonstrates that COP estimation of R12. At conclusive refrigerant temperature of ($T1 = 19$°C and $T2 =$

Table 2 COP values of R12, R134A, R290 and R600a

Sl. No.	Refrigerants	COP
1	R12 (DICHLORODIFLURO METHANE)	2.267
2	R134A (TETRAFLUOROETHANE)	2.79
3	R290 (PROPANE)	
4	R600A (ISOBUTANE)	

Table 3 Performance improvement of COP value used R12, R134a and R290 and R600a

Sl. No.	Refrigerants	COP (%)
1	Rate improved Of COP value R12	15.39
2	Rate improved of COP values of (R134A, R290 and R600A)	23.37

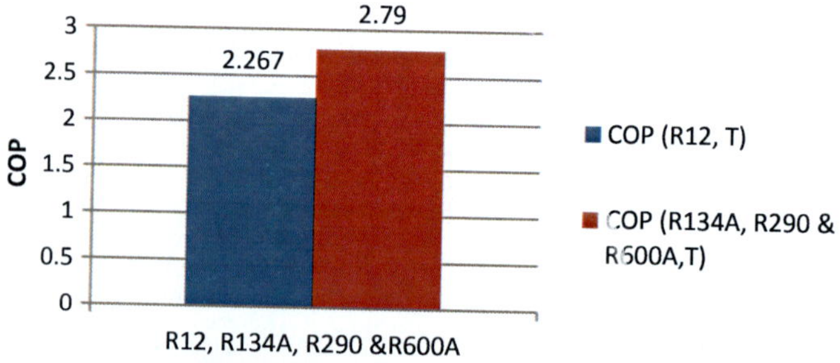

Fig. 2 COP values of R12, R134a, R290 and R600a

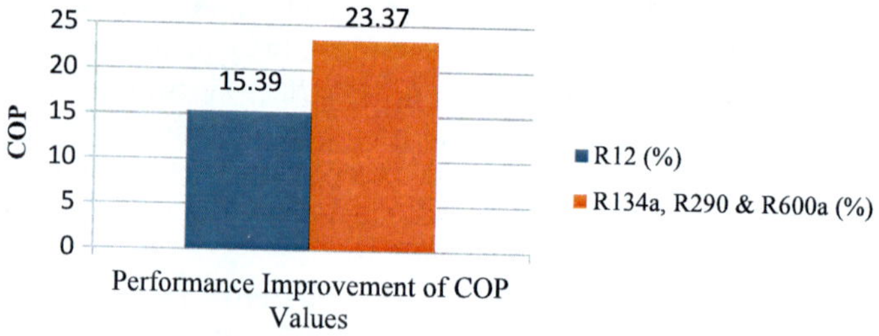

Fig. 3 Performance improvement of COP value used R12, R134a, R290 and R600a

18 °C) shows that COP estimation of R134A, R290 and R600A is 2.79. At the last of performance improvement of COP level is 23.37%. Consequently, the R134A, R290 and R600 refrigerants are superior to R12 refrigerant (Table 3; Figs. 2 and 3).

6 Conclusions

The trial examination of R12, R134a, R290 and R600a in vapor pressure refrigeration framework was completed with the accompanying ends. The major natural effect, for example, ozone layer consumption and a worldwide temperature alteration, is diminished similarly alongside the dimension of coefficient of execution which is higher. In this view, the utilization of R134a, R290 and R600a enhances warm solace and can be utilized as drop-in substitution refrigerant rather than R12.

References

1. Wu W, Skye HM (2018) Progress in ground-source heat pumps using natural refrigerants. Int J Refrig 92:70–85
2. Ghoubali R, Byrne P, Bazantay F (2018) Refrigerant charge optimization for propane heat pump water heaters. Int J Refrig 76:230–244.
3. Dalkilic S, Wongwises S (2010) An execution correlation of vapor-pressure refrigeration framework utilizing different elective refrigerants. Univ Commun Heat Mass Trans 37(9):1340–1349
4. Baskaran A, Koshy Mathews P (2012a) A Performance correlation of vapor pressure refrigeration framework utilizing eco-accommodating refrigerants of low a worldwide temperature alteration potential. Int J Sci Res Publ 2(9):1–8
5. Baskaran A, Koshy Mathews P (2012b) A Performance correlation of vapor pressure refrigeration framework utilizing different elective refrigerant. Univ J Sci Eng Res 3:1–6
6. Baskaran, A, Koshy Mathews P (2013) Energy and exergy analysis of a vapor compression refrigeration system with R134a, R152a and RE170. Doc Des Sci 66(3):1–15
7. Aprea C, Maiorino A, Mastrullo R (2011) Change in vitality execution because of a R422D retrofit: an exploratory examination for a vapor pressure refrigeration plant for a stroll in cooler. Applied Energy 88:4742–4748
8. De Rossi F, Mauro AW, Musto M, Vanoli GP (2011) Long-period nourishment stockpiling family unit vertical cooler: refrigerant charge effect on working conditions amid unfaltering task. Int J Refrig 34:1305–1314
9. James M (2002) Quiet, Outflows and natural effects from cooling and refrigeration frameworks. Int J Refrig 25:293–305
10. Mani M, Selladurai V (2008) International diary of warm science, An Experimental examination of another refrigerant blend as drop-in swap for CFC12 and HFC134a Refrigeration and Air-molding Laboratory. Coimbatore Institute of Technology, Coimbatore, India
11. Mani K, Selladurai V (2008) Experimental examination of another refrigerant blend as dropin swap for CFC12 and HFC134a. Univ J Thermal Sci 47:1490–1495

Design, Optimization and Analysis of Baja Suspension System Using Full Factorial Approach

V. Ajay Ganesh Ram, T. S. Easwar, A. Vishal, A. Andrews, and F. Michael Thomas Rex

Abstract The suspension system is used in vehicles to arrest the vibration. The helical compression springs are used in the suspension system of light vehicles because of its durability, lightweight and easiness of manufacture. Before selecting the spring for any application, its properties should be analyzed. In this paper, a helical compression spring was selected for a Baja vehicle. In general, the Baja vehicles are designed for high durability and strength. Hence, the standard materials and dimensions were taken and optimization was carried out among the standard materials and dimensions using a full factorial approach in MATLAB software. Then, the structural analysis was carried out for the selected springs using ANSYS workbench. The results were compared, and the suitable spring was selected. Further, simulations were carried out and the dynamic response of the selected spring was plotted using MATLAB.

Keywords Helical compression spring · Factor of safety · Analysis

1 Introduction

Coil springs are used in car suspension systems and clutches, as well as valve springs. Springs are also used in mechanical devices, such as toasters, door handles and other types of handles that are constantly depressed. These springs are also used in the suspension system of the buggy. Buggy is a specially designed compact size car which is used in the Baja competition in "Society Of Automotive Engineers (SAE)"; the vehicle must be capable of safe operation over rough land terrain including obstructions such as rocks, sand, logs, steep inclines, mud and shallow water, and it must have adequate ground clearance and traction for the terrain type at the competition. So, there is a need for a specially designed spring with a suitable material to withstand the weight of the vehicle and to isolate the vibration created from the rough surfaces.

V. Ajay Ganesh Ram · T. S. Easwar · A. Vishal · A. Andrews (✉) · F. Michael Thomas Rex
National Engineering College, Tamil Nadu, Kovilpatti, Tuticorin, India
e-mail: andrewsnzt@gmail.com

T. Rajmohan et al. (eds.), *Advances in Materials and Manufacturing Engineering*, Springer Proceedings in Materials 7, https://doi.org/10.1007/978-981-15-6267-9_3

Researchers were used various methods for analysis and optimization of coil spring. H. P. Pawar, D. D. Desale investigated by reducing the number of turns, and the spring was optimized by designing the spring with IS4454 material [1]. S. S. Gailkward, P. S. Kachare analyzed the safe load of helical spring used in two-wheeler horn. The static analysis of the helical compression spring is performed using NASTRAN solver, and the preprocessing of the spring model was done by HYPERMESH software [2]. Pankaj Sharma et al. analyze the suspension spring by designing a quarter car model with 2DOF and analyzed for overshoot and settling time of sprung and unsprung mass in MATLAB [3]. Mohd Izaham Zainal Abidin et al. investigated experimental and numerical investigation of SUP12 Steel Coil Spring and analyzed using experimental and numerical methods, and they have used HYPERWORKS for performing finite element simulation and numerical approaches with practical spring maker and compared the result by tensile testing of SUP12 with JIS specification [4].

In this paper, the helical compression spring was designed for a Baja vehicle by considering the stiffness and factor of safety. The optimization was carried out among a various combination of standard dimensions and standard materials by considering a full factorial design using MATLAB. The selected spring was analyzed in ANSYS workbench to identify the properties. Further, the dynamic response of the selected spring was simulated using MATLAB.

2 Design of Suspension System

2.1 *Standard Data*

To design the suspension system, the total deflection of the spring is kept as fixed as 40 mm. The weight of the Baja vehicle is taken as 2943 N (i.e., 300 kg = 2943 N). Total number of spring in Baja suspension system is 4. When the total weight is acting downward, the spring has to withstand the weight and the total weight is equally shared by four springs. Thus, the force acting on single spring was taken as 735.75 N.

2.2 *Full Factorial Design*

A full factorial design allows all the possible combinations of the input parameters. With the full factorial design, the effect of each parameter in the output can be studied. The standard dimensions of the spring were taken from DIN 2098. Various combinations of the diameters were shown in Table 1. Also, four standard spring materials were selected [5]. Chromium–vanadium alloy steel with an ultimate shear stress of 590 MPa, stainless steel wire austenitic type 301 with an ultimate shear

Table 1 Standard dimensions

Wire diameter (mm)	Coil diameter (mm)	Wire diameter (mm)	Coil diameter (mm)	Wire diameter (mm)	Coil diameter (mm)
4	20	5	50	8	50
4	25	5	63	8	63
4	32	6.3	32	8	80
4	40	6.3	40	8	100
4	50	6.3	50	10	50
5	25	6.3	63	10	63
5	32	6.3	80	10	80
5	40	8	40	10	100
				10	125

stress of 276 MPa, copper base alloy phosphor bronze with an ultimate shear stress of 360 MPa and nickel base alloy Inconel alloy X750 with an ultimate shear stress of 690 MPa were selected as the spring materials for analysis. Thus, for each material, all the standard dimensions were combined and 100 combinations were developed using full factorial approach. A MATLAB code was developed to find out the factor of safety and the number of turns for each combination.

The MATLAB output for the above combinations was shown in Figs. 1 and 2. Figure 1 shows the factor of safety of each combinations, and Fig. 2 shows the number of turns of each combinations. For four-wheeler suspension system, the minimum factor of safety should be 3. Thus, by considering the factor of safety as higher than 3, the optimization was performed. Among 100 iterations, four iterations have a factor of safety of higher than 3 they were listed in Table 2.

Fig. 1 Iteration number (diameter) versus FOS

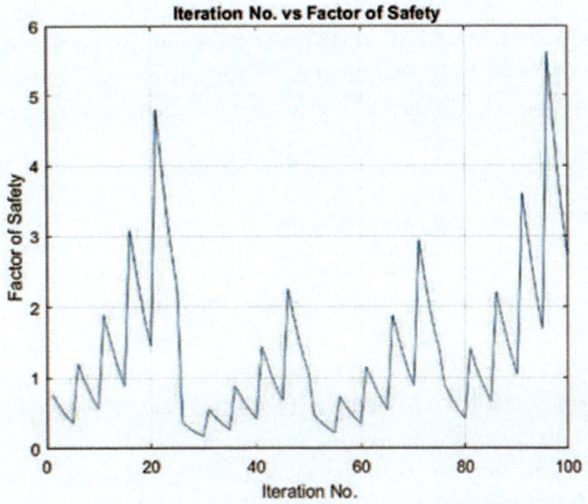

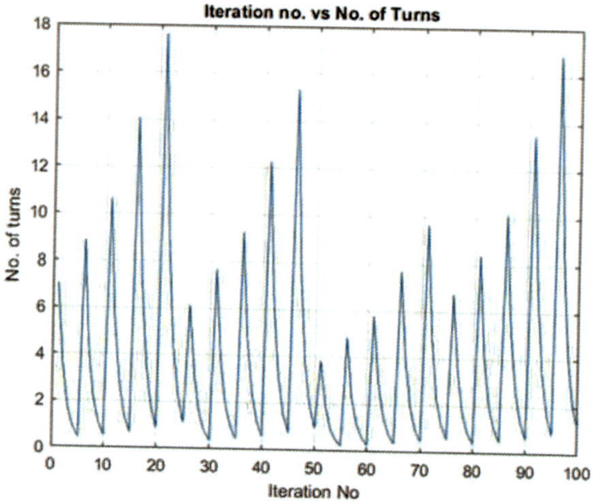

Fig. 2 Iteration number (diameter) versus no of turns

Table 2 Material properties

Spring No	Material	Wire diameter (mm)	Coil diameter (mm)	FOS
1	Chromium–vanadium alloy steels	8	40	3.076
2	Chromium–vanadium alloy steels	10	50	4.806
3	Nickel base alloy Inconel alloy X750	8	40	3.597
4	Nickel base alloy Inconel alloy X750	10	50	5.621

The selected dimensions and materials from Table 2 were used for further analysis. The MATLAB code for suspension design was developed by considering the analytical design procedure. So, to ensure the safety and to verify the results, the structural analysis was carried out in ANSYS workbench.

2.3 *Analysis in ANSYS*

The springs were modeled in SOLIDWORK, and the structural analysis was carried out in ANSYS workbench. Tetrahedral elements were used for this analysis. Figures 3, 4, 5 and 6 show the factor of safety of the selected springs. The total weight of the BAJA vehicle is taken as 2943 N. Four springs were used in the vehicle. So, the load on each spring is taken as 735.75 N.

From the above results, the factor of safety values is compared. The spring number 1 has a minimum factor of safety of 1.2181, the spring number 2 has a minimum

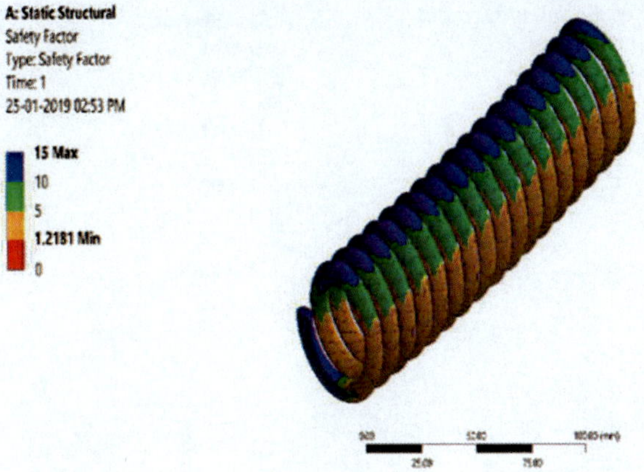

Fig. 3 Safety factor of Spring No 1

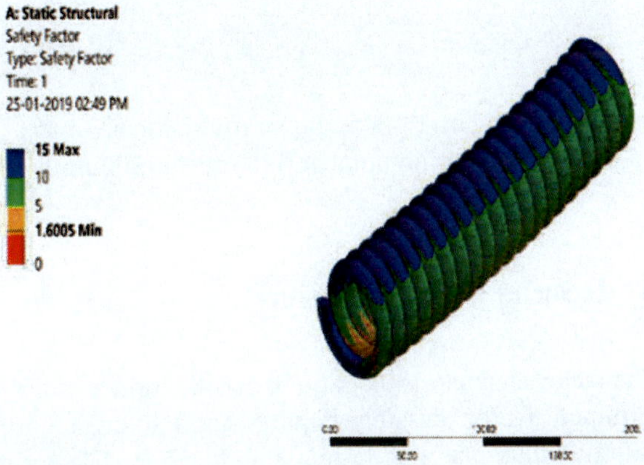

Fig. 4 Safety factor of Spring No 2

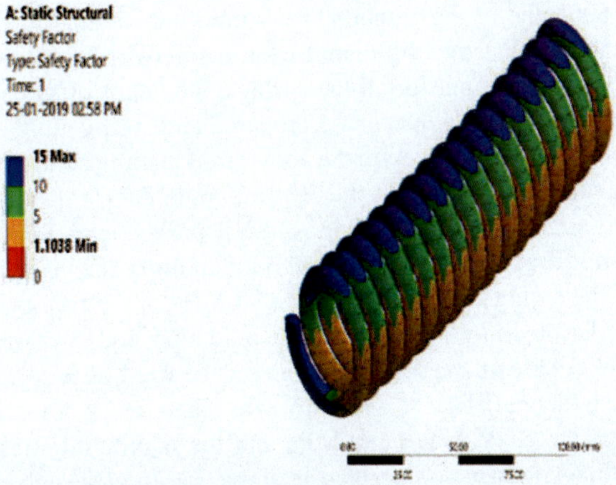

Fig. 5 Safety factor of Spring No 3

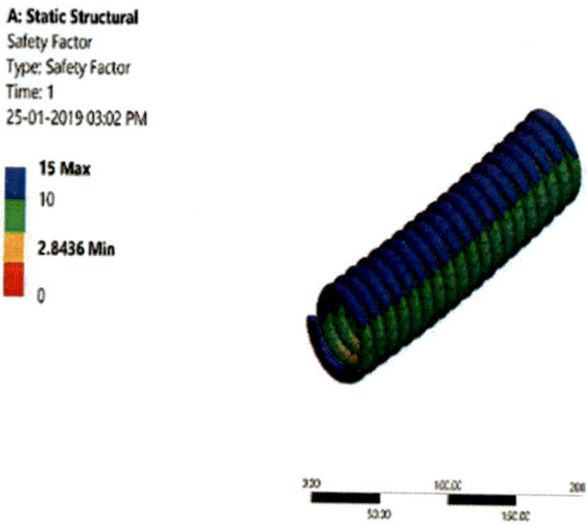

Fig. 6 Safety factor of Spring No 4

factor of safety of 1.6005, the spring number 3 has a minimum factor of safety of 1.1038 and the spring number 4 has the minimum factor of safety of 2.8436.

3 Results and Discussion

The factor of safety is the ratio between ultimate stress and the allowable stress. The minimum factor of safety required for a safe design is one. If the factor of safety is higher than one, the design is highly safe. The spring number 1 has a minimum factor of safety of 1.2181, the spring number 2 has a minimum factor of safety of 1.6005, the spring number 3 has a minimum factor of safety of 1.1038 and the spring number 4 has the minimum factor of safety of 2.8436.

In this design, for all the springs, the factor of safety is higher than one. So, the factor of safety values was compared and spring number 4 was selected as a spring for the Baja vehicle with the above-said loading conditions. Because it has the factor of safety as 2.8436. Further, the maximum deformation and the maximum stress of this spring were analyzed in ANSYS workbench. The maximum deformation of the spring was obtained as 63.968 mm, and it is shown in Fig. 7. The maximum stress of the spring was obtained as 485.3 MPa, and it is shown in Fig. 8.

The Inconel alloys are high oxidation and corrosion-resistant materials. This nickel base Inconel alloy X750 can be used as a spring materials [5]. For nickel base Inconel alloy $\times$ 750 with wire diameter 10 mm and coil diameter 50 mm, the ultimate stress was 1380 MPa and the maximum stress acting on the spring was 485.3 MPa. By comparing the above stresses, the selected spring was safe for this working loads.

Fig. 7 Total deformation of Spring No 5

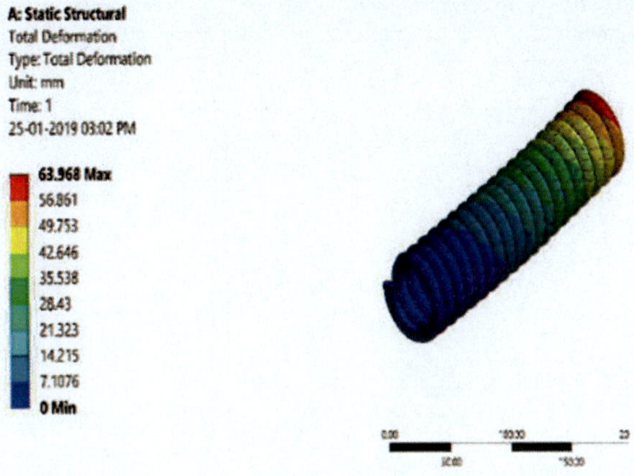

Fig. 8 Equivalent elastic strain of Spring No 5

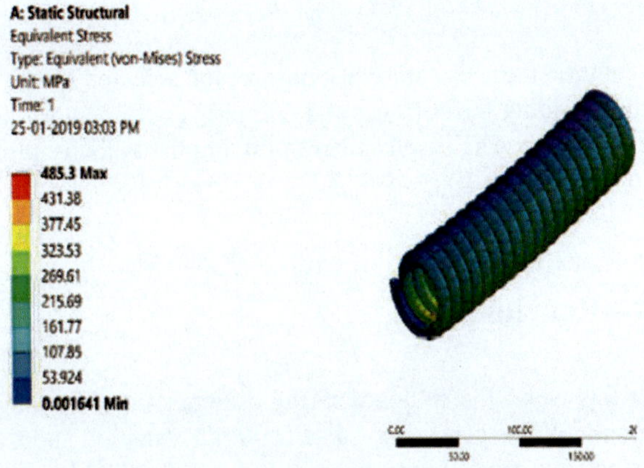

Fig. 9 Time versus displacement

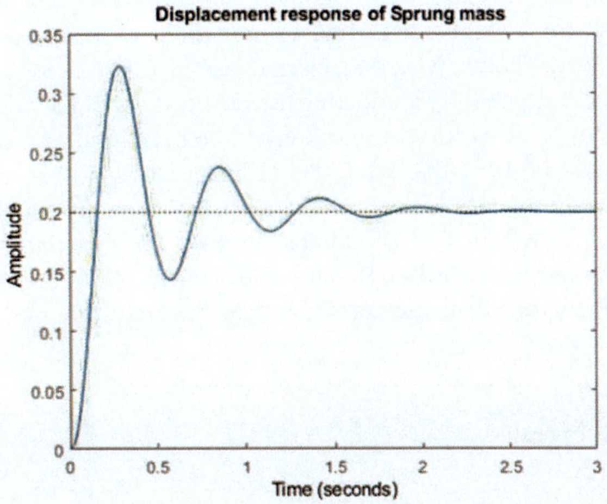

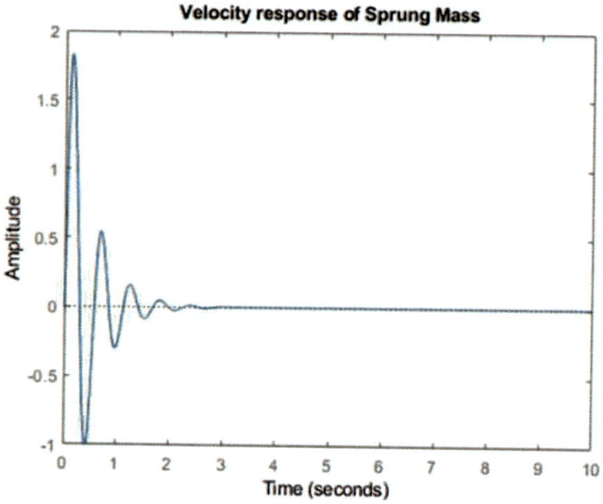

Fig. 10 Time versus velocity

Further, the dynamic response of the selected spring was simulated in MATLAB by providing the stiffness and damping coefficient of the suspension system and the rubber tyre. The displacement plot and the velocity plot was plotted, and it is given in Figs. 9 and 10 [3, 5–7].

4 Conclusion

In this work, the standard spring dimensions and materials were selected to design a suitable suspension system for Baja vehicle. Then, various combinations were selected using full factorial approach. A MATLAB code was developed to calculate the factor of safety of each combination, and the optimization was carried out. Among hundred combinations, four combinations were selected by considering a factor of safety. The analysis was carried out in ANSYS workbench for the five selected combinations. By comparing the results of ANSYS workbench, nickel base Inconel alloy X750 with a wire diameter of 10 mm and coil diameter of 50 mm has a maximum stress of 485 MPa, while the ultimate stress of this material is 1380 MPa. Thus, this spring was selected as the suspension system for Baja vehicle with the above-mentioned loading conditions. Further, the dynamic response of this spring was simulated in MATLAB. The displacement plot and velocity plot were plotted to identify the dynamic behavior of the selected spring in road conditions.

References

1. Pawar HP, Desale DD (2018) Optimization of three wheeler front suspension coil spring. Int Conf Mater Manuf Design Eng 20:428–433

2. Gaikwad SS, Kachare PS Static analysis of helical compression spring used in two-wheeler horn. Int J Eng Adv Technol. ISSN: 2249–8958
3. Sharma P, Saluja N, Saini D, Saini P Analysis of automotive passive suspension system with MATLAB program generation. Int J Adv Technol. ISSN 0976–4860
4. Abidin MIZ, Mahmud J, Latif MJA, Jumahat A (213) Experimental and numerical investigation of SUP12 steel coil spring. Malaysian Int Tribol Conf 68:251–257
5. Joseph E.S Standard handbook of machine design. Tata McGraw Hill Publications
6. Olatunbosun OA, Dunn JW (1991) A simulation model for passive suspension ride performance optimization. In: Automotive Simulation '91. Springer, 131–142
7. https://in.mathworks.com/matlabcentral/fileexchange/18410-analysis-of-suspension-system-using-matlab-simulink-and-simscape

Aerodynamic Analysis of a 3D Small Wind Turbine Blade Using NACA 63415 Aerofoil with MRF

Veludurthi Ajay and Bolleddu Venkateshwarlu

Abstract The flow characteristics over a small wind turbine blade with NACA 63415 aerofoil is analyzed using computational fluid dynamics (CFD). In this analysis, aerodynamic effectiveness over a symmetrical 3D blade is studied experimentally in a low speed wind tunnel. The pressure on the aerofoil surface with chord length of 150 mm was distributed, lift and drag forces were measured at −2, 0, 5, 10, and 15° angle of attacks (AOA) and mean velocity profiles were obtained over the surface. The blade in moving reference frame (MRF) approach is used to verify the pressure and velocity contours on the fan when moving at an optimum speed of 650 rpm. The blade is modelled in CATIA V5R20 and analysis is carried using CFX, ANSYS, and FLUENT 18.1 softwares. It is found that the maximum aerodynamic efficiency (C_l/C_d) is achieved at 5° of angle of attack and it can be considered the safe angle of attack. At this angle, the MRF approach is also carried out and the results showed that the better torque of 14.2 N-m can be obtained at 650 rpm speed. It is also observed that the critical angle of attack lies in between 0° to 10°. If this critical angle value changes from this range, it is resulting in a substanitial decrease of lift.

Keywords Aerodynamic effectiveness · Angle of attack · Drag · Lift · Wind tunnel · Moving reference frame

1 Introduction

In the present scenario, because of an increasing energy crisis, it becomes important to investigate for an alternative source of energy for power generation by replacing fossil fuels. In fact, an economical source of energy for power generation is the wind which can be converted into useful forms using wind turbines. The applications of wind

V. Ajay (✉) · B. Venkateshwarlu
School of Mechanical Engineering, Vellore Institute of Technology, Vellore, Tamil Nadu, India
e-mail: ajaybabu237@gmail.com

B. Venkateshwarlu
e-mail: venkateshwarlu.b@vit.ac.in

T. Rajmohan et al. (eds.), *Advances in Materials and Manufacturing Engineering*, Springer Proceedings in Materials 7, https://doi.org/10.1007/978-981-15-6267-9_4

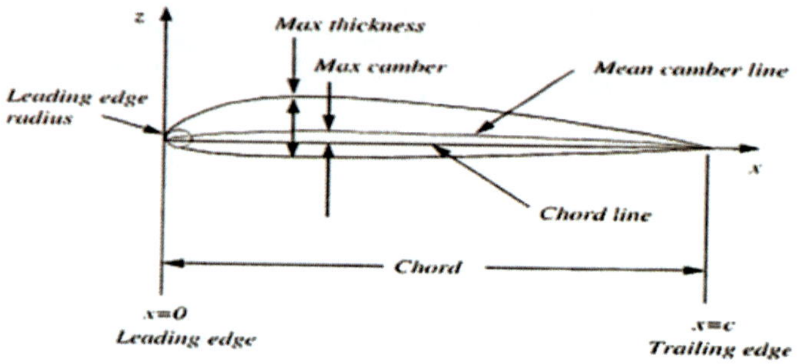

Fig. 1 Nomenclature of aerofoil [1]

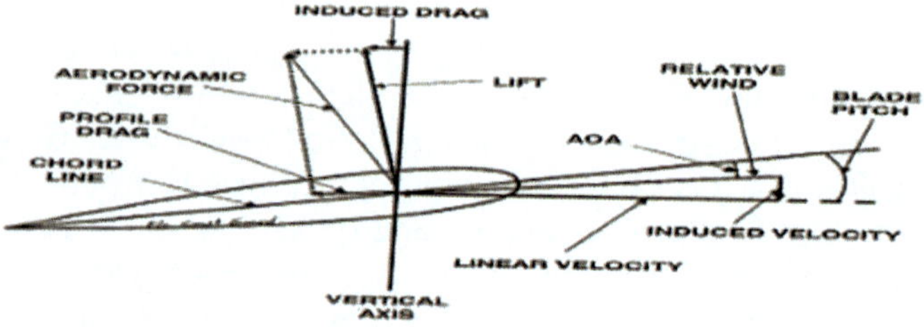

Fig. 2 Blade velocity diagram [2]

turbines for power generation are increasing steadily. Most of the previous research work has been carried out on the aerodynamic analysis of wind turbine blades having aerofoil cross section as shown in Fig. 1. The section with more curvature gives low pressure air and high pressure air goes on the other side of aerofoil section. The phenomenon will be resulting in the lift force which is perpendicular to the flow direction of air as shown in Fig. 2.

2 Theory

In this model, we choose k–ε model based on the Reynolds-Averaged Navier-Stokes equations. This is a two equation model (1) and (2) which shows the turbulence by means of two transport equations [3]. The k–ε model is used to improve the mixing length model and it is used to solve algebraically prescribing turbulent scales in moderate to critical complexity flows. The first transported variable is the turbulence kinetic energy (k) and the second variable is the rate of dissipation of turbulence energy.

For turbulent kinetic energy

$$\frac{\partial}{\partial t}(\rho k) + \frac{\partial}{\partial x_i}(\rho k u_i) = \frac{\partial}{\partial x_j}\left[\left(\mu + \frac{\mu_t}{\sigma_k}\right)\frac{\partial k}{\partial x_j}\right] + G_k + G_b - \rho\varepsilon - Y_M + S_k \quad (1)$$

For dissipation

$$\frac{\partial}{\partial t}(\rho\varepsilon) + \frac{\partial}{\partial x_i}(\rho\varepsilon u_i) = \frac{\partial}{\partial x_j}\left[\left(\mu + \frac{\mu_t}{\sigma_\varepsilon}\right)\frac{\partial \varepsilon}{\partial x_j}\right] + C_{1\varepsilon}\frac{\varepsilon}{k}(G_k + C_{3\varepsilon}G_b) - C_{2\varepsilon}\rho\frac{\varepsilon^2}{k} + S_\varepsilon \tag{2}$$

In these equations, G_k represents the turbulence kinetic generation due to mean velocity gradients, G_b is the turbulence kinetic energy generation due to buoyancy, and Y_M represents the contribution of the fluctuating dilatation in compressible turbulence to overall dissipation.$C_{1\varepsilon,}$ $C_{2\varepsilon,}$ $C_{3\varepsilon}$ are the constants. σ_ε and σ_k are the turbulent prandtl numbers for and k and S_k and S ε are user defined source terms.

2.1 *MRF Approach*

The multi-reference frame model is the simpler of the two approaches for number of zones. In this steady-state approximation method, individual cell zones can move at different rotational speeds. Each cell zone is solved using the moving reference. Local reference frame transformation is performed to enable flow variables in one zone which is used to calculate boundary of the adjacent zones.

In this MRF approach, it does not account for the moving zone relative motion with respect to the adjacent zones, and by fixing the grid in the computation. The analogous freezes the motion of the moving part in the specific position and checks the instantaneous flow field with the rotor in the position. The necessity of the MRF model is to compute the flow field which can be used as an initial condition for transient sliding mesh calculations.

3 Aerofoil Blade Profile

In the present work, the meshing is done for 3D model of a blade using ANSYS ICEM CFD 18.1 software. After completing the modelling of wind tunnel with a blade inside, the splitting is done for the wind tunnel for capturing the wind blade. After giving the edge parameters, the premesh is done. The meshed part of wind turbine blade is shown in Figs. 3 and 4.

Figure 4 shows the blade with 80,800 elements formed on the surface and the blade is meshed with the hexahedral meshing. For having the MRF approach, the 3D blade is taken in to ANSA 16.0 software and periodic is done by rotor at the center portion with the angle of 120° it is rotating and producing three number of blades on the rotor part with including the blades. We create the reference frame for having the rotation of air around the blades (Figs. 5 and 6).

Fig. 3 Meshing of wind turbine blade with wind tunnel

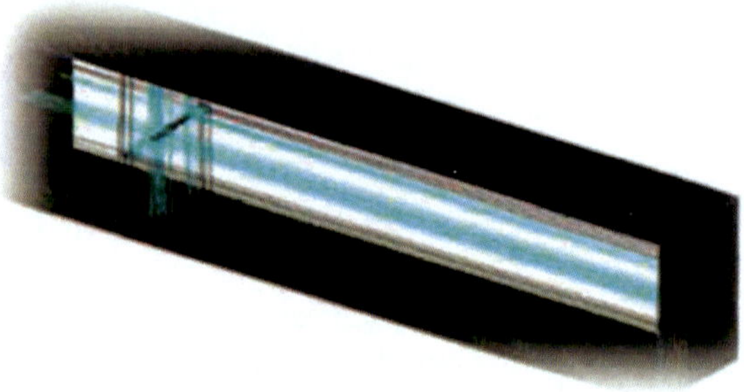

Fig. 4 Meshing of wind blade

Fig. 5 Modelling of blades with rotor along MRF

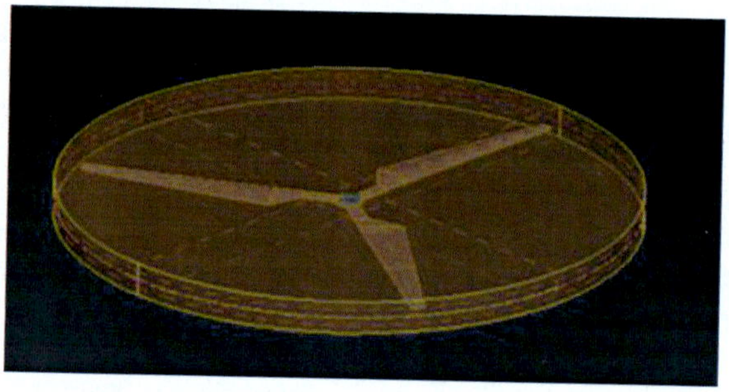

Fig. 6 MRF far field of the blade

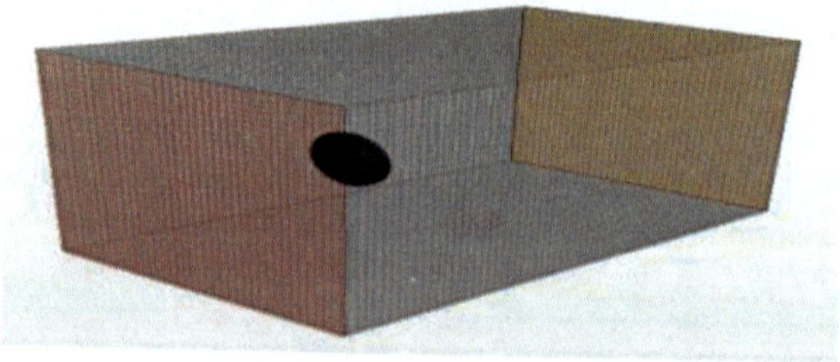

The wind tunnel is designed with four times the rotating diameter of blades, and is meshed with quadrilateral with an element of 1.95 lakhs number and internal meshing is done with a 80,800 elements formed on the blade and MRF section.

4 Formula

Area of aerofoil $A = C \times L = 0.15 \times 0.3 = 0.045\ \text{m}^2$,

Coefficient of drag, $C_d = \frac{F_D}{0.5*\rho*V^2*A}$, Coefficient of drag, $C_l = \frac{F_L}{0.5*\rho*V^2*A}$,

Reynold number $\text{Re} = \frac{\rho v l}{\mu}$, Power available, $\frac{P_{\text{avail}}}{A} = \frac{1}{2} * \rho * V_o^3$.

5 Result and Discussion

In this work, a 3D aerofoil profile using NACA 63415 aerofoil as shown in below figures is analyzed for various angles of attack. The coefficient of lift and coefficient of drag are calculated for this NACA 63–415 for various angles of attack from 0° to 15° and the aerodynamic efficiency is calculated at those corresponding angle of attacks. The 3D blade of contour plots as shown in figures predicts the velocity and pressure variations in various regions near the blade of flow domain are observed. From the calculated results, the coefficient of lift is increasing gradually from 0 to 11° (from 0.1129 to 1.0324), and then decreases from 11 to 15° (from 1.0324 to 0.9615) as we increase the angle of attack from 0 to 15°. The coefficient of drag value is also increasing gradually from 0 to 9° (from 0.0100 to 0.0403), and rapidly increases from 9 to 15° (from 0.0403 to 0.1143) as we increase the angle of attack from 0 to 15° (Figs. 7 and 8).

Aerodynamic efficiency is the ratio of coefficient of lift and coefficient of drag. The aerodynamic efficiency is calculated from the coefficient of lift and coefficient of drag values. From the calculated aerodynamic efficiency values, we can said whether the model will be aerodynamic model or not. For aerodynamic models, the aerodynamic efficiency is more than 5. Coming to this obtained aerodynamic efficiency values, the efficiency is slightly increasing and decreasing at different angle of attacks from 0 to 15°. From this, at what angle of attack the aerodynamic efficiency is more that angle of attack will be called as safe angle of attack.

In this 3D model, the aerodynamic efficiency is more at the angle of attack 5°, then that angle of attack 5 will be called as safe angle of attack. The pressure, density, and velocity distributions are seen at different angle of attacks. The velocity vectors are also seen at different angle of attacks. From this, contours where the stagnation point will occur and how the pressure (low and high pressures) acting on the aerofoil was seen clearly.

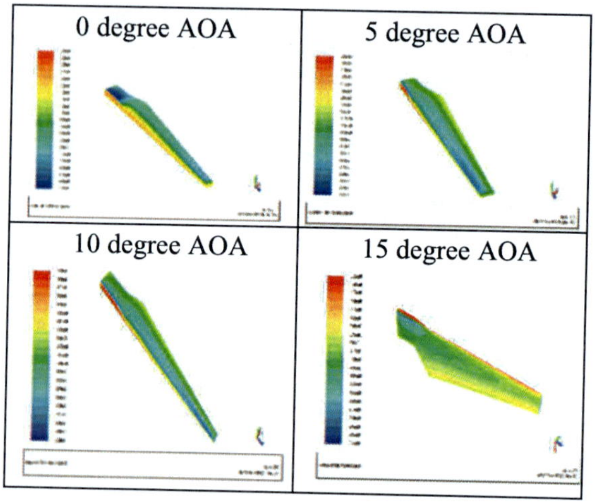

Fig. 7 Pressure distributions along the 3D

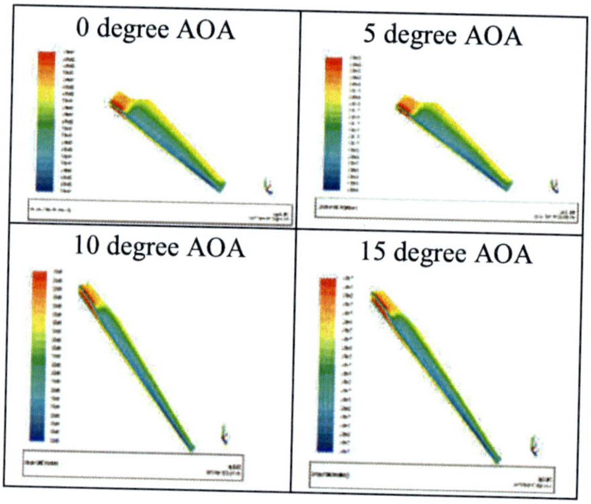

Fig. 8 Velocity distributions along the 3D WTB

The static pressure distributions on the 3D wind turbine blade, i.e., leading edge, trailing edge, lower surface, and upper surface. By this, the pressure distributions are to be identified at different angle of attacks.

The velocity vector is representing the position and motion of a single particle moving in the air. Velocity vectors consisting of arrows to indicate the direction of fluid flow around the wind turbine blade. The velocity moving in the air was moving with magnitude and particular direction. By the velocity vectors, the magnitude and direction of the air flowing around the wind turbine blade can be studied. The velocity vectors on wind turbine blade were shown in the above Fig. 9.

The velocity magnitude of the blade at dynamic condition is pointed at the MRF approach of the blades to the rotor as shown below Fig. 10. The MRF zone setting

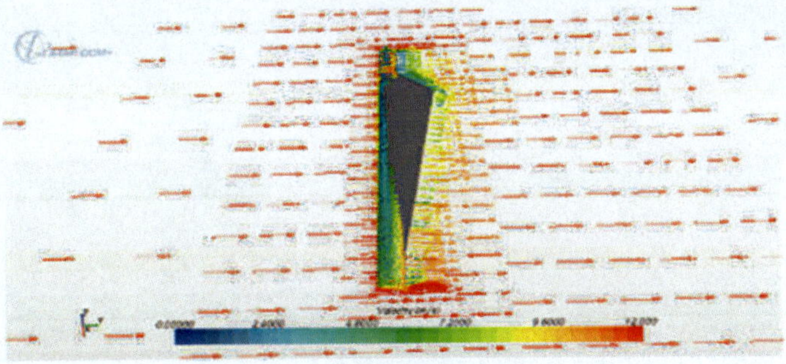

Fig. 9 Velocity vectors along the blade at optimum speed

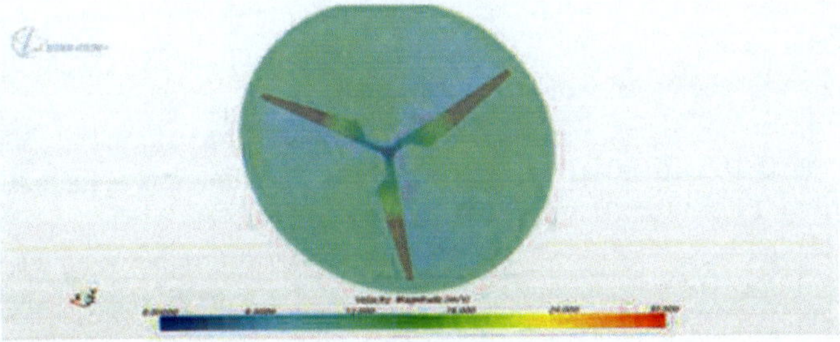

Fig. 10 Velocity magnitude of blade in dynamic conditions with MRF

with a rpm of 650 on the rotor is attained with a velocity of 12 m/s can be used for working in dynamic condition at rotation of blades in the wind tunnel.

The below figures shows the various of pressure and velocity contours in dynamic analysis are (Figs. 11 and 12).

.From the approach, the average torques coefficients of the three blades can be located during rotation. For this approach, the pitch angle at 5° can be fixed the blades to the rotor hub which gives the required torque.

Fig. 11 Pressure distribution of blade

Fig. 12 Velocity distribution of blade

5.1 Graphical Representation on 3D

Figure 13 shows the coefficient of lift (C_l) versus AOA, the turbulent kinetic energy and specific dissipation flow of air effects by changing the AOA values from −2 to 10°; the increases in C_l value from 0.1129 to 1.0324 and then AOA from 11 and 15°; the C_l values was decreased from 1.0324 to 0.9615.

In Fig. 14, the Ccefficient of drag (C_d) versus AOA, the value of C_d is increased gradually from 0.0100 to 0.0413 for AOA of 0 to 10° and then it was increased slowly from 0.0413 to 0.1143 for 10 to 15° AOA.

In Fig. 15, the aerodynamic efficiency value was varying between 11.3098 and 8.4121. From this, the aerodynamic efficiency was gradually increased from 0 to 5° and then gradually decreased from 5 to 15° angle of attack. The aerodynamic efficiency was more at the angle of attack 5°.

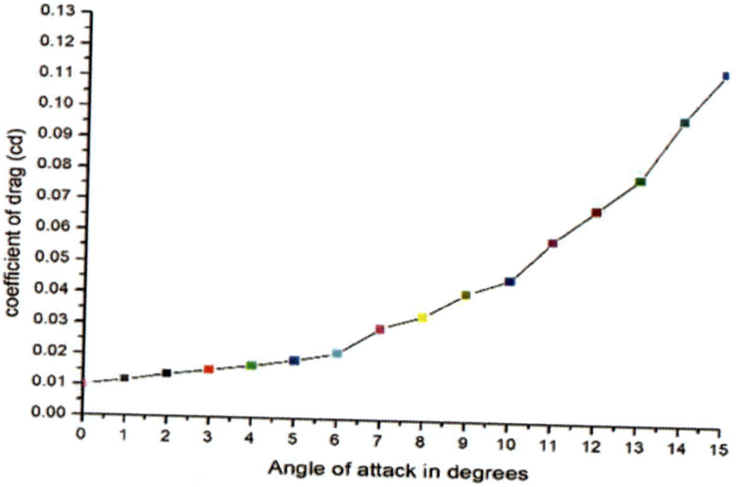

Fig. 13 C_d versus α in wind turbine blade

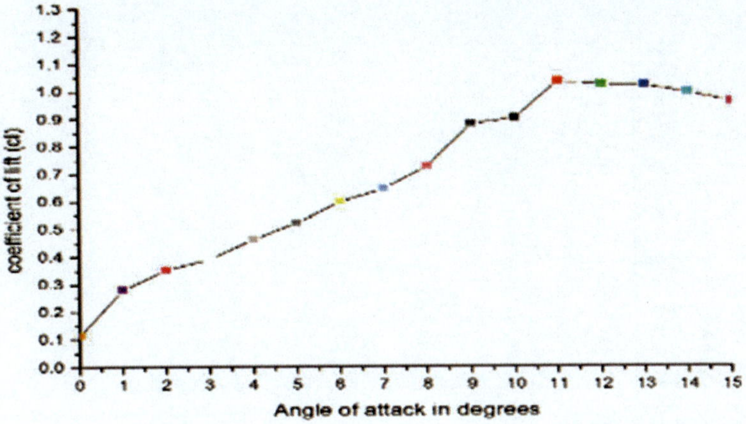

Fig. 14 C_l versus α in wind turbine blade

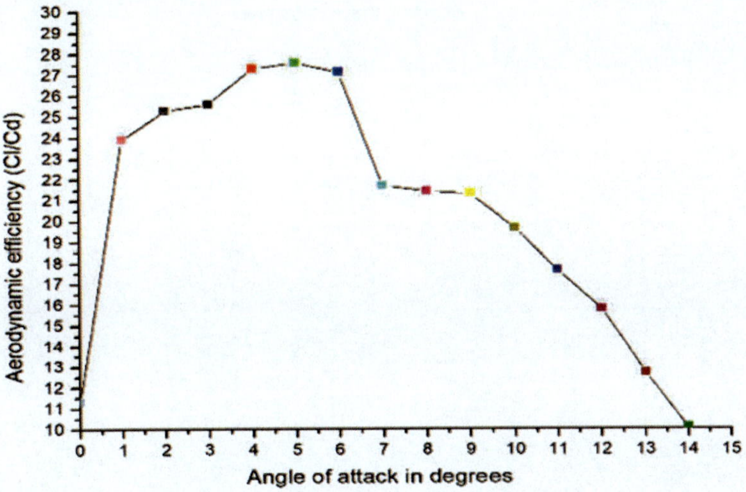

Fig. 15 Aerodynamic efficiency versus α in WTB

5.2 *Graphical Representation of Varing with Wind Speeds*

The coefficient of drag value is estimated at different velocities of speed of 3, 12, 28 m/s. At these conditions, the values of coefficient of drag was increases with increased speed and angle of attacks from −2 to 10 is observed and shown in above Fig. 16.

The coefficient of lift values changed with a velocity of 28 m/s at −2 AOA is very low initially and later it is varied and raised and at 12 m/s; the values are constantly increased to maximum of 0.12 which is shown in above Fig. 17.

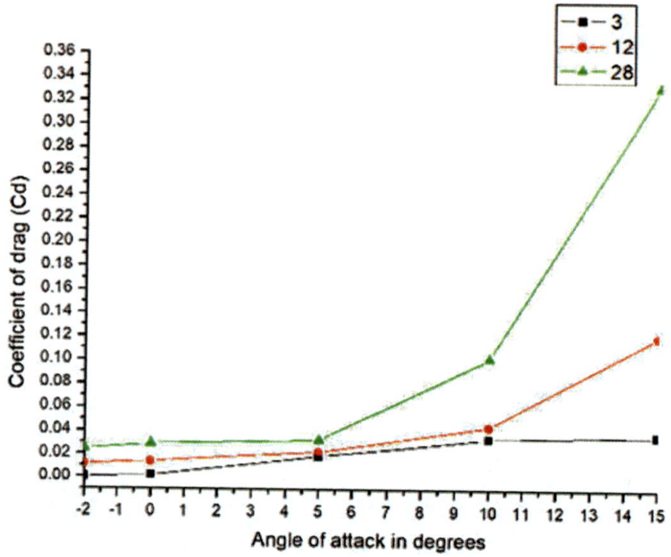

Fig. 16 Varying the air velocities of C_d versus α in WTB

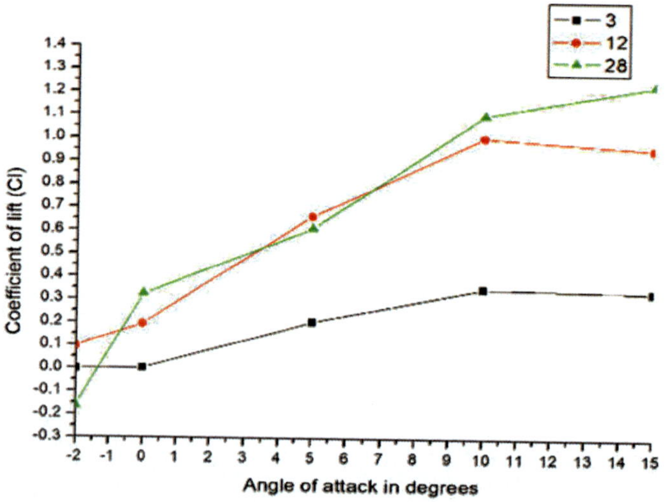

Fig. 17 Varying the air velocities of C_l versus α in 3D WTB

The aerodynamic values are varied for different velocities from 3 m/s to 12 m/s. The cl/cd value is varied for AOA −2–5 is −8.5714–11.027 and is come down after this value, for 12 m/s is 8.27–29.63 and downs the curve and for 28 m/s 8.2–18.71 is going steep for the further AOA are shown in Fig. 18. In this analysis the angle of attack at 5° with 12 m/s is constantly having a good acceptable performance.

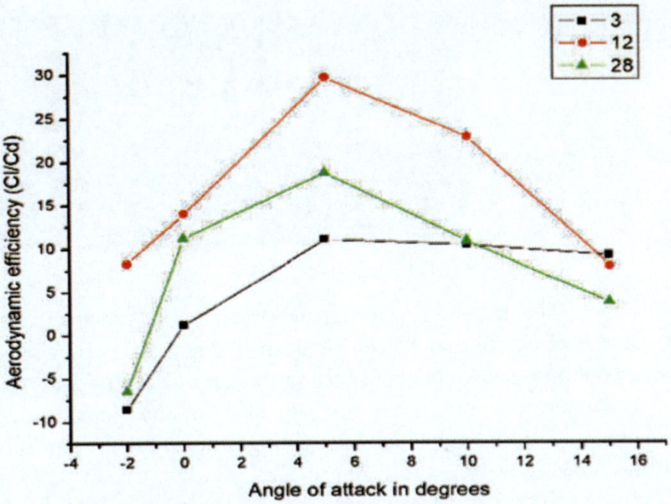

Fig. 18 Varying the air velocities of C_l/C_d versus α in 3D WTB

5.3 Experimentation

The aerofoil was initially fixed at a zero-vs angle of attack in the wind tunnel which is operated at proportionate air velocity. The AOA is differed with streaming behavior of air on suction side of aerofoil velocity and point of separations along with reattachments was noted. Further AOA is adjusted to 5° as by having the constant wind tunnel, velocity of air is allowed to aerofoil which records the flow and separate point is taken. The wind tunnel velocities are set for different angle with nominal velocity. The distribution pressures along the aerofoil are trapped at different AOA of −2, 0, 5, 10, and 15°. The aerofoil was again placed at a 0° angle of attack and the downstream wake was measured using the pitot probe and traverse (Fig. 19).

Drag forces are mainly used to resist the forces of movement of solid that resists the movement of a solid body through a fluid. The drag force usually has pressure and frictional force components acting along the flow direction, which is unified all over

Fig. 19 Wind tunnel unit

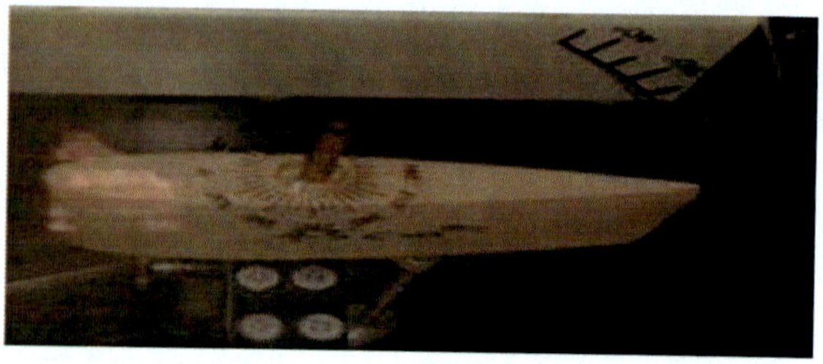

Fig. 20 Specimen placed inside the tunnel

the surface of body. The drag force will depend much on body shape that results in very complicated formula for drag force. Hence, by using drag coefficient approach, the drag force equation is developed. By using this approach, the drag force gives relevance with stagnation points. The above Fig. 20 shows the specimen of 0.15*0.3 of chord and length specimen is cut used for testing in wind tunnel and recorded the following data (Table 1).

In this, keeping the tunnel velocity constant and changing the AOA of the flow in attached Streamers. As the AOA is increased, the flow will separate the upper surface of the aerofoil which results in stall and noted the decrease of the separation of AOA in order for the flow to reattach. The following plots shows the coefficient of lift, drag, and aerodynamic efficiencies to different angles of attack as shown below (Figs. 21, 22 and 23).

The velocity is kept as 12 m/s as constant and varying the angle of attacks of −2°, 0°, 5°, 10°, 15°. Varying and estimated the lift value from 4.42 N to 16.44 N and drag forces varying from 0.66 to 4.65 N for these AOA. Keeping the number of ports at different locations on aerofoils to evaluate the pressure taps at those locations by using the monometers. The pressure obtained which multiplies the area of the aerofoil section which in turn gives the forces on the surface of the aerofoil.

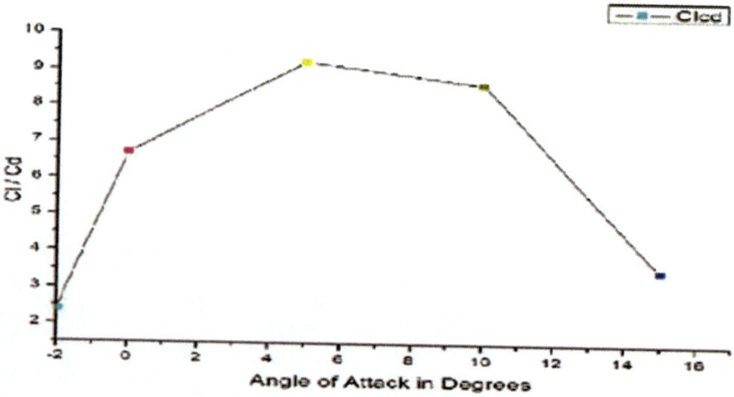

Fig. 21 Experimental dependence on C_l/C_d on AOA

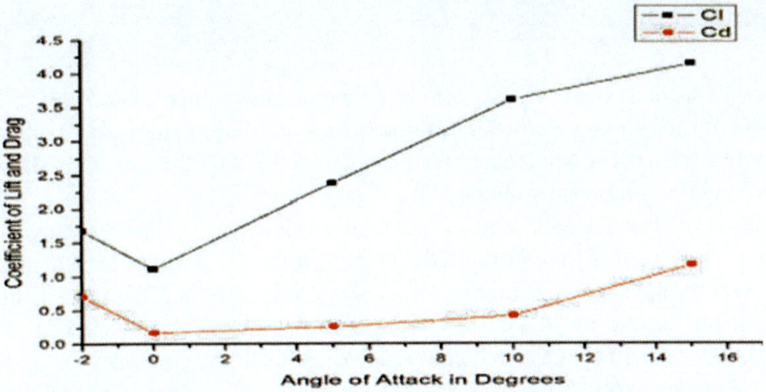

Fig. 22 Experimental dependence on C_l& C_d on AOA

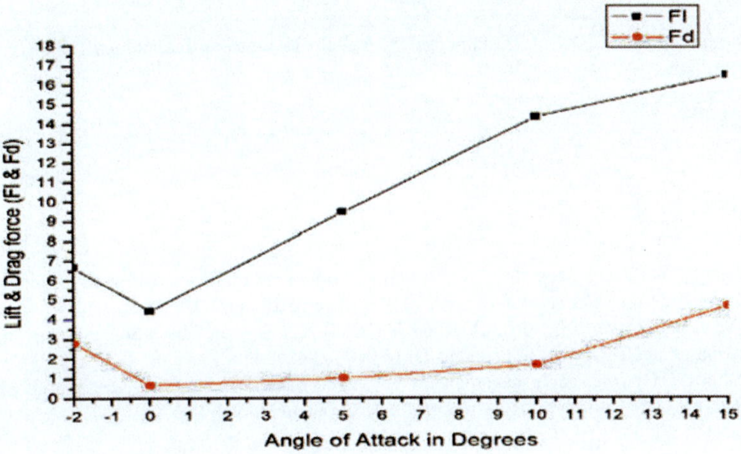

Fig. 23 Experimental dependence on F_l&F_d on AOA

Table 1 Comparison of experimental results with CFD

Experimental						CFD	Error valve
AOA	F_l (N)	F_d (N)	C_l	C_d	c_l/c_d	c_l/c_d	
−2	6.67	2.79	1.6813	0.7028	2.392	8.2758	5.885
0	4.42	0.66	1.1134	0.1662	6.697	13.9568	7.259
5	9.45	1.024	2.3805	0.2579	9.229	29.6339	20.404
10	14.3	1.66	3.6074	0.4181	8.627	22.8095	14.182
15	16.4	4.65	4.1414	1.1714	3.535	7.975	4.44

6 Conclusion

This work is carried out to evaluate the aerodynamic characteristics and performance of 2D and 3D aerofoil-shaped wind turbine blades at different angle of attacks. This investigation primarily compares the results obtained from CFD analysis of 3D model with the experimental results obtained.

It is observed that the maximum aerodynamic efficiency (C_l/C_d) is achieved at 5° of angle of attack for 3D models and this angle can be considered the safe angle of attack. At this angle, the MRF approach is also carried out and the results showed that the better torque of 14.2 N-m can be obtained at 650 rpm speed. It is also experimentally found that the critical angle of attack lies in between 0° and 10°. If this critical angle value changes from this range, it is resulting in a substanitial decrease of lift. The results obtained from 3D models and MRF approach are also showing good agreement with experimental results of aero foil section.

Acknowledgements This work is supported by Vellore Institute of Technology (VIT), Vellore, Tamilnadu, India.

References

1. Pourrajabian A (2015) Aero-structural design and optimization of a small wind turbine blade. Renewable Energy 1–12 https://doi.org/10.1016/j.renene.2015.09.002
2. Ajay V, Venkateshwarlu B, VishnuVardhan T (2018) A review on fabrication and performance evaluation of small wind turbine blades. IJMET 9(1):240–265
3. Chalothorn T Optimal angle of attack for untwisted blade wind turbine. Renewable Energy (ELSEVIER) 1279–1284. https://doi.org/10.1016/j.renene.2008.09.017

Trends on the Abrasive Flow and Electric Discharge Machining of Polymer Matrix Composites

R. Vinayagamoorthy, Shubham Kumar, Suraj Kumar, Mote Sai Sharan, G. M. D. Afzal, and T. V. Rajamurugan

Abstract Machining of polymeric composites is very important to bring the component to the required shape and size. A conventional machining technique uses a tool and is less accurate, creates vibration, generates heat in the machining zone, develops wastage in terms of chips and finally alters the material property. Unconventional methods are advanced techniques which do not use cutting tool and hence eliminate the vibration and reduce material wastage. Among the popular unconventional machining techniques, abrasive flow machining and electric discharge machining are widely used methods for machining polymeric composites. The present research gives a detailed insight of various challenges faced and possible solutions to those challenges during machining of polymeric composites. The survey also addresses about the optimum cutting conditions arrived for different polymeric materials and effective statistical tools for optimization of abrasive flow machining and electric discharge machining.

Keywords Abrasive · Kerf taper · Metal removal rate · Delamination

1 Introduction

Polymer matrix composites (PMCs) are man-made materials to compete the conventional materials in terms of its properties. They are also known to be fiber-reinforced plastics (FRPs) as fibers of different sizes are reinforced in the matrix in the form of regular shapes [1]. During the earlier stage of development, PMCs are made by using a man-made resin and a man-made fiber such as carbon, kevlar and glass as reinforcements [2–4]. These composites are well known for its strengths and have wide applications in various sectors. Since these composites fail in degradability, researchers introduced biofibers derived from plants and animals as reinforcements

R. Vinayagamoorthy (✉) · S. Kumar · S. Kumar · M. S. Sharan · G. M. D. Afzal
SCSVMV Deemed To Be University, Enathur, Kanchipuram 631561, India
e-mail: vin802002@gmail.com

T. V. Rajamurugan
Government College of Engineering, Srirangam, Trichy, India

T. Rajmohan et al. (eds.), *Advances in Materials and Manufacturing Engineering*, Springer Proceedings in Materials 7, https://doi.org/10.1007/978-981-15-6267-9_5

and they enhance the nature of degradability to the composites [5–9]. Many research works have been carried out in the recent past to assess the characters of biofiber-reinforced plastics. Most of them have proved that hybrid composites made by inclusion of a biofiber and a man-made fiber are best in all the characters, and they also produce best machining characteristics [10–16]. Though PMC is usually made in the shape of the final component, they need additional machining operations for finishing and also for making it easy for assembly. Hence, the need for machining of PMC is vital [17,– 19]. Many research studies have been made to investigate the machinability of hole making process in PMC by using conventional drilling and slot making by conventional milling [20–22]. In order to make the machining effectively by reducing the removal waste materials, nowadays, unconventional machining is done on PMC.

Among the unconventional machining techniques, abrasive flow machining (AFM) is one of the vital methods in which abrasive particles are made pass on to the workpiece where the material is to be removed. The machinability of AFM is measured by the kerf taper, delamination factor and metal removal rate as expressed From Eqs. 1 to 3. Kerf taper is measured by using the formula presented in Eq. 1 [23]. The delamination factor is measured by the expression shown in Eq. 2 [24], and metal removal rate is expressed in Eq. 3 [25].

$$\text{kerf taper} = \frac{L_t - L_b}{t} \tag{1}$$

$$\text{delamination factor} = \frac{W_0}{W} \tag{2}$$

$$\text{metal removal rate} = \frac{(L_t + L_b)tL_c}{2T} \tag{3}$$

where L_t denotes the kerf width at the top, L_b denotes the kerf width at the bottom, t denotes the thickness of the composite, W_0 denotes the maximum damaged width, W denotes the nominal width, L_c denotes the length of cut and T denotes the operating time.

2 Abrasive Flow Machining of PMC

Characterization is one of the important processes for any new composite material, and as a next step, machining becomes vital for finishing the material and also to support for assembly [26, 27]. AFM is one of the sustainable techniques used for polishing, de-burring and other finishing of machined components. It is used for finishing both external and internal surfaces which are very difficult to access. The efficiency of AFM technique mainly depends on its process parameters like volume flow rate of the abrasive, flow pressure, number of cycle, time, nature of the material, geometry of the material and properties of the abrasive particles [28, 29]. The major machinability parameters in AJM are surface roughness, kerf geometry and delamination [30]. A study has been made to investigate the efficiency of AFM technique for finishing the plastic gear teeth and reported that AFM technique is a more effective way for finishing the gear tooth as compared to the conventional methods. It has been recorded that, AFM technique increases the surface finish and accuracy of the tooth profile. Also, time consumed for machining is also very less than that of the conventional methods [31]. A research on abrasive water jet machining (AWJM) of carbon-reinforced epoxy laminates investigated the delamination and kerf taper during machining. It has been reported that the delamination goes up with an elevation in abrasive flow rate and pressure, whereas it goes down with a hike in the standoff distance and traverse rate. On the other hand, it is recorded that kerf taper goes down with an elevation in the pressure and goes up with a hike in the standoff distance and traverse rate [32].

Another research on the same material revealed that the standoff distance and feed rate have major dominance over kerf taper, delamination and metal removal rate. Also, it is noted that the surface roughness is affected by the fiber orientation and jet pressure. Among the three fiber orientations, namely 450, 600 and 900, a fiber orientation of 450 produced optimum results [33]. Apart from AWJM, two other methods, namely snow blasting and ice blasting, may also be done for material removal. The difference among these methods is only the jet material, and it is found that water jet is the most appropriate way for an optimum machinability. This has been proved by comparative testing on carbon fiber-reinforced plastics [34]. A comparative study has been made between the AWJM and conventional machining on glass fiber-reinforced composites and reported that AWJM is more suitable for machining under a minimum surface roughness as compared to the conventional machining [35] (Table 1).

3 Electric Discharge Machining of PMC

Another research electric discharge machining (EDM) is a widely used method for creating complicated geometries and also for making very small holes in components. The main advantage of this technique is that during machining, there is no contact between the tool and the workpiece and hence reduces the vibration and chatter. Also, this technique is mostly used for machining components of very small thickness. A study has been conducted to investigate the efficiency of EDM process on carbon–carbon composites and reported that the delamination, surface roughness and the thickness of recast layer vary proportionally to the input power. Also, it is suggested that high-quality holes could be produced only by reducing the temperature in the machining zone and are possible at low discharge energy [48]. The influence of cutting direction and fiber orientation is a major factor that affects the efficiency of EDM process. The metal removal rate goes up with the pulse-on duration during machining normally to the fiber direction, whereas it remains constant

Table 1 Optimum machining condition for AJM

S. No	Fiber/abrasive used	Matrix used	Optimized inputs	Optimized outputs	References
1	Carbon	Epoxy	JP = 75 MPa, SOD = 10 mm, NS = 200 mm/min	DF = 0.4, MRR = 10 mm^3/s, R_a = 2.5 μm	[33]
2	Sundi wood dust/garnet	Epoxy	JP = 150 MPa, SOD = 3.5 mm, NS = 125 mm/min	R_a = 0.1072 μm, PT = 0.0616 s	[36]
3	Graphite, glass	Epoxy	JP = 90 MPa, NS = 75 mm/min Abr % = 10 wt% SOD = 1 mm	KW = 0.864 mm	[37]
4	Carbon, glass	Polyester	FR = 600 g/min, JP = 320 MPa, SOD = 2 mm, NS = 1000 mm/min	R_a = 5.603 μm	[38]
5	Carbon	Epoxy	JP = 175 MPa, NS = 35 mm/s, FR = 200 g/min, SOD = 50 mm	MRR = 193 mm^3/s	[39]
6	Glass/silicon carbide	Epoxy	JP = 0.6 MPa, SOD = 2.5 mm	MRR = 3.8 g/s	[40]
7	Carbon	Epoxy	SOD = 0.59 mm, NS = 252.5 mm/min, t = 5.62 mm	R_a = 1.84 μm, KT = 0.37^0	[41]
8	Carbon	Polye-ether-ether-ketone	JP = 320 MPa, NS = 1.425 mm/s, FR = 5.5 g/s	MRR = 245.09 mm^3/s	[42]
9	Banana	Polyester	JP = 26 MPa, NS = 40 mm/min, SOD = 1 mm	KT = 0.43^0	[43]

(continued)

Table 1 (continued)

S. No	Fiber/abrasive used	Matrix used	Optimized inputs	Optimized outputs	References
10	Glass	Epoxy	FR = 100 g/min, JP = 150 MPa, SOD = 2 mm	R_a = 0.94 μm	[44]
11	Glass, carbon	Epoxy	NS = 137 mm/min, FR = 454 g/min, JP = 300 MPa, SOD = 2 mm	R_a = 3.38 μm, KT = 0.408^0	[45]
12	Carbon	Polye-ether-ether-ketone	JP = 200 MPa, NS = 4 mm/s, FR = 7 g/s, SOD = 2 mm	KT = 0.8^0	[46]
13	Glass/SiC	Epoxy	AS = 200 μm, JP 0.6 MPa, SOD = 0.5 mm	MRR = 0.621 g/min	[47]

JP Jet Pressure, *SOD* Standoff Distance, *NS* Nozzle speed, *t* thickness, *Abr* Abrasive, *FR* Abrasive Flow Rate, *AS* Abrasive Size, *DF* Delamination Factor, *MRR* Metal Removal Rate, R_a Surface roughness, *PT* Process Time, *KW* Kerf Width, *KT* Kerf Taper

when machining parallel to the fiber direction [49]. In a study, carbon fiber-reinforced plastics are subjected to dry EDM and oil EDM for de-burring of drilled holes. It is inferred that the metal removal rate is mainly influenced by the voltage, capacitance and gas pressure in both the EDM. Also, it is found that when oxygen is used as dielectric medium, the metal removal rate is enhanced by three times, and when air is used as dielectric, the metal removal rate is enhanced by two times. The overall performance of dry EDM is found to be better than that of the oil EDM for de-burring of drilled holes [50].

EDM on carbon fiber-reinforced plastics has been made to produce micro-holes of size 110 microns using tungsten carbide electrode. A highest aspect ratio of 10.9 has been achieved for a drill depth of 1.2 mm, and it is reported that this could be possible by targeting the discharge electrons only on the carbon layers. Also, it is shown that a high level of voltage and capacitance must be used for achieving a maximum metal removal, whereas a low level of voltage and capacitance must be used for a lowest tool wear [51]. The electrode material used during EDM has a significant effect on the output parameters. A study on the EDM of carbon-reinforced plastics concluded that copper electrodes are far better than the graphite electrodes in terms of tool wear, and also, it is suggested that tools with positive polarity produce the highest metal removal. Also, a high current density makes the matrix to smear over the surface and affects the metal removal; hence, a minimum current density must be used for a highest metal removal [52]. The optimum conditions arrived during EDM of PMC are presented in Table 2.

Table 2 Optimum machining conditions for EDM

S. No	Fiber/electrode used	Matrix used	Optimized inputs	Optimized outputs	References
1	Carbon	Carbon	PC = 5 A, POD = 100 μs	MRR = 0.75 mm^3/min	[48]
2	Carbon/copper, graphite	Epoxy	POD = 200 μs, Duty factor = 20%	MRR = 2.8 mm^3/min, R_a = 5 μm	[49]
3	Carbon	Epoxy	C = 10 pF, OCV = 220 V, GP = 0.3 MPa	MRR = 16–18 mm^3/min,	[50]
4	Carbon	Epoxy	PC = 1.2 A, POD = 110 μs	MRR = 10.5 mm^3/min	[52]
5	Carbon	Epoxy	PC = 1A, POD = 100 μs, OCV = 120 V, ES = 1000 rpm	MRR = 0.015 g/min R_a = 8 μm	[53]
6	Carbon	Epoxy	OCV = 50 V, PC = 2 A, POD = 70 μ mim	R_a = 3.1 μm	[54]

PC Pulse Current, *POD* Pulse-On Duration, *C* Capacitance, *GP* Gas Pressure, *OCV* Open Circuit Voltage, *ES* Electrode Speed, *MRR* Metal Removal Rate

4 Statistical Tools for Unconventional Machining of PMC

The use of statistical tools is inevitable for any machining processes, and nowadays, design of experiments is used as a common tool for machining [55]. Like the conventional machining analysis, many statistical tools have been utilized to assess the machinability parameters during unconventional machining of PMC. Among different tools, response surface method (RSM) is one of the effective ways for designing an experiment, analyzing the data, optimizing the results and predicting the parameters. RSM may be used in different ways like central composite design (CCD), D-optimal design [56–59] or Box–Behnken Design [60]. A study on the delamination and kerf geometry of carbon-reinforced composites used CCD for designing the experiment. Mathematical models have been created to predict the responses and suit well for the study. Optimization has been done on the bases of desirability function and reported that the output parameters are mainly affected by the process parameters, namely the abrasive flow rate, traverse rate, pressure and standoff distance [30]. In addition, CCD effectively helps to reduce the number of experiments and thus eliminates unnecessary experimentations [1, 61, 62]. A study on the AWJM on wooden-dust-reinforced plastics used Box–Behnken design for analysis and optimization of responses, namely surface roughness and process time. It is shown that the surface roughness is predominantly affected by the pressure and nozzle speed, whereas the process time is affected only by the nozzle speed. Also, it is concluded that the response surface method is found to be an apt way for predicting and analyzing the data [32].

Optimization may be done in several ways. Technological research makes use of design softwares for optimizing the design dimensions and statistical tools for process optimization [63]. Taguchi method is another technique which is most widely used in all engineering problems. The experiments are designed on the basis of orthogonal array and the data analysis, prediction and optimization also done efficiently. A research on the EDM of carbon-reinforced plastics used Taguchi technique and suggested that the discharge current is the most dominating factor for surface roughness [64]. A study on the top kerf width during AWJM of graphite-filled glass-reinforced composites used sequential quadratic programming (SQP) for optimization. It is found that the standoff distance does not have significance on the kerf width and the kerf width could be minimized by using appropriate nozzle diameter [37]. Neural network modeling is another advanced technique for designing and predicting of optimal parameters. A research used neural network modeling in combination with the CCD for prediction of metal removal rate, bottom kerf and top kerf during AJM of glass-fiber-reinforced plastics. It has been reported that the jet pressure affects both the bottom and top kerf predominantly. Also, it is noted that the neural network modeling effectively predicts the process parameters of AJM [40]. Hence, the statistical tools are proven to be best while assessing the performances of natural composites [65, 66, 67].

5 Conclusion

The present survey has made a comprehensive review on various literatures available relating to AFM and EDM of polymeric composites. The major conclusions obtained are as follows:

- The efficiency of AFM depends on volume flow rate of the abrasive, flow pressure, number of cycle, time, nature of the material, geometry of the material and properties of the abrasive particles.
- The major machinability parameters in AJM are surface roughness, kerf geometry and delamination. AFM technique is a more effective way for finishing the gear tooth as compared to the conventional methods.
- EDM is a widely used method for creating complicated geometries and also for making very small holes in components under a reduced vibration and chatter.
- Metal removal rate of EDM is mainly influenced by the voltage, capacitance and gas pressure. When oxygen is used as dielectric medium, the metal removal rate is enhanced by three times, and when air is used as dielectric, the metal removal rate is enhanced by two times.
- The electrode material used during EDM has a significant effect on the output parameters. Copper electrodes are far better than the graphite electrodes in terms

of tool wear, and also, the tools with positive polarity and minimum current density produce the highest metal removal.

References

1. Vinayagamoorthy R, Rajmohan T (2018) Machining and its challenges on bio-fibre reinforced plastics: a critical review. J Reinf Plast Compos 37:1037–1050
2. Azmi AI, Lin RJT, Bhattacharya D (2012) Experimental study of machinability of GFRP composites by end milling. Mater Manuf Process. 27:1045–1050
3. Vinayagamoorthy R, Ankur Sharma, Vignesh I, Navneeth G (2018) Investigation of surface damages in hole making on luffa/jute/glass reinforced plastics. Adv Manuf Process Lect Notes Mech Eng 521–532
4. Vinayagamoorthy R, Venkatakoteswara Rao G (2019) Synthesis and property analysis of green resin based composites J Thermoplast Compos Mater
5. Shanmugam D, Thiruchitrambalam M (2014) Influence of alkali treatment and layering pattern on the tensile and flexural properties of palm leaf stalk fiber (PPLSF)/jute fiber polyester hybrid composites. Compos Inter 21:3–12
6. Rajmohan T, Vinayagamoorthy R, Mohan K (2018) Review on effect machining parameters on performance of natural fibre–reinforced composites (NFRCs) J Thermoplast Compos Mater https://doi.org/10.1177/0892705718796541
7. Vinayagamoorthy R, Manoj IV, Narendra Kumar G, Sai Chand I, Sai Charan Kumar GV, Suneel Kumar K (2018) Challenges on the synthesis, characterization and machining of green fiber plastics: a review. IOP Conf Ser Mater Sci Eng 390:012029
8. Rajesh G, Ratna Prasad AV, Gupta AVSSKS (2015) Mechanical and degradation properties of successive alkali treated completely biodegradable sisal fiber reinforced poly lactic acid composites. J Reinf Plast Compos 34:951–961
9. Vinayagamoorthy R, Koteshwar TN, Madhav VV, Sai TK, Konda V (2019) Drilling associated parametric investigations on chemically treated natural fiber composite. Mater Today Proc 16:277–283
10. Vinayagamoorthy R, Rajeswari N (2014) Mechanical performance studies on Vetiveria zizanioides/jute/glass fiber-reinforced hybrid polymeric composites. J Reinf Plast Compos 33:81–92
11. Sathish S, Kumaresan K, Prabhu L, Vigneshkumar N (2017) Experimental investigation on volume fraction of mechanical and physical properties of flax and bamboo fibers reinforced hybrid epoxy composites. Polym Polym Compos 25:229–236
12. Vinayagamoorthy R (2019) Influence of fiber surface modifications on the mechanical behavior of Vetiveria zizanioides reinforced polymer composites. J Nat Fib 16:163–174
13. Vinayagamoorthy R (2018) A review on the machining of fiber-reinforced polymeric laminates. J Reinf Plast Compos 37:49–59
14. Espert A, Vilaplana F, Karlsson S (2002) Comparison of water absorption in natural cellulosic fibres from wood andone-year crops in polypropylene composites and its influence on their mechanical properties. Compos Part A 35:1267–1276
15. Vinayagamoorthy R, Sivanarasimha S, Vijay P, Vedula G, Karthikeyan S (2015) Experimenal studies on cater absorption and thermal degradation of natural composites. Int J Appl Eng Res 10:663–668
16. Vinayagamoorthy R, Sivanarasimha S, Vinay Kumar KR, Vijay P (2015) Characteristic investigations on loofah, jute and glass fiber reinforced sandwich polymeric composites. Appl Mech Mater 813–814, 14–18
17. Tsao CC, Hocheng H (2008) Evaluation of thrust force and surface roughness in drilling composite material using Taguchi analysis and neural network. J Mater Process Technol 203:342–348

18. Vinayagamoorthy R (2017a) A review on the polymeric laminates reinforced with natural fibers. J Reinf Plast Compos 36:1577–1589
19. Vinayagamoorthy R, Konda V, Tonge P, Koteshwar TN, Premkumar M (2019) Surface roughness analysis and optimization during drilling on chemically treated natural fiber composite. Mater Today Proc 16:567–573
20. Debnath K, Singh I, Dvivedi A (2014) Drilling characteristics of sisal fiber-reinforced epoxy and polypropylene composites. Mater Manuf Process 29:1401–1409
21. Vinayagamoorthy R, Kanakadandi GS, Tuduru NK, Yerasi HR (2016) Modeling and analysis of drilling induced damages on hybrid composites. Indian J Sci Technol 9:1–10
22. Vinayagamoorthy R (2020) Mechanical Performance of glass- and biofibre-reinforced hybrid composites. Glass Fibre Reinf Polym Compos Adv Compos 12:1–16
23. Azmir MA, Ahsan AK (2009) A study of abrasive water jet machining process on glass/epoxy composite laminate. J Mater Process Technol. 20:6168–6173
24. Liu DF, Tang YJ, Cong WL (2012) A review of mechanical drilling for composite laminates. Compos Struct 94:1265–1279
25. Korat MM, Acharya GD (2014) A review on current research and development in abrasive waterjet machining. Int J Eng Res Appl 4:423–432
26. Vinayagamoorthy R, Karthikeyan S, Prem Bhargav RS, Rajivalochan TV (2015) Properties investigations on metallic fiber reinforced sandwich composites. Appl Mech Mater 813–814:101–105
27. Vinayagamoorthy R, Saswath Kaundinya SL, Mani Teja GLSN, Kovuru A (2016) A study on the properties of natural sandwich laminates. Indian. J Sci Technol 9:1–6
28. Pusavec F, Krajnik P, Kopac J (2010) Transitioning to sustainable production—Part I: application on machining technologies. J Clean Prod 18:2174–2184
29. Wang AC, Weng SH (2007) Developing the polymer abrasive gels in AFM process. J Mater Process Technol 192–193:486–490
30. Vigneshwaran S, Uthayakumar M, Arumugaprabu V (2017) Abrasive water jet machining of fiber reinforced composite materials. J Reinf Plast Compos. 37:230–237
31. Kenda J, Duhovnik J, Tavčar J, Kopač J (2014) Abrasive flow machining applied to plastic gear matrix polishing. Int J Adv Manuf Technol 71:141–151
32. Ajit D, Shailendra K (2017) Experimental study of delamination and kerf geometry of carbon epoxy composite machined by abrasive water jet. J Compos Mater 51:3373–3390
33. Prasad Unde D, Gayakwad MD, Patil NG, Pawade RS, Thakur DG, Brahmankar PK (2015) Experimental investigations into abrasive waterjet machining of carbon fiber reinforced plastic. J Compos 2015:1–10
34. David Z, Christian B, Thomas H, Nico W, Werner H., Frank A, Friedrich WB, Thomas H (2012) Repair preparation of fiber-reinforced plastics by the machining of a steeped peripheral zone. SV-J Mech Eng 58:571–577
35. Thiagarajan A, Padmapriyaa G, Raveena R, Velmurugan K, Venkatachalapathy VSK (2017) Investigation of surface roughness on fiber reinforced composite using abrasive water jet machining. Int J Adv Res Manage Eng Technol 1:25–28
36. Jagadish, Sumit B, Amitava R (2016) Prediction and optimization of process parameters of green composites in AWJM process using response surface methodology. Int J Adv Manuf Technol 87:1359–1370
37. Ankur T, Brajesh V, Bhawna D (2015) Optimization of top kerf width produced on graphite filled glass fiber reinforced epoxy composite through abrasive water jet machining. Int J Eng Res Dev 11:33–41
38. Ming IWM, Azmi AI, Chuan LC, Mansor AF (2017) Experimental study and empirical analyses of abrasive water jet machining for hybrid carbon/glass fiber-reinforced composites for improved surface quality. Int J Adv Manuf Technol 95:3809–3822
39. Hocheng H, Tsai NY, Shiue JJ (1997) Feasibility study of abrasive-water jet milling of fiber-reinforced plastics. J Manuf Sci Eng 119:133–142
40. Madhu S, Balasubramanian M (2017) Neural network based optimization of abrasive jet process parameters in machining GFRP composites. Int J Eng Technol Manage. Appl Sci 5:425–435

41. Jagadeesh B, Dinesh Babu P, Nalla Mohamed M, Marimuthu P (2017) Experimental investigation and optimization of abrasive water jet cutting parameters for the improvement of cut quality in carbon fiber reinforced plastic laminates. J Ind Tex 48:178–200
42. Weiyi Lia., Hongtao Zhub., Jun Wanga., Chuanzhen Huang.: Radial-mode abrasive water jet turning of short carbon–fiber-reinforced plastics. Mach Sci Technol 20:231–248
43. Arumuga Prabu V, Thirumalai Kumaran S, Uthayakumar M (2016) Performance evaluation of abrasive water jet machining on banana fiber reinforced polyester composite. J Nat Fib 14:450–457
44. Ibraheema HAI, Iqbalb A, Hashemipourb M (2017) Numerical optimization of hole making in GFRP composite using abrasive water jet machining process. J Chinese Inst Eng 38:66–76
45. Selvam R, Karunamoorthy L, Arunkumar N (2016) Iinvestigation on performance of abrasive water jet in machining hybrid composites. Mater Manuf Process 32:700–706
46. Thongkaew K, Wang J, Li W (2017) An investigation of the hole machining processes on woven carbon-fiber reinforced polymers (CFRPs) using abrasive waterjets. Mach Sci Technol https://doi.org/10.1080/10910344.2018.1449217
47. Suresh R, Sohit Reddy K, Karthik S (2017) Abrasive jet machining for micro-hole drilling on glass and GFRP composites. Mater Today Proc 5:5757–5776
48. Guu YH, Hocheng H, Tai NH, Liu SY (2017) Effect of electrical discharge machining on the characteristics of carbon fiber reinforced carbon composites. J Mater Sci 36:2037–2043
49. Sameh H, Akira O, Sho I (2013) Effect of cutting direction on machining of carbon fibre reinforced plastic by electrical discharge machining process. Int J Mach Mach Mater 13:414–427
50. Islam MM, Li CP, Ko TJ (2017) Dry electrical discharge machining for deburring drilled holes in CFRP composite. Int J Precis Eng Manuf Green Technol 4:149–154
51. Ravinder K, Pramod KA, Inderdeep S (2018) Fabrication of micro holes in CFRP laminates using EDM. J Manuf Process 21:859–866
52. Lau WS, Wang M, Lee WB (1990) Electrical discharge machining of carbon fibre composite materials. Int J Mach Tools Manuf 30:297–308
53. Sameh H, Akira O (2016) Influence of electrical discharge machining parameters on cutting parameters of carbon fiber-reinforced plastic. Mach Sci Technol 20:99–114
54. Lodhi BK, Verma D, Shukla R (201) Optimization of machining parameters in EDM of CFRP composite using Taguchi technique. Int J Mech Eng Technol 5:70–77
55. Kannan P, Balasubramanian K, Vinayagamoorthy R (2015) Defect reduction in ring blank casting through design of experiments. Int Rev Mech Eng 9:536–541
56. Vinayagamoorthy R, Rajeswari N, Balasubramanian K (2014) Optimization studies on thrust force and torque during drilling of natural fiber reinforced sandwich composites. Jordan J Mech Ind Eng 8:385–392
57. Rajmohan T, Palanikumar K (2013) Modeling and analysis of performances in drilling hybrid metal matrix composites using D-optimal design. Int J Adv Manuf Technol 64:1249–1261
58. Vinayagamoorthy R, Rajeswari N, Vijayshankar S, Balasubramanian K (2014) Drilling performance investigations on hybrid composites by using d-optimal design. Int Rev Mech Eng 8:952–961
59. Vinayagamoorthy R, Rajeswari N, Vijayshankar S, Vivekanandan M, Sri Rama Murthy B, Venkata Subramaniam KR (2014) Surface and sub-surface analysis of hybrid polymeric composites during machining operations. Proc Mater Sci 5:2075–2083
60. Vinayagamoorthy R (2017b) Parametric optimization studies on drilling of sandwich composites using the Box-Behnken design. Mater Manuf Process 32:645–653
61. Jung UH, Kim JH, Kim JH, Park CH, Jun SO, Choi YS (2016) Optimum design of diffuser in a small high-speed centrifugal fan using CFD and DOE. J Mech Sci Technol 30:1171–1184
62. Vinayagamoorthy R, Manoj IV, Narendra Kumar G, Sai Chand I, Sai Charan Kumar GV, Suneel Kumar K (2018) A central composite design based fuzzy logic for optimization of drilling parameters on natural fiber reinforced composite. J Mech Sci Technol 32:2011–2020
63. Vinayagamoorthy R, Mothilal T, Madhavan S (2009) Performance enhancement of a diesel engine by providing insulations on engine parts. Int J Des Manuf Technol 3:63–67

64. Srikanth DV, Sreenivasa Rao M (2014) Machining of FRP composites by abrasive jet machining optimization using Taguchi. Int J Mech Mechatron Eng 8:632–636
65. Vinayagamorthy R (2019) Trends and challenges on the development of hybridized natural fiber composites J Nat Fib https://doi.org/10.1080/15440478.2019.1598916
66. Vinayagamoorthy R (2020) Effect of particle sizes on the mechanical behaviour of limestone-reinforced hybrid plastics. Polym Polym Compos 28(6):410–420
67. Vinayagamoorthy R Influence of fibre pretreatments on characteristics of green fabric materials. Polym Polym Compos 096739112094346

Influence of Process Variables on the Ultimate Tensile Strength of Friction Stir Welded AA6061 Matrix Composite

Arun Kumar Shettigar, Subramanya R. Prabhu B., Mervin A. Herbert, and Shrikantha S. Rao

Abstract The present study is focused on the application of the friction stir welding process (FSW) to weld aluminium matrix composites (AMCs). Joints are formed by varying FSW process variables such as tool revolving speed (TRS), tool traverse speed (TTS) and the tool pin geometry (TPG). Influence of these parameters on the ultimate tensile strength (UTS) of the joints is investigated. Process variable optimization is done using Taguchi L18 orthogonal array design. Optimum process variables are determined and confirmed by confirmation tests based on the analysis of variance.

Keywords Friction stir welding · Aluminium matrix composite · Ultimate tensile strength · Optimization

1 Introduction

Ability of joining hard to weld materials such as aluminium, magnesium, alloys and aluminium matrix composites made friction stir welding a novel welding process [1]. Nowadays, FSW process is used generously to join aluminium alloys, magnesium and its alloys, copper, steel and even dissimilar metals. FSW eliminates the common defects which are associated with the fusion welding process such as liquation and solidification cracking, porosity, segregation of reinforcing particles, loss of alloying materials, etc.[2]. Although FSW has numerous advantages, improper selection of process variables results in defective welds as well as poor joint strength. Selection of suitable range for process variables and evaluation of these variables plays an important role in attaining superior mechanical properties. Several studies were done to understand the effect of process variables on metallographic behaviour and

A. K. Shettigar · S. R. Prabhu B. · M. A. Herbert · S. S. Rao
Department of Mechanical Engineering, National Institute of Technology Karnataka, Surathkal, India

S. R. Prabhu B. (✉)
Department of Mechatronics Engineering, Manipal Institute of Technology, Manipal Academy of Higher Education, Manipal, India
e-mail: Subramanya.prabhu@gmail.com

T. Rajmohan et al. (eds.), *Advances in Materials and Manufacturing Engineering*, Springer Proceedings in Materials 7, https://doi.org/10.1007/978-981-15-6267-9_6

joint strength of FS welded parts [2–5]. Most of the study was focused on the classical experimental study where one variable was varying by keeping other variables constant and finding its effect on the joint strength. This technique consumes a lot of money and time.

A statistical technique developed by Taguchi and Konishi has to optimize the industrial problems in order to understand the behaviour of the chosen process variables. It is a simple yet powerful method to identify the best value of design variables. Few studies were performed using orthogonal technique to improve the optimum values of the process variables [4–8]. Lakshminarayanan and Balasubramanian (2008) effectively used Taguchi techniques in FSW of aluminium alloys by considering TRS, TTS and the axial force as variables to understand the effect of each variable on the joint strength. Sundaravel et al. (2010) used Taguchi-based grey relational analysis for multi-objective optimization of FSW process variables during welding of AA5083 alloy. The variables are TRS, TTS and axial force correlating the UTS of the FSW joint.

The influence of the tool geometry on the UTS of the aluminium alloys is evaluated by taking TRS, TTS and the depth of tool plunge as process variables were analysed by Suresh et al. (2013). The dynamic volume to static volume ratio of the FSW tool pin plays a key role on improving the joint strength. Several studies were performed in understanding the effect of tool geometry on the strength of the joint. But a very few studies were done on usage of dual profiled pin in FSW. FSW tool with dual profiled pin offers different dynamic to static volume ratios, alters the material flow and enhances the joint strength.

2 Methodology

Based on process variables, levels and the interaction effects, standard orthogonal designs are chosen for the current study. The FSW process involves TRS, TTS and the TPG as three control factors. For each factors, three levels were chosen based on the trial experiments. The range for the variables were determined by conducting several experiments, and it is fixed by considering the range which gives defect-free welds. Values of three levels for TTS and TRS were fixed by taking maximum, minimum and mid values within the range. The FSW was performed for AA6061–4.5%Cu–5%SiC and AA6061–4.5%Cu–10%SiC composite. The composition of SiC is also selected as one of the control factors with two levels. Mixed level L18 orthogonal array is chosen for the evaluation as the degrees of freedom are less than 18. Table 1 shows the control factors and their levels used in the current study. UTS of the FSW joint is taken as the response of the study. The effect of variables on the responses is determined through the signal-to-noise (S/N) ratio and means. As the higher UTS is desirable, "Larger is Better" criterion is chosen for the study. Using S/N ratio, the shift of the joint quality from the experimental values was evaluated and shown by Eq. (1)

Table 1 Control factors and their levels

Control factor	Level-1	Level-2	Level-3
% Sic	5	10	–
TRS (rpm)	710	1000	1400
TTS (mm/min)	50	63	80
TPG	1 (TC)	2 (DP)	(SP)

$$\eta_y = -10 \log\left[\frac{1}{n}\sum b_{xyz}^2\right] \quad (1)$$

where z is the test number and b_{xyz} is the experimental values of the x quality behaviour in the yth experiment at the nth test. After the analysis, the response values are shown in terms of means of UTS and S/N ratio.

AMCs with 100 × 50 × 5 mm size were used for the study. The matrix is reinforced with SiC particles with 5% AND 10% compositions. The FSW is performed on vertical milling machine (BFW). Figure 1 depicts the FSW setup. The tool is tilted by an angle 1.5° with respect to workpiece to improve the axial force. The parent material moved in the tool traverse direction because of increase of pressure resulted from the tilting of the tool. Three types of TPG were used in the study, namely threaded cylindrical pin (TC), square pin (SP) and the dual profiled pin (DP) with threaded cylindrical and the square profile on a single pin. The dynamic to static volume ratio (DSR) of TC, SP and DP was 1.01, 1.26 and 1.17, respectively. Dynamic volume of the pin is represented by the volume swept by the pin during rotation, whereas static volume refers to the volume of the pin at the rest. Higher DSR results in better mixing of the material at the weld zone as well as higher heat input due to the larger frictional force at the tool workpiece interface [9]. The tensile test specimen is prepared as per the ASTM E8-04 standards from the welded composite, across the weld direction using wire electro-discharge machining. Tensile test is conducted on computerized tensometer.

Fig. 1 FSW setup

3 Results and Discussion

Analysis of variance at 95% level of confidence is used to evaluate the influence of variables, namely TRS, TTS, TPG and the % of Sic particles. Tables 2 and 3 list the ranking of each variable on using Taguchi DOE and evaluation for means and S/N ratio achieved at various variables. The term delta in the Tables 2 and 3 shows the acute alter of UTS resulting from the factorial deviation. Larger difference in the delta values reveals that the response is greatly affected with respect to the control factor under consideration. Tables 2 and 3 indicate that the % composition of SiC is the key factor affecting the UTS followed by TRS. TTS and TPG have exhibited minimal effect on the UTS.

Using analysis of variance (ANOVA) of the means, the relative importance of the control factors was further evaluated. Tables 4 and 5 list the outcome of ANOVA for mean and S/N ratio of the UTS, respectively. The results indicated that % of SiC, TRS, TTS and TPG are consequential variables which influence the UTS. The interaction among the % SiC and the TRS does not have major influence on the UTS. To measure the importance of each term existing in the model, percentage contribution (PC) is used which is defined as the ratio of sum of square to the total sum of square. By looking into the PC of various factors for UTS, the % SiC shows the highest contribution of 72.43%, whereas TRS follows the %SiC with a contribution of 14.70%. The other factors show minimal impact. Thus, %SiC is a key factor to be considered followed by TRS while welding of composite by FSW technique.

The optimum process variables for maximizing the UTS can be evaluated by considering the factors consisting larger S/N values and means [10]. The condition for maximizing UTS is obtained as

Table 2 Mean value

Level	% SiC	TRS (rpm)	TWS (mm/min)	TPG
1	141.1	162.4	158.3	163.5
2	200.2	189.9	171.1	171.0
3	–	159.2	182.1	177.4
Delta	59.1	30.7	23.5	14.0
Rank	1	2	3	4

Table 3 S/N ratio value

Level	% SiC	TRS (rpm)	TWS (mm/min)	TPG
1	42.92	44.06	43.78	44.06
2	45.99	45.43	44.52	44.45
3		43.89	45.07	44.86
Delta	3.06	1.54	1.29	0.80
Rank	1	2	3	4

Table 4 ANOVA for means of UTS

Source	DF	Seq SS	*F*-ratio	*P*-value	P%
% SiC	1	15,727.5	658.38	0	72.69
TRS (rpm)	2	3395.9	71.08	0	15.70
TTS (mm/min)	2	1658.5	34.71	0	7.67
TPG (TC BP SP)	2	587.2	12.29	0.04	2.71
%SiC * TRS (rpm)	2	75.5	1.58	0.264	0.35
Residual error	8	191.1	–	–	0.88
Total	17	21,635.7	778.04		100

Table 5 ANOVA for S/N of UTS of composite joined through friction stir welded

Source	DF	Seq SS	*F*-ratio	*P*-value	P%
% SiC	1	42.2414	639.55	0	72.43
TRS (rpm)	2	8.5738	64.91	0	14.70
TTS (mm/min)	2	5.0333	38.1	0	8.63
TPG (TC BP SP)	2	1.9095	14.46	0.002	3.27
%SiC * TRS (rpm)	2	0.0378	0.29	0.758	0.06
Residual error	8	0.5284	–	–	0.91
Total	17	58.3242	757.31		100

$$\begin{aligned}\text{Predicted S/N} &= \%\ \text{SiC\_2} + \text{TRS\_2} + \text{TTS\_2} + \text{TPG\_3} - 3\times \\ &= 45.99 + 45.43 + 45.07 + 44.86 - 3(44.45) = 47.97 \end{aligned} \quad (2)$$

where x represents the average S/N ratio
Predicted mean = 202.2 + 189.9 + 182.1 + 174.4 −3(170.61) = 237.74 N/mm^2.

4 Confirmation Test

As the above variable combination was not present in the L18 experiment, the predicted value of the UTS was tested by performing experiments at those levels. The average UTS attained from the experiment is 244 N/mm^2. The S/N ratio was within the range of confidence interval. The mean effect plots of all the factors are shown in Fig. 2. The mean value is utilized to evaluate the importance of each control factor. From Fig. 2, it can be seen that the % SiC, TRS, TTS and the TPG have major effect on the UTS. It can also be observed that an increase in the % of SiC the UTS also increases. Increase in UTS is due to the presence of SiC particles which prevents the movement of the dislocation. As the % of SiC increases, the particle-to-particle

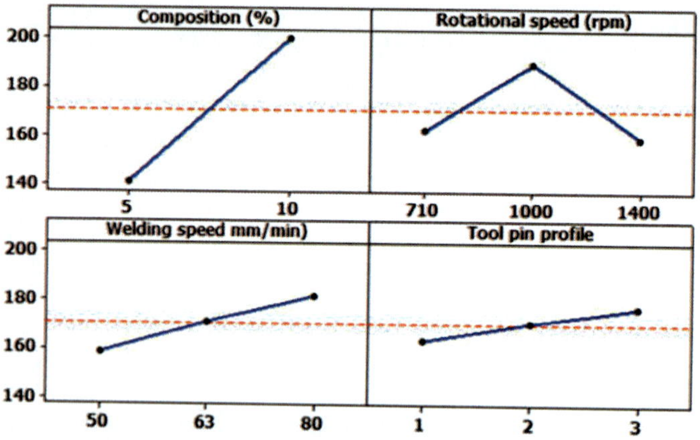

Fig. 2 Main effects plot for means of friction stir welded composites

distance decreases, leading to the increase in the resistance to the motion of dislocation. Hence, during the deformation of the material, either the deformation has to bypass the particle or matrix has to push the particles further. Due to this, pile up of dislocation was taking place and the plastic flow in the matrix was restricted which enhances the composite strength. The dislocation density increases as the quantity of particles increases which improves the UTS. As the TRS increases from 710 to 1000 rpm, the UTS of the joint increases and the further increase in the TRS reduces the UTS. TRS controls the quantity of heat generated at the weld region [11]. At lower TRS, the lesser heat is generated irrespective of the TWS, causing lesser plasticization and the improper material flow which results in lower UTS, whereas at higher TRS, huge amount of heat was generated at the weld area resulting in turbulent material flow and coarsening of grains, which ultimately reduces the UTS. From the study, it was understood that neither the low TRS nor the high TRS gives a better joint strength.

The TTS controls the heat input quantity at the weld region [11]. It has observed that as the TTS increases, the UTS also increases. This is due to the diminishing of softened area as the TTS increases. Lower TTS provides higher heat input and slower cooling, thereby assisting in grain growth. Coarser grain at the weld region reduces the UTS. As the TTS increases, the UTS also increases till a critical TTS value. Beyond this, the weld defects were farmed such as tunnel hole, pin hole, etc.

The TPG has a significant effect on the material flow in the weld region. As the DSR of SP was better than the TC and DP tool, it provides grain refinement. SP tool provides pulsating action during rotation of the tool and sweeps more material for the given static volume, thereby improving grain structure at the weld zone. Smaller grains and uniform distribution of particles in the weld zone enhance the joint strength. The DP tool exhibits close value of UTS as that of SP tool, whereas the UTS of the joint fabricated by the TC tool is much lower. The heat generated during the process with the TC tool is much lesser than the other tools due to the lesser area of contact and the nonappearance of pulsating action

5 Microstructure Analysis of FS Welded Composite

Figure 3 shows the microstructure of stir zone of FS welded composite at TRS of 1000 rpm and the TTS of 80 mm/min using SP tool. The average size of the grains found at the stir zone is 2.13 ± 0.14 microns. Stirring action of the tool at the weld zone develops plastic deformation and generation of frictional heat which results in the grain refinement [12, 13]. Also, stirring action of the tool leads to the redistribution of the particle from segregated heterogeneous distribution at the base metal to the uniform distribution at the stir zone. Generally, the dislocation density is low at the stir zone. Compared to the base material, the stir zone consists of numerous tiny hard SiC particles. Striking of reinforced particle to one other and the abrasive effect of the revolving FSW tool leads to the tiny particles. As the tool rotates and travels along the weld line, it caries the particle along the periphery of the tool pin and redistribute the particles uniformly in the weld region. The load bearing capacity of the composite increases due to the presence of homogeneous distribution of particles [14]. Hence, increase in the % of SiC results in the enhanced joint strength and the hardness. The stir zone showed 25% higher hardness as compared to the base material with a hardness of 105 Hv.

Figure 4 depicted the fracture surface of the FS welded composite joined at a TRS of 1000 rpm and TTS of 80 mm/min using SP tool. The fracture surface indicates

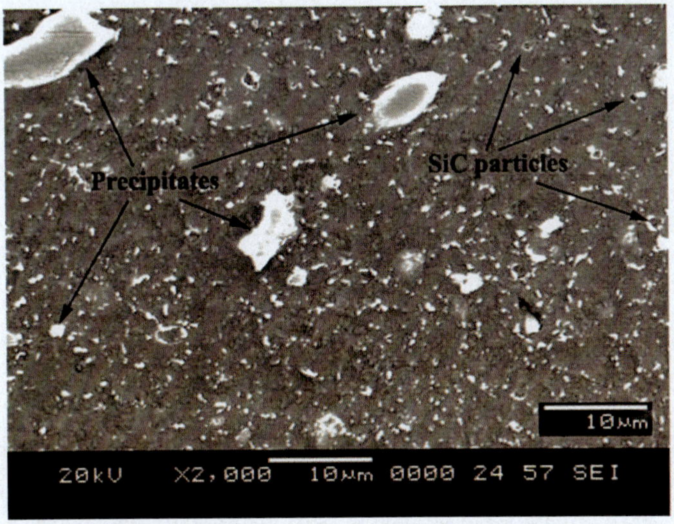

Fig. 3 Nugget zone SEM image of Al–Cu–Mg–Si–10%SiC

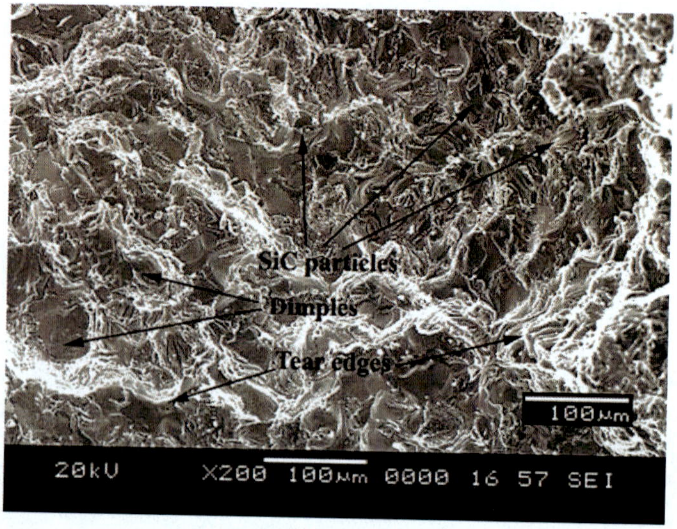

Fig. 4 Facture surface of friction stir welded composite

the ductile fracture, characterized by dimples with tearing edges. All the specimens are fractured at the heat-affected zone which was the weakest zone in the weld area.

6 Conclusions

- FSW of AMCs was successfully accomplished, and the optimum value of the variables which gives maximum UTS has been ascertained. It has been observed that % of SiC has the highest contribution in achieving higher UTS with a contribution of 72.43% followed by tool rotational speed of 14.70% tool traverse speed of 8.63% and the tool pin geometry of 3.27%.
- As the % of SiC increases, the UTS also increases due to the presence of more number of hard particles which restricts the movement of dislocation
- As the tool rotational speed increases, the UTS also increases up to a certain value, and then, it starts to decrease as speed increases further. The highest UTS was obtained at a speed of 1000 rpm.
- UTS has a direct relationship with tool traverse speed. As the speed increases, the UTS also increases, but the increase of speed beyond certain value results in defective welds.
- Square profiled tool offers better UTS due to the grain refinement caused by the pulsating action of the tool.
- An adequate agreement has been attained between the optimum value of the variables obtained from ANOVA and those from the confirmation experiments.

References

1. Mishra RS, Ma ZY (2005) Friction stir welding and processing. Mater Sci Eng R 50:1–78
2. Subramanya P, Amar M, Arun S, Mervin H, Shrikantha R (2018) Friction stir welding of aluminium matrix composites—a review. MATEC Web Conf 144:03002
3. Kumar A, Nayak V, Herbert MA, Rao SS (2014) Microstructure and hardness of friction stir welded aluminum-copper matrix based composite reinforced with 10wt.% SiCp. Mater Res Inn 18:84–89
4. Lakshminarayanan AK, Balasubramanian V (2008) Process parameters optimization for friction stir welding of RDE-40 aluminium alloy using Taguchi technique. Trans Nonferrous Met Soc China 547–554
5. Nami H, Adgi S, Shamabadi (2011) Microstructure and mechanical properties of friction stir welded Al/Mg_2Si metal matrix cast composite Mater Des 32:976–983
6. Periyasamy P, Mohan B, Balasubramanian V, Rajakumar S, Venugopal S (2013) Multi-objective optimization of friction stir welding parameters using desirability approach to join Al/SiC p metal matrix composites. Tran Nonfer Met Soci China 23(4):942–955
7. Sundaravel V, Raju R, Rao SRK (2010) Multiobjective optimization of friction stir welding process parameters on aluminum alloy AA 5083 using Taguchi-based grey relation analysis. Mater Manuf Process 25:1206–1212
8. Suresh S, Shenbaga N, Vinayaga M (2013) Process development in stir casting and investigation on microstructures and wear behavior of TiB_2 on Al6061 MMC. Procedia Eng 64:1183–1190
9. Elangovan K, Balasubramanian V (2008) Influences of tool pin profile and welding speed on the formation of friction stir processing zone in AA2219 aluminium alloy. J Mater Process Techno 200:163–175
10. Tansel I, Gulmez N, Demetgul (2011) Taguchi method-gonns integration: complete procedure covering from experimental design to complex optimization. Expert Syst Appl 38:4780–4789
11. Prabhu S, Shettigar A, Rao K, Herbert M, Rao S (2017) Influence of welding process parameters on microstructure and mechanical properties of friction stir welded aluminium matrix composite. Mater Sci Forum 880:50–53
12. Kumar A, Murugan (2014) Optimization of friction stir welding process parameters to maximize tensile strength of stir cast AA6061-T6/AlN p composite. Mater Des 57:383–393
13. Arunkumar G, Herbert M, Rao S (2013) Microstructural characterization and hardness evaluation of friction stir welded composite AA6061-4.5Cu-5SiC (wt.%). Def Sci J 63(4):1–6
14. Prabhu SR, Shettigar A, Herbert M, Rao S (2018) Multi response optimization of friction stir welding process variables using TOPSIS approach. IOP Conf Ser Mater Sci Eng 376:012134

Mechanical Properties and Microstructural Characteristics of Friction Stir Welded Aluminium Matrix Composite

B. Subramanya R. Prabhu, Arun Kumar Shettigar, Mervin A. Herbert, and Shrikantha S. Rao

Abstract Nowadays, friction stir welding process appears a promising technique to weld difficult materials by conventional welding techniques. Present study aims to analyse the significance of process variables on the mechanical behaviour of aluminium matrix composite joined by friction stir welding (FSW) technique. FSW is carried out at different welding conditions using conventional threaded cylindrical tool (TC). Microstructural study indicates several tiny reinforced particles are uniformly distributed in the nugget region. Recrystallization and grin refinement are observed in the weld area. Nugget region exhibits higher hardness compared to the base material. Joint efficiency of up to 89% is obtained for the FS-welded composite. The fracture surface reveals that the matrix undergoes a ductile fracture whereas reinforced particles exhibit brittle fracture.

Keywords Aluminium matrix composite · Friction stir welding · Nugget region · Microstructure

1 Introduction

Recently, materials with high strength-to-weight ratio gain more interest in the automobile, aviation, marine, and structural industries. This can be realized by combining several constitute of materials, such as reinforcing hard ceramic materials in the aluminium matrix, called aluminium matrix composites (AMCs). Such materials have tolerable mechanical, electrical, and thermal properties [1, 2]. These materials fail to serve their purpose when they are joined by conventional welding process. The drawbacks of conventional welding of MMCs are segregation of ceramic particles

B. Subramanya R. Prabhu
Department of Mechatronics Engineering, Manipal Institute of Technology, Manipal Academy of Higher Education, Manipal, India
e-mail: Subramanya.prabhu@gmail.com

B. Subramanya R. Prabhu · A. K. Shettigar (✉) · M. A. Herbert · S. S. Rao
Department of Mechanical Engineering, National Institute of Technology Karnataka, Surathkal, India
e-mail: arunkumarshettigar@gmail.com

T. Rajmohan et al. (eds.), *Advances in Materials and Manufacturing Engineering*, Springer Proceedings in Materials 7, https://doi.org/10.1007/978-981-15-6267-9_7

during solidification, thermal stress generation due to variation in thermal expansion coefficients, interfacial chemical reaction resulted from high temperature, and high solubility of gas in the molten state that deteriorates the mechanical properties of the joint [3, 4]. To overcome this problem, friction stir welding (FSW) process is adapted. This method has been used to join aluminium alloys, magnesium, steel, copper, metal matrix composites, and dissimilar metals [5–7].

FSW rotating tool is used to mechanically deform and transfer the material under frictional heat. The heat generated at tool and base material depends on the process variables like tool revolution speed and welding speed. Several researches have highlighted that the process parameters also affect the grain and distribution of particles [1, 8–10]. It is also found that numerous fragmented particles are uniformly distributed at weld region. All these factors will assist in increasing the mechanical properties of the materials [2]. Hence, to get desired joint strength, one has to study the process parameters to evaluate the process.

2 Experimental Set-up

In this study, stir cast aluminium matrix composite consists of AA6061-4.5%copper as matrix 10% SiC as reinforcement particle. Table 1 lists the chemical composition of matrix. These composites were sized to a dimension of 100 × 50 × 6 mm by using conventional machining process. Table 2 lists the mechanical properties of the composite. These composites were joined by the vertical milling machine with suitable set-up to convert them to friction stir welding process. Figure 1 shows the FSW process set-up. M2 steel was used to prepare FSW tool and hardened to 55 HRC. A threaded profile pin had M6 × 1 mm pitch, 16 mm shoulder diameter, and 5.7 mm height. Table 3 lists the process variables used in the study. The metallographic characterization was done by using scanning electron microscope (SEM). The size of the grains was measured by using ASTM E112-10 standard taking readings at bottom, middle, and top of the weld nugget region. The ultimate tensile strength (UTS) was measured by tensometer as per ASTM E8M standard.

Table 1 Matrix—AA6061-4.5 wt%Cu—Chemical composition

Cu (wt.)	Mg (wt.)	Si (wt.)	Iron (wt.)	Mn (wt.)	Al (wt.)
4.5	0.8–1.2	0.4–0.8	0.7	0.15	Remaining

Table 2 Sir cast AMC's mechanical properties

AMC	Hardness (Hv)	UTS (N/mm^2)	Tensile elongation (%)
AA6061-4.5 wt% Cu-10 wt% SiC	104 ± 3	252 ± 4	6.4% ± 0.2

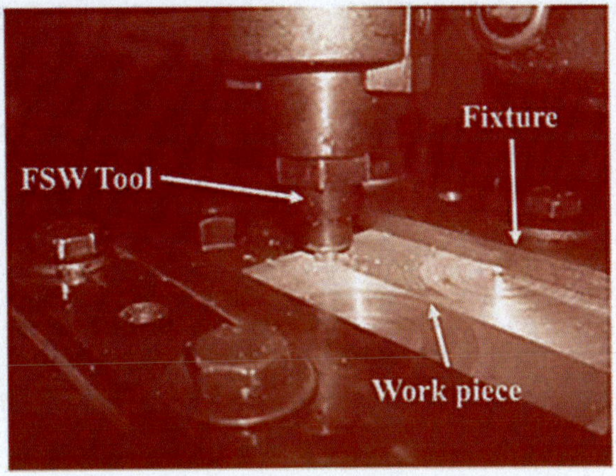

Fig. 1 Set-up used in friction stir welding

Table 3 Process variables and their values used In friction stir welding

Variables	Values
Traverse speed	50, 63 and 80 mm/min
Revolution speed	710, 1000, and 1400 rpm

3 Results and Discussion

3.1 Microstructure of the FS-Welded Joint

The plastic deformation of the material caused by the severe stirring action of the revolving FSW tool results in refinement of grains. The weld area indicates considerably altered microstructure in comparison with the parent material. The analysis of the microstructure FS-welded sample was carried out using SEM.

Figure 2 shows image of one such sample. The microstructure gradually changes from the initial non-deformed, coarse grains of the parent AMCs to equiaxed, fine grains at the mid region of the weld area. Variation in the microstructure divides the weld area into three zones, namely nugget region (NR) which is previously occupied with the FSW tool pin, thermo-mechanically affect region (TMAR) which is adjacent to NR and the heat-affected region (HAR), region between TMAR and the parent material (PM).

Figure 3a–e illustrates the SEM images of AMC‘s FS-welded using TC tool. The joining of composite was carried out with traverse speed of 50 mm/min and revolving speed of 710 rpm. Figure 3b–d represents the grain size distribution at the top, middle, and bottom of the NR, respectively. The average size of the grains at the top, middle, and bottom of the NR is 5.4 ± 0.21, 4.9 ± 0.24, and 4.3 ± 0.23 μm, respectively. It is clear that the grain size changes from top to bottom of the weld zone with bigger grains at the top and smaller ones at the bottom of the NR. Large centrifugal force acts at the top region compared to the bottom of the weld zone

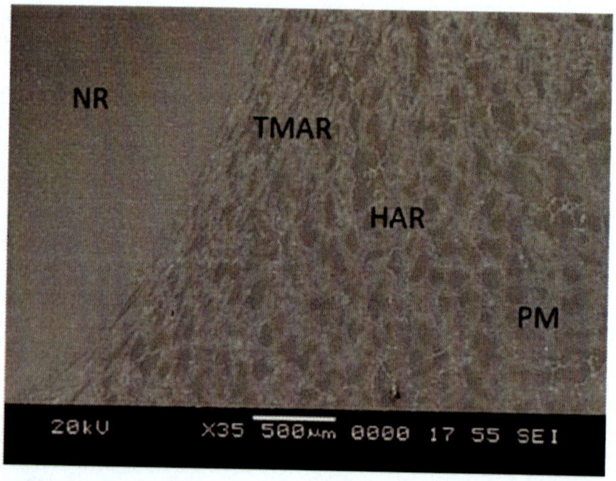

Fig. 2 Scanning electron micrograph of FS-welded composite

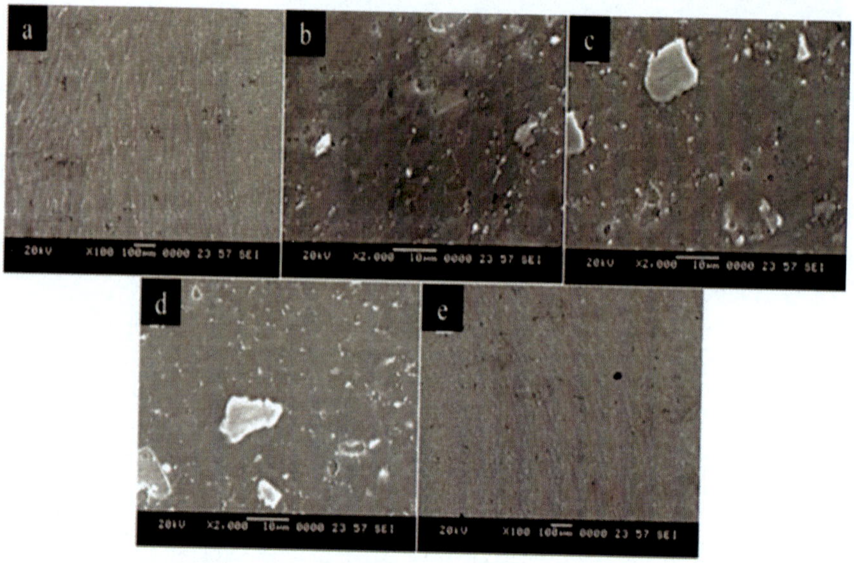

Fig. 3 SEM image of grain size distribution of FS-welded composite joined with a traverse speed 50 mm/min and revolution speed of 710 rpm. **a** TMAR retreating side **b** NR Top, **c** NR middle, **d** NR bottom, and **e** TMAR advancing side

[11]. A small extrusion crushing force was acted on the material causing a larger crystal nucleus during recrystallization. At the same time, the shoulder generates very high frictional heat at the tool workpiece interface. The cooling time reduces results in the grain growth. Whereas due to the smaller length of the pin compared to the workpiece thickness, the tip of the pin reaches slightly above the other end of the workpiece. Due to which insufficient stirring and forging take place at the bottom of the NR result in poor plasticization and the flow of material. At the root of the joint,

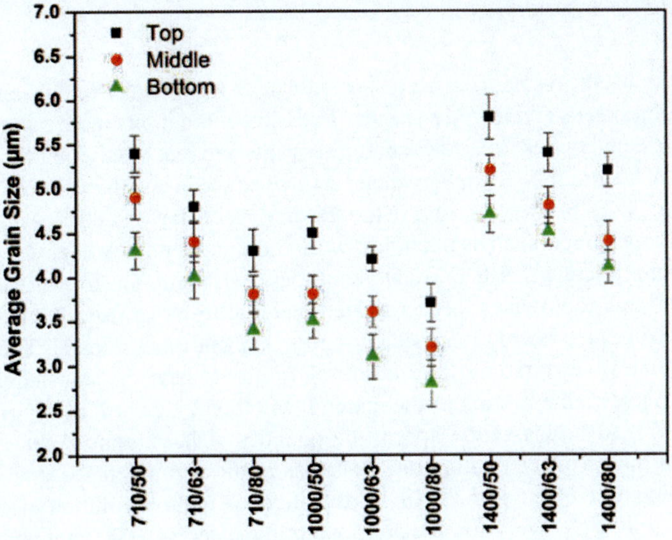

Fig. 4 Average size of the grains at nugget region

the back plate on which workpiece is kept, acts as a heat sink enables heat flow from NR to the back plate through conduction. The extrusion process at the bottom of the NR was formed at slightly lesser temperature than the top of the NR. Due to these finer grains formed at the bottom of the NR [12]. Figure 3a–e represents the grain structure obtained at TMAR on either side of the NR. Highly deformed, elongated, and non-homogeneous coarser grains were found in the TMAR due to the severe stirring effect of the FSW tool. The average reduction in the grain size at the NR is 91.5%.

Figure 4 depicts graphically, the distribution of average grain size formed at NR of MMC during FSW. It is obvious from the plots that, the size of the grains at the top of the NR is larger and gets reduced as one approaches the bottom of the weld NR. The joint formed at traverse speed of 50 mm/min and revolution speed of 1400 rpm exhibited largest average grain size of 5.8 ± 0.25 μm. Higher heat conditions prevailing at lower traverse speed (50 mm/min) with reduced cooling rate resulted in coarsening of grains in the NR. On the other hand, the joint formed with traverse speed of 80 mm/min and revolution speed of 1000 rpm exhibited smallest average grain size of 3.7 ± 0.21 μm. This is because, the increased welding speed (80 mm/min) led to lower heat input due to shorter time available for friction in the process. It also appears that, the prevailing low heat condition also contributed towards inducement of more strain and strain rate, resulting in more dynamic recrystallization which in turn, contributed to grain refinement. Further, the pinning effects [13] of silicon carbide particles which are mostly residing at the grain boundaries, prevent the grain growth and hence the grain size becomes smaller.

3.2 Tensile Properties

Table 4 presents the result of tensile tests of joints FS-welded at different combination of revolution speeds and traverse speeds. The UTS of the joints increases as the tool traverse speed increases from lower to higher value and reaches to a maximum value at a speed of 80 mm/min. Samples welded at traverse speed less than 80 mm/min were fractured at HAR. Further increase in the traverse speed, beyond 80 mm/min, results in the welding defects such as pin holes, worm holes, and tunnel defects. Therefore, the specimens fracture at NR [17] reduces the tensile strength. The forward movement of the revolving tool makes the plasticized material to move from the front to the rear of the tool pin. Revolution speed controls the heat input rate [9]. The traverse speed regulates the duration of exposure of frictional heat per unit length of the joint which affects the heating of the material and thereby growth of the grains was affected [17]. At high traverse speed, the quantity of heat input to the joint area is less, and increases the cooling rate. Increase in the revolution speed from lower value to 1000 rpm increases the UTS. Further increase in the revolution speed results in reduction in UTS. Weld formed at a revolution speed of 1000 rpm and traverse speed of 80 mm/min exhibited highest joint efficiency. Decrease in joint efficiency was observed with further increase in the revolution speed. In the present study, the optimal parameter values for FSW were turned out are traverse and revolution and speed combination of 80 mm/min and 1000 rpm.

Figure 5a–c shows the fracture surface of base material, FS-welded joint across and along the weld direction. It shows the fracture surface of composite joint FS-welded at revolution speed of 1000 rpm and traverse speed of 63 mm/min using TC tool. The microscopic examination of the fracture surface, at high magnifications, shows the large dimples, large voids, associated with matrix particle decohesion, tear ridges, and occurrence of small dimples inside the large dimples due to the ductile failure of the matrix.

Table 4 UTS results of FS-welded composite

Rotational speed (rpm)	Welding speed (mm/min)	Yield strength (N/mm^2)	UTS (N/mm^2)	Elongation %	Joint efficiency (%)
710	50	114 ± 3	167 ± 3	6.1 ± 0.1	66
710	63	132 ± 2	184 ± 3	5.8 ± 0.1	72
710	80	137 ± 3	195 ± 3	5.6 ± 0.1	77
1000	50	156 ± 2	201 ± 3	4.9 ± 0.1	79
1000	63	170 ± 4	211 ± 5	4.8 ± 0.1	83
1000	80	172 ± 3	226 ± 3	4.2 ± 0.1	89
1400	50	119 ± 2	161 ± 2	5.3 ± 0.1	63
1400	63	122 ± 2	174 ± 2	5.1 ± 0.1	68
1400	80	132 ± 3	196 ± 3	4.8 ± 0.1	77

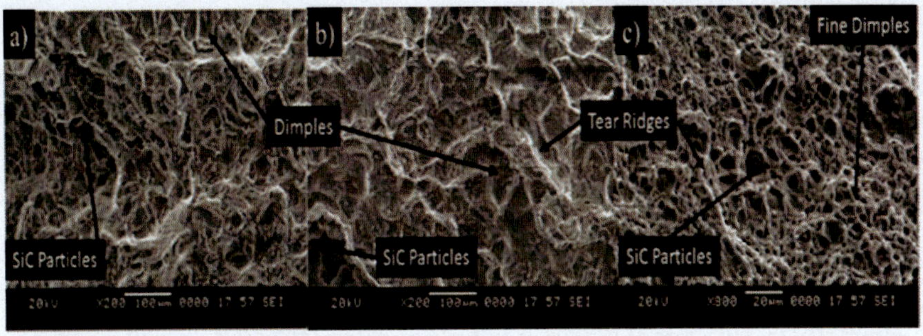

Fig. 5 SEM image of the fractured surfaces of **a** Parent material, **b** FS-welded composite perpendicular to the direction of weld, **c** Along the direction of weld, joined at traverse and revolution and speed combination of 80 mm/min and 1000 rpm

3.3 Hardness

In FSW process, the tool revolution speed is an important process variable which assists in plasticization of the material by generating frictional heat between the interface of revolving tool (pin and shoulder) and the material. Stirring and transfer of the plasticized material around the pin are controlled by the revolution speed [9, 10]. The presence of hard reinforcements in the matrix increases the nucleation sites thereby reduces the grain size in the matrix [14]. Figure 6 depicts the plots

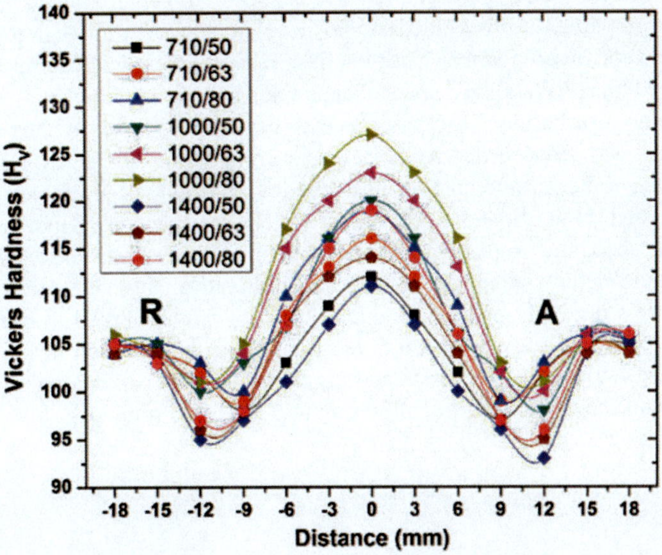

Fig. 6 Hardness across nugget region for the joints prepared at different welding condition (A-Advancing side, R-Retreating side)

depicting the hardness variation across the mid region of FSW joints. FS-welded samples reveal highest hardness in the NR than the parent material because of the finer grain size [2]. High plastic strain caused by the stirring action of the revolving FSW tool results in redistribution of reinforcements from agglomerated and non-uniform arrangements in the parent material to uniform distribution in the NR [15]. This phenomenon has been observed in all the samples at different combinations of traverse speed of 50, 63, and 80 mm/min and revolution speed of 710, 1000, and 1400 rpm. Hardness decreases as the distance is increased from the mid-point, on both sides of NR. HAR shows a lowest hardness value in the weld region. Softening of grains was resulted from the heating effect, caused by revolving FSW tool, reduces the hardness. From Fig. 6, it can be seen that for a given tool traverse speed, hardness value decreases as the revolution speed is increased from 1000 to 1400 rpm. This is due to the increase in the heat input. The higher heat input makes the weld material to experience higher temperature and higher strain rate [9]. This phenomena result in grain growth. Increase in traverse speed from lower to higher value, the hardness of the weld region also increases. This is because of the decrease in the duration of exposure of frictional heat at tool workpiece interface. The more the duration, large amount of heat is supplied, eventually affects the growth of the grains [16].

4 Conclusions

FSW technique was effectively used to join AA6061-4.5Cu-10SiC composite. The influence of tool traverse speed and revolution speed on metallographic behaviour, mechanical properties of FS-Welded composite is studied. The weld region exhibited fine grain size as in comparison with the base material. The microstructure at the NR exhibited fine recrystallized and equiaxed grains of size ranging from 2 to 7 μm. Homogeneous distribution of SiC particles was seen at NR. The size of the reinforced SiC particles was reduced due to the severe stirring action of the revolving FSW tool. The size of the particles also reduced due to the striking of reinforced particles with each other. Higher hardness found at the NR irrespective of tool revolution and traverse speed due to finer grains and smaller-reinforced particles. The tensile test exhibited fractures along the HAR zone on the advancing side, perpendicular to the tensile stress axis, where minimum hardness was observed irrespective of revolution speed and traverse speed. The microscopic examination of the fracture surfaces revealed ductile fracture. The optimal process variables are traverse and revolution and speed combination of 80 mm/min and 1000 rpm.

References

1. Kalaiselvan K, Dinaharan I, Murugan N (2014) Characterization of friction stir welded boron carbide particulate reinforced AA6061 aluminum alloy stir cast composite Mater Des 55:176–182
2. Kumar A, Veeresh Nayak C, Herbert MA, Rao SS (2014) Microstructure and hardness of friction stir welded Aluminum-copper matrix based composite reinforced with 10%SiCp. Mat Res Inn 14(S-6):84–89
3. Storjohann D, Barabash OM, Babu SS, David SA, Sklad PS, Bloom EE (2005) Fusion and friction stir welding of aluminum metal matrix composites. Metall Mats Trans 36(A):3237–3247
4. Subramanya P, Amar M, Arun S, Mervin H, Shrikantha R (2018) Friction stir welding of Aluminium matrix composites—a Review. MATEC Web Conf 144:03002
5. Thomas WM, Nicholas ED, Needham JC, Murch MG, Templesmith P, Dawes CJ (1991) US. Patent 9125978.8.
6. Mishra RS, Ma ZY (2005) Friction stir welding and processing. J Mater Sci Eng R50(12):1–78
7. Shettigar AK, Salian G, Herbert M, Rao S (2013) Characterization and hardness evaluation of friction stir welded composite AA6061–4.5Cu-5SiC (wt.%). Def Sci J 63(4):429–434
8. Periyasamy P, Mohan B, Balasubramanian V, Rajakumar S, Venugopal S (2013) Multi-objective optimization of friction stir welding parameters using desirability approach to join Al/SiCp metal matrix composites. Trans Nonferr Metals Soc China 23:942–955
9. Murugan N, Ashok Kumar B (2013) Prediction of tensile strength of friction stir welded stir cast AA6061-T6/AlNp Composite. Mater Des 51:998–1007
10. Jayaraman M, Sivasubramanian R, Balasubramanian V (2009) Effect of process parameters on tensile strength of friction stir welded cast LM6 aluminium alloy joints. J Mater Sci Technol 25:655–664
11. Xu W; Liu J; Luan G, Dong C (2009) Temperature evolution, microstructure and mechanical properties of friction stir welded thick 2219-O aluminum alloy joint. Mater Des 30:1886–1893.
12. Li JQ, Liu HJ (2013) Characteristics of the reverse dual rotation friction stir welding conducted on 2219-T6 aluminum alloy. Mater Des 148–154
13. Prabhu S, Shettigar A, Rao K, Herbert M, Rao S (2017) Influence of welding process parameters on microstructure and mechanical properties of friction stir welded aluminium matrix composite. Mater Sci Forum 880:50–53
14. Ceschini L, Boromei I, Minak G, Morri A, Tarterini F (2007) Effect of friction stir welding on microstructure, tensile and fatigue properties of the AA7005/10 vol.%Al_2O_3p composite. Comp Sci Technol 67:605–615
15. Guerra M, Schmidt C, McClure JC, Murr LE, Nunes AC (2003) Flow patterns during friction stir welding. Mater Character 49:95–101
16. Subramanya RP, Shettigar S, Herbert M, Rao S (2018) Multi response optimization of friction stir welding process variables using TOPSIS approach. IOP Conf Ser Mater Sci Eng 376:012134
17. Shettigar AK, Prabhu SR, Malghan R, Rao S, Herbert MA (2017) Application of neural network for the prediction of tensile properties of friction stir welded composites. Mater Sci Forum 880:128–131

Influence of Honeycomb Core on Static and Vibration Responses of Sandwich Structures

Sudhansu S. Patro, Ranjan K. Behera, Nitin Sharma, and Kamal Kishore Joshi

Abstract This paper investigates the effect of honeycomb core on flexural and the modal responses of a flat sandwich plate with honeycomb core. Static and modal analyses are carried out for different boundary conditions using finite elements method. The geometry of the same mid-plane kinematics has been modelled by in ANSYS using APDL code, which is based on first-order shear deformation theory (FSDT). The whole model has been formulated using an eight node serendipity SHELL281 element. Initially, the values of natural frequency of sandwich plate are compared and validated with available results. Then, a number of numerical examples have been demonstrated to study the effect of cell size, face sheet thickness, and core thickness on the natural frequencies, central deflection.

Keywords Sandwich plate · Honeycomb core · Natural frequency · Central deflection · FEA

1 Introduction

A sandwich panel has two thin and stiff surface skins at the top and bottom and they are separated by a thick core which is made up of less stiff material. These are commonly used as structural materials due to good strength-to-weight ratio, mechanical properties, and long fatigue life. Aluminum honeycomb is mostly used as core material by aerospace industries. As these structures are subjected to structural vibration, it is very important to test its strength and vibration responses. Ranjbar et al. [1] used numerical methods to predict dynamic responses of vibrating structures. Hao et al. [2] used two-dimensional model with equivalent elastic constant to simplify the numerical methods to perform modal analysis of complex structure like sandwich

S. S. Patro (✉) · R. K. Behera · N. Sharma · K. K. Joshi
School of Mechanical Engineering, Kalinga Institute of Industrial Technology, Deemed To Be University, Bhubaneswar, Odisha 751024, India
e-mail: sudhansu.patrofme@kiit.ac.in

K. K. Joshi
e-mail: kamal.joshifme@kiit.ac.in

T. Rajmohan et al. (eds.), *Advances in Materials and Manufacturing Engineering*, Springer Proceedings in Materials 7, https://doi.org/10.1007/978-981-15-6267-9_9

panels with honeycomb core. He also studied different theories to formulate equivalent elastic properties of sandwich panels. Boudjemai et al. [3] used an equivalent numerical model to evaluate vibration responses of honeycomb panels. Boldrin et al. [4] used both theoretical and experimental methods to investigate dynamic behavior of honeycomb panels. Mahi et al. [5] used hyperbolic shear deformation theory in his study of the free vibration characteristics of flat sandwich panels. Patro et al. [6, 7] and Behera et al. [8, 9] have evaluated the natural frequencies of stiffened composite plates. Kumar et al. [10] used a 2D FEM based on higher order zigzag theory to calculate the free vibration response of sandwich shell. Arunkumar et al. [11, 12] studied about vibro-accoustic responses of honeycomb sandwich panels. Sui et al. [13] studied the effect of honeycomb structures on total weight and the sound transmission loss.

The objective of this paper is to model and simulate the static and dynamic analyses of flat sandwich panels with honeycomb core. First, the sandwich plate with honeycomb core is modelled using an eight-nodded element (SHELL281) available within ANSYS. Successively, the validity as well as convergence of the current model is studied and several numerical examples have been discussed.

2 Methodology

Figure 1 shows the honeycomb core and the sandwich panel, which is studied in the current investigation. The dimensions considered for the panel are: *a*, *b*, *s*, *d*, *t*, *h* and *l*, where *a* and *b* are plate length and width, *h* is half of the core height, *s* is the cell size, *d* is the face sheet thickness, *t* is the thickness of cell wall, and *l* is the width of side.

The methodology employed in the current investigation is presented in this section. First, using sandwich theory [2], the honeycomb core is converted into a orthotropic plate. It is assumed that the orthotropic plate has equal stiffness. Also, in sandwich theory, it is assumed that the transverse shear deformation can be resisted by the core and so, it has in-plane stiffness. Hao et al. [2] reported following equations to evaluate the equivalent material properties of honeycomb core:

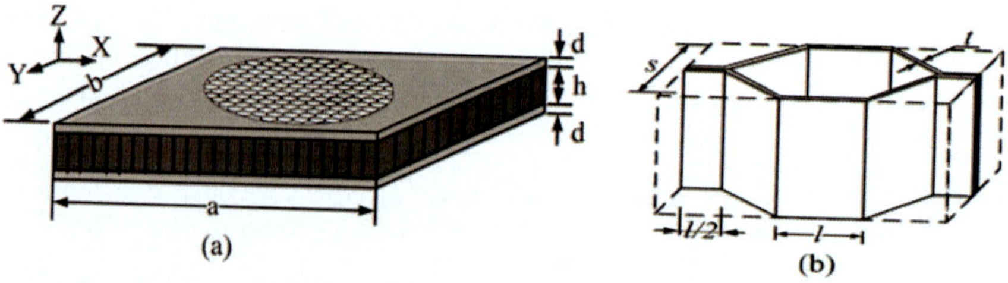

Fig. 1 **a** Dimension of sandwich panel **b** Dimension and configuration of honeycomb core

$$E_x = E_y = \frac{4}{\sqrt{3}}\left(\frac{t}{l}\right)^3 E;\; G_{xy} = \frac{\sqrt{3}}{2}\gamma\left(\frac{t}{l}\right)^3 E;\; G_{xz} = \frac{\gamma}{\sqrt{3}}\left(\frac{t}{l}\right) G;$$
$$\times\, G_{yz} = \frac{\sqrt{3}}{2}\gamma\left(\frac{t}{l}\right) E \quad (1)$$

where E_x, E_y and G_{xy}, G_{yz}, G_{xz} are equivalent material properties (Modulus of elasticity and modulus of rigidity respectively). E is the Modulus of elasticity and G is the modulus of rigidity of the material, and a corrected coefficient (γ) whose value lies between 0.4 and 0.6 [2] is used in the equation.

$$[q \text{ r s}]' = [q_0 + z\,\theta_x r_0 + z\theta_y \text{ r}_0 + z\theta_z]' \quad (2)$$

where q, r and s are the vectors for displacement of any point along x-, , y-, and z-directions, respectively, at time t. The modal parameters used in this modal analysis are evaluated by solving the following equation:

$$[(K_H - K_I) - \omega^2 M]\{\phi\} = 0 \quad (3)$$

where K_H, K_I, M, ω, and $\{\varphi\}$ are the linear global stiffness matrix, matrix of global geometry, mass matrix, natural frequency, and shape vector for corresponding mode, respectively.

3 Results and Discussion

The static and dynamic responses of a flat rectangular sandwich plate with honeycomb core are investigated using ANSYS 15 and presented in this section. To test the accuracy and reliability of the present model, convergence and validation with existing literature are studied. The influence of face sheet thickness, core height, and cell size on deflection and free vibration responses is calculated for different boundary conditions (clamped at all edges (CCCC), simply supported at all edges (SSSS), two edges are clamped and others are simply supported (SCSC)). A flat rectangular sandwich plate of dimension $4 \times 2 \times 0.02$ m^3 with honeycomb core is employed throughout the present analysis unless mentioned otherwise. The following material properties, reported by Arunkumar et al. [11], are considered for the analysis unless mentioned otherwise: $E = 68$ GPa, $\mu = 0.3$, $\rho = 2700$ kg/m^3.

3.1 *Study of Convergence and Validation of Natural Frequencies*

In this section, the natural frequencies are calculated with increasing mesh sizes to investigate the convergence behavior of the model used for the present analysis. First, a flat sandwich rectangular plate with honeycomb core with all edges clamped (CCCC) is modelled. The dimensions of the core are: $h = 10$ mm, $s = 4$ mm, $d = 0.5$ mm, and $t = 0.04$ mm. The changes in natural frequencies for different modes of frequencies are shown in Fig. 2. It can be seen that, the values of natural frequencies for all modes of frequencies are converging nicely with increasing mesh sizes. Based on this, a 20×20 mesh is selected for further investigations.

In order to check the reliability of the methodology followed in the current analysis, the values of natural frequencies computed are compared with the results published by Arunkumar et al. [11] and are presented in the Table1. It can be observed that the results obtained using current methodology are very close to the reference.

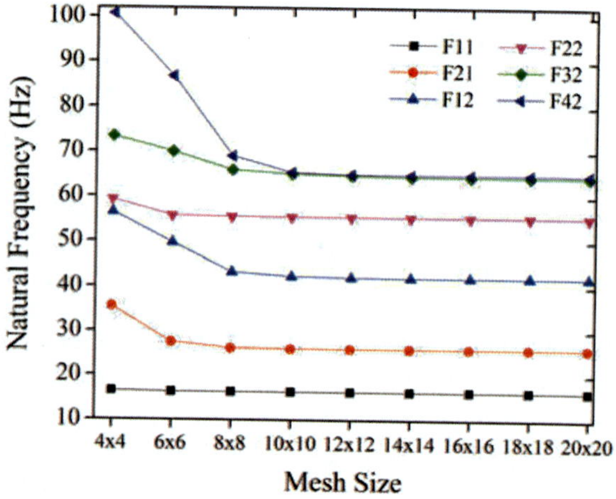

Fig. 2 Convergence of natural frequencies

Table 1 Validation of natural frequency

Natural frequency (Hz)	F11	F21	F12	F22	F32	F42
Arunkumar et al. [11]	16.19	25.81	42.22	55.13	64.55	64.57
Present	16.291	26.013	42.13	55.38	64.527	65.02

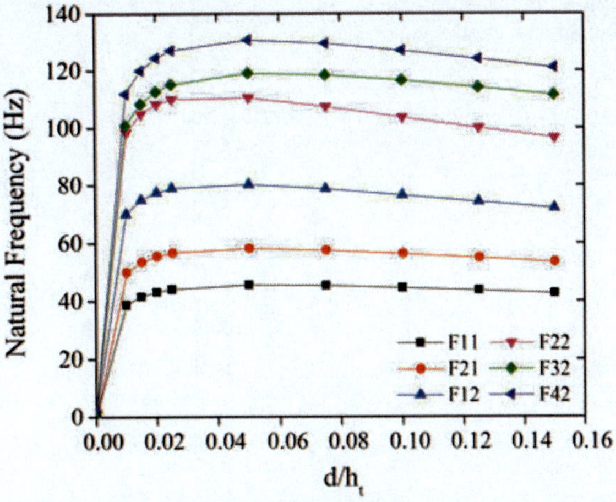

Fig. 3 Influence of face sheet thickness on natural frequency

3.2 *Influence of Face Sheet Thickness on Natural Frequencies*

The effect of change in face sheet thickness on natural frequency of a flat rectangular sandwich plate with honeycomb core with all edges clamped is investigated in the current example. The total thickness ($h_t = 2\text{h} + 2\text{d}$) of the plate is kept constant. The ratio between face sheet thickness and total height (d/h_t) is varied from 0 to 0.15. The variation of natural frequencies for all modes is presented in Fig. 3. It can be observed that initially (from $d/2\text{h} + \text{d} = 0$ to 0.05), the values of natural frequencies are increasing with increasing (d/h_t) ratio. But, after one point ($d/h_t = 0.05$), the values of natural frequencies start decreasing with increasing (d/h_t) ratio.

3.3 *Influence of Core Height on Natural Frequencies*

In this example, the variation in the value of natural frequencies with core height is studied. The plate dimensions and material properties are kept same as used in the previous example. The boundary condition is used in this SSSS. Again, the total thickness of the plate is kept constant. The ratio between core height and total height ($2\text{h}/\text{h}_t$) is varied from 0.5 to 1. The change in values of natural frequencies with the ratio ($2\text{h}/\text{h}_t$) is shown in Fig. 4. It can be observed that, as the ratio ($2\text{h}/\text{h}_t$) increases from 0.5 to 0.9, the values of natural frequencies increases. Afterwards, the values of natural frequencies starts decreasing as the ratio increases.

Fig. 4 Influence of core height on natural frequency

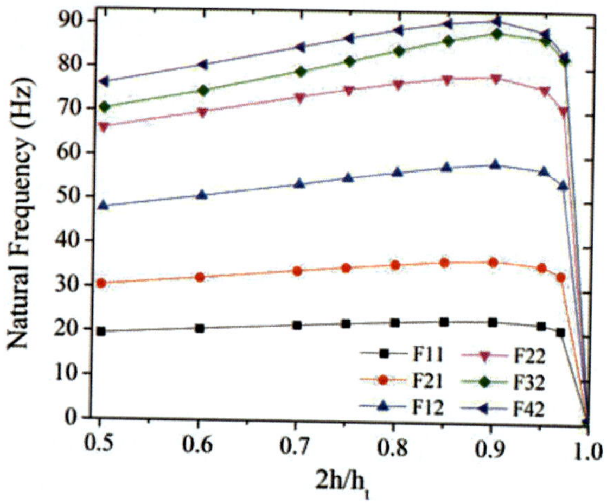

3.4 Influence of Cell Size on Natural Frequencies

The influence of cell size on the values of natural frequencies is studied in this illustration. The dimensions of sandwich plate considered here are total height = 21 mm, core height = 20 mm, face sheet thickness = 0.5 mm, and cell thickness = 0.06 mm. All these dimensions are kept constant and the cell size is varied as 2, 3, 4, 5, 6 mm for this example. First, using Eq. (1), the equivalent mechanical properties are calculated for different cell sizes. The boundary condition used in this example is SCSC. Figure 5 shows the variation of the values of natural frequencies with cell size. It can be observed that, the natural frequencies decreases with increasing value of cell size.

Fig. 5 Influence of cell size on natural frequency

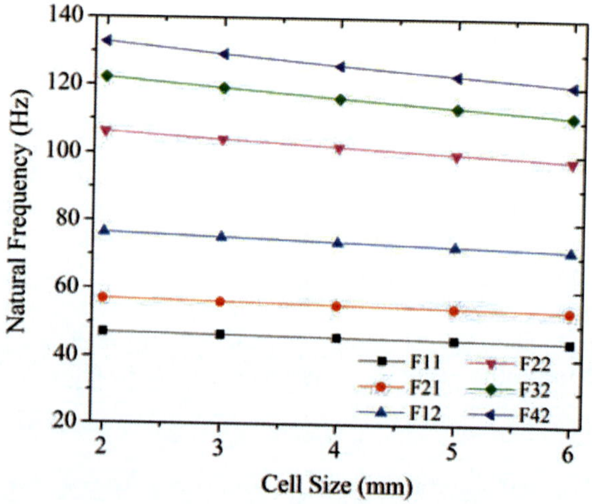

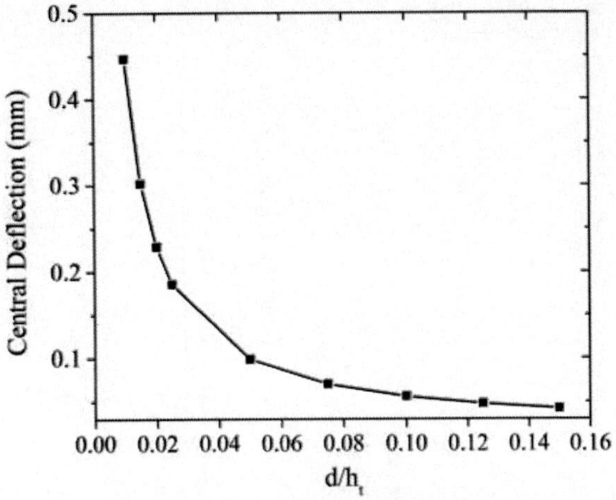

Fig. 6 Influence of face sheet thickness on deflection

3.5 Influence of Face Sheet Thickness on Deflection

The influence of face sheet thickness on the deflection at the center is investigated and presented in this section. The dimensions of the plate with all edges clamped used here are same as the previous examples. First, an uniformly distributed load of intensity of 10.25 N/m is applied at the center line parallel to X axis. The ratio between face sheet thickness and total height (d/h_t) is varied without changing the total height. The central deflections are calculated for different ratios and presented in Fig. 6. It is clear from the figure that, the deflection decreases as the ratio increases. Also, the variation in deflection is higher at the lower value of the ratio.

3.6 Influence of Core Height on Deflection

The influence of core height on the deflection at the center is investigated and presented in this section. Loading pattern is kept same as the previous example. A SSSS-type boundary condition is used in this analysis. The ratio between core height and total height ($2h/h_t$) is varied without changing the total height. The deflections at the center are calculated for different ratios and presented in Fig. 7. It can be observed that, the magnitudes of the deflection increase as the ratio increases.

3.7 Influence of Cell Size on Deflection

The effect of cell size on the deflection at the center is investigated and presented in this part. A SCSC-type boundary condition is used in this calculation. The magnitudes

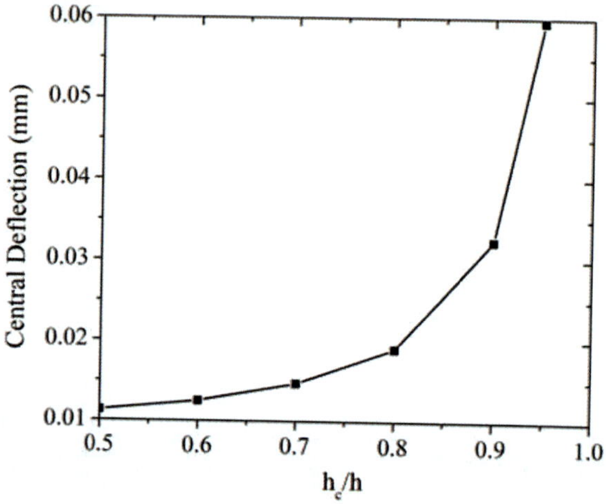

Fig. 7 Influence of core height on deflection

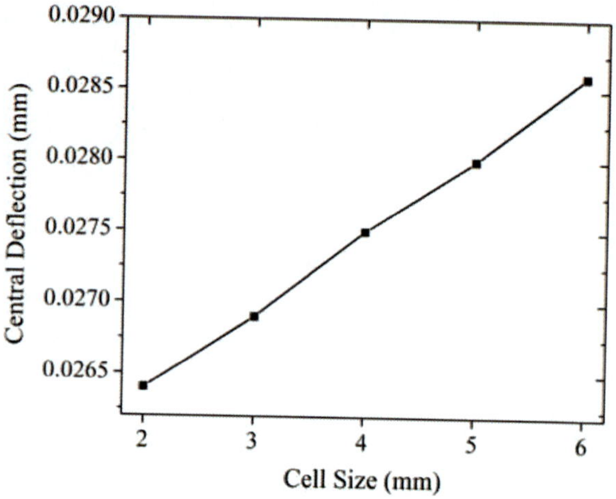

Fig. 8 Influence of cell size on deflection

of deflection for different cell size are calculated and shown in Fig. 8. It is clear from the figure that, the magnitudes of the deflection increase as the cell size increases. The variation is almost linear.

4 Conclusion

The static and vibration responses of sandwich flat plate with honeycomb core are explored in ANSYS 15 using APDL code. First, the results are validated with published results to ensure the accuracy and the reliability of the model used for the analysis. Then, the values of natural frequencies and the magnitudes of the deflection at the center of the plate are obtained for various dimensions of face sheet thickness, core height, and cell size for different boundary conditions. It is observed that, when the ratio between face sheet thickness and total height (d/h_t) is varied from 0 to 0.05, the magnitudes of natural frequencies for all modes increase and then starts decreasing with increasing d/h_t ratio. The similar trend is observed when the ratio between core height and total height ($2h/h_t$) is varied. Also, it is found that the values of natural frequencies decreases with increasing value of cell size. It is also observed that the magnitudes of the deflection decreases as the ratio between face sheet thickness and total height (d/h_t) increases. Whereas, the magnitudes of the deflection increase as the ratio between core height and total height ($2h/h_t$) increases.

References

1. Ranjbar M, Marburg S, Hardtke H (2012) Structural-acoustic optimization of a rectangular plate: a tabu search approach. Finite Elem Anal Des 50:142–146
2. Hao L, Geng L, Shangjun M, Wenbin L (2011) Dynamic analysis of the space craft structure on orbit made up of honeycomb sandwich plates. In: 2011 IEEE international conference on computer science and automation engineering (CSAE), vol 1. IEEE, Shanghai, China, 83–87
3. Boudjemai A, Amri R, Mankour A, Salem H, Bouanane M, Boutchicha D (2012) Modal analysis and testing of hexagonal honeycomb plates used for satellite structural design. Mater Des 35:266–275
4. Boldrin L, Hummel S, Scarpa F, Di Maio D, Lira C, Ruzzene M, Remillat CDL, Lim TC, Rajasekaran R, Patsias S, Lim M, Low S, Jiang L, Liew K (2016) Dynamic behaviour of auxetic gradient composite hexaganal honeycombs. Compos Struct 149:114–124
5. Mahi A, Bedia EAA, Tounsi A (2015) A new hyperbolic shear deformation theory for bending and free vibration analysis of isotropic, functionally graded, sandwich and laminated composite plates. Appl Mathe Model 39(9):2489–2508
6. Patro SS, Behera RK, Sharma N (2018) Prediction of free vibration responses of orthotropic stiffened flat panels. Mater Today Proc 5:20170–20176
7. Patro SS, Sutradhar D, Behera RK, Sharma N (2018) Free vibration analysis of stiffened laminated composite plate in a thermal environment. In: IOP conference series: material science and engineering. 390
8. Behera RK, Patro SS, Sharma N, Joshi KK (2018) Eigen-frequency analysis of stiffened laminated composite plates using finite elements. Mater Today Proc 5:20152–20159
9. Behera RK, Garg K, Patro SS, Sharma N (2018) Eigen-frequency analysis of spherical shell laminated composite plates with and without central cutouts using finite elements. In: IOP conference series: materials science and engineering, 390
10. Kumar A, Chakrabarti A, Bhargava P (2013) Vibration of laminated composites and sandwich shells based on higher order zigzag theory. Eng Struct 56:880–888
11. Arunkumar MP, Jagadeesh M, Pitchaimani J, Gangadharan KV, Lenin Babu MC (2016) Sound radiation and transmission loss characteristics of a honeycomb sandwich panel with composite facings: effect of inherent material damping. J Sound Vibr 383:221–232
12. Arunkumar MP, Pitchaimani J, Gangadharan KV, Lenin Babu MC (2016) Influence of nature of core on vibro acoustic behavior of sandwich aerospace structures. Aerospace Sci Technol 56:155–167
13. Sui N, Yan X, Hunag, T-Y, Xu, J Yuan F-G, Jing Y (2015)A lightweight yet sound-proof honeycomb acoustic metamaterial. App Phys Lett 106:171905

Influence of Point Mass Over FGM Plate for Vibration Signature in Different Boundary Conditions Using FEA

Kamal Kishore Joshi, Ranjan K. Behera, V. R. Kar, and Anugam Chakra

Abstract This paper investigates the effect of mass ratio and volume fraction on natural frequency for functionally graded (FG) square plate having same aspect ratio under clamped, simply supported, and simply-clamped supported condition. The fundamental frequency is calculated using first-order shear deformation theory (FSDT) with point mass attached at the center of the plate. The mechanical properties of the square plate are assumed to vary across the thickness of the plate by power law distribution. Numerical results for different mass ratios and volume fraction index with different support conditions are obtained and some results of fundamental frequencies are also validated with published results. It is observed that with increase in mass ratio and volume fraction index, the non-dimensional frequency of the plate decreases.

Keywords FGM · Point mass · Boundary conditions · Fundamental frequency

1 Introduction

Composite materials are widely used materials due to their adaptability and ease of combination with other material to serve specific purpose. Because of higher strength-to-weight ratio, it finds application in automotive and aerospace industries.

From the literature review, it is observed that considerable amount of work is done by the researchers on composite plates. However; very few literatures are available on static and vibration analysis of functionally graded material (FGM). Talha et.al. [1] applied vibrational approach to derive fundamental equation for FGM plates.

K. K. Joshi (✉) · R. K. Behera · A. Chakra
School of Mechanical Engineering, Kalinga Institute of Industrial Technology, Deemed To Be University, Bhubaneswar, Odisha 751024, India
e-mail: kamal.joshifme@kiit.ac.in

R. K. Behera
e-mail: ranjancet@gmail.com

V. R. Kar
Department of Mechanical Engineering, NIT Jamshedpur, Jamshedpur, Jharkhand 831014, India

T. Rajmohan et al. (eds.), *Advances in Materials and Manufacturing Engineering*, Springer Proceedings in Materials 7, https://doi.org/10.1007/978-981-15-6267-9_10

Results were obtained by using Lagrangian finite element method. Uymaz et al. [2] investigated the effect of aspect ratio and volume fraction index on free vibration in FGM rectangular plate by using linear elasticity theory. Yang et al. [3] analyzes the pre-stressed functionally graded plates under thermal environment for free and forced vibration. Kelly et al. [4] applied third-order shear deformation theory to investigate the effect of thermal buckling vibration on FG plates. Kuilla et al. [5] reviews the development and fabrication methods of graphene-based polymer nano-composites. Prusty et al. [6] apply higher order theory to analyze the effect of high temperature loads on FGM plates. Kapuria et al. [7] implemented zigzag theory-based model in combination with modified rule of mixtures to study static and free vibration response of layered beams made of Al/Sic and Ni/Al_2O_3. Tornabene et al. [8] use generalized differential quadrature method to obtain natural frequency of sandwich shell structures of variable thickness, made of FGM. Patle et al. [9] proposed higher order fuzzy finite element model to determine the frequency and static deflection for laminated structure. Lanzi et al. [10] apply multi-objective optimization technique to design composite stiffened panels which are under post buckling environment. Ali et al. [11] implemented Rayleigh–Ritz method to study the effect of mass ratio and volume fraction on natural frequency for clamped FG plate. Bisen et al. [12] fabricated the natural fiber-reinforced epoxy composite to calculate frequency and deflection both numerically and experimentally.

The objective of the present work is to model and simulate the free vibration behavior of FG plate when point mass is added at the middle position of the plate and at the different position considering different boundary conditions and volume fraction index. The FGM plate is modeled using ANSYS in the framework of FSDT.

2 Theoretical Formulation

2.1 Geometrical Description

Layered FGM square plate of length (a), uniform thickness (h), and width (b) having mass (M), made of metal and ceramic is shown in Fig. 1. A point mass (M_P) is attached at the center position of the plate.

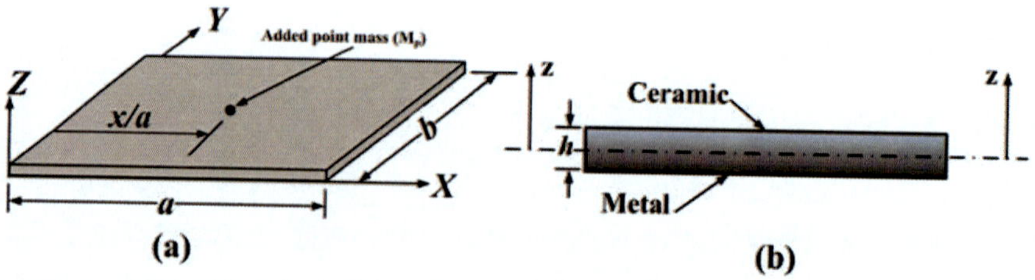

Fig. 1 Layered FGM square plate **a** added point mass and **b** cross-sectional view

2.2 Functionally Graded Material Properties

The material property of functionally graded materials varies continuously in the thickness direction from one surface to another. The property variation is shown here by power law distribution as given in Eqs. (1) and (2).

$$E(z) = (E_t - E_b)\left(\frac{z}{h} + \frac{1}{2}\right)^n + E_b, \ -h/2 \leq\leq h/2 \tag{1}$$

$$\rho(z) = (\rho_t - \rho_b)\left(\frac{z}{h} + \frac{1}{2}\right)^n + \rho_b, \ -h/2 \leq z \leq h/2 \tag{2}$$

where n is the volume fraction index and h is the thickness of plate. The young's modulus $E(z)$ and the density $\rho(z)$ vary along the thickness of plate and are given by Eqs.(1) and (2), respectively, whereas Poisson's ratio (ν) is assumed to be constant.

2.3 Constitutive Relations

The middle plane dynamics for shell or plate structure is calculated based on FSDT and given in Eq. (3):

$$\left.\begin{aligned} u(x,\ y,\ z,\ t) &= u_0(x,\ y) + z\theta_x(x,\ y) \\ v(x,\ y,\ z\ ,t) &= v_0(x,\ y) + z\theta_y(x,\ y) \\ w(x,\ y,\ z,\ t) &= w_0(x,\ y) + z\theta_z(x,\ y) \end{aligned}\right\} \tag{3}$$

where u, v and w are the displacements of the point along the x- , y- , and z-directions. The equation used for carrying out modal analysis for the given structure is given by Eq. (4).

$$[K] - \omega^2[M]\{\varphi\} = 0 \tag{4}$$

where $[M]$, $\{\varphi\}$, ω and $[K]$ are the mass matrix, mode shape factor, natural frequency, and stiffness matrix.

Table 1 Properties of FGM plate

Material	Properties		
	E (Gpa)	ρ (kg/m^3)	ν
Aluminium (Al), metal (m)	70	2707	0.3
Zirconia (ZrO_2), ceramic (c)	151	3000	0.3

Table 2 Comparison of first natural frequency parameter $\lambda = \sqrt{12(1-\nu^2)\rho_m\omega^2 a^2 b^2/\pi^4 E_m h^2}$ for different boundary condition of FG square plate with different mesh size ($n = 5$, $a/b = 1$, $a/h = 5$)

Boundary condition	Mesh size					Umyaz et al.	% difference
	2 × 2	4 × 4	6 × 6	8 × 8	10 × 10		
CCCC	2.1849	2.0899	2.0856	2.0849	2.0847	2.1447	2.87
SSSS	1.4212	1.3946	1.3936	1.3934	1.3934	1.4106	1.23
SCSC	1.8218	1.7673	1.7648	1.7644	1.7642	1.8055	2.34

3 Results and Discussions

3.1 Validation of Natural Frequency Parameter Under Different Boundary Conditions

Shell 281 element is taken to carry out the modal analysis of FG square plate in ANSYS. The material properties used for the analysis are shown in Table 1.

Firstly, convergence study (10 × 10) of non-dimensional frequency parameter (λ) is carried out to check the validity of the model. The present results of analysis of FG square plate are compared with the Uymaz et al. [1] as shown in Table 2. There is a good agreement between the present results and the published results.

3.2 Influence of Added Mass on Fundamental Frequency Parameter Under Different Boundary Condition

The variation of fundamental frequency due to the addition of point mass having different mass ratios, volume fraction index, and boundary conditions are given below. Results indicate that for a particular value of volume fraction index as mass ratio increases fundamental frequency parameter decreases for all the boundary conditions. Further, highest value of frequency parameter is obtained for clamped condition whereas lowest value is for simply supported condition as shown in Table 3, Figs. 2 and 3.

Table 3 Variation of non-dimensional frequency, $\lambda = \sqrt{12(1-\nu^2)\rho_m\omega^2a^2b^2/\pi^4E_mh^2}$ of square FG plate (ZrO_2/Al) with attached point mass ($a = b = 1$, $a/h = 10$)

M/MP	*n*	0.1	0.2	0.5	0.8	1	2	5	10
0.05	CCCC	4.39	4.30	4.09	3.95	3.89	3.74	3.60	3.50
	SSSS	2.58	2.52	2.40	2.32	2.29	2.20	2.14	2.08
	SCSC	3.61	3.53	3.36	3.25	3.20	3.07	2.96	2.88
0.1	CCCC	4.33	4.24	4.03	3.90	3.84	3.69	3.55	3.45
	SSSS	2.56	2.50	2.37	2.30	2.26	2.18	2.11	2.06
	SCSC	3.61	3.49	3.32	3.21	3.16	3.04	2.93	2.85
0.2	CCCC	4.21	4.12	3.92	3.79	3.73	3.58	3.45	3.36
	SSSS	2.51	2.45	2.33	2.26	2.22	2.14	2.07	2.02
	SCSC	3.49	3.41	3.24	3.14	3.09	2.97	2.86	2.78
0.3	CCCC	4.10	4.01	3.81	3.69	3.63	3.49	3.36	3.27
	SSSS	2.46	2.41	2.29	2.21	2.18	2.10	2.04	1.98
	SCSC	3.41	3.33	3.17	3.06	3.02	2.90	2.80	2.72
0.4	CCCC	3.99	3.90	3.71	3.59	3.54	3.39	3.27	3.18
	SSSS	2.42	2.36	2.24	2.17	2.14	2.06	2.00	1.94
	SCSC	3.33	3.26	3.10	3.00	2.95	2.84	2.73	2.66
0.5	CCCC	3.89	3.80	3.61	3.50	3.44	3.31	3.18	3.10
	SSSS	2.38	2.32	2.21	2.13	2.10	2.03	1.96	1.91
	SCSC	3.26	3.19	3.03	2.93	2.89	2.77	2.68	2.60
1	CCCC	3.45	3.37	3.21	3.10	3.06	2.93	2.82	2.74
	SSSS	2.19	2.14	2.03	1.96	1.93	1.86	1.80	1.76
	SCSC	2.94	2.88	2.74	2.65	2.61	2.51	2.41	2.35

3.3 *Influence of Position of Added Mass on Non-dimensional Frequency Under Different Boundary Condition*

The effect of added mass position on frequency parameter is shown in Fig. 4. It is clear that when the mass is placed at the middle position of the plate, it results in lower frequency whereas the frequency parameter increases symmetrically when the mass is moved closer toward the support.

4 Conclusions

- It is observed that, for a given value of volume fraction index, as mass ratio increased, the non-dimensional frequency decreases for all the boundary conditions.

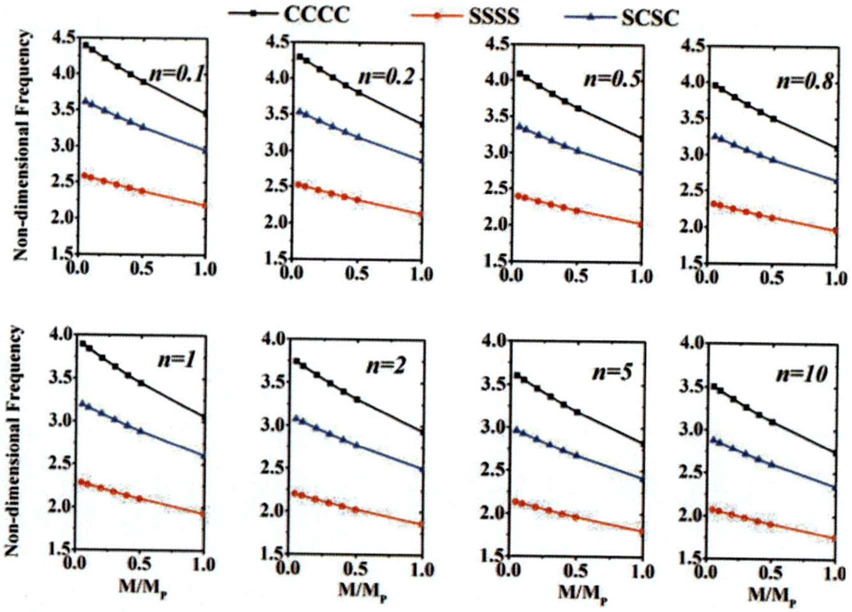

Fig. 2 Variation of non-dimensional frequency for different mass ratios and volume fraction index

Fig. 3 Effect of added mass placed at middle position on non-dimensional frequency (M/MP = 0.2, a/h = 10)

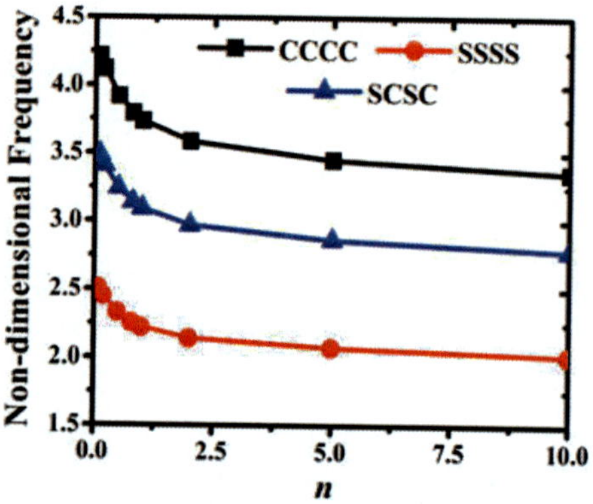

- Results indicates that, for a constant mass ratio and volume fraction index, the highest value of non-dimensional frequency is achieved when the FG plate is in clamped (CCCC) condition followed by simply clamped support (SCSC) and the lowest value is for simply supported condition (SSSS).
- Analysis of square FG plate indicates that, for a fixed mass ratio when point mass is moved closer to the supports, the non-dimensional frequency increases symmetrically for all boundary conditions. When the mass is placed at the middle of the plate, it results in lowest frequency.

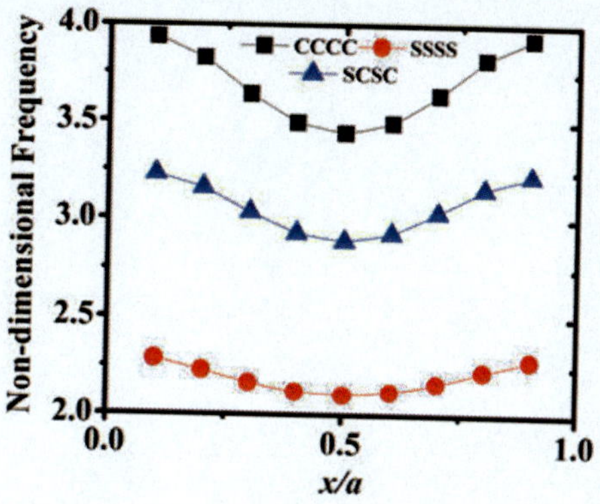

Fig. 4 Effect of point mass at middle of the width on non-dimensional frequency (M/MP = 0

- The value of non-dimensional frequency decreases as the volume fraction index increases for a fixed mass ratio.

References

1. Talha M, Singh BN (2010) Static response and free vibration analysis of FGM plates using higher order shear deformation theory. Appl Math Model 34:3991–4011
2. Uymaz B, Aydogdu M (2007) Three-dimensional vibration analyses of functionally graded plates under various boundary conditions. J Reinf Plast Compos 26:1847–1863
3. Yang J, Shen HS (2002) Vibration characteristics and transient response of shear-deformable functionally graded plates in thermal environments. J Sound Vib 255:579–602
4. Wattanasakulpong N, Gangadhara Prusty B, Kelly DW (2011) Thermal buckling and elastic vibration of third-order shear deformable functionally graded beams. Int J Mech Sci 53:734–743
5. Kuilla T, Bhadra S, Yao D, Kim NH, Bose S, Lee JH (2010) Recent advances in graphene based polymer composites. Prog Polym Sci (Oxford) 35:1350–1375
6. Wattanasakulpong N, Prusty GB, Kelly DW (2013) Free and forced vibration analysis using improved third-order shear deformation theory for functionally graded plates under high temperature loading. J Sandwich Struct Mater 15:583–606
7. Kapuria S, Bhattacharyya M, Kumar AN (2008) Bending and free vibration response of layered functionally graded beams: a theoretical model and its experimental validation. Compos Struct 82:390–402
8. Tornabene F, Fantuzzi N, Bacciocchi M, Viola E, Reddy J (2017) A numerical investigation on the natural frequencies of FGM sandwich shells with variable thickness by the local generalized differential quadrature method. Appl Sci 7:131
9. Patle BK, Hirwani CK, Singh RP, Panda SK (2018) Eigenfrequency and deflection analysis of layered structure using uncertain elastic properties—a fuzzy finite element approach. Int J Approximate Reasoning 98:163–176
10. Lanzi L (2004) A numerical and experimental investigation on composite stiffened panels into post-buckling. Thin Walled Struct 42:1645–1664
11. Ali MI, Kumar S, Ranjan V (2016) Effect of volume fraction index and mass ratio on natural frequency of clamped functionally graded plate with attached point mass. Int Conf Electr Electr Optim Tech ICEEOT 2016:1968–1972
12. Bisen HB, Hirwani CK, Satankar RK, Panda SK, Mehar K, Patel B (2018) Numerical study of frequency and deflection responses of natural fiber (Luffa) reinforced polymer composite and experimental validation. J Nat Fib 00:1–15

Design and Development of Seed Metering Device Implemented in Power Tiller

C. Devanathan, E. Shankar, A. Sivanand, R. Manimaran, and A. Gopinath

Abstract This proposed work is focused on designing and implementation of seed sowing process in power tiller vehicle. This project deals with the design of a mechanical machine, which is affordable to farmers, which can plough and sow seeds simultaneously in order to save time. The sowing operation includes laying fertilizer and seed in proper rows at the regular interval, following the covering the seeds with soil and confirming the proper compaction over the seed. To ensure the seed planting at the regular interval, the seed metering device is used in this work which is considered as the key component of the seed sowing process. A wheel is used to drive the seed metering device. This system could effectively reduce the influence of inhomogeneous sowing caused by the slipping of ground wheels. This project aims in designing a machine whose function is to sow groundnut, crops, etc. This designed machine will be more attractive and very much beneficial to all farmers because it reduces the wastage of seeds, time required for sowing, and less man power is required with reduced time consumption. The observations showed that using machine, it consumes 9 s to sow 0.003 h whereas the manual sowing tool 13.2 s.

Keywords Metering device · Seed sowing · Ploughing · Power tiller

1 Introduction

In India, the agriculture acts as a basic backbone and becomes the main source for livelihood for about 58% of India's population. As compared to the developed countries, the increasing population and low level of land production become the problem in developing countries like Asia and their neighbourhood countries. Farmers looking for applying the new technologies will not disturb the soil quality but will raise the total crop manufacture. In India, cereal yield is 2600 kg/ha approximately, whereas in USA, yield is nearly 6000 kg/ha which is nearly three times higher than India. This is possible due to implementation of automation technologies in agricultural

C. Devanathan (✉) · E. Shankar · A. Sivanand · R. Manimaran · A. Gopinath
Department of Mechanical Engineering, Rajalakshmi Engineering College, Thandalam, Chennai 602105, India
e-mail: devanathan.c@rajalakshmi.edu.in

T. Rajmohan et al. (eds.), *Advances in Materials and Manufacturing Engineering*, Springer Proceedings in Materials 7, https://doi.org/10.1007/978-981-15-6267-9_11

forming. In the USA, the automation technology in agricultural forming has already been implemented the cereal yield is neatly 6000 kg/ha, which is three times more than in India whose yield is just 2600 kg/Hectare approximately.

Now in India, according to the Department of Industrial Policy and Promotion (DIPP), some of the major investment and development are, by earlier 2019, India will start exporting sugar to china, startups related to agri-related foods received more funding, and in Rajasthan, the first mega food park was inaugurated in March 2018.

In India, farmers do agriculture manually, and it involved lot of pain in performing each and every operation. A farmer may sow at preferred seed rate, but inter-row and intra-row spacing of the seeds are likely to be irregular which in turn result in clustering and gaps in the field. The pain and problems involved in doing all operations can be reduced by means of introducing the simple technology. A machine names seed sowing machine which assists the farmers in the sowing of seeds at the desired position and therefore saves money and time.

To achieve the best yield, the spacing between one row to another row, rate of seed sowing, distance between one seed to another seed, and the depth at which the seed is placed can be controlled. These factors can vary from one crop to another. Nagesh B. Adalinge et al. discussed the diverse characteristics of seed sowing machine which will be supportive for the agriculture industries to move in the direction of mechanization [1]. Thorat swapnil V. et al. had designed and fabricated seed sowing machine in which problem encountered during the manual planting was reduced. The machine was developed to plant different types and different sizes of seeds with the required spacing [2]. Kunal A. Dhande et al. developed an automatic operated seed sowing machine in which researchers replaced complex gear system by Hall Effect sensors for easier operation in seed sowing and also lessen the want of labour. The authors concluded that the developed model is very useful for small-scale farmers [3]. Pranil V. Sawalakhe et al. were explored the today's era is marching towards the hasty development of all sectors including the cultivation. To encounter the upcoming food demands, the farmers have to tool the new technique and which will not disturb the soil texture but will intensify the overall crop production [4]. Senthilnathan et al. have fabricated and automated the seed sowing machine using Internet on things (IOT). Agriculture technology is the process of implementing the recent techniques to develop the crops that are being produced [5]. Due to poor quality, the seeds and incompetent farming practices, lack of cold storage and harvest spillage around 30% of the production are wasted.

From the literature, it was found that, powering the farmers with some machines can increase the yield, so in the proposed project work, seed sowing machine is developed which has very less cost and the machine can be easily operated by unskilled farmers. The primary purpose of the invention is to incorporate both seed sowing and ploughing in power tiller. Seed metering device is used for seed sowing process, which ensures seed spacing rate equal and grouping of seeds are avoided.

Table 1 Main components of the model

S. No.	Component name	Type/function
01	Engine	150 cc two stroke IC engine
02	Gauge wheel	Cultivator gauge wheel is used which drives the metering device in order to avoid wastage of seed
03	Plough tool	Chisel type plough tool is used which initially loosens the sand while leaving the rubbish on top. It is usually set at 8″–12″ deep. The maximum depth possible is 18″
04	Metering disc	The disc is made of mild steel, which is placed inside the metering device. Used to carry the seed from storage tank to the outlet pipe one by one to ensure the uniform distribution
05	Metering device	Its basic function is to place the seed in rows at desired depth and ensures the spacing between one seed to another, shield the seeds with sand, and provide proper compaction over the seed
06	Frame	To support the various other parts of the machine

2 Components of Seed Sowing Machine

The proposed model has the different components arranged in certain fashion to perform the required task. The main components and its basic function in the model are given in Table 1.

3 Design of Seed Sowing Machine

The various components of the proposed model were designed using the modelling software. Fig. 1 shows the model of sowing machine in 2D. The CAD model of the metering device is shown in Fig. 2. The original metering device is shown in Fig. 3.

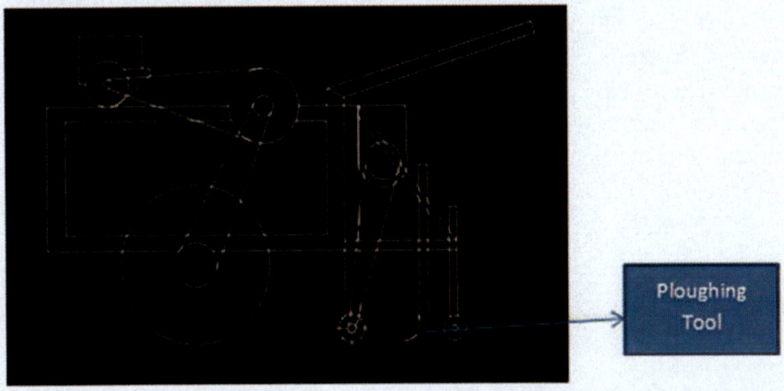

Fig. 1 2D model of sowing machine

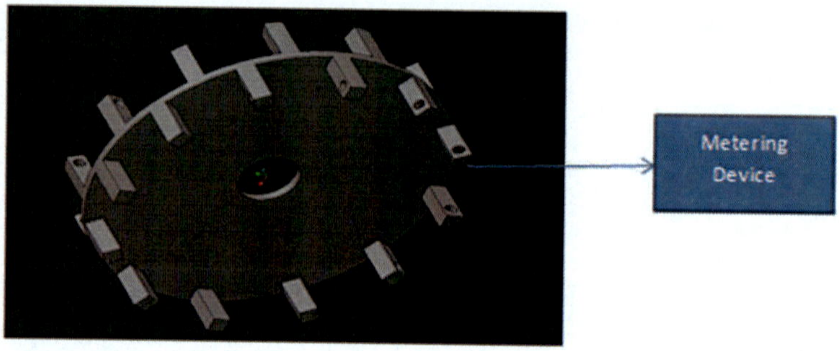

Fig. 2 3D model of metering device

Fig. 3 Metering device

In the designed model, the metering device is an important one which is connected with the shaft, and it forms the surface contact leads to lower pair. The components were designed based on the capacity of the power tiller at which it is modified and fitted. The detailed design calculations were done for rotating the metering device which is driven by chain drive. The metering device was connected to the larger sprocket so that it rotates at lower speed than the driven sprocket. Ploughing tool was fitted at the rear end of the setup which has two ploughing tools. The different components were arranged as per the proposed model, and the machine was ready for checking. The sowing cum ploughing machine is shown in Fig. 4.

The designed seed sowing machine is shown in Fig. 5, and it was tested for its fitness in real-time applications, and the results were discussed in the observations chapter.

3.1 *Observations*

The machine is tested for its performance and compared with the manual sowing. It showed the good improvement in terms of uniform seed sowing and saves human

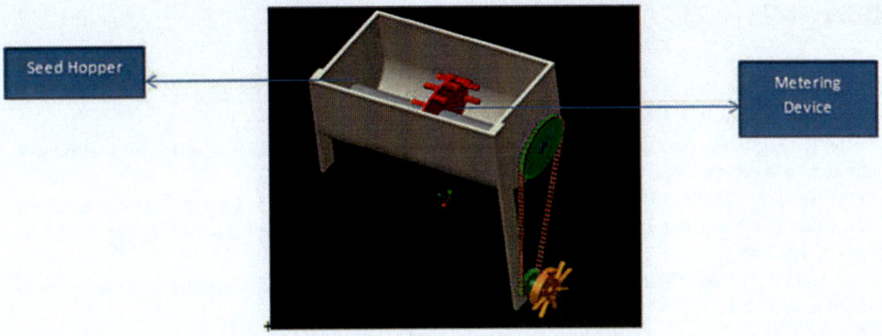

Fig. 4 3D model of seed hopper with metering device

Fig. 5 Seed sowing machine

effort. The developed model was tested in real time, and results were compared with the manual sowing. While using the machine, it took 9 s to sow 30,000 cm^2 area which is equal to 0.003 ha. Whereas in manual operation for the same area, it took 13.8 s. It showed that the considerable reduction in time and improvement in uniformity of sowing.

4 Conclusion and Future Work

The seed sowing and ploughing machine was designed and fabricated in power tiller which has the great prospective to increase the productivity of the seed sowing. The machine was developed using the simple mechanical components with low cost which can be easily affordable by the farmers. Since the machine performs the ploughing and sowing operations simultaneously, it saves lot of energy and reduces the labour cost and labour time. In future, the work can be extended for different types of seeds by changing the design of the metering wheel.

References

1. Adalinge NB, Ghune GP, Lavate GB, Mane RR Design and manufacturing of seed sowing machine. Int J Adv Res Ideas Innov Technol 3(2): 705–708
2. Thorat Swapnil V, Kasturi ML, Patil Girish V, Patil Rajkumar N (2017) Design and fabrication of seed sowing machine. Int Res J Eng Technol IRJET 04:704–707
3. Dhande KA, Sahu OR, Bawane MS, Jiwane AA, Chaware PS Design and development of automatic operated seeds sowing machine. Int J Recent Innov Trends Comput Commun 5(2):277–279
4. Sawalakhe PV, Wandhare A, Sontakke A, Patil B, Bawanwade R, Kurjekar S Solar powered seed sowing machine. Glob J Adv Res 2(4):712–717
5. Senthilnathan N, Gupta S, Pureha K, Verma S (2018) Fabrication and automation of seed sowing machine using Iot. Int J Mech Eng Technol IJMET 9(4):903–912

Pareto Optimization and Metallurgical Characterizations of Dissimilar Friction Stir Welded AA6061–AA7075 Alloys

R. Dinesh Kumar, R. Varthini, and S. Muthukumaran

Abstract Dissimilar friction stir welding of AA6061 and AA7075 is carried out to analyse the influence of tool shoulder profile, travel speed and rotation speed on microstructural and mechanical properties. Placing AA6061 on the retreating side and AA7075 on the advancing side, welding is done with varying shoulder profile such as flat shoulder, concave shoulder and threaded shoulder with taper threaded pin. Numerical models were developed to predict the mechanical characteristics such as ultimate tensile strength, impact toughness, yield strength and elongation under the influence of welding process parameters. Further, the welded joints characterization is done to correlate the microstructure and metal flow with mechanical properties.

Keywords Tool shoulder · Pareto analysis · Precipitate · Strength

1 Introduction

Joining of dissimilar metals is gaining a huge momentum due to its vast application and usage. AA6061 and AA7075, the heat treatable aluminium alloys, are very difficult to be joined by the fusion welding process; and even if welded, they are subjected to defects like porosity, slag inclusion, solidification cracking and distortion. Friction stir welding (FSW) was an accidental invention by The Welding Institute (TWI) of UK in 1991 which is now being an ideal solution for most of the conventionally non-weldable aluminium alloys [1, 2]. Aluminium alloys are efficiently utilized for FSW process which has better-joining strength compared to other fusion welding process.

R. Dinesh Kumar · S. Muthukumaran (✉)
Department of Metallurgical and Materials Engineering, National Institute of Technology, Tiruchirappalli, India
e-mail: smuthu@nitt.edu

R. Dinesh Kumar
e-mail: dineshrd453@gmail.com

R. Varthini
Department of Mechanical Engineering, Government College of Engineering, Srirangam, India
e-mail: varthini.jaku@gmail.com

T. Rajmohan et al. (eds.), *Advances in Materials and Manufacturing Engineering*, Springer Proceedings in Materials 7, https://doi.org/10.1007/978-981-15-6267-9_12

At higher rotational speed, placing AA6061 on advancing side the tensile strength increases [3]. The fracture surface reveals that the low rotating speed leads to poor mixing resulting in the failure of weld nugget [4]. The tool pin with taper thread is found to have better mixing and strength [5]. The microstructural analysis exhibits high-density precipitates of zinc and magnesium causing hardening of processing zone and high level of mechanical strength [6]. Many researchers reported the effect of tool pin profile on mechanical and metallurgical properties [7, 8]. However, there is limited work on the study of the influence of tool shoulder profile on the mechanical and metallurgical property of weld join. This study mainly focusess on the impact of tool shoulder profile over the mechanical property for dissimilar welding of AA6061 with AA7075.

2 Experimental Procedure

Dissimilar aluminium alloys such as AA6061 and AA7075 of size 100 mm × 60 mm × 6 mm are butt-welded by friction stir welding process. The tool is made of D3 material with taper threaded pin having different shoulder profile such as flat shoulder, concave shoulder and threaded shoulder. The shoulder diameter is 18 mm, pin diameter and length are 6 mm and 5.6 mm, respectively, with different shoulder profile such as flat shoulder and taper threaded pin (FTTP), concave shoulder and taper threaded pin (CTTP) and shoulder and taper threaded pin (STTP) as shown in Fig. 1. The chemical composition in weight percentage and mechanical property of

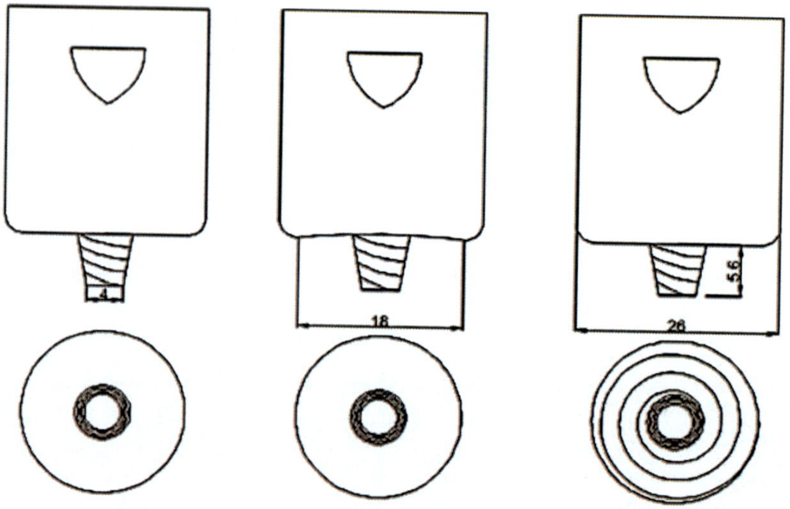

a) Shoulder & Taper threaded Pin b) Concave Shoulder & Taper threaded Pin c) Shoulder & Taper Threaded Pin

Fig. 1 Tool with different shoulder profile

Table 1 Chemical composition of AA6061-T6 parent metal (wt%)

Parent metal	Chromium	Copper	Iron	Magnesium	Zinc	Silicon	Manganese	Aluminium
AA 6061-T6	0.16	0.31	0.48	0.99	0.1	0.67	0.09	Balance
AA 7075-T6	0.22	1.2	0.23	2.1	5.9	0.1	0.12	Balance

Table 2 Mechanical properties of base metal

Material	Yield stress (MPa)	Ultimate strength (MPa)	Elongation (%)
AA6061 T6	276	306	17.3
AA7075-T6	527	571	12

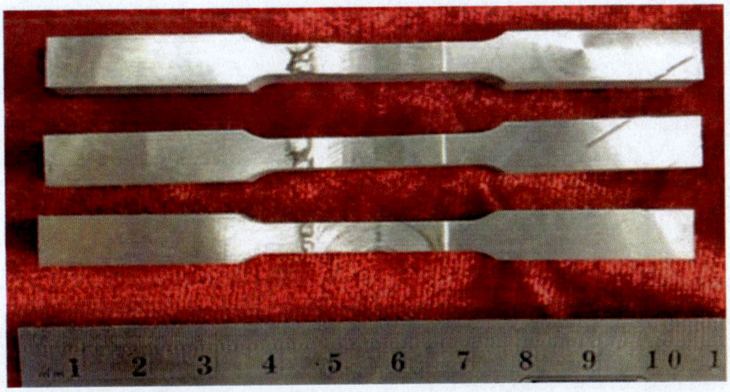

Fig. 2 Specimen for tensile testing

the alloy are shown in Tables 1 and 2. The welded samples are cut for tensile testing based on ASTM E8 standard using wire EDM as shown in Fig. 2.

Further tensile testing is done with strain rate of 1 mm/min. The samples used for Charpy V-notch testing are of dimensions 55 mm × 10mm × 6 mm with V notch at the centre of depth 2 mm and angle 450 included. Micro and macrostructures were observed by mounting the samples in cold setting powder machined, and then, polishing is carried out in different stages like emery polishing with Grade 200, 400, 600, 800, 1000, 1200, 1500 followed by alumina polishing with alumina slurry of 5 microns and diamond polishing with diamond paste of 2 micron size. Modified Keller's reagent does etching with 190 ml of distilled water, 5 ml of HNO_3, 3 ml of HCl and 2 ml of HF is used to reveal the micro and macrostructure. The microstructural observation is carried out by stereo and optical microscope.

3 Results and Discussions

3.1 Modelling Using Response Surface Methodology

A full factorial design was used to create the experimental plan in Minitab 15 statistical software as shown in Table 1. The experimental trials were randomized to avoid any systematic errors and the corresponding responses such as ultimate tensile strength (UTS), impact toughness (IT), yield strength (YS) and elongation (EL) were measured. For each tool profile, a second-order polynomial regression model was developed based on the experimental data shown in Table 3. For this purpose, the developed full factorial design was analysed using response surface methodology in Minitab. In addition, the developed mathematical models are tested for the significance of the regression coefficients using ANOVA.

3.2 Numerical Model Under the Effect of CTTP

Using the experimental data (see trial1-9) shown in Table 3, an empirical model to estimate the UTS, IT, YS and EL of welded joints under the effect of CTTP is developed as follows:

$$\begin{aligned}
\text{Maximise UTS} &= 4.46667A - 0.437083B - 0.0283333A^2 \\
&\quad + 0.000216667B^2 - 0.000625AB + 298.389 \\
\text{Maximise IT} &= 4.45A + 0.005B - 0.045A^2 - 6.89951E \\
&\quad - 21B^2 + 2.03961E - 19AB - 100 \\
\text{Maximise YS} &= -2.63333A - 0.241667B + 0.0333333A^2 \\
&\quad + 0.000108333B^2 - 0.0005AB + 398.444 \\
\text{Maximise EL} &= -1.44419A - 0.0362634B + 0.0187778A^2 \\
&\quad + 0.0000231946B^2 - 0.000214850AB + 55.0445 \\
&\quad 40 \leq A \leq 60 \\
&\quad 1000 \leq B \leq 1400
\end{aligned} \tag{1}$$

3.3 Numerical Model Under the Effect of FTTP

Similarly, a numerical model to estimate the responses under the effect of FTTP using the experimental data (see trial 10–18) shown in Table 3 is developed as follows.

Table 3 Experimental design and the responses

Trial No.	Travel speed (*A*) mm/min	Rotational speed (*B*) (rpm)	Tool profile (C)	UTS (MPa)	Impact toughness (J)	Yield strength (MPa)	Elongation (%)
1	50	1000	CTTP	199	191	5.9456	15
2	40	1400		208	191	9.5300	13
3	60	1400		224	192	12.4956	12
4	40	1200		191	190	7.4889	12
5	60	1000		207	198	10.2000	10
6	50	1200		199	185	6.2000	16
7	40	1000		186	193	5.5156	11
8	60	1200		207	190	10.3856	11
9	50	1400		222	191	10.0289	17
10	60	1400	FTTP	220	172	15.0789	7
11	50	1200		201	168	10.5022	12
12	50	1000		196	171	8.5289	10
13	40	1400		205	172	12.1822	8
14	40	1000		176	169	6.3800	6
15	50	1400		224	175	14.4000	12
16	40	1200		187	169	10.0722	8
17	60	1000		197	172	10.9956	5
18	60	1200		202	168	12.9000	7
19	60	1000	STTP	197	148	15.0356	7
20	40	1000		182	148	12.1389	8
21	40	1400		205	147	16.2222	9
22	50	1200		201	143	14.5422	13
23	60	1200		202	145	17.0089	8
24	60	1400		214	144	17.4000	8
25	50	1400		218	146	16.6522	13
26	50	1000		195	145	12.5000	12
27	40	1200		193	150	15.9000	9

$$\text{Maximise UTS} = 10.9167A - 0.275833B - 0.0916667A^2 + 0.000158333B^2 - 0.00075AB + 34.1111$$

$$\text{Maximise YS} = 1.48333A - 0.185417B - 0.01A^2 + 0.0000875B^2 - 0.000375AB + 238.833$$

$$\text{Maximise EL} = 0.30572A + 0.0177056B + 0.00124450A^2 + 0.00000257B^2 - 0.000214862AB - 19.3779$$

$$40 \le A \le 60$$

$$1000 \leq B \leq 1400 \quad (2)$$

4 Numerical Model Under the Effect of STTP

Under the effect of STTP, a numerical model for the prediction of the responses is presented below.

$$\begin{aligned}
\text{Maximise UTS} &= 7.28333A - 0.1B - 0.0583333A^2 \\
&\quad + 0.0000791667B^2 - 0.00075AB + 35.2222 \\
\text{Maximise IT} &= 4.45A + 0.0325B - 0.045A^2 - 1.25E - 5B^2 \\
&\quad - 4.06829E - 20AB - 118 \\
\text{Maximise YS} &= -2.01667A - 0.00458333B + 0.02333A^2 \\
&\quad + 0.000008333B^2 - 0.000375AB + 202.944 \\
\text{Maximise EL} &= 0.708575A + 0.0691094B + 0.010528A^2 \\
&\quad - 0.0000206388B^2 - 0.000214862AB - 16.0958 \\
&40 \leq A \leq 60 \\
&1000 \leq B \leq 1400 \quad (3)
\end{aligned}$$

The above-developed models were then tested statistically for adequacy using ANOVA. Generally, the coefficient of determination (R^2) is used widely to determine the reliability of the formulated model. The statistical results showed higher values of calculated and adjusted R^2 values (greater than 0.85) and p-values less than 0.05 implying that the models are highly reliable and significant, respectively.

5 Multi-Objective Optimization Using Augmented Epsilon Constraint Method (AUGMECON)

In order to determine the optimal responses, the numerical models developed in the previous section were subjected to a multi-objective procedure—AUGMECON. This posteriori or generation method, proposed by [9], is a development over the conventional epsilon constraint method, where each objective function is prioritized based on the decision maker's preference. AUGMECON was coded in GAMS 23.5.1 and solved on a 64-bit server with a 3.20 GHz Intel® Core(TM) i5-4570 CPU and 4 GB of RAM using BARON which is a NLP solver.

The Pareto fronts of the developed mathematical models (Eqs. 1–3) are shown in Fig. 3a–c, respectively. The ideal solution (IP) in each model is also plotted in Fig. 3.

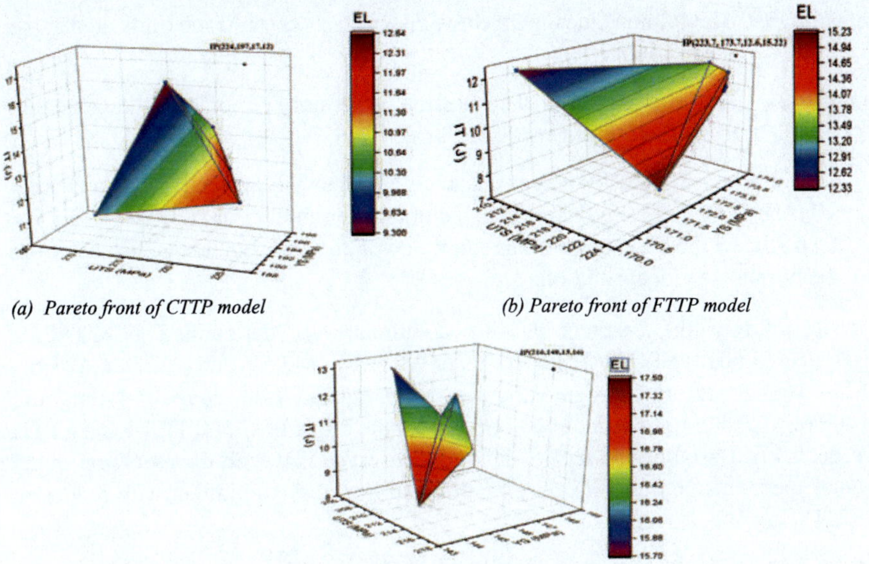

(a) Pareto front of CTTP model

(b) Pareto front of FTTP model

(c) Pareto front of STTP model

Fig. 3 3D Pareto front of the developed model with the four objectives—UTS, IT, YS, EL.

The decision maker chooses appropriate design variables for the desired objective functions based on the non-dimensional solutions obtained. From Fig. 1, it can be observed that, for smaller values of IT, the Pareto surfaces (i.e. feasible region) under all three cases become narrower, and a higher EL is found in these regions. For larger IT, a simultaneous reduction in other responses was also observed. Further, the responses UTS, YS and IT reach a maximum value under the effect of CTTP, and EL reaches a maximum under the effect of STTP.

5.1 Heuristics to Determine the Best Compromise Solution

In order to select the best compromise solution among the developed Pareto solutions, the following heuristics based on the work of [10] is proposed.

Step 1Normalize each objective function in the Pareto front as shown in Eq. (4).

$$f_i'(x) = \frac{f_i(x) - \min(f(x))}{\max(f(x)) - \min(f(x))} \tag{4}$$

where

$f_i'(x)$ Normalized value of the objective function $f(x)$.

$\max(f(x))$ Maximum value the objective function $f(x)$ can reach (indicated in the ideal solution).

$\min(f(x))$ Minimum value of the objective function $f(x)$ among the generated Pareto solutions ($i = 1, 2, \ldots n$).

Step 2Determine the Euclidean distance between the normalized ideal Pareto solution (1, 1, 1, 1) and the normalized objective functions (UTS, YS, IT, EL).
Step 3Select the best compromise solution with the least Euclidean distance from the normalized ideal solution.

Accordingly, the best compromise solutions were determined as (224.225, 191.391, 13.507, 11.816), (223.453, 173.549, 11.738, 14.057) and (216.908, 143.67, 12.3, 16.329) for the design variables (58 mm/min, 1400 rpm), (53 mm/min, 1400 rpm) and (51 mm/min, 1400 rpm) under the effect of CTTP, FTTP and STTP, respectively. Based on the results, it can be observed that with the reduction in the travel speed, a significant increase in elongation is observed; albiet with reduction in UTS and YS.

6 Metallographic Analysis

6.1 *Macrostructure and Microstructural Analysis*

Macrostructural image captured by a stereo microscope shown in Fig. 4f. with 2 × magnification reveals the different weld region such as nugget zone, thermo-mechanically affected zone (TMAZ), heat affected zone (HAZ) and base metal of

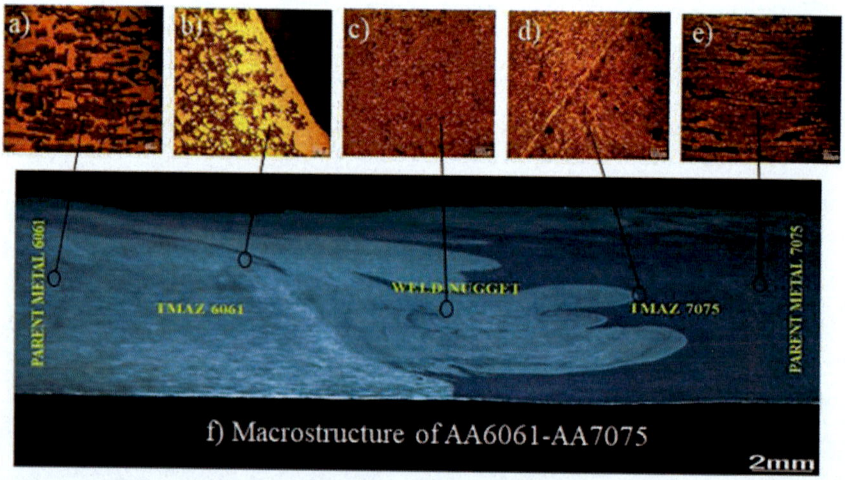

Fig. 4 Microstructure and macrostructure of AA6061 and AA7075

both AA6061 and AA7075 alloys. The microstructure of the weldments observed in 400 × magnification Fig. 4a shows the parent metal of AA6061 with coarser grains. Figure 4b shows the interlayer formed at TMAZ of 6061 having mixture of coarser and finer grains. Figure 4c results with fine equi-axised grains in the weld nugget with a proper mixture of both the materials. Figure 4d exhibits both finer grains and elongated grains in the TMAZ of 7075 side, the interlayer is also clearly visible, and finally, Fig. 4e is observed to have elongated grains revealing the rolled nature of 7075 base metal.

7 Conclusion

Dissimilar FSW of AA6061 and AA7075 was successfully carried out by varying tool shoulder profile, travel speed and rotation speed. The following conclusions are derived:

1. Numerical models to predict the responses—ultimate tensile strength, impact toughness, yield strength and elongation under the effects of process parameters are developed
2. Concave shoulder and taper threaded pin (CTTP) result in better ultimate tensile strength, yield strength and impact toughness with a maximum value of 224 MPa, 197 MPa and 17 J under the and Elongation reaches a maximum of 17.5% under the effect of Shoulder and Taper Threaded Pin.
3. When travel speed is reduced, a significant increase in elongation is observed; albiet with reduction in ultimate tensile strength and yield strength.
4. Micro and macrostructure reveals the proper mixing of AA6061 and AA7075 in the weld nugget.

References

1. Threadgill PL, Leonard AJ, Shercliff HR, Withers PJ (2009) Friction stir welding of aluminium alloys. Int Mater Rev 54(2):49–93
2. Dinesh Kumar R, Ilhar Ul Hassan MS, Muthukumaran S, Venkateswaran T, Sivakumar D (2019) Single and multi-response optimization and validation of mechanical properties in dissimilar friction stir welded AA2219-T87 and AA7075-T73 alloys using T-GRA. Exp Tech 43:245–259. https://doi.org/10.1007/s40799-019-00305-3
3. Guo JF, Chen HC, Sun CN (2014) Friction stir welding of dissimilar material between AA6061 and AA7075 Al alloys effect of process parameter. Mater Des 56:185–192
4. Rodriguez RI, Jordon JB, Allison PG (2015) Microstructure and mechanical properties of dissimilar friction stir welding of AA6061 to 7050 aluminium alloy. Mater Des 83:60–65
5. Ravikumar S, Seshagiri RV (2014) Effect of process parameters on mechanical properties of friction stir welded disiimilar materials between AA6061-T651 and AA7075-T651 alloy. Int J Adv Mech Eng 4:101–114
6. Ghangas G, Singhal S (2018) Effect of tool pin profile and dimensions on mechanical properties and microstructure of friction stir welded Armor alloy. Mater Res Express 5(6)

7. Rengarajan S, Rao VS (2015) Characteristics of AA7075-T6 and AA6061-T6 Friction stir welded joints. 14-CSME-89, E.I.C. Accession 3749
8. Nandhini R, Kumar RD, Muthukumaran S, Kumaran S (2019) Arabian J Sci Eng. https://doi.org/10.1007/s13369-019-03770-5
9. Mavrotas G (2009) Effective implementation of the ε-constraint method in multi-objective mathematical programming problems. Appl Math Comput 213(2):455–465
10. PrasannaVenkatesan S, Kumanan S (2012) Multi-objective supply chain sourcing strategy design under risk using PSO and simulation. Int J Adv Manuf Technol 61:325–337

Emission Reduction in Four-Stroke S.I Engine Using EGR and Catalytic Converter

Abhijith Reji, P. Anu Nair, A. K. Saurav, Insam Ismail, and Nidhin Babu

Abstract The objective of this investigation is to reduce the NOx, CO and HC emissions by using catalytic converter and exhaust gas recirculation. The EGR is to reduce the NOx emissions, and catalytic converter reduces the CO and HC emissions. The experimental study was carried out in a four-stroke single cylinder spark ignition engine. The experimental results show that appreciable reduction of NOx by 40% for every 2.5% EGR introduction, while CO and HC emission level increases to certain level. The combined effect of EGR and catalytic converter simultaneously reduce the CO, HC and NOx emissions. The emission parameters were analysed and compared with the pure gasoline operation

Keywords EGR · Catalytic converter · Emission

1 Introduction

Internal combustion engines produce undesirable emissions during the combustion process. It is the dream of the researcher in the field of IC engines to develop fuel and engine with less quantity of harmful emissions (i.e. NO_x, HC and CO) are generated, and these could be let into the surrounding without a major impact of the environment. The major causes of emission are dissociation of nitrogen, non-stoichiometric combustion and impurities in fuel and air. The emission of the SI engine is worse than the CI engine.

Several technical developments have taken place to combact the stringent emission standards world over that are ranging from new engine design for low emissions to fuels specification and advancement in three-way catalytic converter, oxidation catalytic converter and EGR technology [1]. The current emission standards would

A. Reji · A. K. Saurav · I. Ismail · N. Babu
Department of Mechanical Engineering, GISAT, Payyapady, Kottayam, Kerala 686516, India

P. A. Nair (✉)
Department of Mechanical Engineering, Mar Baselios Christian College of Engineering & Technology, Kuttikkanam, Peermade, Kerala 685531, India
e-mail: anunairp@mbcpeermade.com

T. Rajmohan et al. (eds.), *Advances in Materials and Manufacturing Engineering*, Springer Proceedings in Materials 7, https://doi.org/10.1007/978-981-15-6267-9_13

need to incorporate all the technologies together, and metal substrate type catalytic converters are in the market. Metallic substrates possess the technical properties necessary to conform with both current and future legislations, particularly with regard to durability, light off and space requirements. Due to the low heat capacity of the metal in relation to its weight, the amount of heat which the exhaust gas must transfer to the substrate is less in order to reach the light-off temperature [2]. The metallic catalytic converter can reach the operating temperature about twice as fact when compared to ceramic converter. This leads to significant decreases in emissions whenever the vehicle is started under cold starting condition. As the thermal conductivity of the metal substrate is higher, there is less thermal shock stress resulting in long life to the converter. Anu Nair et al. [3] conducted an experiment to study with the use of an exhaust gas recirculation (EGR) system on a given spark ignition engine for studying the performance and emission characteristics.

Generally, three ways of catalytic converter are employed in SI engines for reducing the CO, HC and NOx emission. Temperature is an important parameter for better chemical reactions [4]. However, catalyst ageing tendencies are higher at very high temperatures [5]. The problem of reduction of NOx emission using catalytic converter has attracted many people, and they have done lot of research on this problem.

This paper presents about experimental investigation carried on a four-stroke single cylinder spark ignition engine using EGR and catalytic converter. The emission characteristics were studied on SI engine with and without catalytic converter, exhaust gas recirculation, and it is compared for different speeds with constant load. The noble metals are generally used as catalyst in catalytic converter. It is very expensive, so attempts are made to use low cost catalytic material—copper oxide in catalytic converter, and tests are conducted on the engine.

2 Experimental Methods

The experiment was conducted out in a Hero Honda, four-stroke spark ignition engine. The oxidation catalytic converter is fixed in exhaust pipe, and EGR valve is connected with inlet manifold of engine as shown in Fig. 12. The engine was run at various speeds of constant load without connecting the EGR and catalytic converter was studied. Then, experiment is conducted on engine using EGR and catalytic converter for various speeds at constant load. The exhaust emissions CO, HC, CO_2 and O_2 are measured by NDIR analyser, and NOx emissions is measured by exhaust analyser. The emission characteristics are studied on SI engine for the following attempts at different speeds at constant load.

- Engine operated without EGR and catalytic converter
- Engine operated with EGR for different speeds
- Engine operated with catalytic converter for different speeds
- Engine operated with EGR and catalytic converter for different speeds.

3 Results and Discussions

Engine parameters such as CO, NOx, HC and conversion efficiency are presented against different speeds for all the attempts. Figure 1 shows the variation of CO emission with different engine speeds. It can be observed that CO emission increases marginally with increase in the engine speed. CO will be increased by about 2% volume for every 1.5% of EGR introduction. This is due to O_2 concentration which is slightly reduced while increasing the EGR ratio at higher speeds.

Figure 1 shows the HC emission versus different engine speeds. This results show that HC will be increased by about 200 ppm for every 1.5% of EGR introduction. Generally, if EGR ratio increases, it is possible that combustion temperature is reduced with EGR on account of lower oxygen concentration.

Figure 2 shows the variation of NOx emission with different engine speeds. The results show that NOx emissions appreciably are reduced at higher speeds, when compared to sole fuel operation. This reduction is due to introduction of exhaust gas regulation which reduces the combustion temperature. From these investigations, it can be confirmed that, the use of EGR reduces the NOx emissions.

Variation of CO and HC with different engine speeds is shown in Figs. 3 and 4. The initial temperature of the bed is low; hence, the marginal reduction of CO is achieved. The bed temperature increases, as reduction of CO level increases from 2.5% volume to 3.5% volume. It can be noticed that the appreciable reduction in the CO emission level is obtained where optimum engine speed is reached.

Figure 5 shows the variation of NOx emissions at various engine speeds. It can be observed that the catalytic converter can help to reduce the NOx emission marginally.

Fig. 1 CO versus speed with EGR

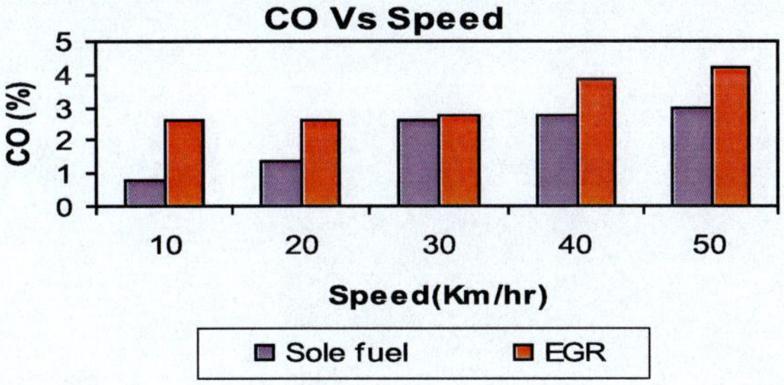

Fig. 2 HC versus speed with EGR

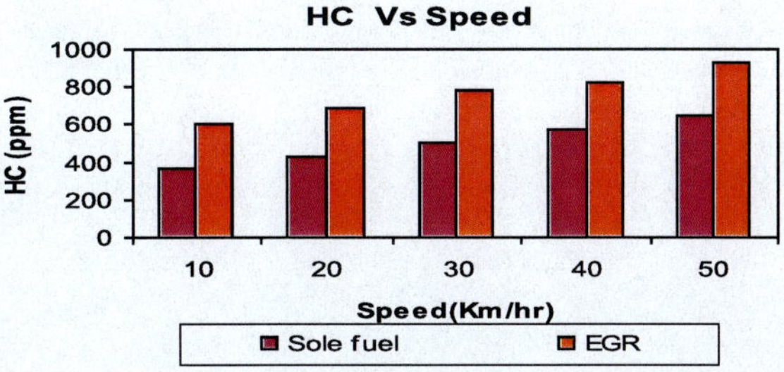

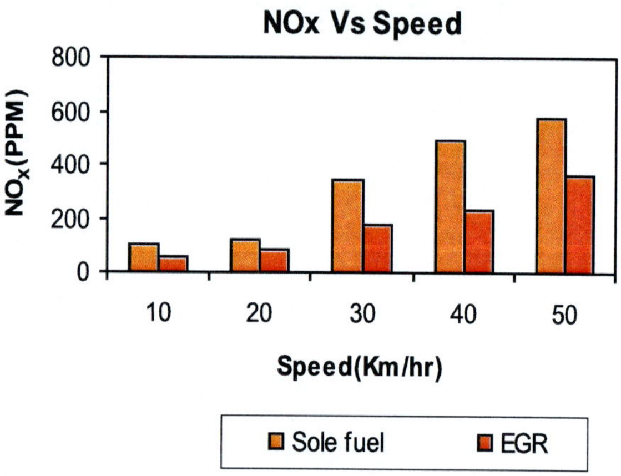

Fig. 3 NOx versus speed with EGR

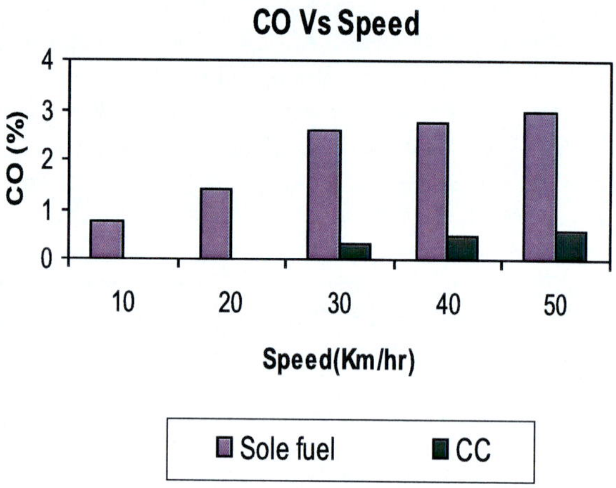

Fig. 4 CO versus speed with catalytic converter

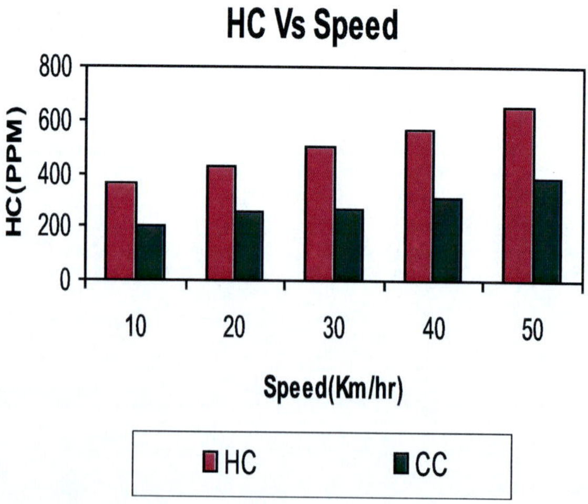

Fig. 5 HC versus speed with catalytic converter

The NOx level does not vary appreciably with use of catalytic converter. This is due to the copper-coated catalyst to oxidize the CO and HC emissions [6]. The emissions of HC and CO are much higher for an engine without catalytic converter insert.

Figures 6 and 7 show the effect of EGR with catalytic converter fitted on the exhaust on CO % by volume and HC in ppm in exhaust, respectively. Initially, the CO emission is marginal reduction due to lower bed temperature of the catalytic converter. After 30 km/h, the CO level shows decreases about 2.5% by volume. This is due to prominent effect of increased bed temperature, due to metal supported catalyst converter to reach faster the light of temperature. The temperature of the catalytic converter bed is increased due to resistance offered to the flow of exhaust gases and due to chemical activity in the bed of the catalytic converter. The effect of the catalytic converter reduces HC level uniformly which decreases about 100 ppm at maximum engine speed when compared to standard engine. The combined effect of

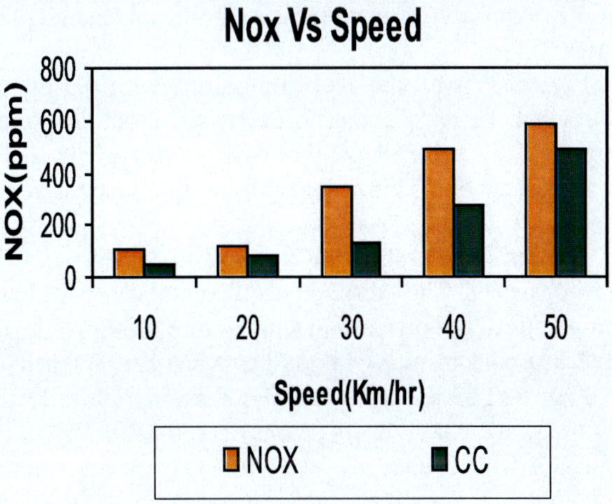

Fig. 6 NOx versus speed with catalytic converter

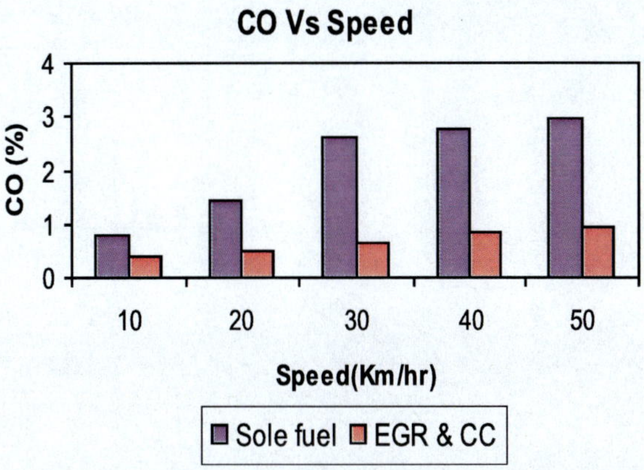

Fig. 7 CO versus speed with EGR and catalytic converter

Fig. 8 HC versus speed with EGR and catalytic converter

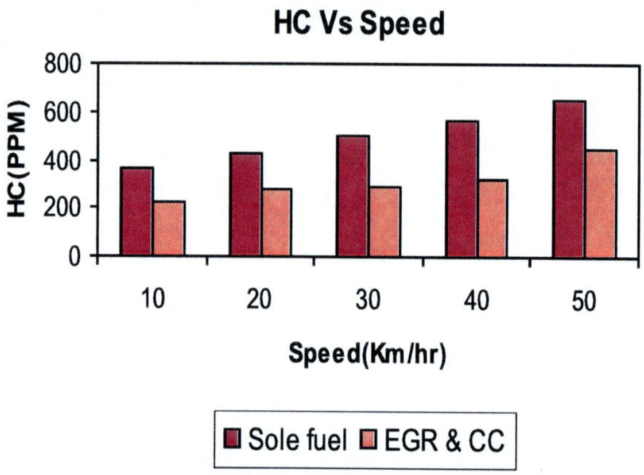

EGR with catalytic converter is a significant change in the emission level of exhaust gases.

Figure 8 gives the NOx emissions with various engine speeds. The results show that the NOx emission decreases about 300 ppm at maximum speed of the engine. The combined effect of EGR with catalytic converter reduces the NOx emission appreciably. This will result in simultaneous reduction in CO, HC and NOx emissions.

Figures 9 and 10 show the conversion efficiency of CO/HC emission with different engine speeds. The conversion efficiency is low at lower speeds due to lower bed temperature. The increase in temperature favours oxidation of Carbon monoxide and Hydrocarbons resulting in good conversion efficiencies [3].

Figure 11 gives the NOx emissions and their reduction rate with EGR ratio. The results show that NOx emissions decreases linearly with increase in the EGR ratio and NOx will be reduced by 10% for every introduction of 1.5% of EGR [3] (Fig. 12).

Fig. 9 NOx versus speed with EGR and catalytic converter

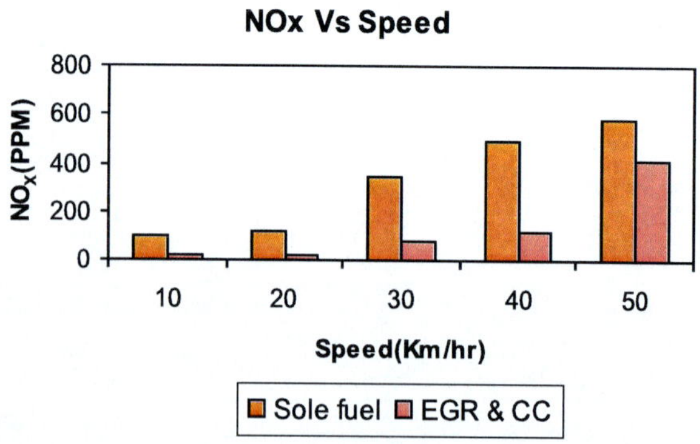

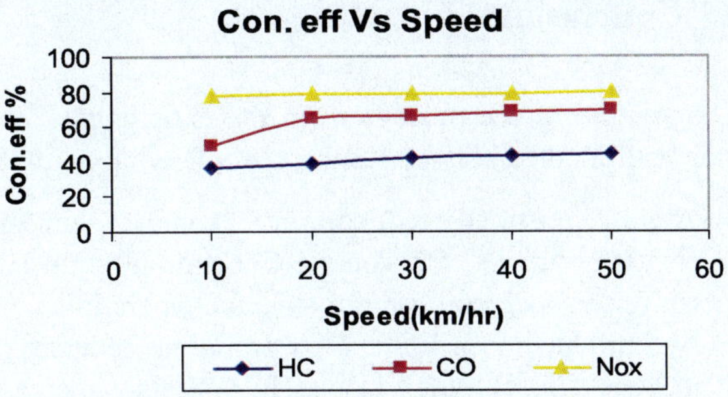

Fig. 10 Conversion efficiency versus speed for CO/HC emission

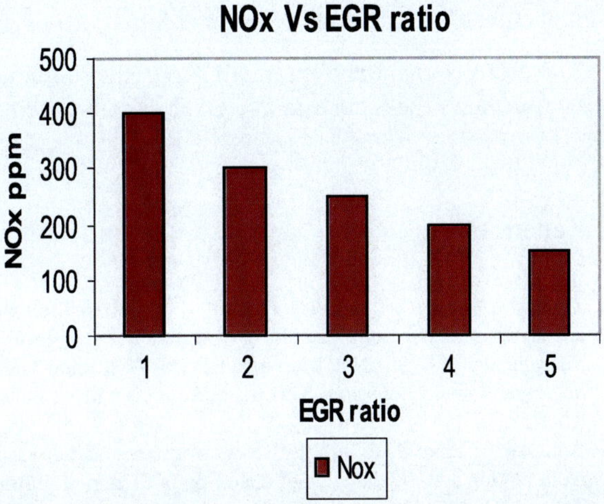

Fig. 11 NOx emission versus EGR ratio

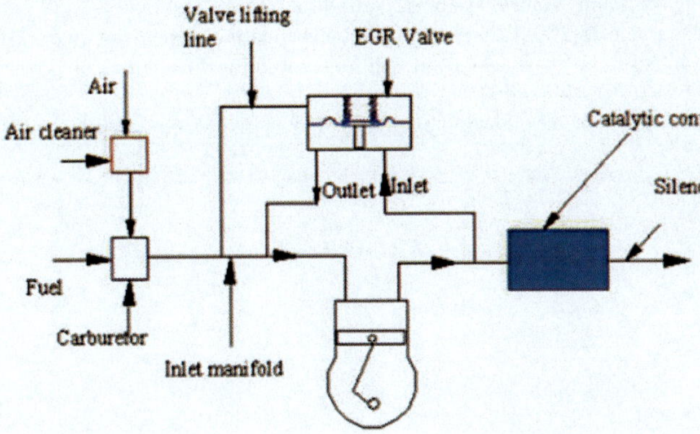

Fig. 12 Experimental set-up

4 Conclusions

The significant conclusions from the experimental investigation are summarized below. From the obtained results, the following points were observed.

- There is drastically reduction if CO and HC emissions of engine with catalytic converter fitted. 2.5% volume CO and 100 ppm HC are reduced with catalytic converter when compared to sole fuel operation.
- By introduction of EGR, NOx emissions reduction rate increases with increase in EGR ratio. But the CO and HC emissions increased marginally.
- The combined effect of EGR and catalytic converter shows the simultaneous reduction of CO, HC and NOx emissions

Therefore, from the above investigations, the combined effect of EGR and catalytic converter shows the reduction of CO, HC and NOx emissions.

References

1. Zheng M, Reader GT, Hawley JG (2004) Diesel engine exhaust gas recirculation-review on advanced and novel concepts. Energy Convers Manage 45(6):883–900
2. Subramanian KA, Shailesh Mariappa RP, Das LM, Babu MKG (2009) Simultaneous reduction of NO_x and smoke emissions of a DI diesel engine with partial HCCI mode. SAE Paper No. 2009-01-0259
3. Anu Nair P, Elias S, John V (2015) Performance and emission characteristics of exhaust gas recirculation system and ethanol operation in SI engine. Euro J Adv Eng Technol 2(8):82–86
4. Twigg MV (2011) Haren Gandhi 1941–2010: Contributions to the development and implementation of catalytic emissions control systems. Platin Met Rev 55(1):43–53
5. Saravanan N, Nagarajan G (2008) An experimental investigation on performance and emissions study with port injection using diesel as an ignition source for different EGR flow rates. Int J Hydrogen Energy 33(16):4456–4462
6. Heywood JB (1988) Internal combustion engine fundamentals. McGraw-Hill, New York

Experimental Analysis of Fuel Spray Impingement Against the Tip on the Performance of CI Engine

G. Gopeekrishnan, P. Anu Nair, Midhun Das, Sandeep Santhosh, and Tom Mathew

Abstract In this study, the conventional DI diesel engine is modified by installing the impingement head in the form of triangular tips in the cylinder head. The tips are installed in the path of fuel spray issued from the nozzle hole. Each fuel spray after impingement is broken into two zones resulting in more uniform and wide distribution of fuel in combustion chamber. The improved performance in terms of lower-specific fuel consumption and lower smoke level is obtained.

Keywords DI engine · Impingement · Emission

1 Introduction

The theme of the project is to impinge the fuel spray from nozzle hole against the impinging wall head. The impinging wall head is in the form of small 30° triangular bar of 1 mm size named as impingement tip and is installed in the cylinder head in the path of fuel spray from nozzle hole as shown in Fig. 1.

The spray from the nozzle hole impinges on impingement tip, breaking the fuel spray as shown in Fig. 2, which results in more uniform and wide distribution of fuel inside the combustion chamber [1, 2]. Also the use of impingement tip results in diffused nature of spray and fuel evaporation at initial stage because of the fuel impingement on hot tip. The purpose of this experimentation is only to verify the possibility of improving diesel combustion process by the impingement of spray against the tip [3, 4]. The experimentation is carried out on three configurations, namely original engine, engine fitted with three tips and engine fitted with single tip.

G. Gopeekrishnan · M. Das · S. Santhosh · T. Mathew
Department of Mechanical Engineering, GISAT, Payyapady, Kottayam, Kerala 686516, India

P. Anu Nair (✉)
Department of Mechanical Engineering, Mar Baselios Christian College of Engineering & Technology, Kuttikkanam, Peermade, Kerala 685531, India
e-mail: anunairp@mbcpeermade.com

T. Rajmohan et al. (eds.), *Advances in Materials and Manufacturing Engineering*, Springer Proceedings in Materials 7, https://doi.org/10.1007/978-981-15-6267-9_14

Fig. 1 Triangular impingement tips installed on the cylinder head

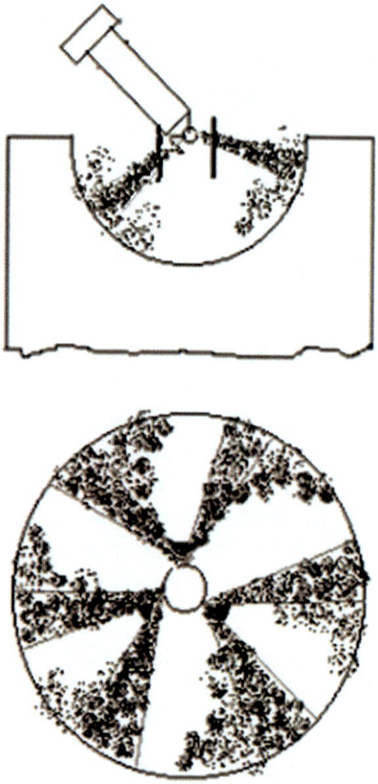

Fig. 2 Fuel spray after impinging against the edge of triangular impingement tip brakes into two zones

2 Experimental Set-up and Experimentation

A single-cylinder, vertical, water-cooled, DI, stationary, Kirloskar make AVI-type diesel engine was used for experimentation. The schematic representation of the experimental test set-up used is shown in Fig. 3. The engine specifications are given in Table 1.

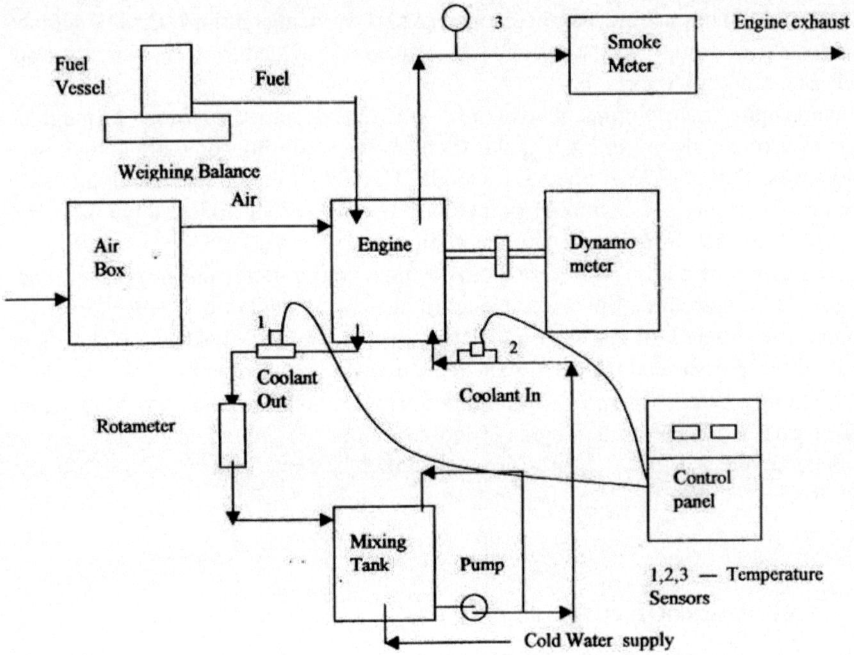

Fig. 3 Experimental set-up diagram

Table 1 Specification of engine

Specification	Details
Engine type	Single-cylinder, four-stroke, direct injection, water-cooled, AVI-type Kirloskar make diesel engine
Rated power (1500 rpm)	3.7 kW
Bore and stroke, mm × mm	80 × 110
Compression ratio	16.50
Fuel injector	Multihole fuel injector with three holes, centrally mounted on combustion chamber
Combustion chamber	Central hemispherical cavity on the top of piston

The test set-up for performing the trials was provided with rope brake dynamometer for measuring brake load with two 50 kg spring balances. The fuel consumption was measured by digital weighing balance and stopwatch. The exhaust gas temperature was measured at a distance of 10 cm from the flange of exhaust manifold with the help of thermocouple. The smoke level in exhaust was measured in terms of smoke number k with Neptune make Opax—2000 II smoke meter. The value of smoke number *k* ranges from 0 to 10. The cooling water circuit was specially designed for maintaining the engine at constant temperature irrespective of load. The outlet coolant temperature was maintained at 70 °C by using mixing tank. The coolant flow was circulated by the pump, and rotameter was provided for measuring the flow of coolant through the engine.

The impingement tips installed in the cylinder head were in the form of triangular bars. The angle of isosceles triangular tip faces the incoming spray breaking it into two zones. This angle is termed as tip angle. The size of the tip is the length of side opposite to tip angle. The impingement tips in the form of 30° triangular bars having size of 1 mm were used for experimentation.

The experimentation was carried out on three configurations as original engine, engine fitted with three tips and the engine fitted with single tip. The performance parameters selected for this experimentation were BSFC and smoke level. The effect of injection pressure on the performance parameters was also seen.

The observations for various configurations were taken by changing the load on the engine and keeping the engine speed constant. The engine speed was kept at 1500 rpm, and the injection pressure was varied for three values of 180, 200 and 220 bar.

3 Results and Discussions

3.1 *Effect on BSFC*

The engine fitted with single impingement tip shows lower fuel consumption than original engine at all loads (Fig. 4). It gives 7% decrease in BSFC at full load and 21.6% at 20% load, whereas the engine fitted three tips show higher BSFC at port

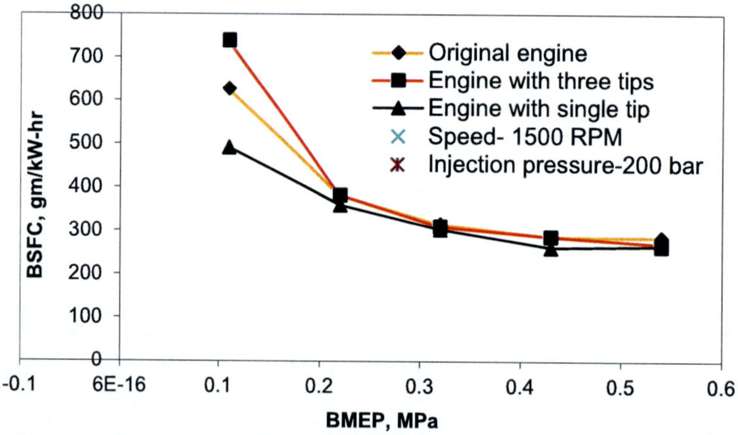

Fig. 4 BFSC versus BMEP

load and lower BSFC at full load. The decrease in BSFC is 5.3% at full load for three tips.

From the results obtained, it is found that the engine fitted with single tip gives lower-specific fuel consumption compared to original engine. This gives an indication for improvement in fuel consumption by reducing the fuel sticking to the tips.

The reduction in fuel consumption with the use of impingement tips may be because of spray diffusion or early evaporation of fuel droplets from hot impingement tips. Their exact contribution towards the improvement in diesel combustion could not be predicted since it requires analysis of $P - \theta$ diagram and combustion photography.

Comparing the results obtained, the present spray impingement system demonstrates less fuel consumption from part load to full load compared to NICS [5]* and OSKA-DH [6]* system. The present system gives 21.6% decrease in BSFC at part load and 7% decrease at full load. The NICS system gives 3% increase in BSFC at part load and 20% increase at full load. The OSKA-DH system shows 10% decrease in BSFC at part load and 2% decrease at full load.

3.2 *Effect on Smoke Level*

The smoke level increases with the increase in load for each configuration (Fig. 5). The sharp rise in smoke level was observed at full load. This is quite obvious, with the increase in load, the quantity of fuel injected increases leading to incomplete combustion. The engine fitted with single tip shows lower smoke level at all loads compared to original engine, whereas for the engine fitted with three tips, the smoke level was slightly higher at middle engine loads and lower at part load and full load compared to original engine. The reduction in smoke level for the engine fitted with

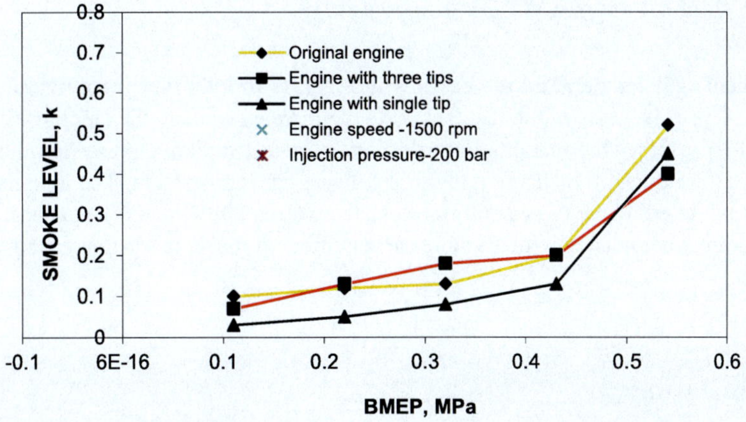

Fig. 5 Smoke level versus BMEP

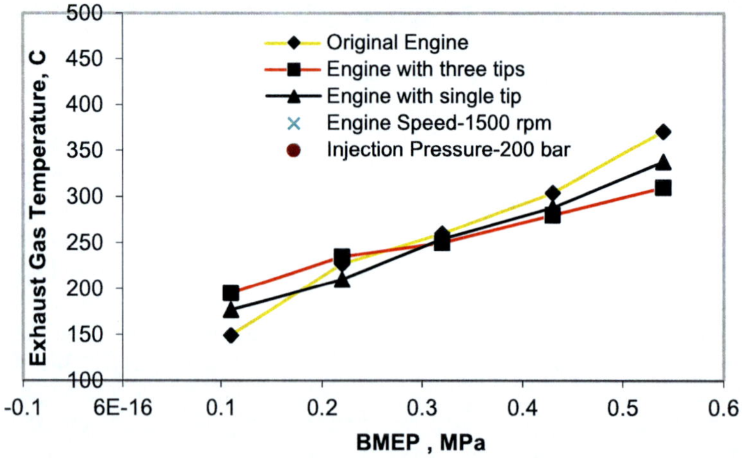

Fig. 6 Exhaust gas temperature versus BMEP

impingement tips is mainly because of brake up of core and diffused nature of broken spray.

Comparing the results obtained, the present spray impingement system demonstrates lower smoke level compared to NICS [5]* and OSKA-DH [6]* system. The present system gives 60% decrease in smoke level at part load and 23.07% decrease at full load. The NICS system gives 62.5% increase in smoke level at part load and 12.5% increase at full load. The OSKA-DH system shows 15.38% increase in smoke level at full load.

3.3 Effect on Exhaust Gas Temperature

The exhaust gas temperature increases with increase in load for all configurations (Fig. 6). This is because of the fact that, with the increase in load, the amount of fuel injected increases which results in more degree of incomplete combustion, hence more afterburning of the charge and more exhaust gas temperature. The interesting trend is exhibited by the three configurations from 60 to 100% load. The exhaust gas temperature is decreasing with the increase in number of fuel zones in the combustion chamber.

4 Conclusions

The significant conclusions from the experimental investigation are summarised below. From the obtained results, the following points were observed:

1. The engine fitted with impingement tip gives marginally lower smoke level and fuel consumption compared to original engine.
2. The observation of the tips after the total running of 12 h shows no carbon deposition with brown colour appearance which is an indication of good diesel combustion. The diesel engine run was normal.
3. The reduction in both parameters smoke level and specific fuel consumption together with no evidence of carbon deposition on tips clearly indicates the improvement in diesel combustion process with the use of impingement tip.
4. The marginally lower performance of engine with three tips as compared to single tip is because of the more amount of fuel sticking to the tips.
5. Finally, the use of impingement tip clearly gives the clue for improvement in combustion process which requires further detailed study on spray diffusion with tip.

References

1. Henein NA, Bhattacharyya A, Schipper J, Kastury A, Bryzik W (2006) Effect of injection pressure and swirl motion on diesel engine-out emissions in conventional and advanced combustion regimes. SAE Paper No. 2006-01-0076
2. Fiveland SB, Assanis DN (2000) A four-stroke homogeneous charge compression ignition engine simulation for combustion and performance studies. SAE Paper No. 2000-01-0332
3. Christensen M, Hultqvist A, Johansson B (1999) Demonstrating the multi fuel capability of a homogeneous charge compression ignition engine with variable compression ratio. SAE Paper No. 1999-01-3679
4. Celikten I (2003) An experimental investigation of the effect of the injection pressure on engine performance and exhaust emission in indirect injection diesel engines. Appl Therm Eng 23(16):2051–2060
5. Fang T, Coverdill RE, Lee CF, White RA (2008) Low temperature premixed combustion within a small bore high speed direct injection (HSDI) optically accessible diesel engine using a retarded single injection. Int J Autom Technol 9(5):551–561
6. Fang T, Coverdill RE, Lee CF, White RA (2009) Influence of injection parameters on the transition from PCCI combustion to diffusion combustion in a small bore HSDI diesel engine. Int J Autom Technol 10(3):285–295

Solving the Flexible Job Shop Scheduling Problem Using an Effective Jaya Algorithm

Rylan Caldeira and A. Gnanavelbabu

Abstract Flexible job shop scheduling problem (FJSSP) is considered to be NP-hard, in which the allocation of operations to machines is not predetermined. Due to the complex nature of this problem, various metaheuristics have been increasingly employed to address the problem, obtaining satisfactory solutions in a reasonable computational time. In this work, an upcoming heuristic named Jaya algorithm (JA) is employed to minimize the makespan. The JA enjoys the advantages of absence of any algorithm-specific parameter to be tuned to achieve optimal solutions and reduced computational effort. Although JA has powerful exploration capabilities, it lacks exploitation capability. To enhance this shortfall, an effective local search technique is integrated to improve the local search ability of the JA. Hence, the proposed approach possesses superior diversification and intensification search abilities. The performance of the effective Jaya algorithm is compared with other well-known reported algorithms with Kacems benchmark instances. Experimental results revealed that the proposed algorithm gave the best results for all five problem instances.

Keywords Flexible job shop scheduling · Local search · Makespan · Jaya algorithm

1 Introduction

Competitiveness is the backbone of sustenance of an industry. One of the factors through which the latter can be achieved is through an effective scheduling system. Scheduling is the process of allocating resources in a manufacturing plant. It is considered to be a complex but most essential process for the efficient functioning of the manufacturing system. Scheduling is required to meet customer demands on or before the due date, thereby improving customer satisfaction. Proper scheduling will also enable the available resources to be utilized effectively. FJSSP is a classification

R. Caldeira (✉) · A. Gnanavelbabu
Department of Industrial Engineering, CEG Campus, Anna University, Chennai, India
e-mail: rylan.caldeira@gmail.com

T. Rajmohan et al. (eds.), *Advances in Materials and Manufacturing Engineering*, Springer Proceedings in Materials 7, https://doi.org/10.1007/978-981-15-6267-9_15

of the JSSP which is a NP-hard problem [1] in which the allocation of operations to machines is not predefined. In FJSSP, machines are capable of processing all or some jobs. This makes it more practical and complex as compared to JSSP. The FSSP consists of two subproblems: (1) routing subproblem which deals with the assignment of each operation to a machine from a set of capable machines; and (2) sequencing subproblem in which developing an arrangement of the assigned operations on the assigned machines satisfying precedence constraints and optimizing a predefined performance measure. Depending upon the flexibility to process operations, FJSSP can be classified as total FJSSP wherein each operation can be processed on any of the available machines and partial FJSSP in which some of the operations can be processed on certain available machines.

The paper is ordered as follows: after the literature review in Sects. 2, 3, it defines the FJSSP. Section 4 describes the Jaya algorithm and the local search technique. The computational results are presented in Sect. 5. Finally, Sect. 6 concludes the paper suggesting the directions for future research.

2 Literature Review

Over the past years, various methods and techniques were employed to address the FJSSP. The earliest contributions are being that of Brucker and Schlie [2]. In recent years, due to its complexity, a large number of evolutionary algorithms and hybrid techniques were used by researchers to achieve near-optimal solutions in a reasonable time. Shen et al. [3] addressed the FJSSP with sequence-dependent set-up times considering the performance measure of minimizing makespan and developed an improved TS algorithm to solve large instances of the problem. Zandieh et al. [4] employed SA along with an improved imperialist competitive algorithm to enhance the performance by efficiently exploring the solution space for the FJSSP under condition-based maintenance. Zhang et al. [5] addressed a multi-objective FJSSP and employed a multi-population genetic algorithm considering the makespan and the machine workload as objectives. Wang et al. [6] proposed an effective ant colony optimization (ACO) algorithm which included improved selection rules for machines, initializing uniformly distributed ant mechanism, a pheromone guiding mechanism, node selection method and a pheromone updating mechanism. Zhang et al. [7] solved the FJSSP employing a chaotic differential evolution algorithm with the performance measure as makespan. Gao et al. [8] addressed the FJSSP considering fuzzy processing times and new job insertions and proposed a two-stage artificial bee colony (TABC) algorithm to solve it with an objective to minimize fuzzy makespan. Zeng and Wang [9] proposed a novel artificial immune algorithm (AIA) integrated with PSO to maintain diversity and prevent being trapped in the local optima. Buddala and Mahapatra [10] solved the FJSSP considering an integrated approach using teaching–learning-based optimization (TLBO) along with a local search, to improve the solution quality and a mutation strategy, to maintain the diversity of the population. Gao et al. [11] addressed the FJSSP with new job insertion and employed Jaya algorithm to solve it with an objective to minimize machine workload.

From the above literature, it can be seen that various metaheuristics are increasingly employed to solve the FJSSP. An upcoming metaheuristic named Jaya algorithm, proposed recently by Rao [12] ,is employed in this work. Jaya algorithm is considered to have superiority over other algorithms in terms of absence of any algorithm-specific parameters which need to be tuned to achieve optimal solutions and reduced computational effort due to less number of function evaluations [13]. Hence, an attempt is made to investigate the performance of Jaya algorithm in solving the FJSSP. The proposed Jaya algorithm is integrated with an effective local search technique to improve its local search ability and thereby improving the solution quality.

3 Problem Formulation

A FJSSP comprises *m* machines on which *j* jobs are to be processed. The processing sequence of each job may be different. The FJSSP involves the allocation of operation of jobs to machines from a set of available machines in the predetermined processing sequence considering the specified processing times so as to minimize the makespan. Each job has *o* number of predetermined operations $O = \{O_{j1}, O_{j2}, O_{j3}, \ldots, O_{jm}\}$ to be processed on m machines. Each operation of the jobs can be processed on several machines with different processing times resulting in $R = \{R_1, R_2, R_3, \ldots, R_r\}$ processing routes. The transportation time and machine set-up times are included in the processing time of the jobs. For the complete processing of a job, all the operations need to the completed in the predefined sequence resulting in a constrained optimization problem. Pre-emption is not allowed, i.e. interruption of an operation cannot be done once started. A machine can process only a single operation at a time. All jobs and machines are available at time $t = 0$. The objective is to minimize the makespan. The detailed mathematical model for the FJSSP can be referred to in [10].

4 Jaya Algorithm

Jaya algorithm is recently proposed by Rao [14]. The algorithm tends to shift the current solution in the direction of the best solution of a population and away from the worst solution for every computational step. Unlike other metaheuristics, the algorithm has an advantage of the absence of any algorithm-specific parameters which need to be tuned to obtain an optimal solution, resulting in easy implementation. Since the algorithm has only a single equation, the numbers of function evaluations are also less as compared to other metaheuristics. The algorithm is mathematically described in the following manner. For each step, k, identify the best and worst solution as F_{best} and F_{worst}, respectively, then the remaining solutions are modified as follows:

$$F_{\text{new}k} = F_{\text{old}k} + x_1 \times (F_{\text{best}} - |F_k|) - x_2 \times (F_{\text{worst}} - |F_k|) \quad (1)$$

where x_1 and x_2 are the two random numbers between 0 and 1. The terms $x_1 \times (F_{\text{best}} - |F_i|)$ and $x_2 \times (F_{\text{worst}} - |F_i|)$ represent the tendency of the solution to improve its value by moving in the direction of the best solution and away from the worst solution. If the new solution results in a better value, it is selected. The drawback of Jaya algorithm is that it lacks local search capability. Hence, an effective local search is proposed to enhance exploitation capability and improve the solution quality. Population initialization rules are employed to generate the initial population.

4.1 Local Search Technique

The following local search technique is performed after the Jaya algorithm during each iteration.

(I) Select the set of worst solution from the current population, i.e. solution which has minimum makespan value.
(II) For every solution, identify an individual operation among the set of operations.
(III) Iteratively allocate machines other than the present machine allocated for that operation.
(IV) If a solution generates a better value of makespan, replace it with the original solution.

5 Computational Results

The performance of the proposed Jaya algorithm is compared with eight other metaheuristics on five Kacem's [12, 15] benchmark problems. The algorithm was coded in MATLAB 2011. The various test problem sizes range from four parts and five machines to 15 parts and ten machines. Each of the test problems is solved five times resulting in 25 instances of problems-solving, to explore the inherent randomness present in the procedural in working of the algorithm. Average makespan values are tabulated in Table 1. The convergence of Kacem's 10 * 15 problem is shown in Fig. 1 for 30 iterations. A Gantt chart obtained using the proposed approach is shown for Kacem's 8 * 8 problem instance in Fig. 2. The x-axis shows the processing time, and y-axis indicates the machine.

Table 1 Results of benchmark problems

S. No.	Problem size	Xing's algorithm [16]	AIA [9]	HDE [17]	HTSA [18]	MBBO [19]	NIMASS [20]	IACO [21]	TLBO [10]	Proposed JA
1	4 × 5	11	11	11	11	11	11	11	11	11
2	8 × 8	14 1	14	14	14	14	14	14	14	14
3	10 × 7	11	11	11	11	NA	NA	11	11	11
4	10 × 10	7	7	7	7	7	7	7	7	7
5	15 × 10	11	11	11	11	12	11	11	12	11

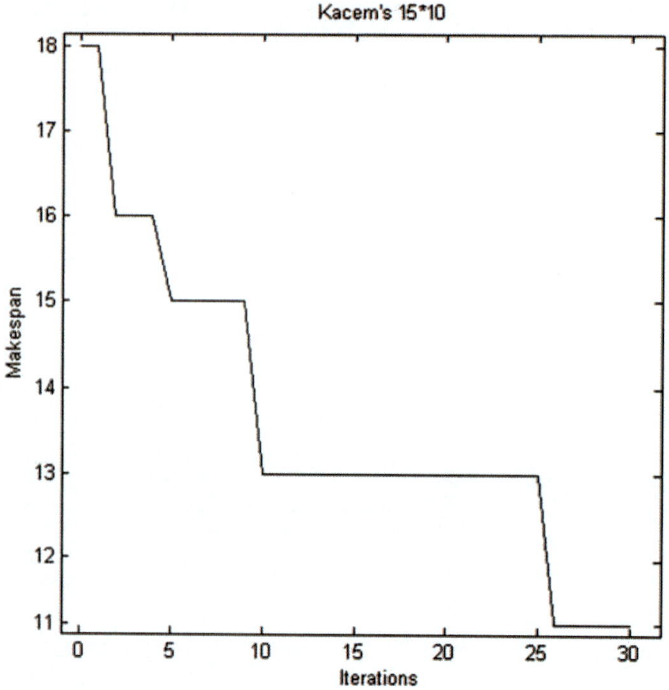

Fig. 1 Convergence of Kacem's 15 * 10 instance

6 Conclusion

This work considers makespan as the performance measure in the FJSSP. Experimental results from Table 1 show that the proposed Jaya algorithm gives the best results across all problem instances. The highlight of the Jaya algorithm is the absence of any algorithm-specific tuning parameter and reduced computational effort making its implementation easy and efficient. Previous metaheuristics employed to solve FJSSP contain algorithm-specific tuning parameters making a selection of the right parameter for the optimal solution a difficult task. An apparent extension of this work is to incorporate the Jaya algorithm with algorithms like simulated annealing and tabu search. In addition, a time-based comparison of these heuristic can be explored.

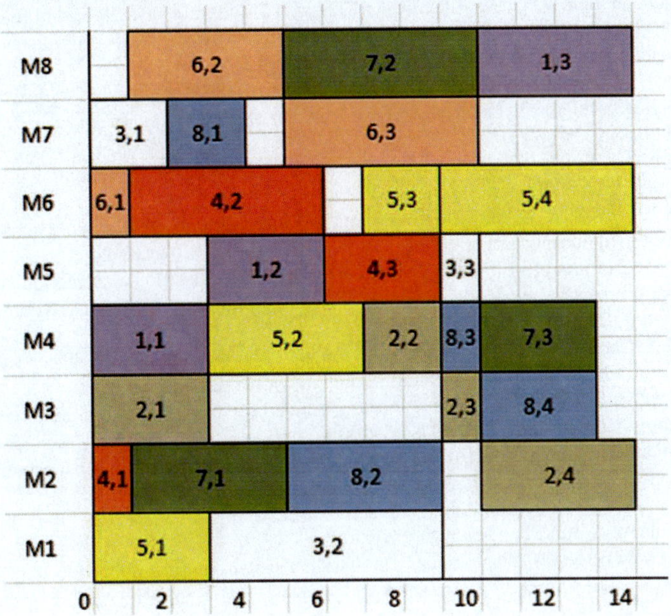

Fig. 2 Gantt chart of Kacem's 8 * 8 instance

References

1. Garey MR, Johnson DS, Sethi R (1976) The complexity of flow shop and job shop scheduling. J Math Res 1(2):177–129
2. Brucker P, Schlie R (1990) Job shop scheduling with multi-purpose machine. Computing 45(4):369–375
3. Shen L, Dauzère-Pérès S, Neufeld JS (2018) Solving the flexible job shop scheduling problem with sequence-dependent setup times. Euro J Oper Res 265(2):503–516
4. Zandieh M, Khatami AR, Rahmati SHA (2017) Flexible job shop scheduling under condition-based maintenance: improved version of imperialist competitive algorithm. Appl Soft Comput 58:449–464
5. Zhang W, Wen JB, Zhu YC, Hu Y (2017) Multi-objective scheduling simulation of flexible job-shop based on multi-population genetic algorithm. Int J Simul Modell 16(2):313–321
6. Wang C, Jiang P (2018) Manifold learning based rescheduling decision mechanism for recessive disturbances in RFID-driven job shops. J Intell Manufact 29(7):1485–1500
7. Zhang H, Yan Q, Zhang G, Jiang Z (2016) A chaotic differential evolution algorithm for flexible job shop scheduling. In: Theory, methodology, tools and applications for modelling and simulation of complex systems. Springer, Singapore, pp 79–88
8. Gao KZ, Suganthan PN, Pan QK, Tasgetiren MF, Sadollah A (2016) A Artificial bee colony algorithm for scheduling and rescheduling fuzzy flexible job shop problem with new job insertion. Knowl Based Syst 109: 1–16
9. Zeng R, Wang Y (2018) A chaotic simulated annealing and particle swarm improved artificial immune algorithm for flexible job shop scheduling problem. EURASIP J Wireless Commun Netw 2018(1):101
10. Buddala R, Mahapatra SS (2018) An integrated approach for scheduling flexible job-shop using teaching–learning-based optimization method. J Ind Eng Int 1–12

11. Gao K, Sadollah A, Zhang Y, Su R, Li KGJ (2016) Discrete Jaya algorithm for flexible job shop scheduling problem with new job insertion. In: 14th international conference on control, automation, robotics and vision. IEEE, Thailand, pp 1–5
12. Kacem I, Hammadi S, Borne P (2002) Pareto-optimality approach for flexible job-shop scheduling problems: hybridization of evolutionary algorithms and fuzzy logic. Math Comput Simul 60(3–5):245–276
13. Caldeira RH, Gnanavelbabu A (2019) Solving the flexible job shop scheduling problem using an improved Jaya algorithm. Comput Ind Eng 137:1–16
14. Rao R (2016) Jaya: A simple and new optimization algorithm for solving constrained and unconstrained optimization problems. Int J Ind Eng Comput 7(1):19–34
15. Kacem I, Hammadi S, Borne P (2002b) Approach by localization and multiobjective evolutionary optimization for flexible job-shop scheduling problems. IEEE Trans Syst Man Cybern Part C Appl Rev 32(1): 1–13
16. Xing LN, Chen YW, Yang KW (2009) An efficient search method for multi-objective flexible job shop scheduling problems. J Intell Manuf 20(3):283–293
17. Yuan Y, Xu H (2013) Flexible job shop scheduling using hybrid differential evolution algorithms. Comput Ind Eng 65(2):246–260
18. Li JQ, Pan QK, Liang YC (2010) An effective hybrid tabu search algorithm for multi-objective flexible job-shop scheduling problems. Comput Ind Eng 59(4):647–662
19. Yang Y (2015) A modified biogeography-based optimization for the flexible job shop scheduling problem. Math Probl Eng 2015
20. Xiong W, Fu D (2018) A new immune multi-agent system for the flexible job shop scheduling problem. J Intell Manuf 29(4):857–873
21. Wang L, Cai J, Li M, Liu Z (2017) Flexible job shop scheduling problem using an improved ant colony optimization. Sci Progr 2017

Wear Analysis of Epoxy Resin Composites Reinforced with Seashell

A. Sivanand, C. Devanathan, E. Shankar, and P. PrasannaKumar

Abstract Make in India initiative becomes a major drive for Indian Composites Market growth. For renewable energy, gas and oil, chemical industries, smart city construction (building and construction) and water management, the components made of composites are imminent. To support the effort, in this project work, two different composites were prepared and tested for its wear resistance and specific wear resistance using pin on disc apparatus. Composites were made with pure epoxy and 20% seashell mixed with epoxy resins. This type of composites was mostly used with large number of resins, motor housing, telephones and electrical fixtures, etc. To carry out the wear analysis load, sliding velocity and sliding distance had been taken as different parameters. From the results, it was learnt that the parameters considered were indirectly proportional to wear resistance and specific wear resistance. When comparing the two composites, seashell mixed composites had offered better wear resistance than the pure epoxy resin.

Keywords Epoxy resin · Seashell · Natural composites · Pin on disc

1 Introduction

Epoxy resins are more costly than polyesters but their comprehensive range of properties can make them the better choice for critical applications like aircraft. The major benefits of using epoxy include its capability to tailor made to different products with low shrinkage, strong mechanical properties, which can be used at higher temperature as high as 1750 °C and are compatible with common reinforcement. The viable importance in epoxy resins was initially made by the German Patent publication with number 676 117 by I G Farben1 in 1939.

An epoxy resin belongs to a family of polymer materials under the support of thermoset. Epoxy resins are formed by a reaction of an epoxide with a hardener or polyamine that has tremendous cross linking to create a very tough and stiff

A. Sivanand (✉) · C. Devanathan · E. Shankar · P. PrasannaKumar
Department of Mechanical Engineering, Rajalakshmi Engineering College, Thandalam, Chennai 602105, India
e-mail: sivanand.a@rajalakshmi.edu.in

T. Rajmohan et al. (eds.), *Advances in Materials and Manufacturing Engineering*, Springer Proceedings in Materials 7, https://doi.org/10.1007/978-981-15-6267-9_16

polymer. Epoxy resin has been a dominant matrix material used in the development of advanced composite materials because of the excellent properties. Zion Market Research recently had published the survey report, according to which the IN 2016, worldwide epoxy resins market accounted for 7.9 Billion USD in 2016 and is anticipated to extent 11.5 Billion USD by 2022, growing at a compound annual growth rate (CAGR) of around 6.8% between 2017 and 2022. Manchester University scientist Leeds acquired an inspiration from seashell, calcium carbonate to create a strong and tough polymer.

Oladele et al. [1] studied the mechanical properties and wear resistance of epoxy polymers filled with calcined particles of African land snail shells using Taber Abraser and UTM and concluded that higher amount of loading of filler and smaller particle size in bio composites have increased the mechanical properties and resistance against wear. Larger particle size showed no significant enhancement in mechanical properties.

Bello et al. [2] reviewed the currently ongoing research in epoxy resin hybrid nano composites for engineering applications. Kumar et al. [3] investigated the tribological properties of the composites fabricated by biowaste horn fibre and epoxy resin and optimized using Grey relational analysis. The specimen prepared by optimum parameters has lesser density and higher coefficient of friction, and therefore, they can be used in brake pads, clutch discs, etc.

Vignesh et al [4], studied the mechanical properties of seashell particles with various particle sizes which are mixed with unsaturated polyester resins. For smaller particle size, the void is very thin, and in turn, it avoids the voids, so that the materials are very strong. Odusanya et al. [5] considered the water absorption characteristic of unsaturated polyester composites. Researcher tried 250 μm ground seashell at different percentage of resin polyester and studied flexural strength, impact and hardness properties. It was observed that the properties were greatly increased for 10% loading of seashell filler. Karthick et al. [6] made an attempt to analyse tribological properties of seashell nano powder introduced in PMMA composites at different volume fraction, and the results showed that 12% addition of seashell powder offered more wear resistance. Further increase in reinforcement particles developed more frictional force which results in the decrease in wear resistance.

In the proposed work, two different composites were prepared with 100 and 80% epoxy with 0 and 20% seashell reinforcement. The composites were tested in pin on disc apparatus for specific wear rate and wear rate.

2 Experimental Procedure

Raw materials such as seashell and isophthalic epoxy resin were taken and mixed at required proportion, then catalyst methyl ethyl ketone peroxide and accelerator cobalt naphthenate are added to ensure the proper and uniform mixing. The resin and seashell powder are shown in Figs. 1 and 2. The mixture is then poured in the designed mould and allowed for 6–8 h to prepare the composites. The used mould is

Fig. 1 Isophthalic epoxy resin

Fig.2 Powdered seashell

Fig. 3 Composite mould

shown in Fig. 3. The cured composites were removed from the mould and prepared for wear testing in pin on disc apparatus. Figure 4 shows the composites prepared for further testing.

The basic concept of pin on disc is shown in Fig. 5. The experimental set-up involves a hardened steel disc. The initial weight of the specimen was measured, and after the test the final weight was measured with a single-pan electronic weighing machine with an accuracy of 0.001 g. Material loss was determined by weight loss method. The actual machine set where the experiments were conducted is shown in Fig. 6.

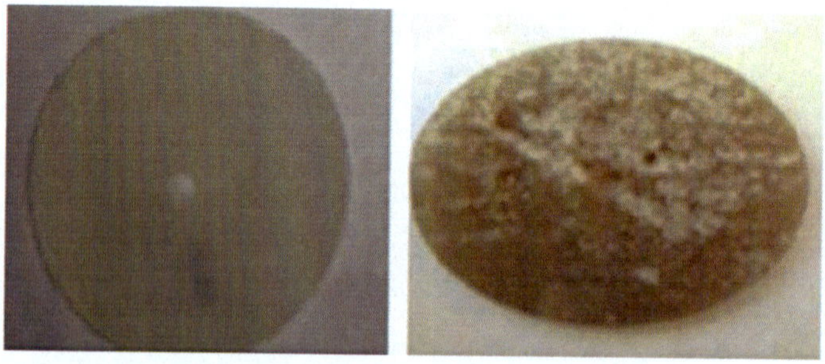

Fig.4 Prepared composites

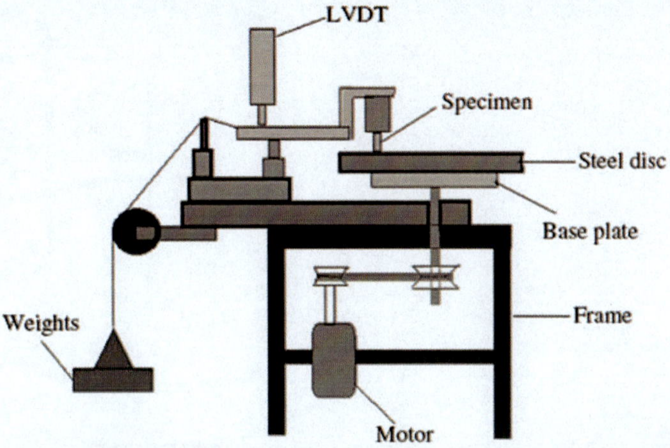

Fig. 5 Concept of pin on disc set-up [9]

Fig. 6 Pin on disc testing machine

3 Results and Discussion

A tribometer is an instrument that measures tribological quantities, such as coefficient of friction, friction force and wear volume, between two surfaces in contact.

The volume of wear in any arrangement will, in general, depend upon the number of system factors such as the machining characteristic, sliding velocity, sliding distance and properties of the material. The value of any wear test method lies in

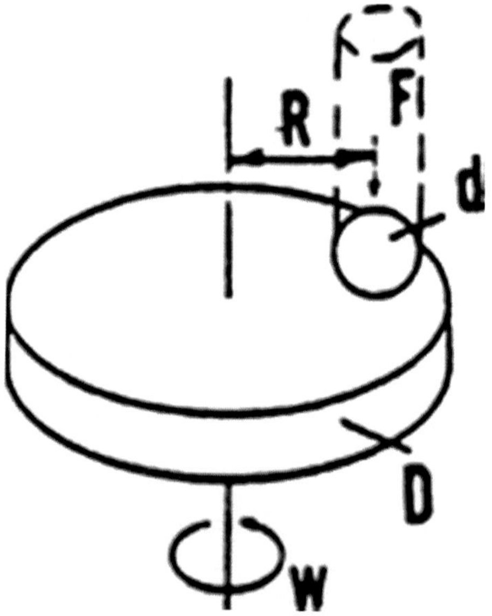

Fig. 7 Pin on disc test system

predicting the relative ranking of material combinations. The specimen for pin on disc apparatus was prepared as per ASTM G99-05 which is shown in Fig. 7.

In the present work, to analyse the specific wear rate and wear rate for different process parameters such as load, sliding velocity and sliding distance, these parameters are taken at three different levels in order to conduct the pin on disc experiments. L9 orthogonal array was selected for conducting the experiments. Table 1 shows the different parameters and their levels.

The weight loss of the specimen was measured after the wear test in order to calculate the specific wear rate by the following equation:

$$\text{Specific Wear Rate(SWR)} = \Delta m/\rho.t.\text{VS.FN}\ \left(\text{mm}^3/\text{Nm}\right) \quad (1)$$

Wear rate is calculated by using the following equation:

$$\text{Wear Rate (WR)} = m/2\Pi.r.n.t \quad (2)$$

Table 1 Process parameters for wear test

S. No.	Parameter	Level 1	Level 2	Level 3
1	Load (N)	10	20	30
2	Sliding velocity (m/s)	1	2	3
3	Sliding distance (m)	1500	1000	500

The specific wear rate and wear rate were calculated for various parameters, and its effect was plotted as graph. Figures 8, 9 and 10 depict the effect of load, sliding velocity, sliding distance on wear rate and specific wear rate, respectively, for pure epoxy resin without addition of seashell.

From Figs. 8, 9 and 10, it was clear that both wear rate and specific wear rate decrease when the parameter value increases. All the three considered parameters had inverse effect on output parameters. The current experimental results were very similar to the results obtained by Emad omrani et al., while checking for polymer matrix composites reinforced with natural fibres [7]. This trend is identical for

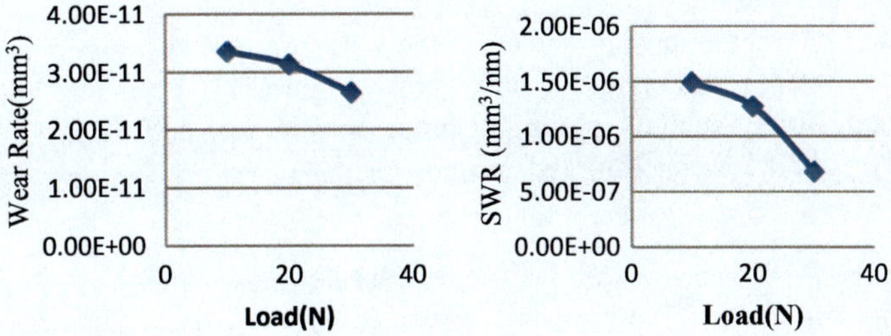

Fig.8 . Effect of load on WR and SWR for pure epoxy

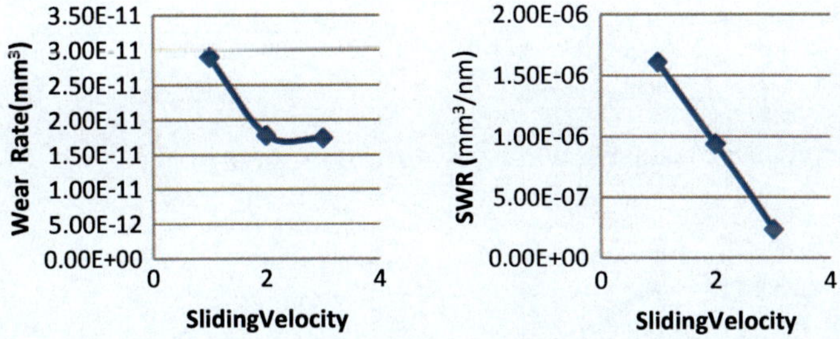

Fig. 9 . Effect of sliding velocity on WR and SWR for pure epoxy

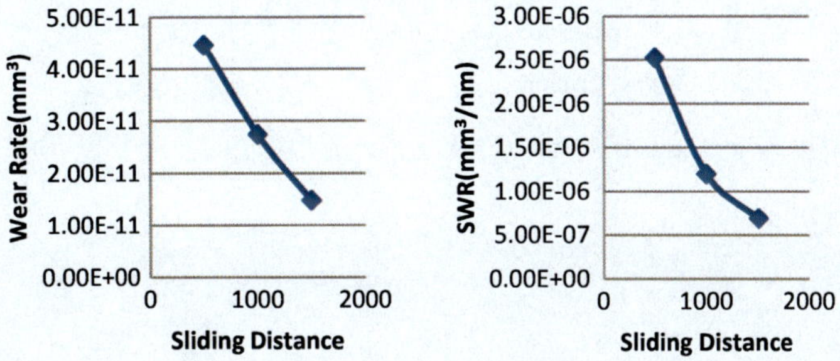

Fig.10 Effect of sliding distance on WR and SWR for pure epoxy

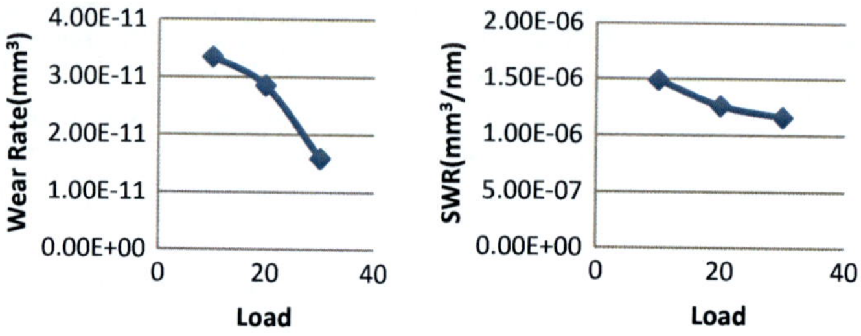

Fig.11 . Effect of load on WR and SWR for 20% seashell

20% seashell mixed composites, whereas the 20% seashell reinforced composites were showed better results than the pure epoxy. Figures 11, 12 and 13 showed the effect of load, sliding velocity, sliding distance on wear rate and specific wear rate, respectively, for 20% seashell mixed composites.

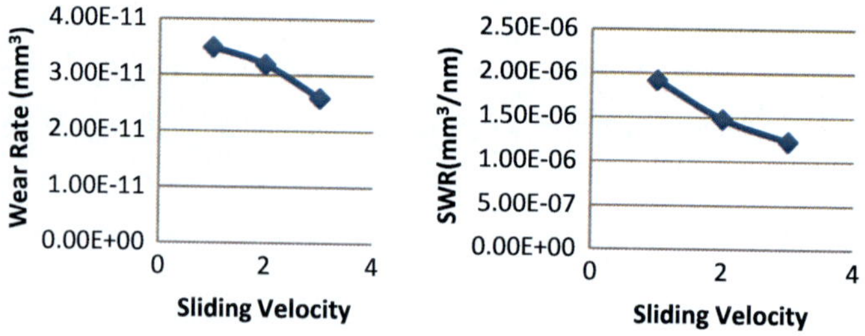

Fig.12 Effect of sliding velocity on WR and SWR for 20% seashell

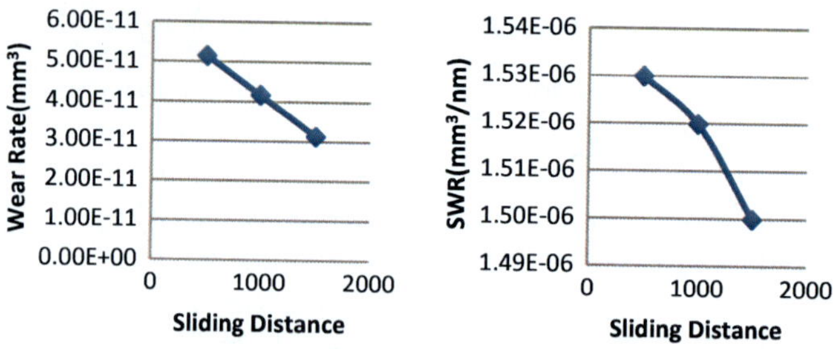

Fig.13 Effect of sliding distance on WR and SWR for 20% seashell

4 Conclusion

In the stated work, two different types of composites were prepared: one with pure resin with addition of reinforcement and the other one with 20% addition of seashell. These composites were tested on pin on disc apparatus to analyse various tribological factors like specific wear rate and wear rate when changing various input parameters like load, sliding velocity and sliding distance. The test results revealed that increasing the above said parameters leads to decrease in the specific wear rate and wear rate. This trend is common for both pure epoxy resin and 20% seashell mixed composites. When comparing the values among these two composites, the seashell mixed composites had given the better results than the pure epoxy resin composites.

References:

1. Oladele IO, Olajide JL, Amujede M (2016) Wear resistance and mechanical behaviour of epoxy/mollusk shell biocomposites developed for structural applications. Tribol Ind 38(3):347–360
2. Bello SA, Agunsoye JO, Hassan SB, Zebase Kana MG, Raheem IA (2015) Epoxy resin based composites, mechanical and tribological properties: a review. Tribol Ind 37(4):500–524
3. Kumar D, Rajendra Boopathy S, Sangeetha D Investigation on tribological properties of horn fibre reinforced epoxy composites. Int J Mech Mech Eng IJMME-IJENS 16(03): 79–87.
4. Vignesh K, Anbazhagan K, Ashokkumar E, Manikandan R, Jayanth A Experimental analysis of mechanical properties of sea shell particles-polymer matrix composite. Int J Mech Ind Technol 3(1): 13–21
5. Odusanya AA, Bolasodun B, Madueke CI (2014) Property evaluation of sea shell filler reinforced unsaturated polyester composite. Int J Sci Eng Res 5(11):1343–1349
6. Karthick R, Sirisha P, Ravi Sankar M (2014) Mechanical and tribological properties of PMMA-sea shell based biocomposite for dental application. In: 3rd international conference on materials processing and characterization (ICMPC 2014). Procedia materials science, vol 6, 1989–2000
7. Omrani E, Menezes PL, Rohatgi PK (2016) State of the art on tribological behavior of polymer matrix composites reinforced with natural fibers in the green materials world. Eng Sci Technol Int J 19:717–736

A Brief Review: Study of Machinability Aspects of Hard Metals Using Micro Textured Inserts

Indraneel Soppa and Swastik Pradhan

Abstract Nowadays, new technologies are being developed in the field of machining to enhance the machinability criteria of the cutting tools. In the current years, sustainable machining approach is used to improve performance of cutting inserts. The recent advancement is made in the field of micro texture manufacturing on the rake face and flank face of the cutting inserts. Fabricating micro texturing on the cutting inserts leads improving of tribological properties. Current study provides fabrication of micro texture, study of machinability aspect and comparison between coated the uncoated micro texture tool. Present literature study shows that micro texture cutting tools with inclusion of solid lubricants which is enhancing the performance of the cutting inserts lead to increase in tool life.

Keywords Micro texture cutting insert · Tool wear · Chip morphology · Cutting temperature · Surface roughness

1 Introduction

Dry machining of hard metals is difficult because of its poor thermal conductivity and high reactivity. It causes severe tool wear, like adhesion wear and flank wear, and deformation in tools takes place while machining of hard metals. So, researches are used in cutting fluids to eliminate the effect on the tool. Its causes a harmful effect to the operator and storing and disposal of cutting fluid is taking twice the cost of total machining cost. So researches developed sustainable machining conditions like surface texturing, solid lubrications cryogenic and minimum quantity lubrication (MQL) conditions [1]. Recently, researches developed the micro and nano textures on rake face with or without filling of the solid lubricant in the cutting tool inserts; it also called as self-lubricating tools [2]. This technique is sustainable machining which refers alternatively flooded machining. The present literature survey is carried

I. Soppa · S. Pradhan (✉)
School of Mechanical Engineering, Lovely Professional University, Phagwara, Punjab 144411, India
e-mail: swastik.rock002@gmail.com

T. Rajmohan et al. (eds.), *Advances in Materials and Manufacturing Engineering*, Springer Proceedings in Materials 7, https://doi.org/10.1007/978-981-15-6267-9_17

out with respect to the last four-year enhancements in the field of the micro textured insert.

2 Literature Survey

2.1 Surface Texture Manufacturing on Cutting Insert

Texturing on the surface of the inserts will act self-lubricating, and it can improve tribological performance of the inserts [3]. There are several ways to create textures on the surface [4]. These various researchers manufacture micro texture using wire electric discharge machining [5–11], laser surface texturing (LST) [12–25], different hardness tester and scratch tester micro texture are created on the surface on the rake surface [26–29]. In this survey, researchers use different types of textures like parallel, perpendicular, holes, elliptical, chevron, squares pyramid, dimples and volcano shapes are created on the rake or flank face.

2.2 Cutting Forces

Cutting force acting on the cutting tool by its cutting conditions reduction of forces would exhibit extension of tool life by following researchers. Machining of Ti-6Al-4 V cutting forces like feed and thrust forces acting on tool is reduced in perpendicular texture tool than non-groove inserts [5].Comparison of experimental and simulation results is almost equal in perpendicular texture tool [6].Between coated and uncoated tool, cross texture insert is coated with TiAlN there reduced feed and thrust forces by 22% and 35%, respectively [7]. Reduction of forces in micro texture is coated with TiAlN tool [24]. Parallel-type deep submillimetre texture tool is reducing cutting force [16]. Comparison of cutting forces between AlCrN and AlTiN coated micro texture tools shows that main cutting forces are exhibited maximum reduction up to 4.8% when compared to plain tool [17]. Machining of Al-Cu/TiB_2 composites with linear texture perpendicular to chip flow has reduced the principal cutting force [8]. Machining of hardened steel with a composite micro textured ceramic tool shows a good reduction in all the three forces [9]. Force is increased with respect to machining speed and depth of cut in parallel texture insert under cryo-treatment [11]. Machining of 4340 hardened steel based on L27 Taguchi layout cutting forces is reduced in micro texture design created at a distance of 100 μm, width of 100 μm, pitch of 360 μm, dimple diameter of 100 μm and depth of 140 μm [12]. Machining of AISI 316 forces investigated under dry and wet conditions with micro texture tool with parallel to chip flow reduces forces by 5% in dry conditions and in wet conditions 4.7% is reduced when compared to conventional tool [13]. Machining of ZK60 Mg alloy under cryogenic conditions reduces maximum of 22% force in comparison with textured and untextured tool [15]. Machining of Al7075-T6 aerospace alloy cutting force and thrust force reduces to 9% and 19%, respectively [18]. Machining in A6061 alloy is that volcano type texture is lower in rough cutting force and in finishing volcano type texture increases the force because of coolant [23]. By machining of AISI 1040 alloy steel with micro textured coated and uncoated solid lubrication of MoS_2 on HSS tool, the forces get reduced in the solid lubricated coated tool [26]. Machining of AISI 1040 steel with HSS tool filled and unfilled micro texture tool also reduces the forces [28]. Micro texture coated with MoS_2 also reduced the forces [29]. Machining of mild steel and aluminium with HSS tool affects the axial, radial and main cutting forces by 41%, 38% and 22% and in the case of aluminium 20%, 18% and 13%, respectively, as compared to that of the conventional tool [27]. By FEM simulation, cutting forces are reduced averagely 11.88% to plain tool [30]. In microgroove tool, average of 16.68% reduction in main force than plain tool [31] based upon microgrooves rake angles also plays a role in reduction in forces [32].

2.3 *Cutting Temperature*

Cutting temperature on the cutting tool plays a major role. By reducing the cutting temperature of cutting tool reduces the tool wear, shear strength of tool and tool–chip interface. Cutting temperature rise with respect to cutting speed but by using perpendicular texture tool a reduction of 29% in cutting temperature was shown during machining [5]. Comparison between coated and uncoated tools than TiAlN coated tool shows 32% reduction in temperature with respect to remaining texture [7]. Angle to chip flow texture has generated lower cutting temperature while in parallel to chip texture temperature is increased because more chip interface is observed [9]. Using MQL reduced the temperature in the micro texture design of 1 mm deep hole with diameter 0.4 and 0.6 on different sides of the insert up to 40% and 30%, respectively [10]. Under cryogenic conditions, maximum 50% reduction of temperature is observed in micro texture tool [15]. Temperature of the workpiece increases with respect to feed and speed while MoS_2 filled insert has reduced the workpiece temperature [28]. HSS tool with micro textured has reduced temperature up to 3.28% in mild steel while in case of aluminium there is a maximum reduction up to 12% [27]. Temperature is decreased in microgrooved tool due to less tool–chip interface [30]. Average cutting temperature microgrooved tool has 19.03% less than plain tool [31].

2.4 *Surface Roughness*

While machining, good surface roughness of the material should be achieved; otherwise, manufacturing cost would increase. So the surface roughness on the cutting tool is studied by the researcher. Desirable surface finished is achieved in micro hole texture when compared to the normal tool [10]. Higher machining speed and depth of cut affect the surface roughness of the material in the case of parallel texture and cryo-treatment [11]. There are three types of microgroove texture like parallel vertical and a micro hole to cutting edge. The surface roughness of Gcr15 material is desirable in the case of micro hole texture PCBN inserts [14]. Better surface finish is achieved in under micro texture under cryogenic conditions. There is a 20% improvement in surface finish compared to the conventional tool [15]. In terms of micro texture on the flank face, there is no improvement [21]. Micro texture tools of coated and uncoated achieved good machined surface when compared to conventional tool [24]. Due to the presence of graphite in micro texture, R_a value reduces approximately 40–20% at higher speed [25]. Surface roughness is improved 15% and 25% in mild steel and aluminium, respectively, using micro texture tool with respect to cutting speed [27]. Good surface finish got in lubrication filled micro texture. Surface finish affects while increasing feed [28].

2.5 Chip Morphology

Chip morphology plays a major role in the cutting tool by reduction of tool–chip interface will reduce tool wear and improve tool life and machining force by the following researcher. Different types of chips produced while machining of Ti-6Al-4 V that less saw tooth chips are formed at a lower speed. Due to less friction, curled chips are formed in the case of perpendicular texture tool [5]. Serrated chips are formed in TiAlN coated texture tool [7]. Snarled chips are produced in the long and short chip in all types of insert. Diameter of snarled chips is reduced in perpendicular texture tool [8]. Serrated chips are formed in case of traditional tool while in micro textured breaking of serrated chips and chip width is also reduced [9]. Snarled ribbon chips are formed in the normal tool while in micro texture tool coil-type chips are formed [10]. Small and curled chips are formed due to micro texturing with CaF_2 lubrication, and the tool–chip interface is less so friction is less [12]. Smaller thickness with curling in textured coated tool was observed due to the reduction in tool–chip interface area [17]. Chip formed in dimple texture gets adhesion and increases the rake angle [18]. Texture inserts chips show texture impressions on the back side of chips [20]. Powder chips are produced in green ceramic composites due to which microgrooves abrasion wear is prevented by powder chips [21]. Curlier chips in rough turning and smooth chips are formed in volcano type texture tool and act as chip breaker [23]. Curled chips are formed, and curled radius of chips is reduced in coated texture tools [24]. Due to the presence of MoS_2 in micro texture tool, it acts as thin-film lubrication and reduces tool–chip interface, and there is no change in chip thickness [26]. Micro texture acts as a chip breaker and reduces the tool–chip interface that forms a continuous chip in aluminium and mild steel [27].Serrated chips are formed, and less chip thickness is observed in lubricated insert [28].

2.6 *Tool Wear and Adhesion*

Based upon tool wear and adhesion, the tool life is dependent if the tool life is increased more parts are to be manufacture by the following researcher. Tool wear in micro texture tools less crater wear appeared than a normal tool [5]. The perpendicular texture tool has low wear because of less friction. Maximum flank wear is reduced in TiAlN coated perpendicular texture tool [7]. Low amount of adhesion of perpendicular texture tool and no formation of built-up edge and tool wear are reduced under the proper lubrication [8]. Adhesion of *Fe* is taking place in the traditional tool while micro texture tool of less deposition of *Fe* is observed [9]. Micro holes have reduced the formation of built-up edge and reduction of flank wear that increased tool life [10]. Due to abrasion mechanism between the rake surface of textured cutting tool and chip coming out at the time of machining the texture flank wear exists [11]. Deposition of *Fe* takes place in a conventional tool, and less deposition of *Fe* takes place due to lubrication heat dissipated on the tool [13]. Adhesion wear is found in dry conditions, and a smaller amount of adhesion is 27% of reduction in adhesive wear in parallel texture tool shown in cryogenic conditions [15]. Formation on built-up edge in tool due to abrasions and coating delamination takes place in AlCrN coated tool, and due to micro texture there is a reduction in flank wear [17]. Micro dimples are filled with built-up layer also formed bulge [18]. Tool life is increased by using of low pressure jet coolant, and flank wear is reduced in texture on rake surface [19]. Combination of dimple and square pyramids texture on rake and flank surface reduced the flank wear [20]. Low rake face wear exists, and abrasion wear appears in flank face and parallel-type texture are good to wear resistance [21]. Volcano and microgroove texture have a similar type of adhesion, and crater wear appears as filling of microgroove [23]. That coated textured tools are a reduction in flank wear, and no cracks appeared in coated texture tool [24]. Lubrication filled texture decreased the tool–chip interface, and lower tool wear exists [29].

3 Conclusion

The current literature study presents the study of different machining effect on the machinability of the workpiece using the micro texture tool insert. Most of the researcher manufactured the micro texture using laser surface texturing to achieve more accuracy of desirable dimensions. Using micro texture on tool rake face that reduces the tool and chip interface so that effect of cutting forces and cutting temperature, tool wear is reduced by increasing the quality of the machined surface as compared to conventional tool inserts. The overall performance of the coated micro texture inserts is good as compared to uncoated micro texture inserts. Addition of solid lubrication, minimum quantity lubrication (MQL) and cryogenic condition to micro textured inserts will enrich the performance of the cutting by increasing the tool life and surface finish.

References

1. Ghosh S, Rao PV (2015) Application of sustainable techniques in metal cutting for enhanced machinability: a review. J Clean Prod 100:17–34
2. Wu Z, Deng JX, Lian YS, Wang ZJ, Zhao J Experimental study on the cutting temperature with micro-texturing self-lubricated tools. In: Advanced materials research, pp 715–720. Trans Tech Publ
3. Gajrani KK, Sankar MR (2017) State of the art on micro to nano textured cutting tools. Mater Today Proc 4:3776–3785
4. Arslan A, Masjuki H, Kalam M, Varman M, Mufti R, Mosarof M, Khuong L, Quazi M (2016) Surface texture manufacturing techniques and tribological effect of surface texturing on cutting tool performance: a review. Crit Rev Solid State Mater Sci 41:447–481
5. Arulkirubakaran D, Senthilkumar V, Kumawat V (2016) Effect of micro-textured tools on machining of Ti–6Al–4V alloy: an experimental and numerical approach. Int J Refract Metal Hard Mater 54:165–177
6. Arulkirubakaran D, Senthilkumar V, Dinesh S (2017) Effect of textures on machining of Ti-6Al-4V alloy for coated and uncoated tools: a numerical comparison. Int J Adv Manuf Technol 93:347–360
7. Arulkirubakaran D, Senthilkumar V (2017) Performance of TiN and TiAlN coated micro-grooved tools during machining of Ti-6Al-4V alloy. Int J Refract Metal Hard Mater 62:47–57
8. Arulkirubakaran D, Senthilkumar V, Chilamwar VL, Senthil P (2019) Performance of surface textured tools during machining of Al-Cu/TiB2 composite. Measurement
9. Kumar CS, Patel SK (2018) Effect of WEDM surface texturing on Al2O3/TiCN composite ceramic tools in dry cutting of hardened steel. Ceram Int 44:2510–2523
10. Rao CM, Rao SS, Herbert MA (2018) Development of novel cutting tool with a micro-hole pattern on PCD insert in machining of titanium alloy. J Manuf Process 36:93–103
11. Palanisamy D, Balasubramanian K, Manikandan N, Arulkirubakaran D, Ramesh R (2019) Machinability analysis of high strength materials with Cryo-Treated textured tungsten carbide inserts. Mater Manuf Process 34:502–510
12. Sharma V, Pandey PM (2017) Geometrical design optimization of hybrid textured self-lubricating cutting inserts for turning 4340 hardened steel. Int J Adv Manuf Technol 89:1575–1589
13. Vasumathy D, Meena A, Duraiselvam M (2017) Experimental study on evaluating the effect of micro textured tools in turning AISI 316 austenitic stainless steel. Procedia Eng 184:50–57
14. Pan C, Li Q, Hu K, Jiao Y, Song Y (2018) Study on surface roughness of Gcr15 machined by micro-texture PCBN tools. Machines 6:42
15. Dinesh S, Senthilkumar V, Asokan P (2017) Experimental studies on the cryogenic machining of biodegradable ZK60 Mg alloy using micro-textured tools. Mater Manuf Process 32:979–987
16. Li N, Chen Y, Kong D, Tan S (2017) Experimental investigation with respect to the performance of deep submillimeter-scaled textured tools in dry turning titanium alloy Ti-6Al-4V. Appl Surf Sci 403:187–199
17. Mishra SK, Ghosh S, Aravindan S (2018) Characterization and machining performance of laser-textured chevron shaped tools coated with AlTiN and AlCrN coatings. Surf Coat Technol 334:344–356
18. Sasi R, Subbu SK, Palani I (2017) Performance of laser surface textured high speed steel cutting tool in machining of Al7075-T6 aerospace alloy. Surf Coat Technol 313:337–346
19. Obikawa T, Nakatsukasa R, Hayashi M, Ohno T High performance cutting using micro-textured tools and low pressure jet coolant. In: AIP conference proceedings. AIP Publishing, pp 070019
20. Alagan NT, Zeman P, Hoier P, Beno T, Klement U (2019) Investigation of micro-textured cutting tools used for face turning of alloy 718 with high-pressure cooling. J Manuf Process 37:606–616
21. Liu Y, Deng J, Wu F, Duan R, Zhang X, Hou Y (2017) Wear resistance of carbide tools with textured flank-face in dry cutting of green alumina ceramics. Wear 372:91–103

22. Li J, Tao B, Huang S, Yin Z (2018) Built-in thin film thermocouples in surface textures of cemented carbide tools for cutting temperature measurement. Sens Actuators A 279:663–670
23. Kang Z, Jun MBG, Fu Y (2018) Performance of cemented carbide cutting tools with volcano-like texture on rake face. In: ASME 2018 13th international manufacturing science and engineering conference. American Society of Mechanical Engineers, pp V004T003A009–V004T003A009
24. Mishra SK, Ghosh S, Aravindan S (2019) Performance of laser processed carbide tools for machining of Ti6Al4V alloys: a combined study on experimental and finite element analysis. Precis Eng
25. Karthikeyan S, Suresh Kannan I, Kumar KR (2018) Impact of laser micro textured cutting tool inserts filled with graphite in reducing chip adhesion during the turning process of aluminium alloy AA2025
26. Gajrani KK, Reddy RPK, Sankar MR (2016) Experimental comparative study of conventional, micro-textured and coated micro-textured tools during machining of hardened AISI 1040 alloy steel. Int J Mach Mach Mater 18:522–539
27. Jesudass Thomas S, Kalaichelvan K (2018) Comparative study of the effect of surface texturing on cutting tool in dry cutting. Mater Manuf Process 33:683–694
28. Gajrani KK, Sankar MR, Dixit US (2018) Environmentally friendly machining with MoS 2-filled mechanically microtextured cutting tools. J Mech Sci Technol 32:3797–3805
29. Gajrani KK, Sankar MR (2018) Sustainable machining with self-lubricating coated mechanical micro-textured cutting tools. Refe Mod Mater Sci Mater Eng
30. Maity K, Pradhan S (2018) Investigation of FEM simulation of machining of titanium alloy using microgroove cutting insert. Silicon 10:1949–1959
31. Zou Z, He L, Jiang H, Zhan G, Wu J (2019) Development and analysis of a low-wear micro-groove tool for turning Inconel 718. Wear 420:163–175
32. Kaya E, Kaya İ (2018) Efficiency investigations of textured cutting tools in orthogonal cutting of Ti6Al4V alloy: a numerical approach. Euro J Sci Technol: 164–168

Optimization of Process Parameters in EDM Using Standard Deviation and MOORA Method

J. Anitha and Raja Das

Abstract In the current study, the application of standard deviation and multi-objective optimization based on ratio analysis (MOORA) is applied to optimize the process parameters in Electro discharge machine (EDM). The impact of the process parameters, namely current, pulse on time, duty cycle and voltage on the performance measures like the material removal rate (MRR) and surface roughness (Ra) is studied in this experimental work. MOORA in combination with standard deviation is used as an improvement procedure. Standard deviation is applied to choose the weights that are used for normalizing the performance measures which are obtained from the experimental outcomes. The weights obtained using standard deviation are as follows: MRR is 0.53 and Ra is 0.47. The best combination of process parameters to maximize the MRR and minimize the Ra are current 15 units (level 3), duty cycle 50 (level 1), pulse on time 100 units (level 3) and voltage 50 units (level 3).

Keywords Multi-objective method-MOORA · EDM · Material removal rate · Surface roughness · Standard deviation

1 Introduction

Electric discharge machining (EDM) is a widely and efficiently used non-conventional machining process. The main advantage of this process is that it can engine any electrically conductive solid regardless of its hardness. It is mainly used for manufacturing automobile components, aerospace components, injection molds, plastic molds and surgical instruments. Difficult-to-machine materials such as alloys based on nickel, titanium and hardened tool steel are used in this process. In the die making and tool making, AISI D2 steel has a wide variety of application, as it can be hardened and strengthened to give higher strength and wear resistance as compared to other low-carbon steels [1–4].

J. Anitha (✉) · R. Das
Vellore Institute of Technology University, Vellore, India
e-mail: anithasastry2077@gmail.com

T. Rajmohan et al. (eds.), *Advances in Materials and Manufacturing Engineering*, Springer Proceedings in Materials 7, https://doi.org/10.1007/978-981-15-6267-9_18

The manufacturers frequently face problems in decision making like judging a wide range of alternatives and selecting the best one, which are generally conflicting in nature. There is no single description method of selection to find the best option, but the decision maker must consider a lot of options and finally select the best option. The criteria for selection should be simple, systematic, logical and use mathematical tools. The decision maker should consider several selection criteria and their interrelations [5]. Nowadays, several MODMs are prevailing to agree with varying criteria and choice, and this paper mainly explores the application of new MODM method, namely the multi-objective optimization on the basis of ratio analysis (MOORA) method to optimize the process parameters in EDM. This method is quite simple, easy to calculate, and helps the decision maker to remove undesirable alternatives and select the suitable alternatives and helps in strengthening the selection procedure.

Many researchers used the MOORA method in different applications to prove its potentiality, applicability and flexibility. Brauers [6] first introduced MOORA method for estimating the stakeholder's society design, also for privatization in control and cybernetics [7], evaluation of inner climate [8] multi-objective contractors ranking [9], evaluating the road design [10, 11] and project management by MULTIMOORA for transition economies [12]. Gadakh [13, 14] applied this method for optimization in milling and WEDM process. Optimization using standard deviation and MOORA method by Muniappan et al. [15]. Chakraborty [16, 17] has used this method in solving six different problems like (a) a flexible manufacturing system, (b) an industrial robot, (c) an automated system, (d) a rapid prototyping process, (e) a computerized controlling machine and (f) the most appropriate non-traditional machining process for given workpiece irrespective of its material and shape feature combination. MOORA is a JAYA algorithm which is based on the concept that the solution obtained for the problem moves away from the worst solution and moves toward the best solution. MOORA is a very simple algorithm which shows outstanding performance in various applications.

2 Experimental Environment and Method

2.1 Experimental Setup

Experiments were conducted under the following machining conditions:

- Processing Machine: Electronica Electra plus PS 50ZNC.
- Workpiece material: A rectangular shaped AISI D2 tool steel, having a thickness of 4 mm (with negative polarity) and density 7.7 g/cc is used.
- Electrode material: It is electrolytic copper with a positive polarity and 30 mm in diameter.
- Flushing: Side flushing technique with 0.3 kg f/cm^2 pressure.
- Dielectric fluid: A viable grade EDM oil which has a specific gravity of 0.76 and freezing point of 94 °C.

Table 1 Various input values along with units at different levels

Input values	Unit	Level 1	Level 2	Level 3
Ip	A	5	10	15
Ton	μs	50	75	100
T		50	66.5	83
V	V	40	45	50

The four input parameters along with units are shown in Table 1.

2.2 Optimization Problem

If MRR $= f_1$(Ip, Ton, T, V) and Ra $= f_2$(Ip, Ton, T, V), then the multi-objective optimization problem is

Maximize f_1(Ip, Ton, T, V) and
Minimize f_2(Ip, Ton, T, V)

Subject to $5 \leq Ip \leq 15$
$50 \leq Ton \leq 100$
$50 \leq T \leq 83$
$40 \leq V \leq 50$
And Ip, Ton, T, V € R

2.3 Weight Calculation Using Standard deviation Concept

Standard deviation is related to calculation of unbiased assignment of weights.

Step 1 Calculate

$$X_{ij}^{1} = \frac{X_{ij} - \min_{1<j<n} X_{ij}}{\max_{1<j<n} X_{ij} - \min_{1<j<n} X_{ij}} \tag{1}$$

where max X_{ij} is the maximum, and min X_{ij} is the minimum value for the measure (*j*)

Step 2 Calculate standard deviation using equation (2)

$$\mathrm{SDV}_j = \sqrt{\frac{\sum_{i=1}^{m}\left(X_{ij} - \overline{X_j^i}\right)^2}{m}} \tag{2}$$

where *Xij* is the average of the values for the *j*th measure, where $j = 1, 2, 3$ … *n*.

Step 3 Calculate weights using Eq. (3)

$$w_j = \frac{\mathrm{SDV}_J}{\sum_{j=1}^{n} \mathrm{SDV}_j} \tag{3}$$

where w_j are the weights for $j = 1, 2, 3 \ldots n$.

2.4 *MOORA Method*

MOORA is one of the MCDM methods which is used to select the best option from number of selections. This method was first presented by Brauers to solve intricate decision-making problems in the manufacturing industry. The MOORA method (Kalibatas et al. [8]; Lootsma [18]) begins with a decision matrix that represents various substitutes and objectives.

Step 1 Describing the problem and establishing the objectives are the first step. In the present work, MRR must be maximized and Ra must be minimized.

Step 2 Next step is to create a decision matrix based on the experimental results of the number of output parameters. Represent the decision matrix using *X*.

$$X = \begin{bmatrix} x_{11} & x_{12} \ldots & x_{1n} \\ x_{12} & x_{12} \ldots & x_{2n} \\ \vdots & \vdots & \vdots \\ x_{m1} & x_{m2} & x_{mn} \end{bmatrix}. \tag{4}$$

Step 3 The performance of the *i*th alternate on *j*th feature is normalized using Eq. (5)

$$x_{ij}^* = \frac{x_{ij}}{\sqrt{\sum_{i=1}^{m} x_{ij}^2}} \tag{5}$$

where $j = 1, 2, 3 \ldots n$.

Step 4 In the multi-objective optimization process, these normalized values are added if it is a case of maximization and subtracted if it is a case of minimization. Hence, the optimized values are calculated using Eq. (6)

$$y_i = \sum_{j=1}^{g} x_{ij}^* - \sum_{j=g+1}^{n} x_{ij}^* \tag{6}$$

where *g* are the number of features which have to be maximized, and the number of features which has to be minimized is $(n - g)$

Step 5 It is quite often detected that some features are more significant than the other features. To give more significance to such features, it is multiplied with corresponding weight. The optimized problem is calculated using Eq. (7)

$$y_i = \sum_{j=1}^{g} w_j x_{ij}^* - \sum_{j=g+1}^{n} w_j x_{ij}^* \tag{7}$$

for $j = 1, 2, 3 \ldots n$, where w_j is the weight of the jth attribute, which is determined using the standard deviation method.

Step 6 The y_i values can be positive or negative depending on the maximizing attributes and minimizing attributes in the decision matrix. Rank the values from the highest to the lowest. The best alternative has the highest value of y_i, and the worst value has the lowest y_i.

3 Results and Discussion

3.1 Allocation of Weights

In the current work, the weights are allocated to the output parameters MRR and Ra. The range is calculated for these output parameters using Eq. (1), later standard deviation is calculated using Eq. (2), and weights are calculated based on Eq. (3). The corresponding weight for calculation of MRR is 0.53 and for Ra is 0.47.

3.2 Analysis

Table 2 shows the experimental values of MRR and Ra. The standardized values of MRR and Ra are shown in Table 3. The normalized values of X_i and the ranks calculated using MOORA method are shown in Table 4. The Material removal rate ranges between 0.0270 and 0.2575 and the surface roughness in the range 0.0774–0.1864. From these observations, experiment 4 has the best rank. The values for run 4 are Ip 15 units (level 3), pulse on time 100 units (level 3), duty cycle 50 (level 1) and voltage 50 units (level 3).

4 Conclusion

In this study, the combined methods of standard deviation and MOORA are applied for optimizing the machining parameters, so that MRR is maximized and Ra is minimized. The conclusions are as follows:

Table 2 Experimental values of MRR and Ra

Run	Ip	Ton	*T*	*V*	MRR	Ra
1	10	75	66.5	45	10.35	5.55
2	15	50	83	50	29.163	8.43
3	15	50	50	50	33.103	7.43
4	15	100	50	50	51.09	8.1
5	15	100	50	40	51.004	10.93
6	15	100	83	40	33.023	12.49
7	15	50	50	40	29.737	10.49
8	10	75	66.5	40	8.936	8.2
9	10	75	66.5	50	11.007	6.35
10	15	75	66.5	45	33.084	9.68
11	10	75	66.5	45	11.01	6.25
12	10	50	66.5	45	9.182	5.87
13	10	75	66.5	45	9.352	6.75
14	5	75	66.5	45	5.361	6.07
15	5	50	83	50	14.12	5.19

Table 3 Standardized values of MRR and Ra

EXP_MRR	EXP_Ra	Standardized MRR	Standardized Ra
10.35	5.55	0.0722	0.0995
29.163	8.43	0.0204	0.0063
33.103	7.43	0.0524	0.0033
51.09	8.1	0.3872	0.0012
51.004	10.93	0.3849	0.1778
33.023	12.49	0.0516	0.4036
29.737	10.49	0.0241	0.1306
8.936	8.2	0.0897	0.0023
11.007	6.35	0.0647	0.0423
33.084	9.68	0.0522	0.0627
11.01	6.25	0.0646	0.0482
9.182	5.87	0.0865	0.0737
9.352	6.75	0.0844	0.0228
5.361	6.07	0.1427	0.0596
14.12	5.19	0.0347	0.1330
Standard deviation		0.3284	0.2906

Table 4 Normalized values of MRR and Ra and MOORA ranking

Run	Normalized MRR Ra		Weighted MRR Ra		Max–min	Rank
1	0.0984	0.1762	0.0522	0.0828	0.0306	9
2	0.2773	0.2676	0.1470	0.1258	0.0212	5
3	0.3148	0.2359	0.1669	0.1109	0.0560	3
4	0.4859	0.2571	0.2575	0.1209	0.1367	1
5	0.4851	0.3470	0.2571	0.1631	0.0940	2
6	0.3141	0.3965	0.1665	0.1864	0.0199	8
7	0.2828	0.3330	0.1499	0.1565	0.0066	7
8	0.0850	0.2603	0.0450	0.1223	0.0773	15
9	0.1047	0.2016	0.0555	0.0947	0.0393	11
10	0.3146	0.3073	0.1668	0.1444	0.0223	4
11	0.1047	0.1984	0.0555	0.0933	0.0378	10
12	0.0873	0.1863	0.0463	0.0876	0.0413	12
13	0.0889	0.2143	0.0471	0.1007	0.0536	13
14	0.0510	0.1927	0.0270	0.0906	0.0635	14
15	0.1343	0.1648	0.0712	0.0774	0.0063	6

a. Combined standard deviation and MOORA method are used to select the optimized parametric combination in electric discharge machining of AISI D2 steel workpiece with copper electrode.
b. The best rank is given to run 4 in the experiment. The corresponding parameters for run 4 are Ip 15units (level 3), pulse on time 100 units (level 3), duty cycle 50 (level 1) and voltage 50 units (level 3).
c. Standard deviation is applied to find the relative importance of MRR and Ra. The weight ratios for MRR are 0.53 and Ra 0.47.
d. The optimized results obtained are used in validation study, and the results obtained can be effectively used for better productivity which helps the manufacturers to compete in the world market.

References

1. Abbas NM, Solomon DG, Bahari MF (2007) A review on current research trends in electrical discharge machining (EDM). Int J Mach Tools Manuf 47(7):1214–1228
2. Ho KH, Newman ST (2003) State of the art electrical discharge machining (EDM). Int J Mach Tools Manuf 43(13):1287–1300
3. Pradhan MK (2010) Experimental investigation and modelling of surface integrity, accuracy and productivity aspects in EDM of AISI D2 steel. Ph.D. Dissertation
4. Pradhan MK, Biswas CK (2008) Neuro-Fuzzy model on material removal rate in electric discharge machining in AISI D2 steel. In: Proceedings of the 2nd international and 23rd all india manufacturing technology, design and research conference. IIT Madras, India

5. Rao RV (2007) Decision making in the manufacturing environment: using graph theory and fuzzy multiple attribute decision making methods. Springer, London
6. Brauers WK (2003) Optimization methods for a stakeholder society, a Revolution in Economic thinking by Multi-objective optimization series: no convex optimization and its applications, 73. Kulwer academic publishers, Boston, 342
7. Brauers WKM, Zavadskas EK (2006) The MOORA method and its applications to privatization in some transition economy. Control Cybern 35:445–469
8. Kalibatas D, Turskis Z (2008) Multicriteria evaluation of inner climate by using MOORA method. Inf Techol Con 37:79–83
9. Brauers WKM, Zavadskas EK, Turkis Z, Viutiene T (2008) Multi-objective contractors ranking by applying the MOORA method. J Bus Econ Manage 9:245–255
10. Brauers WKM, Zavadskas EK (2008) Pedschus, Turkis, Z: Multi-objective decision making for road design. Transport 23:183–193
11. Brauers WKM, Zavadskas EK Pedschus Turkis Z (2008) Multi-objective optimization of road design alternatives with an application of the MOORA method. In: Proceedings of the 25th international symposium on automation and robotics in construction. Lithuania, pp 541–548
12. Brauers WKM, Zavadskas EK (2010) Project management by MULTIMOORA as an instrument for transition economies, Technological and economic development of economy. Balt J Sustain 16(1):5–24
13. Gadakh VS, Shinde VB, Khemnar NS Optimization of welding process parameters using MOORA method. Int J Adv Manuf Technol. https://doi.org/10.1007/s00170-013-5188-2
14. Gadakh VS (2011) Application of MOORA method for parametric optimization of milling process. Int J Appl Eng Res 1(4):743–758
15. Muniappan A, Raj JA, Jaykumar V, Prakash S, Sathyaraj R (2018) Optimization of WEDM process parameters using standard deviation and MOORA method. In: 2nd international conference on advances in mechanical engineering (ICAME 2018)
16. Chakraborty S (2010) A decision making framework for selecting non-traditional machining process using the MOORA method. In: Proceedings of the 4th international conference on advances in mechanical engineering, Sept (2010)
17. Chakraborty S (2010) Applications of the MOORA method for decision making in manufacturing environment. Int J Adv Manuf Technol 54(9–12):1155–1166
18. Lootsma FA (1999) Multicriteria decision analysis via ratio and difference judgement. Springer, London

Wear Behaviour and Mechanical Properties of AA2024/Al_2O_3/SiC/Gr HMMC Using Advanced Squeeze Casting Technique

L. Natrayan and M. Senthil Kumar

Abstract Hybrid metal matrix composites (HMMC) play a vital role in meeting the global needs in automotive applications for low-cost, high-performance, and quality materials with excellent mechanical and thermal properties. The study aims to explore the wear and mechanical properties of AA2024-reinforced Al_2O_3/SiC/Gr with different weight fractions using liquid metallurgy route particularly in advanced squeeze casting technique. Different wt% of Al_2O_3/SiC/Gr reinforcement is added to AA2024 and the effect of this combination on the material is evaluated based on its tensile strength, hardness, wear rate and compressive strength. The fabricated specimens were prepared and tested as per the ASTM standard. The results show that increased reinforcement up to 9 wt% has better hardness and tensile strength. On the other hand, the reinforcement metal matrix with Al_2O_3/SiC/Gr particles up to 9 wt% shows less wear rate and high compressive strength. Wear worn surface shows that fragments from the inner part of the material forms a layer on the composite surface.

Keywords Squeeze casting · Hardness · Tensile strength · Compressive strength · Wear · Worn surface · Weight percentage · HMMC

1 Introduction

Most of the engineering applications require high strength light-weight materials besides offering greater resistance to corrosion and wear during their operation [1]. Adding reinforcements improves its properties by transferring the load onto the reinforcements [2]. The materials for engineering applications are selected based on their properties like strength, wear, density, lightweight, melting point, cost, etc. Aluminium (Al) and mild steel have better properties compared to other materials. Steel is non-preferable for cylinder liners as it's density is three times of Al with low CTE and thermal conductivity. Having good thermal conductivity makes Al a better

L. Natrayan (✉) · M. Senthil Kumar
School of Mechanical Engineering, VIT, Chennai, India
e-mail: natrayanphd@gmail.com

T. Rajmohan et al. (eds.), *Advances in Materials and Manufacturing Engineering*, Springer Proceedings in Materials 7, https://doi.org/10.1007/978-981-15-6267-9_19

choice. Though the CTE of steel and Al are similar, Al piston satisfies the requirement of lightweight material. Al and its alloys exhibit low hardness and the mechanical and tribological properties degrade at elevated temperatures leading to their disadvantage in engineering applications. Similarly, proper selection of reinforcements like Al_2O_3, SiC, Gr, B_4C, Si_3N_4, TiC, TiB_2, TiO_2, Ti and W helps the AMC in achieving the desired mechanical properties [3]. Though studies about MMCs on the potentials of agro unwanted waste ashes are still skimpy (compared to the synthetic reinforcement), the accessible outcome represents that Al-based synthesized ceramics compounds such as silicon carbide and alumina have first -rate properties in correlation to the agro waste ash augmented category [4]. When hard ceramic particles like silicon carbide (SiC) is used with Al alloy, the abrasive nature of the ceramic improves the strength, stiffness, corrosion resistance, wear and hardness of the composites [5]. Al_2O_3 is a low-cost core material [6]. The mixture of Al_2O_3 with the Al alloy produces lightweight materials due to excellent particle bonding. The lightweight materials are used in energy consumption materials where it improves the load carrying capacity than the usual metal matrix composites. The present investigation is undertaken to determine the wear and mechanical properties of hybrid metal matrix composite by adding different wt% of reinforcement particles such as Al_2O_3/SiC/Gr using squeeze casting technique.

2 Material Selection

In this research, AA2024 was selected as a base material. Table 1 shows the chemical properties of AA2024. Reinforcement with different wt% of Al_2O_3/SiC/Gr was used. Table 2 shows the reinforcement wt% taken for this experimental study. The reinforcement with the average particle size of 10 μm was prepared by the advanced squeeze casting technique.

Table 1 Of AA 2024 chemical composition

Element	Mn	Cr	Fe	Cu	Si	Mg	Zn	Ti	Al
Weight %	0.01	0.05	0.17	0.33	0.71	1.12	0.1	0.01	Bal.

Table 2 Sample preparation

S. No.	AA2024 (g)	Al_2O_3 (g)	SiC (g)	Gr (g)	Reinforcements (wt%)
S-1	1000	0	0	0	0
S-2	970	10	10	10	3
S-3	940	20	20	20	6
S-4	910	30	30	30	9
S-5	880	40	40	40	12

2.1 Base Material

AA2024 is mainly used for automobile engine applications. It has a lower melting point of 660 °C compared to other aluminium alloys, and the melted metal hardens at a constant temperature with high flexibility. It retains high tensile strength, good deformation, and good wear resistance.

2.2 Reinforcements

In the present study, different wt% of silicon carbide (SiC), aluminium oxide (Al_2O_3) and graphite (Gr) have been selected as the reinforcements for their capability of increasing the wear resistance and strength of the composite.

The hybrid metal matrix composites were developed by considering Al_2O_3/SiC as primary reinforcement and graphite (Gr) as secondary reinforcement due to its solid lubrication capability in a wide range of temperatures. Low density, high reactivity with aluminium, high stability and thermal conductivity are the other favourable characteristics of Gr.

3 Experimental Procedure

In the experimental set-up shown in Fig. 1, 1 kg of AA2024 alloy was taken in a graphite crucible and was fired in a furnace at 750 °C, and stirred during the molten state. The reinforcements Al_2O_3/SiC/Gr were preheated at 400 °C in a furnace and were mixed with the molten aluminium during stirring. The preheated Mg 1 wt% was added to increase the wettability [7]. The molten solution was stirred for 10 min at 250 rpm to achieve the homogeneous dispersion. The experiment was conducted in an argon atmosphere to reduce oxidation. The molten metal solution was poured into the die cavity. 100 MPa of squeeze pressure was applied using a hydraulic plunger. Pressure holding time was 20 s; pouring temperature took 800 °C, and the die temperature of 250 °C was maintained constantly [8]. Die-metal interface promotes rapid heat transfer, and then the sample was ejected from the die. The specimens prepared according to the ASTM standards were characterized for their hardness, tensile strength, compressive strength and wear.

Fig. 1 Squeeze casting machine

4 Results and Discussion

4.1 Hardness

Vickers hardness test machine was used to inspect the macrohardness of HMMC as per ASTM E92 standard. Diamond-shaped indenter of 10 mm diameter with 0.5 kg was used for 10 s at five different locations to the all specimens, and the average was calculated [9]. Three trails of experiments were conducted. The specimen surface was prepared by grinding and polishing for a well-defined indentation. The relationship between Al_2O_3/SiC/Gr wt% and the hardness of the composite is shown in Fig. 2. From the results, it was found that hardness of Al2024/Al_2O_3/SiC/Gr increased from 137 to 167 HV with wt% increasing up to 9.

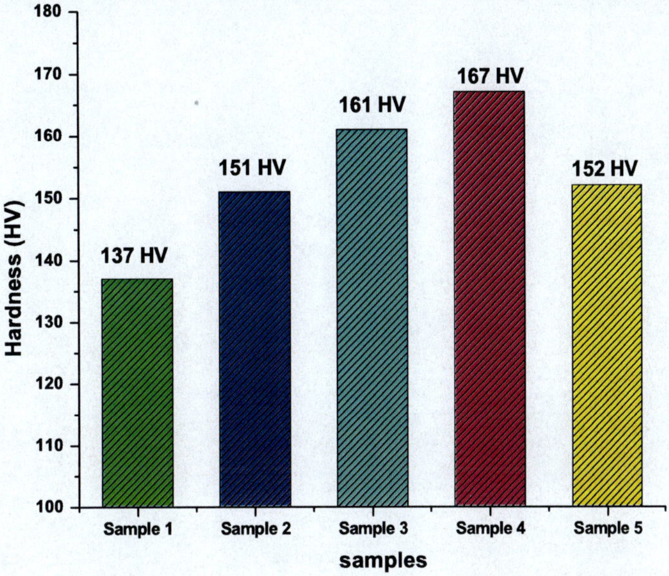

Fig. 2 Hardness of HMMC

4.2 *Tensile Strength*

Tensile test samples were prepared as per the ASTM E8M-08 standard, and cross-head speed of 0.5 mm/min was used for the measurement of tensile strength [10]. Figure 3 shows the relationship of reinforcement particles and tensile strength of the fabricated composite. Weight fraction of the composite reinforcement is vital in UTS. Figure 3 shows the tensile values of the fabricated composites. AA2024 base metal gives 420 MPa. Sample 4 shows a tensile strength of 469 MPa when it is exposed to reinforcement AA2024/Al_2O_3/SiC/Gr which is higher than the base metal. When the Al_2O_3/SiC/Gr reinforcement up to 9 wt% was added with the AA2024 molten metal, the tensile strength increased by 10%. The bonding boundary of the matrix was observed to be good.

4.3 *Compressive Strength*

Compressive test samples were machined as per the ASTM C39 (Cylinder shape) standard, and cross-head speed of 0.5 mm/min was used for the compressive strength measurement [10]. Figure 4 shows that the compressive strength of the fabricated composite increased by 7% with increased wt% of Al_2O_3/SiC/Gr reinforcement up to 9 wt%. Adding more than 9 wt% of Al_2O_3/SiC/Gr particles in molten metal decreased

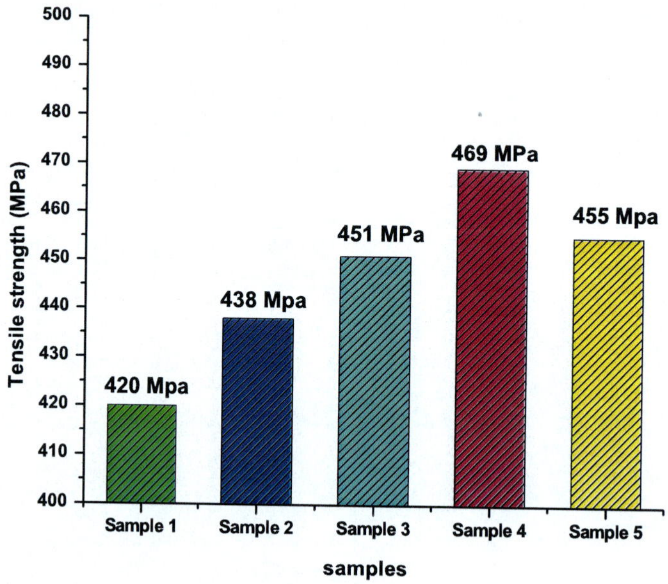

Fig. 3 Tensile strength of HMMC

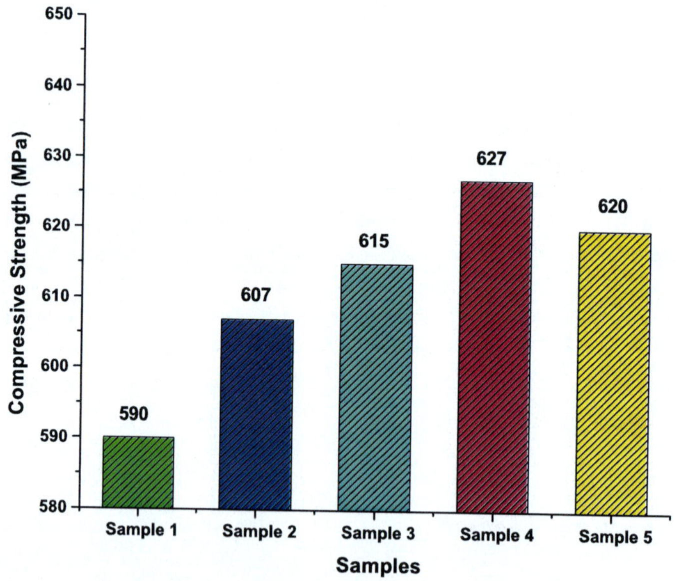

Fig. 4 Compressive strength of HMMC

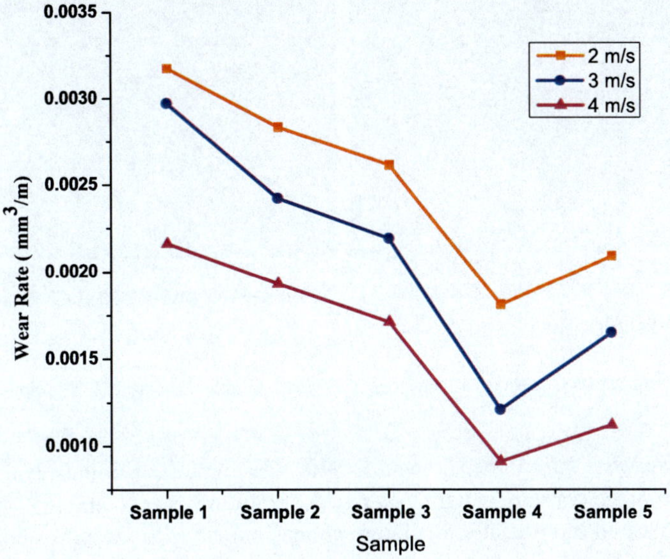

Fig. 5 Wear rate of HMMC with different velocity

the compressive strength of fabricated composite due to fracture of reinforcement that weakens the bonding boundary of the matrix.

4.4 Wear

Wear specimens were machined as per ASTM G99-05 standard. The selected dimensions of the pin sample were 15 mm height and 6 mm diameter. Samples were tested at the ambient condition, where the relative humidity was about 60–65%. Wear test was performed at different sliding speeds of 2, 3, 4 m/s for loads 20 N and sliding distance 1200 m, respectively. Measurements were performed in an unlubricated condition. The dry sliding wear behaviour of HMMC at three different velocities is shown in Fig. 5. Wear rate decreased with increase in the reinforcement up to 9 wt% for different velocities. The wear rate of sample 5 was high with 12 wt% reinforcement. The wear rate was found to be the lowest for sample 4 tested at a constant velocity of 4 m/s.

4.5 Worn Surfaces

Figure 6a, b shows the worn surfaces of Al2024/Al_2O_3/SiC/Gr. During the wear test, reinforcement thrown out of the composite pin gets logged on the disc. This converts

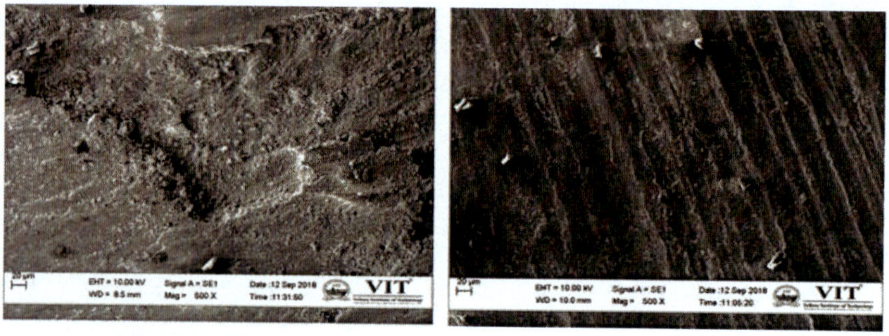

(a) Sample 2 (b) Sample 4

Fig. 6 SEM micrograph of the worn surfaces of HMMC samples **a** sample 2, **b** sample 4

the adhesive wear into abrasive wear. Due to this a greater amount of material is removed from the pin sample with lower wt% of reinforcement (sample 2—Fig. 6a).

An increase in interfacial temperature creates oxidation layer on the composite. The oxide layer acts as a protective layer that prevents sliding interfaces (Fig. 6b). Increase in sliding velocity from 2 to 4 m/s decreased the wear rate of the composite by 28%.

5 Conclusion

AA2024 with different wt% of Al_2O_3/SiC/Gr reinforcement was fabricated with advance squeeze casting technique, and fabricated HMMC was tested for hardness, tensile strength, compressive strength, SEM worn surface and wear rate. The specimens were characterized, and the subsequent inferences are drawn.

- The hardness of the HMMC improved significantly with the addition of the reinforcement particle. It increased from 137 HV to 167 HV in HMMC.
- The tensile properties of HMMC increased with increasing reinforcement up to 9 wt% due to good bonding with the matrix. From the experimental results, the tensile strength of sample 4 increased by 10% compared to the other samples.
- When 9 wt% of reinforcement was added into the molten metal, the compressive strength increased while adding 12 wt% of reinforcement decreased the compressive strength as the bonding boundary of the matrix weakened due to the fracture of the reinforcement.
- Wear rate engaged by various factors such as the velocity of 2, 3, 4 m/s, load of 20 N and sliding distance of 1200 m. Wear rate decreased in direct proportion to an increase in the velocity. Wear rate decreased with increase in Al_2O_3/SiC/Gr up to 9 wt%. Adding 12 wt% of Al_2O_3/SiC/Gr showed a high wear rate.

- Wear worn surface observed that wear track was found on the edges of sample 2 and the removal of fragments from the inner composite surface was also high. The surface damage and delamination were less in sample 4.

References

1. Surappa MK (2003) Aluminium matrix composites: challenges and opportunities. Sadhana 28(1–2):319–334
2. Natrayan L, Kumar MS (2020) A potential review on influence of process parameter and effect of reinforcement on mechanical and tribological behaviour of HMMC using squeeze casting method. J Crit Rev 7(2):1–5
3. Bodunrin MO, Alaneme KK, Chown LH (2015) Aluminium matrix hybrid composites: a review of reinforcement philosophies; mechanical, corrosion and tribological characteristics. J Mater Res Technol 4(4):434–445
4. Natrayan L, Kumar MS (2020) Influence of silicon carbide on tribological behaviour of AA2024/Al_2O_3/SiC/Gr hybrid metal matrix squeeze cast composite using Taguchi technique. Mater Res Express 6(12):1265f9
5. Yogeshwaran S, Prabhu R, Natrayan L, Murugan R (2015) Mechanical properties of leaf ashes reinforced aluminum alloy metal matrix composites. Int J Appl Eng Res 10(13), ISSN 0973-4562
6. Natrayan L et al (2018) Optimization of squeeze cast process parameters on mechanical properties of Al_2O_3/SiC reinforced hybrid metal matrix composites using taguchi technique. Mater Res Express 5:066516. https://doi.org/10.1088/2053-1591/aac873
7. Jiang J, Xiao G, Che C, Wang Y (2018) Microstructure, mechanical properties and wear behavior of the rheoformed 2024 aluminum matrix composite component reinforced by Al_2O_3 nanoparticles. Metals 8(6):460
8. Natrayan L, Senthil Kumar M (2019) Optimization of tribological behaviour on squeeze cast Al6061/Al_2O_3/SiC/Gr HMMCs based on taguchi method and artificial neural network. J Adv Res Dyn Control Syst 11(7):493–500
9. Natrayan L, Kumar MS (2020) Optimization of wear behaviour on AA6061/Al_2O_3/SiC metal matrix composite using squeeze casting technique-Statistical analysis. Mater Today Proc 27:306–310
10. Kumar S et al (2018) Experimental investigations on mechanical and microstructural properties of Al_2O_3/SiC reinforced hybrid metal matrix composite. IOP Conf Ser Mater Sci Eng 402: 012123.

Experimental Investigation on Machining Properties Carbon Fibre-Reinforced Epoxy Composites with the Addition of Nano SiC

R. Ravi Kumar, Seshadri Sridharan, and Arun Kumar Srirangan

Abstract Composites of carbon fibre reinforced in polymer matrix competes with aluminium–lithium alloys in aerospace applications because of its better balance between strength and the weight. The major issue in CFRP composites was reduced fracture toughness. To increase the fracture toughness of the composites, nanoSiC was added to epoxy resin as reinforcement. In this study, the machining properties of CFRP composites reinforced with nanoSiC were studied with weight fraction of SiC, spindle speed, and feed rate as process parameters and the torque was obtained as the response. The analysis and optimization of parameters were done using RSM technique. It has been observed that the parameter with 1% weight fraction of SiC, 50 mm/min feed rate, and spindle speed of 1500 rpm has given the maximum torque of 0.34 Nm.

Keywords Epoxy resin · Carbon fibre · Nano-SiC · Taguchi L_{18} orthogonal array · Response surface methodology

1 Introduction

From the time, immemorial people are drawn towards the materials which have distinct properties among them like less weight and more stiffness, so they have been used in novel applications and they were called as composites [1]. Definition for composites states that composites are materials composed of more than one constituent which are mixed along the infinitesimal level with negligible solublity. It is basically heterogeneous, having two or more physically distinct components, and when combined together, they become stronger and stiffer and perform superior to each of the distinct components. In general, composites are combination of two parts: one is matrix which was known to be continuous nature while the other part is reinforcement seen present along the matrix as an intermittent phase [2]. Here, the matrix is epoxy resin and the reinforcement is carbon fibre which contains carbon atoms

R. Ravi Kumar · S. Sridharan · A. K. Srirangan (✉)
School of Mechanical Engineering, SASTRA University, Thanjavur, Tamil Nadu, India
e-mail: arun1988@gmail.com

T. Rajmohan et al. (eds.), *Advances in Materials and Manufacturing Engineering*, Springer Proceedings in Materials 7, https://doi.org/10.1007/978-981-15-6267-9_20

whose diameter ranges from 4 to 5 μm. In the bundle form, the carbon fibres are five times stronger and two times stiffer than the steel. Its four different views like plain view, twill view, etc. Production of carbon fibres consists of an process in which the carbon atoms which are crystalline are bonded together. To achieve the requirements of maintaining high fraction of a relationship between strength and the volume, the alignment of carbon atoms on the fibre plays an important role, and it should be noted that bonding of the carbon atoms should parallel along the direction of the fibre [3]. Carbon fibres weigh only about two-thirds the weight of the steel and it has a tensile strength of 5.9 GPa and tensile modulus of 300 GPa so that they are used in structural applications in the field of aerospace and motor sports [4]. Epoxy resins are polymer compounds which contain epoxide as their functional group. They react with polyfunctional amines and acid anhydrides though a process called catalytic homopolymerisation. These amines and anhydrides are called as hardners or curatives which help to stabilize the epoxy resins. It is seen that the high strength of carbon fibre comes with reduced fracture toughness since the brittleness is directly proportional to hardness.In order to increase the fracture toughness of the material, nanosilicon carbide is added to it. Chang-uk Kim et al., on their study of mechanical properties of carbon fibre composites by adding reinforcements such as carbon nanotubes and graphene nanoparticles found that more amount of reinforcement of nanoparticles increases the mechanical properties of composites [5]. J. Herwan et al., on their investigation on the study of load handling efficiency of pin-jointed CFRP composites found that the Poly Acrylo Nitrile (PAN) nanofibres spun by electro spinning technique and deposited on the composites increased its load capacity [6]. PrashanthBanakar and H. K. Shivananda on their investigation about mechanical properties aluminium nanoparticle-reinforced glass fibre composites found that composites were showing high compressive properties along the transverse direction of loading [7]. A. K. Birru et al. proposed the technique of ultrasonification for uniform dispersion of nanoparticles in epoxy matrix [8]. D. Chandramohan et al. reported that while carrying out the drilling experiments on silicon carbide-reinforced metal matrix composites, feed rate was the most dominant parameter [9]. From the literature review, it is felt that the works on parameter optimization of machining properties of carbon fibre composites were found to be very little so the optimization of drilling parameters on CFRP composites reinforced with nanoSiC was done.

2 Experimental Procedure

The experiment was carried out using carbon fibre-reinforced epoxy composites by adding nanosilicon carbide at the wt% of (0, 1 and 3).The composites were fabricated by "Hand Lay-up" method. The materials chosen were carbon fibre of 3 m length (3 sheets of 1 m each), epoxy resin (Araldite LY550), and hardener(Araldite LY951) at an empirical ratio of 60% of epoxy resin to 40% of carbon fibre. Based on carbon fibre weight, nanoSiC was added to 60% of the epoxy resin in three levels (0, 1 and 3%) to make three composite work pieces. The specimens were prepared by hand

Table 1 Composition of specimen

Specimen number	Carbon fibre (%)	Epoxy resin (%)	Weight fraction of nanoSiC (%)
1	40	60	0
2	40	60	1
3	40	60	3

Table 2 Process parameters and their levels

Parameters	Units	Level 1	Level 2	Level 3
Weight fraction of SiC	%	0	1	3
Spindle speed	rpm	1000	2000	3000
Feed rate	mm/min	50	100	150

layup method. Table 1 shows the composition of specimens. The composites thus made were drilled in the vertical machining centre. The drill bit used was TiAlN-coated tungsten carbide of shank diameter 5 mm. The process parameters taken were weight fraction of SiC (%),spindle speed (rpm), and feed rate (mm/min). The response parameter analysed was torque (Nm). The experimental was designed using Taguchi L18 orthogonal array and the optimization was done using response surface methodology. Table 2 shows the process parameters and their levels (Tables 3, 4 and 5).

3 Results and Discussions

3.1 *Analysis of Quadratic Equation*

The quadratic equation which was obtained by using "response surface methodology" is given below.

The equation is expressed in terms of torque:

$$\begin{aligned}\text{TORQUE} = {} & -0.86350 + 0.09000 * c + 0.012165 * f \\ & + 6.22750E - 004 * v + 3.1500E - 003 * c * f \\ & - 1.07500E - 004 * c * v - 5.72500E - 003 * f * v\end{aligned}$$

where

- **c** = weight fraction of SiC in %
- f = feed rate in mm/min
- v = spindle speed in rpm.

Table 3 Design matrix table by Taguchi L_{18} orthogonal array

S. No.	Weight fraction of SiC (%)	Spindle speed (rpm)	Feed rate (f) (mm/min)
1	1.5	2000	100
2	1.5	2000	100
3	3	1000	50
4	0	3000	50
5	1.5	2000	100
6	1.5	2000	100
7	1.5	2000	150
8	3	3000	150
9	0	2000	100
10	3	1000	150
11	1.5	3000	50
12	1.5	2000	100
13	1.5	2000	100
14	1.5	2000	50
15	3	3000	50
16	0	1000	50
17	0	3000	150
18	1.5	1000	100

3.2 3D Surface Plot

Figure 1 shows the three-dimensional interrelationship between process parameters ("Feed, speed and weight fraction of SiC") with respect to torque. From the figure, it is seen that, torque increases with respect to the weight fraction of SiC and feed rate. The change in slope of the plot is as same as feed and SiC weight fraction. The change in slope for speed remains the same and increases with respect to weight fraction. Here, as the individual factor weight fraction of SiC is dominant, whereas in interaction factor, speed and feed are dominant. The change in slope for speed remains the same and increases with respect to feed. Here, speed and feed rate are the dominant factors which influences the torque. Table 6 shows the values obtained from RSM-based desirability analysis (Table 7).

4 Conclusions

The composites were effectively fabricated by handlay up method and optimization of machining parameters was done to find the optimal combination. Results of ANOVA show that weight fraction of SiC was dominant on response parameter obtained from

Table 4 Experimental values

S. No.	Weight fraction of SiC (%)	Feed rate (mm/min)	Spindle speed (rpm)	Torque (Nm)
1	1.5	100	2000	0.69
2	1.5	100	2000	0.65
3	3	50	1000	0.45
4	0	50	3000	0.63
5	1.5	100	2000	0.62
6	1.5	100	2000	0.62
7	1.5	150	2000	0.85
8	3	150	3000	0.94
9	0	100	2000	0.61
10	3	150	1000	2.5
11	1.5	100	3000	0.45
12	1.5	100	2000	0.44
13	1.5	50	2000	0.6
14	1.5	100	2000	0.65
15	3	50	3000	0.89
16	0	50	1000	0.4
17	0	150	3000	0.59
18	1.5	100	1000	0.6

Table 5 ANOVA values

Source model	Sum of squares	df	Mean square	*F* value	*p*-value prob > *F*	
A-SiC	2.985538	6	0.497756	8.54549	0.000668	Significant
B-FEED	0.81225	1	0.81225	13.9055	0.002498	
C-SPEED	0.73984	1	0.73984	12.7067	0.003458	
AB	0.446513	1	0.12321	2.116122	0.16492	
AC	0.655513	1	0.446513	7.668818	0.01594	
BC	0.756918	1	0.208013	3.572519	0.081242	
Residual	0.751834	13	0.655513	11.25838	0.005167	
Lack of fit	0.005083	8	0.058224	92.46576	5.26E-005	Not significant
Pure error	3.472225	5	0.093973			
Corr total	0.797738	19	0.0001017			
R^2	0.797738					

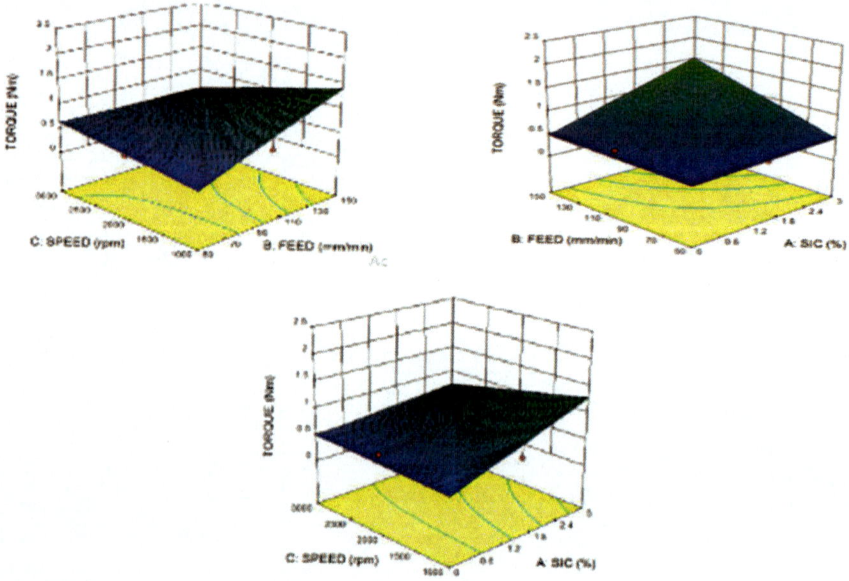

Fig. 1 3D surface plots showing relationship of torque with various combinations of speed, feed, and weight fraction of SiC

Table 6 Desirability table

S. No.	Weight fraction of SiC (%)	Feed rate in (mm/min)	Speed (rpm)	Torque (Nm)	Selection	Desirability
1	1.168002	50.31276	1442.545	0.340493	Selected	0.98
2	1.465661	50.225588	1598.327	0.395027	Not selected	0.96

Table 7 Confirmation test

S. No.	Weight fraction of SiC (%)	Feed (f) (mm/min)	Speed (s) (rpm)	Torque (Nm)		Error (%)
				Experimental	Predicted	
1	1	50	1000	0.37	0.34	3
2	1	50	1000	0.4	0.39	2.564

the machining process. The optimal combination was found to be spindle speed of 1000 rpm, feed rate of 50 mm/min, and weight fraction of SiC as 1% which give the minimum torque of 0.34 Nm.

References

1. Hikosaka Y, Matsuzaki R (2013) Enhancement of interfacial toughness of carbon/epoxy composite joints by in- mould surface preparation. Express Polym Lett 7(3):293–303
2. Subita B, Pardeep K (2013) Effect of graphite filler on mechanical behaviour of epoxy composites. Int J Emerg Technol Adv Eng 3(2):138–147
3. Rothon RN (2007) Particulate-filled polymer composites, Rapra Technology Ltd.
4. Stabik J, Dybowska A (2010) Polymer composites filled with powders as polymer graded materials. J Achiev Mater Manuf Eng 43(1)
5. Kim C, Kim S, Park J, Song J (2013) Fabrication and evaluation of mechanical properties of CF/GNP composites. Procedia Manuf 2: 368–373
6. Lisunova M, Hildmann A, Hatting B, Datsyuk V, Reich S (2013) Nanofibres of CA/PAN with high amount of carbon nanotubes by core-shell electrospinning. Compos Sci Technol 70(11):1584–1588
7. Vinay HB, Govindaraju HK, Banakar P (2017) Experimental study on mechanical properties of polymer based hybrid composite. Mater Today Proc 4(10):67–76
8. Birru AK, Reddy GK, Ajay G, Kumar NK (2014) Effect of reinforced silicon carbide nanoparticles in epoxy composites. Mater Today Proc 2(9):4348–4352
9. Chandramohan D, Presin Kumar AJ (2017) Experimental data on the properties of natural fiber particle reinforced polymer composite material. Data Brief 13: 460–468

Process Parameter Optimization Using TOPSIS for Electric Discharge Machining of Incoloy 800HT

Paul Joshua Samuel, K. C. Aswinkumar, Arunkumar Ganesan, and Arun Kumar Srirangan

Abstract Electric discharge machining (EDM) was carried out on Incoloy 800HT by using copper as electrode in dielectric medium kerosene. In order to find out the optimal process parameter, the experiments were designed using L_9 orthogonal array. The input parameters considered were voltage, current, and pulse-on time. The output parameters taken are material removal rate (MRR), surface roughness (SR), and radial overcut (ROC); then, these parameters are optimized using TOPSIS methodology. TOPSIS is used to suggest one or more among some alternatives, having many attributes. The principle is "The chosen alternative should have the shortest distance from the positive ideal solution and the longest distance from the negative ideal solution." Other parameters like MRR, SR, and ROC were calculated based on the experiments. White light spectroscopy was used to find the surface roughness of the machined surface of the sample, and these output values were optimized using TOPSIS method. The preference values were obtained after the optimization processes. The readings for the experiment which had the highest preference value were selected as the best value. Sample 7 had the highest value, and sample 3 had the least value.

Keywords Incoloy 800HT · EDM · TOPSIS · MRR · ROC · SR · White light spectroscopy

1 Introduction

Nowadays, research shows the failure in structural analysis and manufactured products, particularly those comprising of welded joints, when they are subjected beyond the design load capacity. This situation created a new way to find out the best optimized parameters. The machining of superalloys using lathe, etc., will lead to the increase in production time and also increase in production cost. In order to optimize the production rate and production cost, some of the special machining methods have

P. J. Samuel · K. C. Aswinkumar · A. Ganesan · A. K. Srirangan (✉)
School of Mechanical Engineering, SASTRA University, Thanjavur, Tamil Nadu, India
e-mail: arun1988@gmail.com

T. Rajmohan et al. (eds.), *Advances in Materials and Manufacturing Engineering*, Springer Proceedings in Materials 7, https://doi.org/10.1007/978-981-15-6267-9_21

to be carried out in an effective manner and also with better accuracy. Therefore, the special process like EDM is the better way of machining super alloys.

Incoloy 800 series of superalloys were invented by the special metal corporation group of companies by having the objective as maintaining the ultimate chemical properties for resistance to oxidation and high-temperature strength, other types of high corrosion and carburization [1]. By monitoring, they noticed that maintaining aluminum and titanium on the upper portion of the specified material range will increase the creep and stress rupture properties and also increases the maximum allowable design stress [2]. Incoloy 800HT belongs to nickel–chromium-based superalloys. Nickel plays a major role in this superalloy since nickel is an austenite stabilizer [3]. Incoloy 800HT is best suitable for high-temperature application such as heat exchanging equipment, hot ducts, and superheaters in power plants. Incoloy 800HT has more carbon content than Incoloy 800 and Incoloy 800H [4, 5]. Nickel is acting as a resistance to stress corrosion cracking. Chromium is acting as a withstanding element during the addition of FeF_2, FeF_3, and NiF_2 to corrosion media [6]. In EDM, an electrical spark is generated between the workpiece and electrode. This spark indicates the flow of electricity and produces the heat from 8000 to 12,000 °C [7]. Therefore, the melting of any of the superalloy is possible at this temperature. The spark should be controlled and localized on the specific area so that it affects only the surface of the material. The spark which is taking place in the dielectric medium that is kerosene should be carefully controlled so that the water acts as a coolant and removes the eroded material particles [8]. TOPSIS is an algorithm of multi-objective optimization invented by Hwang and Yoon in 1981 which is based on a principle that the desired alternative should be in the shortest distance from the ultimate solution and it should be in the largest distance from the non-beneficial solution. TOPSIS is the most genuine algorithm in the world of optimization because of the compromise it makes with the criteria [9].Jitendra kumar et al. on their investigation on face milling operation of titanium alloys used the TOPSIS for optimal combination of parameters which gave them better surface finish [10]. Manivannan et al. employed TOPSIS algorithm for the improvement of machining parameters of micro-EDM process [11]. Senthil et al. advocated the application of TOPSIS on optimization of EDM parameters [12]. From the literatures, it is seen that there is little amount of work on TOPSIS optimization of electric discharge machined Incoloy 800HT samples.

2 Experimental Setup

Incoloy 800HT superalloy plates of dimension 20 × 20 × 4 mm were machined using electrical discharge machining process. Incoloy 800HT base material chemical composition is based on ASTM-B409. Electric discharge machining was chosen to machine the workpiece (Incoloy 800HT) with copper material as the electrode. The input parameters such as pulse-on time, voltage, and current for the experiments were designed using L_9 orthogonal array. Current was set with the values ranging from 4, 8, and 12 amps, voltage was set with the values ranging from 2, 4, and 6 V, and pulse-on time was set with the values ranging from 30, 60, and 90 μs. By setting these values as inputs, respectively, MRR was calculated, SR was calculated using white light spectroscopy, and radial overcut (diameter) was measured. The obtained values are represented in Table 1. The machined specimens are shown in Fig. 1.

Table 1 L_9 orthogonal array experiments and their corresponding results

Exp. no.	Current (A)	Voltage (V)	Pulse-on time (μs)	MRR (mm^3/min)	Radial overcut (mm)	Surface roughness (μm)
1	4	2	30	1.21746	0.02005	34.53
2	4	4	60	0.62132	0.02625	36.15
3	4	6	90	0.40302	0.0536	43.66
4	8	2	30	6.21746	0.3267	44.87
5	8	4	60	2.569269	0.14605	45.28
6	8	6	90	1.62888	0.1018	39.05
7	12	2	30	9.67254	0.1059	23.98
8	12	4	60	5.64231	0.35625	62.71
9	12	6	90	3.736356	0.2145	46.99

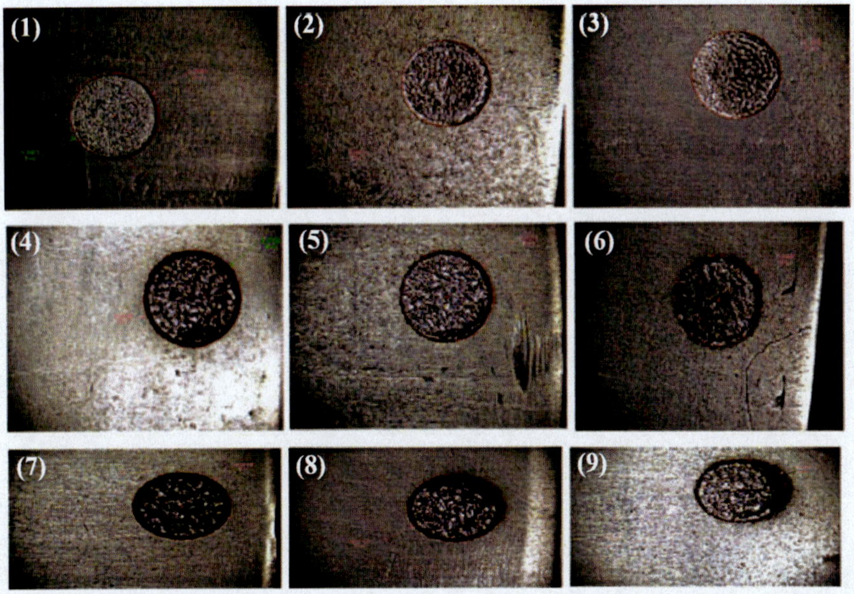

Fig. 1 Machined samples (1–9)

3 Methodology

TOPSIS is one of the methods for solving the decision-making problem. The output responses considered for optimizing are the ROC, MRR, and SR.

Step I: The first step of the TOPSIS methodology is to obtain normalized matrix.

$$D_m = \begin{bmatrix} q_{11} & q_{12} & q_{13} & \cdots & q_{1y} \\ q_{21} & q_{22} & q_{23} & \cdots & q_{2y} \\ q_{31} & q_{32} & q_{33} & \cdots & q_{3y} \\ \vdots & \vdots & \vdots & \ddots & \vdots \\ q_{x1} & q_{x2} & q_{x3} & \cdots & q_{xy} \end{bmatrix}$$

Step II: The obtained normalized matrix is shown below

$$R_{mn} = \frac{q_{mn}}{\sqrt{\sum_{m=1}^{x} q_{mn}^2}}; \quad n = 1, 2, \ldots, y$$

$$R_{9*3} = \begin{bmatrix} 0.088488 & 0.028938 & 0.267125 \\ 0.045159 & 0.037886 & 0.279658 \\ 0.029292 & 0.077359 & 0.18551 \\ 0.4519 & 0.471517 & 0.347116 \\ 0.186741 & 0.21079 & 0.350288 \\ 0.118391 & 0.146925 & 0.302092 \\ 0.703024 & 0.585824 & 0.337755 \\ 0.410097 & 0.514165 & 0.485127 \\ 0.271567 & 0.309582 & 0.363516 \end{bmatrix}$$

Step III: The assumed value for w_m ($m = 1, 2, \ldots, y$). The expression used to obtain weighted normalized decision matrix $U = [u_{mn}]$ is $U = w_m r_{mn} \cdot \sum_{n=1}^{y} w_n = 1$

$$U_{9*3} = \begin{bmatrix} 0.022122 & 0.007234 & 0.066781 \\ 0.01129 & 0.009471 & 0.069914 \\ 0.007323 & 0.01934 & 0.046378 \\ 0.112975 & 0.117879 & 0.086779 \\ 0.046685 & 0.052697 & 0.087572 \\ 0.029598 & 0.036731 & 0.075523 \\ 0.175756 & 0.146456 & 0.084439 \\ 0.102524 & 0.128541 & 0.121282 \\ 0.067892 & 0.077395 & 0.090879 \end{bmatrix}$$

Step IV: The equation to obtain positive and negative ideal solutions are shown below:

$$A^{+} = \{U1^{+}, \ldots, Uy^{+}\}, \text{ where } U^{+} = \{\max(U_{mn}) \text{ if } n\hat{I}J;\ \min(U_{mn}) \text{ if } n\hat{I}J'\}$$

$$A^{-} = \{U1^{-}, \ldots, Un^{-}\}, \text{ where } U^{-} = \{\min(U_{mn}) \text{ if } n \in J; \max(U_{mn}) \text{ if } n \in J'\}$$

Step V: Segregation between alternatives was identified from the "ideal" solution and is given by (Table 2):

$$S_m^{+} = \left[\sum (U_n^{+} - U_{mn})^2]^{1/2}\right] \quad m = 1, 2, \ldots\ldots.x$$

Similarly, the negative ideal alternative:

$$S_m^{-} = \left[\sum (U_{mn} - U_n^{-})^2]^{1/2}\right] \quad n = 1, 2, \ldots\ldots x$$

Table 2 Positive and negative ideal solutions

Positive	Negative
0.015372676	0.015887
0.017670763	0.015039
0.019221641	0.012089
0.015086321	0.008013
0.013692819	0.007131
0.015034186	0.009589
0.000902824	0.029425
0.020869229	0.005801
0.013345444	0.005401

Table 3 Alternatives taken from positive and negative ideal solutions

$S1^+$	$S2^+$	$S3^+$	$S1^-$	$S2^-$	$S3^-$
0.140605	0.007018	0.037102	0.005858	0.124688	0.097025

Step VI: The expression to calculate relative nearness is given as: $P_m = \frac{S_m^-}{S_m^+ + S_m^-}$; $m = 1, 2, \ldots, x$ $0 < P_m < 1$ (Tables 3 and 4).

4 Results and Discussion

See Fig. 2.

4.1 *White Light Spectroscopy*

The surface roughness sample was analyzed on machine surface by spectroscopy. Each transition in absorption spectrum 2D spectrum shows a pair of diagonal peaks. The negative peak is exactly on the diagonal (BLUE) corresponds to the photo-bleaching of direct bandgap transition, and white blue-shifted positive peak is along the axis from an excited absorption. Surface roughness is 23.98 μm for sample 7, and surface roughness is 43.66 μm for sample 3 (Figs. 3 and 4).

5 Conclusion

Multi-attribute optimization has been performed using TOPSIS to determine the most significant set of process variables. The response parameters considered were MRR, SR, and ROC, and white light spectroscopy was used to find the surface roughness of the machined sample.

The optimal parameters were found to be current (amps) = 12, voltage (v) = 2, and pulse-on time (μs) = 30 from TOPSIS. Therefore, results obtained from the highest possible solution for the set of input parameters depend upon the required performance characteristics. The results of the research work can be suggested to the industries for the improvement in quality and processing.

Table 4 Rankwise distribution of preference values

Exp. no	1	2	3	4	5	6	7	8	9
Preference value	0.508234	0.459773	0.386108	0.346888	0.34246	0.389428	0.970231	0.217495	0.288105
Rank	2	3	5	6	7	4	1	9	8

Fig. 2 Rankwise preference distribution

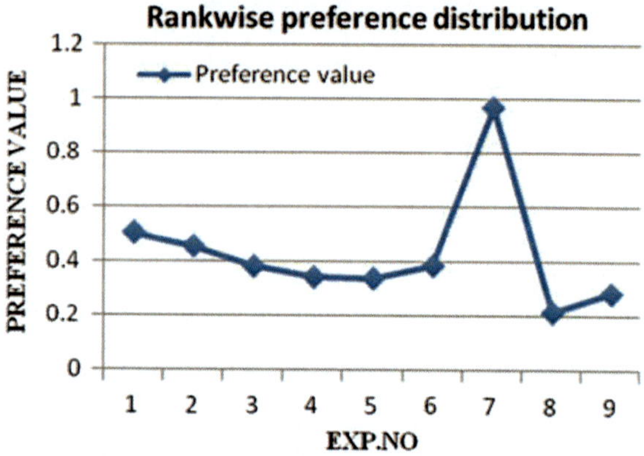

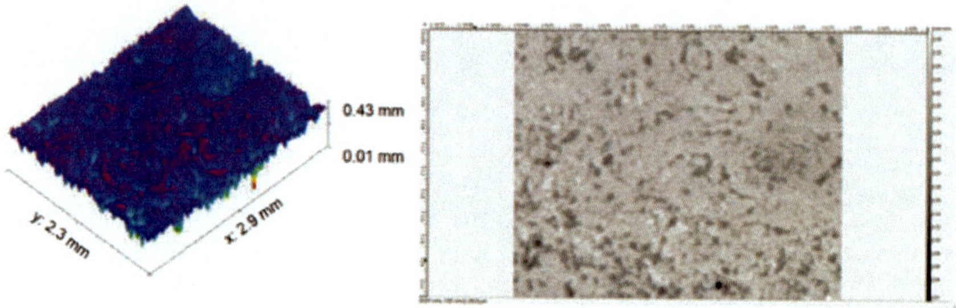

Fig. 3. 3D and 2D surface roughness images and structure of sample 7

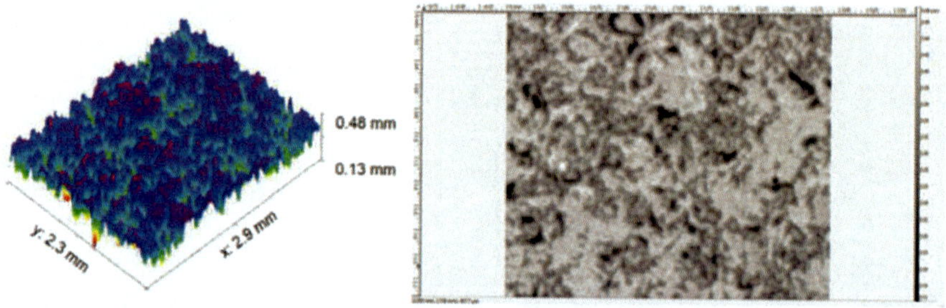

Fig. 4. 3D and 2D surface roughness images and structure of sample 3

References

1. Srisrual A, Pitaksakorn K, Promdirek P (2018) Oxidation behaviour of Incoloy 800HT in pure oxygen. Mater Today: Proc 5:9238–9243
2. Tan Yu, Yang J, Wang W, Shi R, Liang K (2016) Effects of PbO on the oxide films of incoloy 800HT in simulated primary circuit of PWR. J Nucl Mater 473:119–124
3. Sinha S, Ballav R, Kumar A (2017) Investigation of material removal rate and tool wear rate on electrical discharge machining of Incoloy 800HT by using response surface methodology. Mater Today: Proc 4(2017):10603–10606

4. Rahmel A, Grabke HJ, Steinkusch W (1998) Carburization—introductory survey. Mater Corros 49:221–225
5. Rajesh Purohit RS, Rana RKD, Banoriya D, Singh SK (2015) Optimization of electric discharge machining of M2 tool steel using grey relational analysis. Mater Today: Proc 2:3378–3387
6. Vishnu P, Santhosh Kumar N, Manohar M (2018) Performance prediction of electric discharge machining of Inconel-718 using artificial neural network. Mater Today: Proc 5:3770–3780
7. Baroi BK, Kar S, Patowari PK (2018) Electric discharge machining of titanium grade 2 alloy and its parametric study. Mater Today: Proc 5:5004–5011
8. Richard J, Giandomenico N (2018) Electrode profile prediction and wear compensation in EDM-milling and micro-EDM-milling. Proc CIRP 68:819–824
9. Joshi SN, Pande SS (2011) Intelligent process modeling and optimization of die-sinking electric discharge machining. Appl Soft Comput 11:2743–2755
10. Kumar J, Singh GS (2016) Optimization of machining parameters of titanium alloy steel using: TOPSIS method. Int J Sci Res Sci Eng Technol 1019–1022
11. Pradeep Kumar RM (2016) Multi-response optimization of Micro-EDM process parameters on AISI304 steel using TOPSIS. J Mech Sci Technol 30: 137–144
12. Senthil P, Vinodh S, Sing AK (2014) Parametric optimisation of EDM on Al-Cu/TiB2 in-situ metal matrix composites using TOPSIS method. Int J Mach Machinab Mater 16(1): 80–94

Analysis on Thermal Properties of Polyethylene–Vinyl Acetate (PEVA) Matrix with Polytetrafluroethylene (PTFE) Particle-Reinforced Composites

R. Mahesh Kumar, N. Rajini, K. Mayandi, Nadir Ayrilmis, M. Srisuryadharan, P. Venkatesh, and M. Vijayakumar

Abstract To improve the thermal properties of polyethylene–vinyl acetate matrix, polytetrafluroethylene particles were introduced into the matrix. By hosting small amount of PTFE particles into the matrix, thermal properties had improved particularly with 20% PTFE. It was evident that thermal conductivity of the composite increased with the addition of fillers. The melt mixing in an injection molding process was used for mixing the copolymer and filler that ensured proper blending. Lee's disk apparatus was utilized to resolve the thermal conductivity of the composites. Count of 20% by weight composition of PTFE particle with PEVA increased the thermal conductivity of the samples to anticipated range.

Keywords Polyethylene–vinyl acetate · Polytetrafluroethylene · Thermal properties · Mechanical properties

1 Introduction

All composites consist of two or more large ingredients that have different shapes and chemical structures that are insoluble with each other. Polyethylene–vinyl acetate (PEVA) is a copolymer that approaches elastomeric ('rubber-like') materials in elasticity and softness [1]. The material has good clarity and shine, hot-melt adhesive, waterproof properties, and protection from UV radiation [2]. Its properties can be effectively changed by adjusting the vinyl acetate comonomer proportion from 10 to 40% with the rest of the ethylene. Thermal investigation of polyethylene–vinyl acetate (PEVA) and its composites confirms a vital job.

R. Mahesh Kumar · N. Rajini · K. Mayandi (✉) · M. Srisuryadharan · P. Venkatesh · M. Vijayakumar
Department of Mechanical Engineering, Kalasalingam Academy of Research and Education, Krishnankoil 626126, Tamil Nadu, India
e-mail: k.mayandi@gmail.com

K. Mayandi · N. Ayrilmis
Department of Wood Mechanics and Technology, Faculty of Forestry, Istanbul University-Cerrahpasa, Bahcekoy, Sariyer, 34473 Istanbul, Turkey

T. Rajmohan et al. (eds.), *Advances in Materials and Manufacturing Engineering*, Springer Proceedings in Materials 7, https://doi.org/10.1007/978-981-15-6267-9_22

PEVA displays incredible business prospective in the packing industry. PEVA is used in the photovoltaic industry as an encapsulation material for crystalline silicon solar powered cells in the construction of photovoltaic units [3].

PTFE maintains high strength, self-lubrication, and toughness at temperatures less than 5 K and good flexibility at temperatures over 194 K. Processing PTFE can be troublesome and costly. PTFE does not flow, even at melting temperature but rather behaves as a gel because of the absence of high melt viscosity and crystalline phase. PTFE (Teflon™) is best known for its utilization in coating non-stick frying pans and other cookware, as it is hydrophobic and has genuinely high heat resistance. Its frictionless characteristics allow improved flow of very thick fluids and for uses in applications such as brake hoses.

2 Experimental

2.1 Materials

PTFE polymer in powder form size varying from 5 to 15 μm was bought from M/s. Modoplast Company Pvt. Ltd, Kolkata, West Bengal, India. PEVA copolymer resins in pellet form in 2 mm diameter were procured from Sree Meenakshi Plastics, Madurai, Tamilnadu, India. Injection mold made of EN8 mild steel was used to prepare sample specimens in polymer injection molding machine WINDSOR, and SP130 available at Surya Polymer, Kovilpatti, Tamilnadu, India, is used for making specimen plates [4].

2.2 Composites Preparation

The samples prepared were: Sample 1: PEVA, Sample 2: PEVA + 10% by weight composition of PTFE powder particle, and Sample 3: PEVA + 20% by weight composition of PTFE powder particle. Initially, the moisture available in the PEVA pellets and PTFE powder was removed by heating at 100 °C in void oven for 24 h individually. Then, the materials as per requirement were taken in three batches with different weight compositions which were fed one by one into the hopper, and molded test plate samples as per required dimensions were made [5].

3 Characterization

3.1 Thermal Conductivity

A material by itself which conduct heat through it is called thermal conductivity of the material. Thermal conductivity value for the three samples: PEVA, PEVA + 10%PTFE, and PEVA + 20%PTFE with unknown conductivity was determined by placing the sample one by one between two samples of known conductivity (usually brass plates). Heat was supplied under steady-state condition for 30 min and allowed to move from top to bottom to avoid any convection [6]. As per ASTM standard, E1530 samples having size 50 mm diameter and 3 mm thickness were made from water jet cutting method for performing thermal conductivity test at Central Institute of Polymer Engineering and Technology (CIPET), Chennai.

3.2 Optical Microscopy

Light from a mirror was reflected up through the specimen to be viewed, into the targeted focal point, which produced the principal magnification. The image produced from the focal point was then magnified again by the eyepiece lens, which acted as a simple amplifying glass. The specimens were kept in the focal point, and the magnified images were recorded in the screen.

3.3 Thermogravimetric Analysis

Thermogravimetric analysis (TGA) under controlled atmosphere was used to determine the quantity and change in the weight of a material with time rate. The powder sample was kept in the TA instruments in a platinum pan in nitrogen atmosphere 100 ml/min supply. Perkin Elmer thermal analysis between temperature ranges from room temperature to 700 °C until zero mass reduction with temperature increment of 10 °C/min was conducted successfully.

4 Results and Discussion

4.1 Thermal Conductivity Report

The thermal conductivity '*k*' values obtained from Lee's disk apparatus in W/mK for the different samples under steady-state condition with mean temperature 55 °C

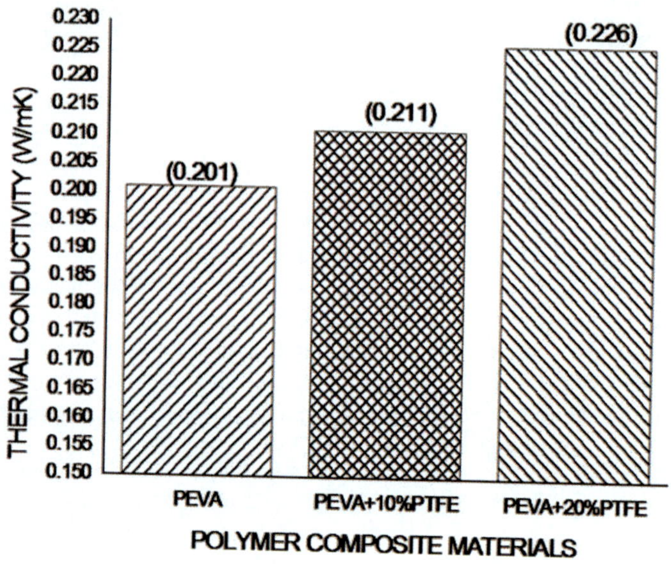

Fig. 1 Thermal conductivity for various compositions of composite materials and pure PEVA matrix

of heat flow from the top of the specimen obtained from Central Institute of Plastics Engineering and Technology were given as bar chart. The '*k*' value for Sample 1—PEVA is 0.201 W/mK and '*k*' value for Sample 2—PEVA + 10%PTFE is 0.211 W/mK, which has improvement of 5% and '*k*' value for Sample 3—PEVA + 20%PTFE is 0.226 W/mK which has improvement for 12.5% [6] (Fig. 1).

4.2 *Optical Microscopy Report*

The optical microscope image analysis studies were done using Motic optical microscope instrument. All the images was taken at International Research Centre, KARE, India. This microscope images confirm the dispersen of the PTFE particles in the PEVA matrix. The particle distribution was effective and achieved successfully on behalf of melt mixing to obtain tailor-made composite samples by injection molding technique [7] (Figs. 2, 3 and 4).

4.3 *Thermogravimetric Report*

The TGA report obtained from Sophisticated Analytical Instruments Facility, STIC, Cochin University, India. The TGA curves were drawn using pure PEVA and PEVA composites, it was shown in Fig 5. The TGA experimental was conducted between

Fig. 2 Microscope image of pure PEVA matrix

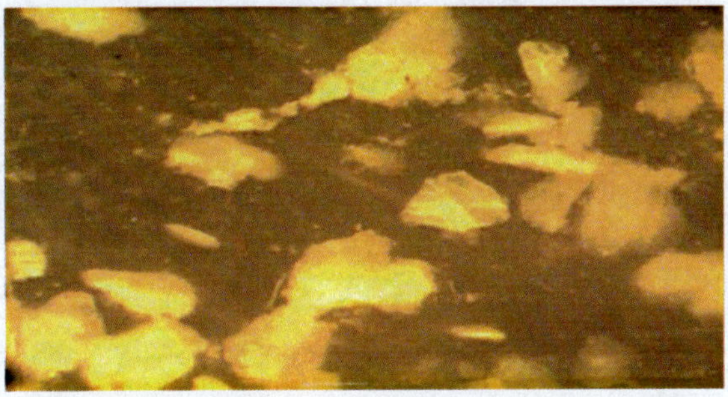

Fig. 3 Microscope image of 10% PTFE particle fillers in PEVA matrix

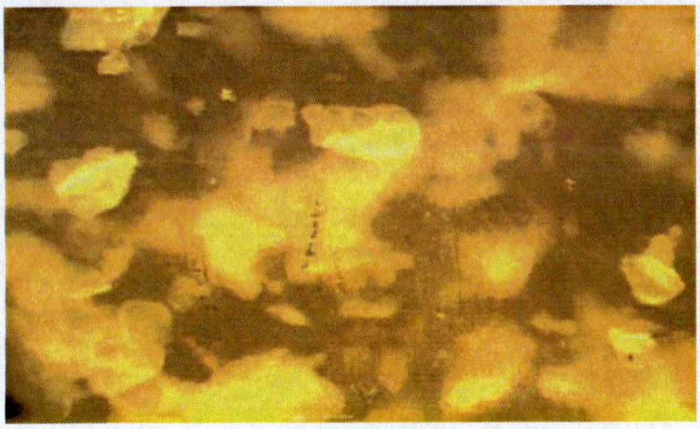

Fig. 4 Microscope image of 20% PTFE particle fillers in PEVA matrix

a temperature range of 50 °C to 650 °C and the increasing temperature values is set at10 °C/min. For PEVA, there is an evidence for the thermal degradation in two steps [8]. The first stage completed at around 390 °C, with long radius of curvature that suggested a slow rate of degradation process with high thermal stability. The second stage has sharp bend with less radius of curvature and degradation process ended at fast rate (within the interval of 465–550 °C). It is evident from the graph that due to the addition of PTFE particle, curve appears to be three-step thermal degradation process [3]. For composites PEVA + 10%PTFE and PEVA + 20%PTFE, degradation

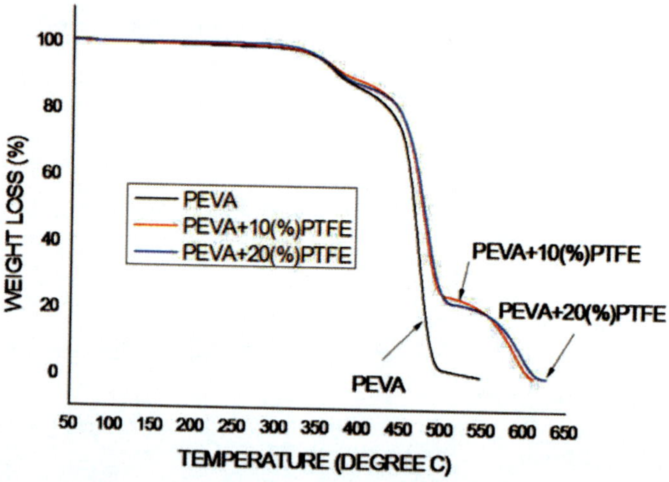

Fig. 5 TGA test report of PEVA & PEVA composites

Table 1 TGA data for PEVA and PEVA/PTFE particle composites

Samples	$T_{-5\ wt\%}$ degradation (°C)	$T_{-50\ wt\%}$ degradation (°C)	T_{max} (°C)
Sample 1 PEVA	390	465	550
Sample 2 PEVA + 10% PTFE	390	480	610
Sample 3 PEVA + 20% PTFE	390	480	625

process increases up to 610 and 625 °C, respectively, for zero percentage weight loss (Fig. 5).

From Table 1 given below, it was observed that the 5% weight loss degradation temperatures for all the three samples are same. It was also noticed that 50% weight loss temperature and peak temperature values obtained during degradation were more for composites blended with PTFE powder samples. The composite with 10%PTFE and 20%PTFE had very high degradation temperature 610 and 625 °C, respectively.

5 Conclusions

The observation from the result stated that PEVA polymer composites blended with 20% Wt. composition and had improved properties. From the references, it was observed that inclusion of fillers improved thermal properties and upgraded the heat transfer rates. It is also noted that PEVA composites blended with 10 and 20% Wt.

PTFE particles had improved thermal properties overall. Thus, the composite innovation technique was found to be successful for thermal analysis of PEVA polymer matrix with PTFE particle filler.

References

1. Manas D, Manas M, Stanek M, Danek M (2008) Improvement of plastic properties. Archiv Mater Sci Eng 32:69–76
2. Dikobe DG, Luyt AS (2009) Morphology and properties of polypropylene/ethylene vinyl acetate copolymer/wood powder blend composites. Exp Polym Lett 3:190–199
3. Badiee A, Ashcroft IA, Wildman RD (2016) The thermo-mechanical degradation of ethylene vinyl acetate used as a solar panel adhesive and encapsulant. Int J Adhes Adhes 68:212–218
4. Tham DQ, Tuan VM, Thanh DTM, Chinh NT, Giang NV, Trang NTT, Hang TTX, Huong HT, Duang NTK, Hoang Y (2015) Preparation and properties of ethylene vinyl acetate copolymer/silica nanocomposites in presence of EVA-G-Acrylic Acid. J Nanosci Nanotechnol 15:2777–2784
5. Bahmanyar M, Ahmad Ramazani SA, Baniasadi H (2015) Preparation and properties of Ethylene-vinyl acetate/linear low-density polyethylene/graphene oxide nanocomposite films. Polym-Plast Technol Eng 54:1152–1158
6. Al Mizan R, Islam MA (2018) Augmentation of thermal conductivity for polypropylene/ethylene-vinyl acetate based composites filled by organic and inorganic particles. Univ J Mater Sci 6:49–57
7. Omprakash H (2012) Nautiyal: molding of EVA soles using expanding and reducing agents. Int J Eng Sci Technol 4:3050–3058
8. Rimez B, Rahier H, Van Assche G, Artoos T, Van Mele B (2008) The thermal degradation of poly (vinyl acetate) and poly (ethylene-co-vinyl acetate), Part II: Modeling the degradation kinetics. Polym Degradation Stab 93:1222–1230

A Study on Wire Electrical Discharge Machining Process Parameters and Performances

P. S. Gowthaman and S. Jeyakumar

Abstract Nowadays, the manufacturing industries require the advanced machining techniques for micro-machining. It produces reliable and quality product. However, it is extremely difficult in conventional machining to process an increase in hardness of material, high utilization of energy, and decrease in cutting speed. Wire electrical discharge machining (WEDM) is a precision machining technique for cutting any electrically conductive materials. They are possible to cut the materials like titanium, nickel, cobalt, tungsten, ceramics, and composite material. It has a competence to produce better surface, complex profiles and control tolerances. This paper reviews the overview study of WEDM process parameters and performances of various materials. The conclusion is drawn with the future scope as well.

Keywords Wire electrical discharge machining (WEDM) · Material removal rate (MRR) · Surface roughness (SR) · Cutting rate (CR)

1 Introduction

A wire electrical discharge machining (WEDM) is advanced machining techniques. It espouses to be the most excellent options in present-day application, which is used to remove materials from the workpiece by a series of electric sparks. A wire of about (0.05–0.30 mm) diameter is used. The wire is usually made up of copper, brass, and tungsten. The gap between the wire and workpiece is typically from (0.025–0.075 mm). The WEDM can be proficient to cut thickness as 300 mm. The WEDM experiences a very high temperature of around 10,000 °C [1]. The applications of WEDM are as follows: aerospace, automotive, biomedical, electronics, marine industry, nuclear, tool and die industry. It was found that the WEDM process is subjected to axial mechanical stress in order to keep it straight motion during the cutting with respect to programmed path [2]. The WEDM has a capability for the manufacturing of special gears, bearing cage, cam wheels, tools and dies, gauges, and

P. S. Gowthaman (✉) · S. Jeyakumar
Department of Mechanical Engineering, P.S.N.A. College of Engineering & Technology, Dindigul 624622, Tamil Nadu, India
e-mail: psgowtham36@yahoo.com

T. Rajmohan et al. (eds.), *Advances in Materials and Manufacturing Engineering*, Springer Proceedings in Materials 7, https://doi.org/10.1007/978-981-15-6267-9_23

prototype [3]. This paper summarizes the various research activities conceded out in the area of wire-EDM process. The WEDM reviews involve the working principle and process parameters. Furthermore, the experimental work on WEDM performances is concluded and influences the various factors that affect the machining performance.

2 Wire Electrical Discharge Machining

The wire electrical discharge machining (WEDM) is an exceptional machining process. It endeavors the tolerance of ±0.005 mm and surface finish of 0.1 μm. It was found that the series of interrupted electric discharges is during the cutting of materials in WEDM [4]. The temperature increases due to the high level of liberated discharge energy. Figure 1 illustrates the schematic diagram of WEDM.

2.1 *Principle of WEDM*

WEDM is a specialized thermal machining technique. It is based on the principle of electro-thermal mechanism. It transforms electrical energy into thermal energy for cutting the materials. The high frequency is produced between tool and workpiece, and it leads to generation of more ions and loss of dielectrics. In WEDM, the motion of the wire is very slow and moves the prescribed path [5]. A process discharges the single spark, and it is qualified for next spark [6].

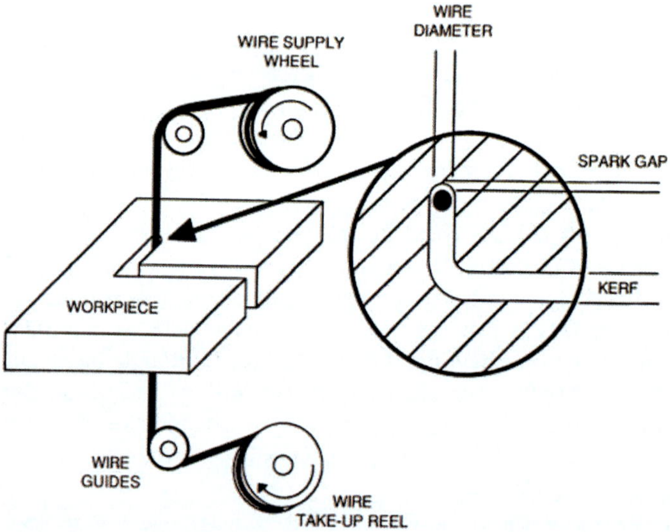

Fig. 1 Schematic diagram of WEDM (*Source* Scott et al. [7])

2.2 *Process Parameters*

In WEDM, the important performance measures are: material removal rate, surface roughness, and cutting rate. The parameters which affect the following performances are: pulse frequency, pulse duration, discharge current, wire speed, wire tension, and dielectric flow rate.

2.2.1 Power Supply

The power supply is an essential part of WEDM system. It employs to produce sparks between the gap. The surface finish is dominated by discharge current, wire speed, and pulse duration [7]. They enhanced the voltage and pulse on time is an uttermost compelling parameters of controlling MRR [8].

2.2.2 Dielectric Fluid

The dielectric fluid serves the functions as:

(i) It gives the higher material removal rate and better surface finish.
(ii) Provides an effective cooling medium and good degree of fluidity.
(iii) Removing the chips from the sparking area.

The dielectric unit plays a dominant role in WEDM. The deionized water is served as a dielectric purpose. It is used for the upcoming reasons: low viscosity, immense cooling rate, and no fire hazard. It was reported lower MRR, and streaks are observed over the surface area [9].

2.2.3 Wire Electrode

The appropriate selection of wire is an crucial task in WEDM. The tool should possess the desirable properties: good hardness, resistance to wear, high conductor of electricity, high melting, and heat resistant. It was found that the performance of WEDM is improved by tensile strength, good conductivity, and material coating [10]. A coating wire can increase the cutting speed and dimensional accuracy. The coated wire is overheated and evaporated during the high energy discharge. They examined that the wear rate of brass wire is more than that of zinc-coated wire [11]. The wire tension plays a prominent role in machining and job accuracy. The higher value of wire tension increases the cutting speed and accuracy; hence, it exceeds the strength and it leads to wire breakage. The wire rupture is another obstacle that occurs in WEDM, as a result of surplus spark energy, reduction of gap voltage, and poor flush conditions.

2.2.4 Workpiece

In WEDM process, the workpiece material impacts the MRR, SR, and CR. The higher MRR is attained through the lesser melting and lower thermal conductivity, while the lower melting point endures the improper heat treatment and distortion [12]. They explored that the heat treatment of material undergoes the stability, shockability hardness, and wearability [13]. It was investigated that the thicker workpiece provides a more chances of sparks to occur, while the surface conditions are affected [14].

3 Experimental Studies

The WEDM is generally achieved in a multi-cutting mode. The workpiece is processed for rough cutting with higher energy of spark, longer dielectric fluid, and low wire tension to go through the primary shape. The amount of total cutting passes depends on accuracy and surface finish. The MRR, SR, and CR of the machined surface get a main attention in the study of WEDM.

3.1 Material Removal Rate (MRR)

A material removal rate (MRR) is an essential performance measure to determine the machinability. The MRR is expressed as the amount of material removing from workpiece under the working time (mm^3/min). It mainly depends upon electrode material, melting temperature, and wear rate. The limited sparking time between the wire and workpiece consequences the rapid cutting and higher MRR [15]. It was investigated that the effect of various factors are: wire speed, pulse duration, and pulse frequency for maximization of MRR [16]. In dry-wire-EDM, the maximum quantity of mixing flow rate impinges the machining performance and it results in variation of MRR [17]. They performed WEDM on EN-31 alloy steel, and it concludes that the pulse width, wire speed, and servo voltage play the major role in MRR [18].

3.2 Cutting Rate (CR)

The cutting rate is otherwise called as cutting speed. It is termed as the ratio of length of workpiece cut to the duration of cutting time. It was reported that pulse-on time, pulse duration, peak current, servo voltage, wire speed, and electrical capacitance are the essential parameters of cutting speed [19]. It was explored that the increase in peak current and electrical capacitance undergoes the increase in cutting speed [20]. Table 1 illustrated the WEDM performances on different materials.

Table 1 Overview of WEDM on different materials

Authors	Material	Contributions
Maher et al. (2016)	AISI 1050 steel	Developed the ANFIS model and determined the white layer thickness
Mouralova et al.(2019)	Inconel-625	Studied the machining characteristics of Inconel-625
Rauma sen et al.(2018)	Maraging steel (300 Grade)	Investigated the process parameters using neural network and Jaya algorithm
Muhammed Azam et al. (2016)	HSLA steel	Developed the mathematical model and explored the machining parameters
Nithin Raj et al (2017)	Cu/Ni-Si/Tic composite	Evaluated the process performances using the Taguchi method
Himadri majumder et al. (2018)	Ni–Ti shape memory alloy	Highlighted the general regression neural network of Ni–Ti alloy
Abimannan Giridharan et al. (2017)	AISI 4340 steel	Experimented the erosion mechanism and machining properties
Somvir singh et al. (2018)	Udimet-L605	Optimized the machining parameters using PSO
Pramanik et al. (2018)	2205 Duplex stainless steel	Investigated the surface morphology of machined components
Meinam Annebushan et al. (2018)	Alumina (MWNCT) composite	Identified the machining properties and erosion phenomena of MWNCT

3.3 Surface Roughness (SR)

The surface roughness is a prominent parameter that arouses the performance of machining component. It primarily depends on accuracy, friction, surface contact, lubrication, and deformation. It was reported the machining surface is affected by high discharge energy and long pulse duration [21]. They explained the competent selection of heat treatment, and machining parameters distressed the surface properties [22].

4 Conclusion

The WEDM is a proficient non-conventional machining process. WEDM offers the benefits as: sharp internal corners, higher material utilization, and elimination of cutting forces.

Today's companies using WEDM techniques make extrusion dies and tools, but its application prolongs much wider. The WEDM is used for low-volume production of complex parts such as gears, splines, and medical equipment. Furthermore, the current research trends included in applications as: micro-wire-EDM (μ-WEDM), wire electrical discharge grinding (WEDG), and ultrasonic-assisted WEDM. Therefore, the research of WEDM in new applications needs a more focus to make prudent.

References

1. Bisaria H, Shandilya P (2018) Experimental investigation on wire electrical discharge machining of Nimonic-263 superalloy. J Mater Manuf Process. ISSN 1042-6914
2. Plaza S, Ortega N, Sanchez JA, Pombo I, Mendikute A (2009) Original models for the prediction of angular error in Wire-EDM taper cutting. Int J Adv Manuf Technol 44:529–538
3. Lusi N, Muzaku K, Oedy B, Soepangker P (2016) Parametric optimization of wire electrical discharge machining process on AISI H13 Tool steel using weighted principal component analysis and Taguchi method. ARPN J Eng Appl Sci 11(2)
4. Puri AB, Bhattacharya B (2003) An analysis and optimization of the geometrical accuracy due to wire lag phenomenon in WEDM. Int J Mach Tools Manuf 43:151–159
5. Kuriakose S, Shanmugam MS (2005) Multi-objective optimization of wire-electro discharge machining process by non-dominated sorting genetic algorithm. J Mater Proces Technol 170:133–141
6. Benedict GF (1987) Nontraditional manufacturing process. CRC Press, London UK
7. Scott D, Boyina S, Rajurkar KP (1991) Analysis and optimization of parameter combination in wire electrical discharge machining. Int J Prod Res 29(11):2189–2207
8. Chaudhary T, Siddique AN, Chanda AK (2018) Effect of wire tension on different output response during wire electrical discharge machining on AISI 304 stainless steel. J Def Technol
9. Kunieda M, Furudate C (2001) High precision finish cutting by Dry WEDM. CIRP Ann 90(1):121–124
10. Kruth JP, Lauwers B, Schacht B (2004) Composite wire with high tensile core for wire EDM. CIRP Ann 53(1):171–174
11. Kapoor J, Sehijpal S, Khamba JS (2020) Recent developments in wire electrodes for high performance WEDM. In: Proceedings of the world congress on engineering, vol 2. London
12. Kim DI, Jung SC, Lee JE, Chang SH (1998) Parametric study on the design of composite foam-resin concrete sandwich structure for precision machine tool structure. Compos Struct 43:155–163
13. Kamody DJ (1993) Cryo-treatment. US Patent 5 259,200
14. Tomura S, Kunieda M (2009) Analysis of Electromagnetic force in wire-EDM. J Precis Eng 33(3):255–262
15. Sultan T, Kumar A, Gupta RD (2014) Material removal rate, electrode wear rate and surface roughness evaluation in Die sinking EDM with hollow tool through response surface methodology. Int J Manuf Eng
16. Mahapatra SS, Patnaik A (2006) Optimization of wire electrical discharge machining (WEDM) process parameters using genetic algorithm. Ind J Eng Mater Sci 13:494–502
17. Suresh Kumar M, Boopathi S (2017) Grey Relational optimization of powder mixed near-dry Wire cut Electrical discharge machining of Inconel 718 alloy. Asian J Res Soc Sci Hum 7(3)
18. Jaganathan P, Naveen Kumar T, Siva Subramanian R (2012) Machining parameters optimization of WEDM process using Taguchi method. Int J Sci Res 2(12)
19. Manjaiah M, Narendaranath S, Basavarajappa S (2014) Review on non conventional machining of shape-memory alloys. Trans Non-ferrous Metals Soc Chin 24(1):12–21

20. Parmeshewararao, Sarcar M (2009) Evaluation of optimal parameter for machining brass with wire cut EDM. J Sci Indus Res 68(1):32–35
21. Dey A, Pandey KM (2018) Wire electrical discharge machining characteristics of AA 6061/Cenosphere Aluminium matrix composite using RSM. J Mater Today Proc 5:1278–1285
22. Srinivasa Rao P, Ramji K, Satyanarayana B (2018) Surface integrity of wire EDMed aluminium alloy a comprehensive experimental investigation. J King Saud Univ 30:368–376

Investigation of Tensile and Morphological Properties of Kevlar/S-Glass/Jute Fibre-Reinforced Epoxy Hybrid Composite

G. Manoj Kumar, S. Kavin Raj, K. S. Ajai Bhalaji, and J. Karthik

Abstract The term composite materials are deeply suited into our engineering culture. Nowadays, composite materials are gaining huge attention. This work focuses on the development and personation of hybrid Kevlar/S-Glass/jute-reinforced epoxy composite. A total number of fibre layers are fixed and both weight percentage of the matrix and different fibre layers are varied. Three specimens were fabricated using hand layup technique. The specimens are fabricated according to the ASTM standard by using epoxy resin LY556 as the matrix material and hardener—HY 951. Out of three specimens, the best composition is selected by undergoing various tests like tensile test, SEM, etc. The reinforcement includes fibres like Kevlar, S-glass and jute. Since the world is moving on towards the composite materials and smart materials, the ultimate aim of this paper is to develop a hybrid composite materials to achieve helmet of superior quality, low cost and lightweight. This helmet is suitable for various applications like bike helmet, bicycle helmet, baseball helmet and cricket helmet.

Keywords Kevlar fibre · S-glass fibre · Jute fibre · Bike helmet

1 Introduction

All helmets try to protect the user's head by absorbing mechanical energy. It also try to protect the head against penetration. From the case studies, it is observed that the helmet gets cracked it through the middle and also at front side and finally was broken due to impact during accidents.

When the helmets hit the road, the plastic shell gets shattered and the inner foam gets squished and crumbled and finally the helmet gets disintegrated on impact.

G. Manoj Kumar (✉) · S. Kavin Raj · K. S. Ajai Bhalaji · J. Karthik
Department of Mechanical Engineering, Mepco Schlenk Engineering College, Sivakasi, Tamil Nadu, India
e-mail: manojkumargmech@gmail.com

S. Kavin Raj
e-mail: kavinkacer@gmail.com

T. Rajmohan et al. (eds.), *Advances in Materials and Manufacturing Engineering*, Springer Proceedings in Materials 7, https://doi.org/10.1007/978-981-15-6267-9_24

Though the person driving wearing helmet is not died, but had a very serious injury, no helmet will protect from a falling girder!—(construction site). Composites helmet would be a protective helmet that could save your life. In the broad sense, the term composite materials refers to all solid materials composed of more than one component, where in those components are in separate phases [1]. Composite materials are used here because they have excellent properties like high specific stiffness, specific strength and near net shape of the products that can be produced. Composite materials provide capabilities of part integration [2].

2 Materials and Methods

The conventional helmet is made up of acrylonitrile–butadiene–styrene (ABS) and high density polyethylene (HDPE). Modern helmets that are polymer-based and sometimes polycarbonate [3, 4] are used based on the application. This work involves the use of composite materials like Kevlar fibre, S-glass fibre and jute fibre. The helmet material as a whole consists of four layers; the bottom most layer consists of polystyrene foam which provides a cushioning effect to the user's head. Epoxy resin (LY-556) and hardener (HY-951) were used as a matrix material (Fig. 1).

2.1 *Processing*

The composite specimen and the helmet were fabricated using hand layup technique. The composite specimen was prepared for all the three compositions as per the die specification using hand layup technique [5]. After curing at a room temperature of 15–16 h, developed composite specimen was taken out. The below mentioned values represent the layers of the fibres of the respective specimen (Table 1).

The above-mentioned values denote the weight percentage of the concerned fibre (Fig. 2).

Fig. 1 Kevlar fibre mat (gold colour), S-glass fibre mat (white colour), jute fibre mat (brown colour), epoxy resin and hardener

Table 1 Designation of fibres

S. no.	Designation	Composition
1	1 kevlar fibre + 3 glass fibre + 1 jute fibre + epoxy resin	10% Kevlar + 32% glass + 6% jute + 52% epoxy
2	1 kevlar fibre + 2 glass fibre + 2 jute fibre + epoxy resin	10% Kevlar + 20% glass + 10% jute + 60% epoxy
3	1 kevlar fibre + 1 glass fibre + 3 jute fibre + epoxy resin	10% Kevlar + 12% glass + 20% jute + 58% epoxy

Fig. 2 Composite specimen of different compositions

2.2 *Property Calculations*

The properties of the Kevlar, S-glass, jute fibre and the matrix material were obtained [6, 7] (Table 2).

The calculation procedure is proceeded for all the three compositions and finally the density, Young's modulus and Poisson's ratio are found and tabulated [8] (Table 3).

Table 2 Properties of matrix and reinforcements

Properties	Kevlar	S-Glass	Jute	Epoxy/hardener
Density (ρ)	1.4 g/cm^3	2.5 g/cm^3	1.46 g/cm^3	1.2 g/cm^3
Young's modulus (E)	124 GPa	89 Gpa	30 GPa	3.42 Gpa
Poisson's ratio (μ)	0.36	0.22	0.3	0.35

Table 3 Properties of the developed composite specimen

S. no.	Composition	Density of composite (kg/m^3)	Young's modulus (GPa)	Poisson's ratio
1	Composition 1	1058.59	14.750	0.24
2	Composition 2	1017.284	12.34	0.252
3	Composition 3	1000.409	11.352	0.2517

Fig. 3 Jigsaw cutting machine

2.3 Cutting of Laminates in to Samples of Desired Dimensions

The jigsaw blade was used to cut the laminates into various samples of desired dimensions. Sample specimens were made for tensile test, impact test, hardness test and bending test as per the ASTM standards. The number of test specimens is twelve and corresponding properties are taken from the experimental procedure (Fig. 3).

2.4 Fabricated Specimens

The specimens are fabricated as per the ASTM standard into desired shape. The fabricated specimens are given below (Fig. 4).

3 Numerical Analysis of the Composite Specimens Using ANSYS

3.1 Modelling of the Composite Specimen

The composite specimen was modelled using SOLIDWORKS 2018. The rectangular specimen was modelled with (120 mm × 12 mm × 3 mm) dimensions. All these specimens were developed as per ASTM standards.

Fig. 4 Specimen samples for hardness test, impact test, bending test and tensile test for all three compositions

3.2 *Analysis of the Specimens Using ANSYS*

3.2.1 Tensile Test

The tensile test specimen was actually modelled in dog bone shape as per the ASTM standards. Finally, the whole body was meshed (edge sizing mesh, body sizing mesh). The tetrahedral type of mesh element was applied to the corresponding tensile test specimen. The element size of the mesh was given as 4 mm. The deformation results obtained from ANSYS were then compared with results obtained from the experimental procedure (Figs. 5, 6 and 7).

The same tensile test analysis procedure was repeated for other two specimens, and the total deformation results from ANSYS were tabulated and then compared with the experimental results (Table 4).

Fig. 5 3D model

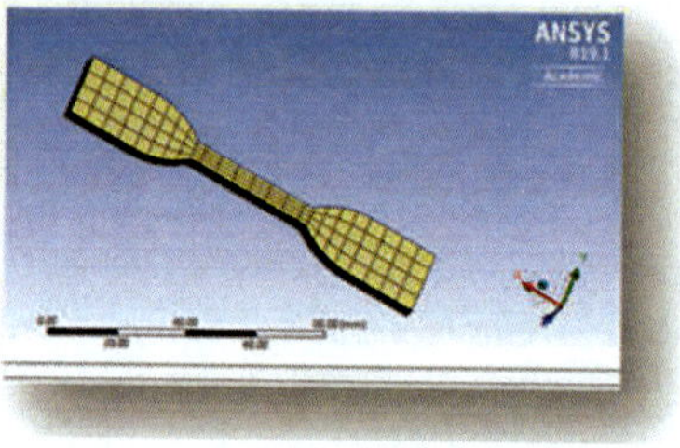

Fig. 6 Mesh generation

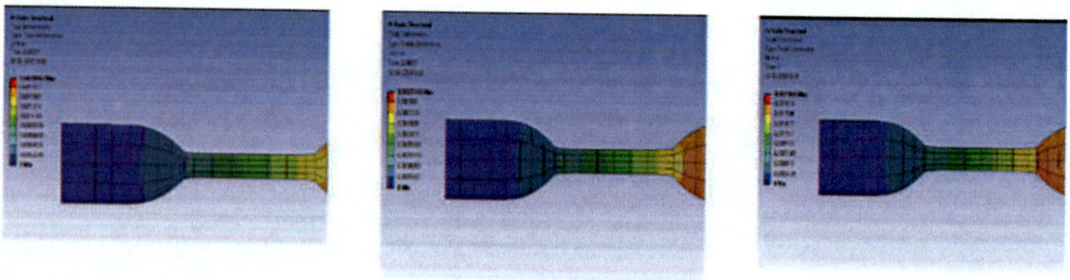

a) Specimen 1 b) Specimen 2 c) Specimen 3

Fig. 7 Tensile test analysis for specimens in ANSYS

Table 4 Tensile test results obtained from ANSYS

Composition	Load applied (N)	Deformation (mm)
Composition 1	7.75	2.006
Composition 2	7.28	2.234
Composition 3	6.5	2.1865

4 Experimental Analysis

4.1 Tensile Test

Static tensile properties such as tensile strength, tensile modulus and Poisson's ratio were determined as per ASTM standard. The tensile test was carried out using universal testing machine (UTM) where one end of the specimen was fixed and the other end of the specimen was subjected to gradual load till the fracture of the sample. The stress–strain graph for the test carried out on respective specimens is shown in Figs. 8, 9 and 10 (Table 5).

The experimental results of all the three compositions were compared with results of ANSYS. It was observed that the deformation of the tensile test specimen in the ANSYS was more or less similar and matches with the experimental results.

Table 5 Tensile test results obtained from experimental procedure

Composition	Load applied (N)	Deformation (mm)	% of elongation	Tensile strength (MPa)
Composition 1	7.75	3.93	10.28	166.31
Composition 2	7.28	3.28	11.11	111.3
Composition 3	6.5	2.94	6.72	105.22

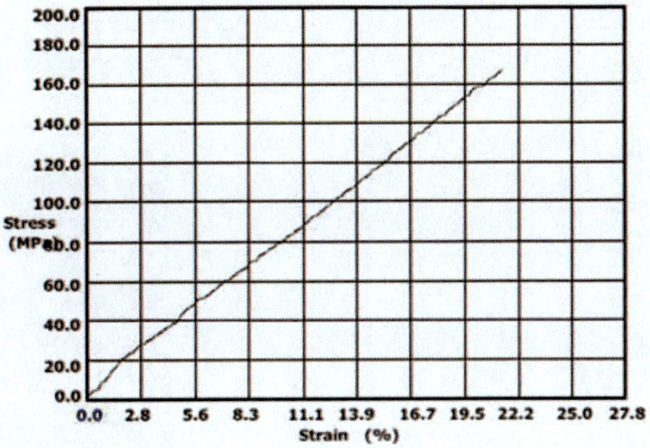

Fig. 8 Tensile test (specimen 1)

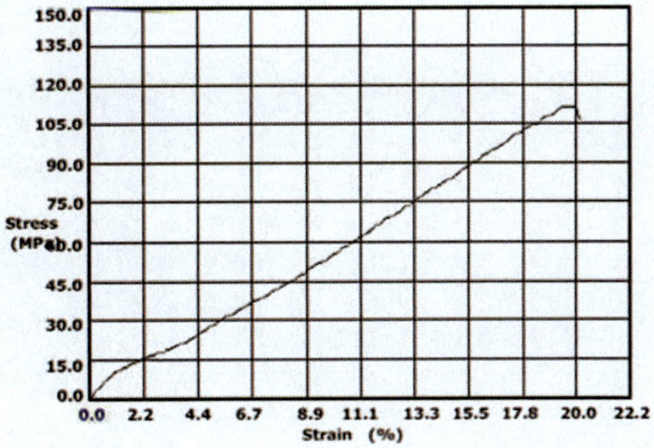

Fig. 9 Tensile test (specimen 2)

4.2 SEM Results

The surface details of the composites were studied using scanning electron microscopy (SEM). The images from SEM will be useful to study the fibre–matrix interaction [9, 10] (Figs. 11, 12, 13, 14, 15 and 16).

It also helps to study how breaking and debonding of fibres take place. The SEM images help to study the morphology of the different specimen after and before testing.

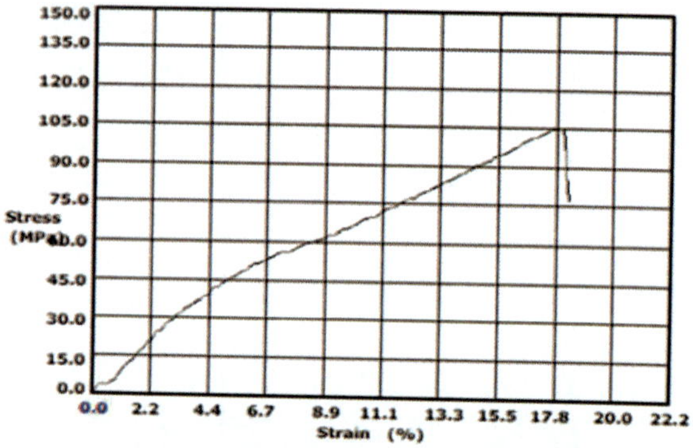

Fig. 10 Tensile test (specimen 3)

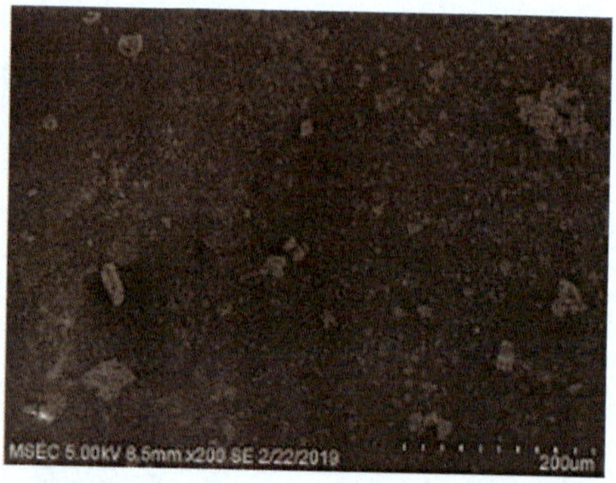

Fig. 11 SEM of specimen 1

Fig. 12 SEM of specimen 2

Fig. 13 SEM of specimen 3

Fig. 14 Tensile (SEM) of specimen 1

Fig. 15 Tensile (SEM) of specimen 2

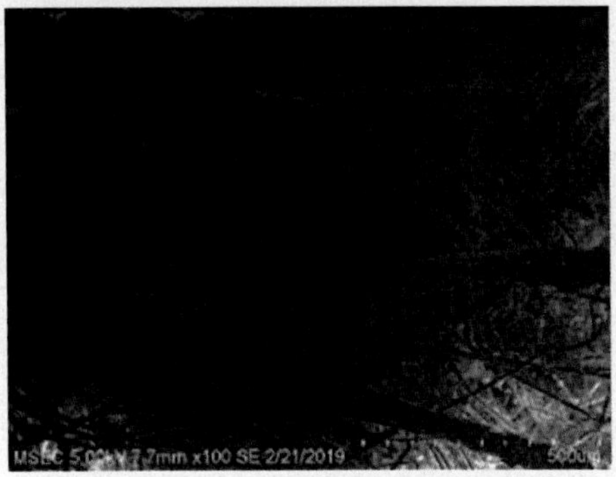

Fig. 16 Tensile (SEM) of specimen 3

5 Conclusions

This paper presents the fabrication of hybrid composite (Kevlar, S-glass, jute fibre) with epoxy resin using hand layup technique. From the tests, the following conclusions are drawn:

From the tensile test, it is observed that the specimen of composition 2 has the tensile strength of 111.3 MPa which is better than specimen 3 having tensile strength of 105.22 MPa. Even all the ANSYS results of tensile test with the respective specimens agree with the results of experimental tests of the specimens.

With the results obtained, the properties of the specimen tested suit for various applications like safety helmet for industrial workers and sports persons.

References

1. Strong AB (2008) Fundamental of Composites manufacturing: materials, methods and applications. Society of Manufacturing Engineers
2. Design and analysis of industrial safety helmet using composite material. In: 1st national conference on recent innovations in mechanical engineering (NCRIME-2018 17 I Page)
3. Salman SD, Sharba MJ, Leman Z, Sultan MTH, Ishak MR, Cardona F (2015) Physical, mechanical, and morphological properties of woven kenaf/polymer composites produced using a vacuum infusion technique. Int J Polym Sci 2015:894565. https://doi.org/10.1155/2015/894565
4. Zhang D, Sun Y, Chen L et al (2014) Influence of fabric structure and thickness on the ballistic impact behavior of Ultra high molecular weight polyethylene composite laminate. Mater Des 54:315–322
5. Sanjay MR, Yogesha B (2016) Studies on mechanical properties of jute/E-glass fibre reinforced epoxy hybrid composites. J Miner Mater Character Eng 4:15–25
6. Salman SD, Leman Z, Sultan MTH et al (2015) Kenaf/synthetic and Kevlar/cellulosic fiber-reinforced hybrid composites: a review. Bio Resour 10:8580–8603
7. Tham CY, Tan VBC, Lee HP (2008) Ballistic impact of a KEVLAR® helmet: Experiment and simulations. Int J Impact Eng 35(5):304–318. https://doi.org/10.1016/j.ijimpeng.2007.03.008
8. Kaw AK (2006) Mechanics of composite materials, 2nd edn. Mechanical Engineering, vol 29. ISBN 0-8493-1343-0
9. Salman SD, Sharba MJ, Leman Z et al (2015) Cardona, physical, mechanical, and morphological properties of woven kenaf/polymer composites produced using a vacuum infusion technique. Int J Polym Sci 2015:894565
10. Bajpai PK, Ram K, Gahlot LK, Jha VK (2017) Fabrication of glass/jute/epoxy composite based industrial safety Helmet. Mater Today: Proc 5: 8699–8706

Optimization of Roller Burnishing Parameters of Al(SiC)p Metal Matrix Composite with TiAlN-Coated Roller Using Response Surface Methodology

E. Shankar, T. Sampath Kumar, M. R. Stalin John, and C. Devanathan

Abstract In this paper, the optimization of roller burnishing parameters of silicon carbide particles-reinforced aluminum composites of metal matrix base using response surface methodology (RSM) was carried out. For the burnishing roller material, tungsten carbide was coated with TiAlN using physical vapor deposition (PVD) technique. Experiments were conducted in dry condition by changing speed of the burnishing tool and number of passes. The input parameters were changed at different levels in order to evaluate its influence on output responses such as roughness and hardness of the surface. The optimization was carried out using response surface methodology.

Keywords Roller burnishing · TiAlN roller · RSM · Lathe · MMC

1 Introduction

Nowadays, burnishing process is very popular to improve the surface finish and enhance the surface hardness after machining operations like turning. It avoids the need of separate superfinishing machines like grinding, polishing, honing, etc. Due to its large contact area, roller burnishing produces good surface finish in comparison with ball burnishing. There are many parameters which affect the burnishing performance like burnishing feed, force, speed, number of passes, diametric size of ball, tool material, lubricant used, initial roughness and hardness of the workpiece, burnishing direction, etc. Materials like hardened steel EN 24, brass, cast Al–Cu,

E. Shankar (✉) · C. Devanathan
Department of Mechanical Engineering, Rajalakshmi Engineering College, Thandalam, Chennai 602 105, Tamil Nadu, India
e-mail: deepanshankar@gmail.com

T. Sampath Kumar
School of Mechanical Engineering, Vellore Institute of Technology, Vellore Campus, Vellore 632014, India

M. R. Stalin John
Department of Mechanical Engineering, SRM Institute of Science and Technology, Kattankulathur, Chennai 603203, India

T. Rajmohan et al. (eds.), *Advances in Materials and Manufacturing Engineering*, Springer Proceedings in Materials 7, https://doi.org/10.1007/978-981-15-6267-9_25

mild steel, and Al (SiC)p composites have been burnished and its reports were available. The consequence of burnishing operations on the roughness and hardness of the surface was found [1–5]. It was found that the TiN coating on EN31 roller during burnishing has positive impact on the output parameters.

Based on the literature analysis, it was found that less work is reported on burnishing of metal matrix composites and optimizing its parameters. This paper uses response surface methodology to determine the optimal burnishing conditions to get multiobjective function for instance minimum surface roughness and maximum surface hardness.

2 Experimental Procedure

Al(SiC)p MMC was chosen as the workpiece material. The conventional stir casting method was used to prepare the work material. 10 wt% of SiC reinforcement is added to molten Al6061 matrix at a temperature of 750 °C and stirred at speed of 500 rpm for 10 min and then it was poured into a cavity and permitted to solidify. The cast workpiece was turned using center lathe. The Vickers hardness machine (FTM make TV-50 model) and contact type profile meter (Mitutoyo make SJ210 model) was used to obtained the initial surface hardness (42.3 HV), and initial surface roughness (2.6 μm), respectively. Burnishing tool with WC roller material was purchased from bright burnishing tools. Titanium aluminum nitrate coating was done on the tungsten carbide roller by PVD technique. Figure 1 shows photographic view of the roller burnishing tool.

The cutting speed (N), the lubricant (L), and number of times the tool passes (n) were the three controlling factors selected. The output responses were surface roughness and hardness of the surface, and the output values are shown in Table 1. Figure 2 shows photographic view of the burnishing operation using roller burnishing tool.

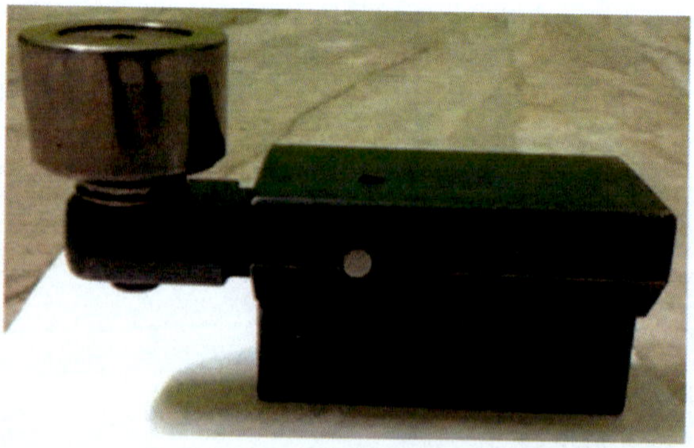

Fig. 1 Roller burnishing tool

Table 1 Experimental matrix of test

No. of passes	Speed in rpm	Dry condition(D)			
		Uncoated roller		Coated roller	
		HV10 (HV)	Ra (μm)	HV10 (HV)	Ra (μm)
1	200	69	0.650	98	0.890
1	300	67	0.774	79	1.220
1	500	63	1.658	75	1.440
2	200	71	0.998	101	0.880
2	300	75	1.128	86	1.005
2	500	67	1.4	79	0.990
3	200	83	1.246	103	0.850
3	300	83	1.344	98	0.660
3	500	75	1.678	86	0.742

Fig. 2 Burnishing operation

3 Case Study

3.1 Case Study 1: Dry Condition and Uncoated Roller

3.1.1 Mathematical Models

Using response surface methodology, the roughness and hardness of the surfaces were modeled for the input factors of burnishing speed and number of times the tool passes.

The response equations are

$$\begin{aligned} \text{HV10} &= 55.9 - 0.90\,\text{pass} + 0.0819\,\text{speed} + 2.33\,\text{pass} * \text{pass} \\ &\quad - 0.000133\text{speed} * \text{speed} - 0.00429\,\text{pass} * \text{speed} \end{aligned} \tag{1}$$

$$\mathrm{Ra} = 0.083 + 0.339\,\mathrm{pass} + 0.00103\,\mathrm{speed} + 0.0497\,\mathrm{pass} * \mathrm{pass} + 0.000004\,\mathrm{speed} * \mathrm{speed} - 0.001019\,\mathrm{pass} * \mathrm{speed} \quad (2)$$

3.1.2 Response Surface Plots

Figures 3 and of 4 show the response surface plots between number of pass, speed and surface roughness and surface hardness, respectively. The hardness was high at higher speeds. This is due to work hardening of the workpiece on the surface. Surface quality was very poor at higher speed. It may be due to flaking of over-hardened layer of aluminum and exposure of SiC particles. It may also be due to the chatter and vibrations.

Fig. 3 RSM plot between Ra and passes

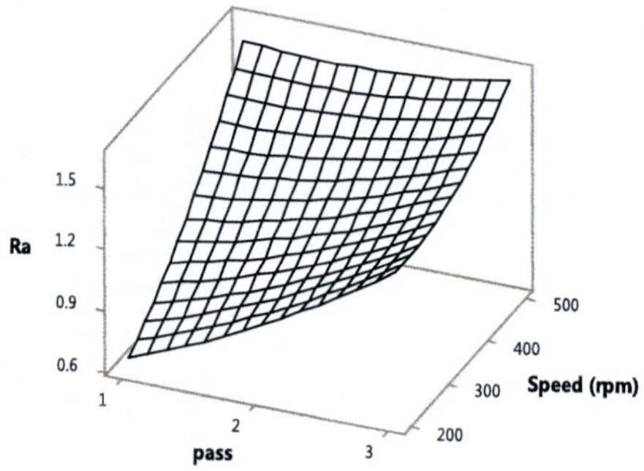

Fig. 4 RSM plot between HV and passes

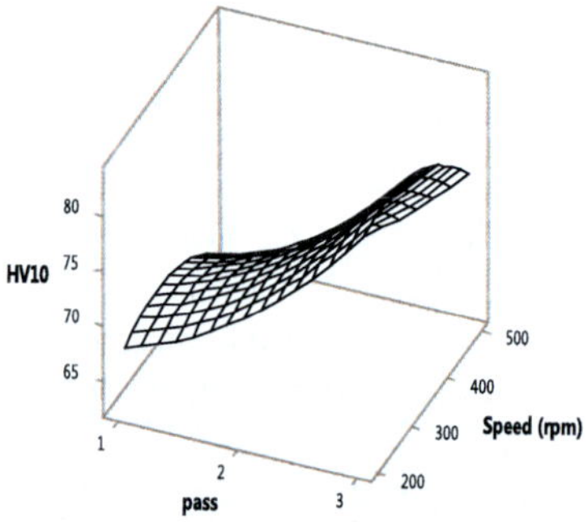

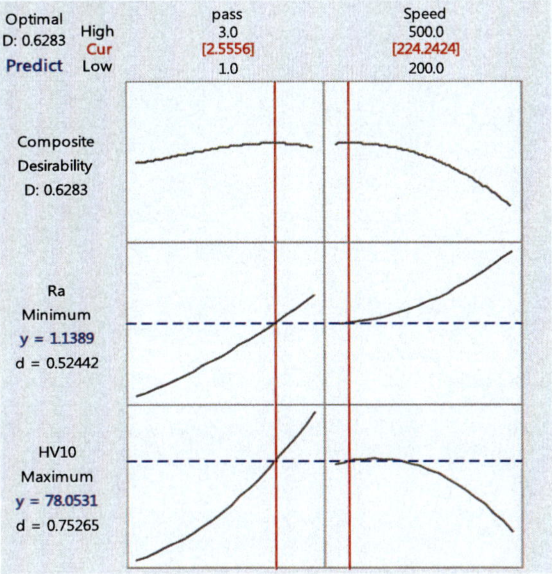

Fig. 5 Response surface optimization for the case study 1

3.1.3 Optimization of Response

The factors are optimized using response surface methodology (RSM) with an objective function. The objective functions were minimizing the surface roughness and maximizing the surface hardness. Minitab software is used for optimization. The optimized factors were speed of the burnishing tool = 224.2424 rpm and number of times the tool passes = 3. Figure 5 indicates the optimized plot of the burnishing factors. With the desirability factor of 0.6238, it was obtained that the optimized surface roughness is 1.1389 μm and hardness of the surface is 78.0531 HV.

3.2 *Case Study 2: Dry Condition and Coated Roller*

3.2.1 Mathematical Models

The response equations are

$$\begin{aligned} \text{HV10} = {} & 141.0 - 0.7\,\text{pass} - 0.294\,\text{speed} + 1.17\ \text{pass} * \text{pass} \\ & + 0.000306\,\text{speed} * \text{speed} + 0.0057\,\text{pass} * \text{speed} \end{aligned} \tag{3}$$

$$\begin{aligned} \text{Ra} = {} & 0.418 + 0.079\,\text{pass} + 0.00354\,\text{speed} + 0.0087\,\text{pass} * \text{pass} \\ & - 0.000001\,\text{speed} * \text{speed} - 0.000989\,\text{pass} * \text{speed} \end{aligned} \tag{4}$$

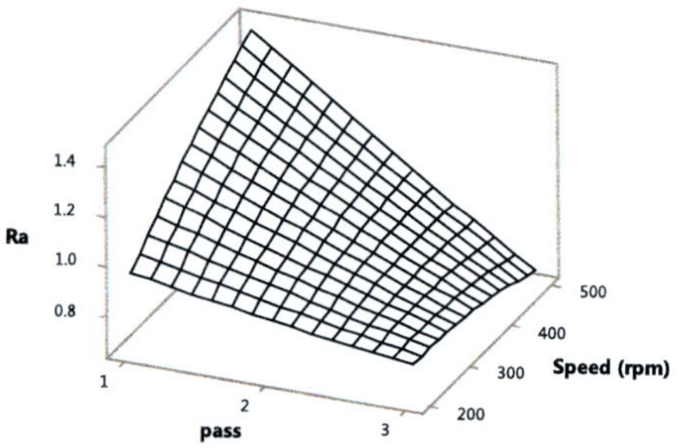

Fig. 6 RSM plot between Ra and passes

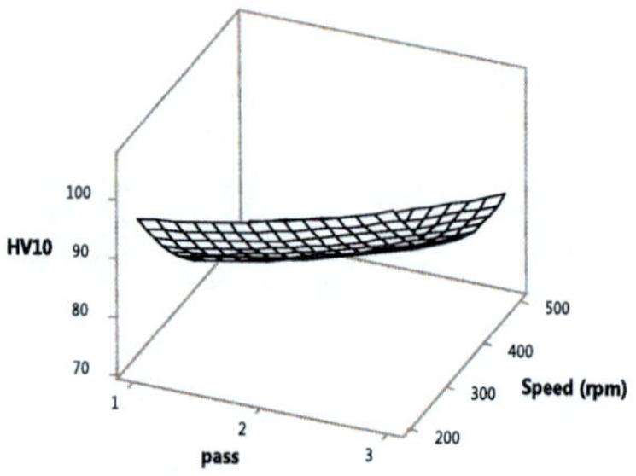

Fig. 7 RSM plot between Ra and passes

3.2.2 Response Surface Plots

Figures 6 and 7 show the response surface plots between speed, number of pass, against surface roughness and hardness of the surface, respectively. It was found that for repeated number of passes the quality of surface was good. This is because of the movement of substance from peaks into the valleys of the surface irregularities on the machined surface. During first pass, roughness of surface increases with increase in speed. At higher passes, the surface finish was not affected by rotational speed of the workpiece. Hardness was higher at the third pass where the surface finish was also good.

3.2.3 Optimization of Responses

The best input factors are found using RSM and the values are: speed of the burnishing tool = 200 rpm and number of times the tool passes = 3. The optimized surface roughness and surface hardness are found out using RSM with the desirability factor of 0.9116 and the values are 0.7918 μm and 106.127 HV correspondingly. Figure 8 shows the response surface optimization plots for the case study 2.

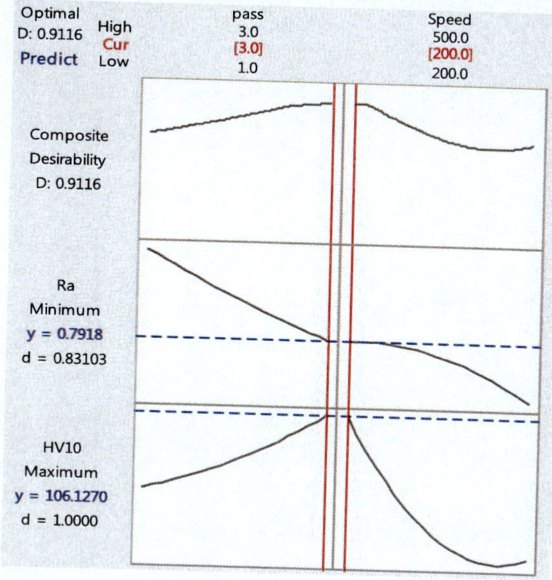

Fig. 8 Response surface optimization for the case study 2

4 Conclusions

Roller burnishing was done on Al(SiC)p MMC using TiAlN-coated roller. Based on the results, the following conclusions can be made.

1. Surface quality improves with increase in number of passes.
2. Optimization was obtained using response surface methodology.
3. Comparing both the case studies, the coated burning tool had given the better surface finish and hardness when compared with non-coated burnishing tool.
4. In future, the work can be extended to check for different conditions like kerosene and castor oil as lubricants and results can be compared with the dry condition.

References

1. Hamadache H, Laouar L, Zeghib NE, Chaoui K (2006) Characteristics of Rb40 steel superficial layer under ball and roller burnishing. J Mater Process Technol 180:30–136
2. Mahmood Hassan A, Al-Dhifi SZS (1999) Improvement in the wear resistance of brass components by the ball burnishing process. J Mater Process Technol 96:73–80
3. Hassan AM, Maqableh AM (2000) The effects of initial burnishing parameters on non-ferrous components. J Mater Process Technol 102:115–121
4. El-Tayeb NSM, Low KO, Brevern PV (2007) Influence of roller burnishing contact width and burnishing orientation on surface quality and tribological behaviour of aluminum 6061. J Mater Process Technol 186:272–278
5. Luca L, Neagu-Ventzel S, Marinescu I (2005) Effects of working parameters on surface finish in ball-burnishing of hardened steels. Precis Eng 29:253–256

Optimization of Machining Parameters During Turning of AISI 316L Stainless Steel Under Nanocutting Fluid Environment

S. D. Sathishkumar and T. Rajmohan

Abstract Nanofluids are a new class of fluids obtained by dispersing nanoparticles in the conventional base fluids. In this research work, optimization of material removal rate in turning operation is done by using nanoparticle-filled lubricant on an all geared center lathe at different cutting velocities and feeds. In the present work, the machining performance of nanofluids is compared with the conventional cutting fluids. The experimental investigation is performed on AISI 316L stainless steel rod by plain turning operation under flooded nanofluid conditions using ceramic inserts. Nanofluids are prepared by mixing carbon nanotubes with SAE20W40 engine oil by 0.25 wt.%. The optimization of machining parameters to maximize material removal rate is done by using Taguchi's technique. The experimental results indicate that spindle speed and depth of cut are the dominant variables on responses.

Keywords AISI 316L stainless steel · Carbon nanotubes · Material removal rate · Nanofluids · Optimization · Taguchi

1 Introduction

Ramana et al. [1] have studied the performance of coolants in turning of titanium alloy under different machining conditions. The experimental results reveal that surface roughness is improved under MQL machining when compared to conventional machining conditions. It is observed that feed rate has more effect on the surface roughness compared to other factors. Vishnu and Kumar [2] have determined the influence of nanofluids in turning of EN-19 steel alloy at various cutting parameters. The surface roughness is improved by obtaining optimum conditions of 1500 rpm spindle speed, feed rate of 0.2 mm/min, 1.5 mm depth of cut, and

S. D. Sathishkumar (✉) · T. Rajmohan
Department of Mechanical Engineering, Sri Chandrasekharendra Saraswathi Viswa Maha Vidyalaya Deemed to be University, Enathur, Kanchipuram 631561, India
e-mail: sathishsd.2@gmail.com

T. Rajmohan
e-mail: rajmohanscsvmv@yahoo.com

T. Rajmohan et al. (eds.), *Advances in Materials and Manufacturing Engineering*, Springer Proceedings in Materials 7, https://doi.org/10.1007/978-981-15-6267-9_26

gold-coated cutting tool by using Taguchi's technique. Khalil et al. [3] have studied the performance of nanolubricant in turning of AISI 1050 steel under MQL conditions. The experimental results reveal that nanolubricant with SDBS exhibits better performance in reduction of tool wear when compared to dry and pure nanolubricant conditions. Singh et al. [4] have examined the role of nanofluids in turning of AISI 304 stainless steel under MQL conditions. The experimental results reveal that MWCNT nanofluid exhibits better performance in reduction of surface roughness and cutting temperature. Ali et al. [5] have made an investigation to optimize the machining parameters during turning of Inconel 718 by using surfactant added Al_2O_3 nanofluids under MQL condition. It is observed that MQL nanofluids have improved the machining performance with ecological benefits. Siddique et al. [6] have studied the impact of input parameters on material removal rate in turning of AISI 4140 alloy steel under dry conditions. They have obtained the optimum machining parameters to maximize the material removal rate. Narayanan et al. [7] have studied the performance of coolants in face milling of EN 31 steel using tungsten carbide tipped tool by MQL technique. The aluminum oxide-based nanofluids are selected as the cutting fluid. The experimental results reveal that the material removal rate is improved by using 60 nm nanoparticles at 3% concentration filled in the base fluid. Naidu et al. [8] have made an experimental study to optimize the machining parameters in turning of EN-36 steel alloy by using Taguchi's technique. The experimental results reveal that the optimum conditions are obtained under vegetable oil-based nanofluids. Vishnu [9] has made an attempt to study the effect of coolants on material removal rate in turning of EN 19 alloy steel under flooded condition. It is observed that vegetable oil showed better performance than nanofluids. Vishnu et al. [10] have investigated the performance of vegetable oil with MQL in turning of EN 353 alloy steel by using carbide cutting tools. It is observed that vegetable oil exhibited great cooling effect under MQL and flooded conditions. From the literature, it is clear that the use of nanocutting fluids has enhanced the machining characteristics of hard materials by their improved thermal properties. Therefore, the present work deals with the optimization of material removal rate during plain turning of AISI 316L stainless steel using flooded lubrication conditions under nanocutting fluids.

2 Materials and Methods

2.1 Materials Used

AISI 316L stainless steel is chosen as the workpiece material. The chemical composition of the stainless steel material is shown in Table 1. In this research work, ceramic insert is selected as the cutting tool material. Multi-walled carbon nanotubes are chosen as the nanoparticle. The nanofluids were prepared by mixing 10 g of carbon nanotubes into the SAE20W40 oil-based fluid (4000 ml) by using esterification setup.

Table 1 Chemical composition of AISI 316L stainless steel

Element	C	Si	Mn	S	P	Ni	Cr	Mo	N	Fe
Wt. %	0.018	0.350	1.664	0.014	0.043	10.176	16.115	2.197	0.10	Rest

Table 2 Properties of nanofluids

Thermal property	Without MWCNTs	With MWCNTs
Flash point (°C)	235	251
Fire point (°C)	263	255
Pour point (°C)	−12	−15
Viscosity @ 100 °C	16.07	17.22
Viscosity @ 40 °C	135.36	143.28
Thermal conductivity @ 40 °C	0.140	0.162

Table 3 Machining parameter and levels

Input parameters	Symbol	Units	Level 1	Level 2	Level 3
Cutting fluid type	A		SAE20W40	SAE20W40 + MWCNT	
Spindle speed	B	rpm	140	315	500
Feed rate	C	mm/rev	0.05	0.102	0.143
Depth of cut	D	mm	0.40	0.60	0.80

The thermal properties of nanofluids are determined in Delta lab, Chennai, India. Table 2 provides the thermal properties of engine oil with and without nanoparticles.

2.2 *Experimental Design*

In this study, the Taguchi technique was employed for getting the data in a controlled way. According to the Taguchi quality design concept, a L18 orthogonal array was used for the experiments. It reduces the number of experiments. The methodology is valuable when design parameters are qualitative and discrete. Table 3 shows four machining parameters as control factors and their levels.

2.3 *Experimental Procedure*

The experiment is conducted for wet turning operation on AISI 316L stainless steel as work material and ceramic inserts as tool material. All geared lathe machine is used for conducting the experiments. The experimental setup is shown in Fig. 1. The

Fig. 1 Experimental setup

experiments are performed on the workpiece for a length of 130 mm. The response variable which is measured in this study is material removal rate. The experimental results are given in Table 4. The material removal rate is measured by using the following formula.

Let $t =$ machining time, $D_{\text{avg}} =$ average diameter of the cutting section, and $d =$ depth of cut.

So, experimental formula for M.R.R.

$$= \frac{\pi \times L \times d \times Davg}{t} \text{mm}^3/\min \tag{1}$$

Table 4 Experimental results

Trail no.	Cutting fluid type	Spindle speed (rpm)	Feed rate (mm/rev)	Depth of cut (mm)	Machining time(min)	Material removal rate, MRR in mm^3/min	S/N ratio of material removal rate
1	SAE20W40	140	0.05	0.4	8.85	915.57	59.2338
2	SAE20W40	140	0.102	0.6	9	1323.24	62.4328
3	SAE20W40	140	0.143	0.8	9.10	1694.66	64.5817
4	SAE20W40	315	0.05	0.4	4.52	1662.54	64.4154
5	SAE20W40	315	0.102	0.6	4.48	2461.38	67.8236
6	SAE20W40	315	0.143	0.8	4.50	3165.61	70.0091
7	SAE20W40	500	0.05	0.6	2.78	3719.74	71.4103
8	SAE20W40	500	0.102	0.8	2.88	4628.61	73.3090
9	SAE20W40	500	0.143	0.4	2.90	2230.75	66.9690
10	SAE20W40 + MWCNT	140	0.05	0.8	9.05	1386.33	62.8373
11	SAE20W40 + MWCNT	140	0.102	0.4	8.17	743.83	57.4295
12	SAE20W40 + MWCNT	140	0.143	0.6	9	985.62	59.8742
13	SAE20W40 + MWCNT	315	0.05	0.6	4.30	1994.54	65.9969
14	SAE20W40 + MWCNT	315	0.102	0.8	4.25	2583.05	68.2427
15	SAE20W40 + MWCNT	315	0.143	0.4	4.33	1222.39	61.7442
16	SAE20W40 + MWCNT	500	0.05	0.8	2.57	3966.47	71.9681
17	SAE20W40 + MWCNT	500	0.102	0.4	2.67	1835.53	65.2752
18	SAE20W40 + MWCNT	500	0.143	0.6	2.63	2702	68.6337

3 Results and Discussions

3.1 Optimization of Machining Parameters Using Taguchi's Technique

The category the larger-the-better is always preferred to calculate the S/N ratio for quality characteristics of material removal rate. The equation for calculation S/N ratio for larger-the-better characteristic is given in Eq. 2 (in decibels).

Table 5 Response table for means

Level	Cutting fluid type	Spindle speed (rpm)	Feed rate (mm/rev)	Depth of cut (mm)
1	2422	1175	2274	1435
2	1936	2182	2263	2198
3		3181	2000	2904
Delta	487	2006	274	1469
Rank	3	1	4	2

$$\eta = -10 * \log\left(\Sigma\left(1/Y^2\right)/n\right) \tag{2}$$

where Y is the value of responses under study and n is the number of tests.

By applying the Eq. 2, the S/N ratio values for each experiment of L18 are calculated. Based on the response table for means which is shown in Table 5, the optimal solution is obtained at type of cutting fluid of SAE20W40 engine oil, 500 rpm of spindle speed, feed rate of 0.05 mm/rev, and 0.80 mm of depth of cut. These optimum cutting conditions are an untrailed experiment. Therefore, it is necessary to predict the S/N ratio using Minitab software. The predicted S/N ratio for this condition is 73.7186. By using this, considering the larger-the-better equation, the predicted material removal rate is 4852.10 mm^3/min which is the optimum value.

4 Conclusions

This study presents the optimization of process parameters such as spindle speed, feed rate, depth of cut, and type of cutting fluid in wet turning of AISI 316L stainless steel using Taguchi's method. The following conclusions can be obtained based on the experimental results and analysis as follows:

- The material removal rate is maximum for all cutting conditions when using MWCNT-filled cutting fluid.
- The material removal rate increases with the increase in spindle speed.
- Based on the ANOVA results, the most influential parameter which affects the response is spindle speed followed by depth of cut.
- Based on the Taguchi method, the optimal solution for the cutting parameters is determined.

References

1. Ramana MV, Vishnu AV, Rao GKM, Rao DH (2012) Experimental investigations, optimization of process parameters and mathematical modeling in turning of titanium alloy under different lubricant conditions. IOSR J Eng 2(1):86–101

2. Vishnu AV, Kumar GK (2017) Analysis of surface roughness in turning using Nano fluids: prediction model and cutting parameters optimization. Int J Res Appl Sci Eng Technol 5(12):340–347
3. Khalil ANM, Ali MAM, Azmi AI (2015) Effect of Al_2O_3 Nano lubricant with SDBS on tool wear during turning process of AISI 1050 with minimal quantity lubricant. Proc Manuf 2:130–134
4. Singh RK, Sharma AK, Bishwajeet, Mandal V, Gaurav K, Sharma A, Kumar A, Dixit AR, Mandal A, Das AK (2018) Effect of multi-walled carbon Nano tubes based Nano fluid on surface roughness and cutting temperature in turning operation using minimum quantity lubrication. In: International Conference on Mechanical, Materials and Renewable Energy and IOP Conference Series: Materials Science and Engineering, p 377
5. Ali MAM, Khalil ANM, Azmi AI, Salleh HM (2017) Optimization of cutting parameters for surface roughness under MQL, using Al_2O_3 Nano lubricant, during turning of Inconel 718. In: International Research and Innovation Summit and IOP Conference Series: Materials Science and Engineering, p 226
6. Siddique MS, Akhtar MN, Iqbal MW, Beg N, Haq MZ (2017) Turning parameter optimization for Material removal rate of AISI 4140 Alloy steel by Taguchi method. Int Res J Eng Technol 4(6):848–854
7. Narayanan MR, Nallusamy S, Kumar SL (2018) Characterization of machining parameters on EN 31 with Al_2O_3 Nano particles using Taguchi Technique. Int J Mech Eng Technol 9(1):1173–1183
8. Naidu GG, Vishnu AV, Naveen Kumar S, Sagar TP, Srinivas MG (2017) Analysis on the effect of machining parameters on MRR using different coolants. Int J Adv Res Innov Ideas Educ 2(2):28–32
9. Vishnu AV (2017) An investigation on comparison of vegetable oil and Nano fluids coolants in turning of EN 19 Alloy steel. Int J Pure Appl Res Eng Technol 6(4):1–15
10. Vishnu AV, Rajashekar M, Teja PS, Upadhyay SP, Sumanth S, Akash M (2017) Optimization of process parameters for material removal rate under flooded and MQL conditions. Int J Adv Res Innov Ideas Educ 2(2):48–56

Surface Integrity Studies on WEDM of Magnesium Matrix Nano-SiC Reinforced Composites

S. Vijayabhaskar, T. Rajmohan, K. Lalitesh, and S. Sai Vivek

Abstract Magnesium with the addition of ceramic nanoparticles exhibits improved mechanical properties such as an increase in strength and hardness without compromising the ductility which makes it attractive for lightweight applications. But the presence of hard ceramic particles makes it difficult to machine through the conventional process. Wirecut electrical discharge machining (WEDM) is considered as one of the favorable processes to machine such nanocomposites with high accuracy. The surface quality of the machined surface is one of the attributes which needs special attention. In the present investigation, the surface integrity of wirecut electrical discharge machined magnesium matrix nanocomposite has been studied. Magnesium metal reinforced with SiC nanoparticle at three different weight percentage (wt %) viz 0, 0.5, and 1 that is fabricated through ultrasonic cavitation has been used in this investigation. The surface microstructures were examined with a scanning electron microscope (SEM), the surface roughness, and microhardness at various depths that have been measured.

Keywords Magnesium · Nanocomposites · Surface roughness · WEDM · Microhardness

1 Introduction

Magnesium matrix composite is one of the most promising materials to meet the increasing demands for high performance and lightweight materials for structural applications. The development of magnesium matrix nanocomposites with the growing availability of metallic and ceramic nanoparticles offers enhanced properties like tensile, compressive, hardness, fatigue, wear, and corrosion properties when compared to conventional magnesium composites reinforced with micro-sized particles [1]. Ultrasonic cavitation is found to be an efficient method for the effective

S. Vijayabhaskar (✉) · T. Rajmohan · K. Lalitesh · S. Sai Vivek
Department of Mechanical Engineering, Centre for Composite Science and Tribology, Sri Chandrasekharendra Saraswathi Viswa Mahavidyalaya, Enathur, Kanchipuram 631561, India
e-mail: mailvijayabhaskar@gmail.com

T. Rajmohan et al. (eds.), *Advances in Materials and Manufacturing Engineering*, Springer Proceedings in Materials 7, https://doi.org/10.1007/978-981-15-6267-9_27

dispersion SiC nanoparticles in magnesium melt and to obtain a uniform distribution of the reinforcement particles along with grain refinement during solidification process of magnesium matrix nanocomposites [2]. The presence of hard ceramic reinforcement in the nanocomposite makes it difficult to machine through traditional processes resulting in other problems like reduced tool life and the damages at a subsurface level [3–5]. WEDM is considered as one of the favorable unconventional machining processes to machine such materials having varying hardness with high accuracy [6].

WEDM is basically a thermal process involving a complex metal removal mechanism, in which the plasma channel formation between the electrode/wire and workpiece results in the removal of material from the workpiece [7], leading to work hardness, surface roughness, surface structural changes, heat affected zone, white layer, cracking, and metallurgical transformations. Surface integrity includes these properties which play a key role in determining the operational behavior of the material [8]. Surface integrity is that the surface condition of a workpiece once being changed by a machining process, which in turn alters the material properties [9]. In this investigation, WDM experiments have been carried out to study the machined workpiece surface integrity, including the microhardness and microstructures, under a wide range of machining conditions.

2 Experimental

2.1 Materials

Magnesium metal of purity 99% is used as the matrix material, and silicon carbide nanoparticles of average particle size 50–80 nm were used as the reinforcement. The nanocomposites with varying wt.% (0, 0.5, 1) of nano-SiC were fabricated using ultrasonic cavitation method to overcome the problems associated with the dispersion of nanoparticles in the molten metal. Magnesium in the form of billet was loaded into the furnace and heated to around 610 °C, and then, the SiC nanoparticles were added to the melt after preheating to 800 °C. The ultrasonication process was carried out by dipping the probe into the melt, and the whole melt environment is protected by argon gas. The Mg melt with SiC nanoparticles was poured into a steel mold which is preheated to 500 °C.

2.2 Experimental Design

In the present investigation, Taguchi L9 orthogonal array was used to plan the experimental design. This method is considered as one of the effective, simple, and methodical approaches to improve the design for performance, cost, and quality [10]. The

Table 1 Machining parameters and their levels

S. no	Parameters	Unit	Levels		
			1	2	3
1	Pulse ON	μs	1	6	10
2	Voltage	V	40	55	70
3	Feed rate	mm/min	1	6	10
4	Wt. % SiC	%	0	0.5	1

method is appreciated when design factors are discrete and qualitative. The most influencing parameters were identified from the past research literature, and their respective levels were selected and presented in Table 1.

2.3 *Experimental Procedure*

All the experiments were carried out on ecocut four-axis CNC WEDM machine. Negatively polarized brass wire of diameter 0.25 mm is used as electrode, and deionized water is used as the dielectric medium. Three specimens of size (65 × 35 × 10) mm were machined on WEDM by varying the parameters such as pulse on time, voltage, feed rate, and wt. % of SiC considering the output responses such as surface roughness (SR) and material removal rate (MRR). The surface roughness is measured through computerized surface roughness tester, where the average of two measurements were taken for each experiment, and the MRR is calculated using the relation:

$$\text{MRR} = k t V_c \rho \tag{1}$$

where k is kerf width, t is thickness of the workpiece, V_c is the cutting speed, and ρ is the density of the material. The observed experimental results are shown in Table 2. Figure 1 shows the machined samples.

In this investigation, microhardness is measured by using Micro Vicker's Hardness Tester (Model MH6). A force of 100 g for a duration of 10 s is applied on the machined surface, and measurements were made at four different depths perpendicular to the machined surface, i.e., 50, 100, 150, and 200 μm. The measured hardness values are presented in Table 3.

Table 2 Experimental results

Exp. no	Cut	Pulse on (μs)	Voltage (V)	Feed rate (mm/min)	Wt. % of SiC (%)	Kerf width (mm)	MRR (g/min)	SR (μm)
1	1	1	40	1	0	0.339	0.047	3.55
2	2	10	70	6	0	0.347	0.025	3.36
3	3	6	55	10	0	0.344	0.033	3.16
4	1	1	55	6	0.5	0.357	0.026	3.15
5	2	6	70	1	0.5	0.344	0.023	3.39
6	3	10	40	10	0.5	0.338	0.047	3.72
7	1	10	55	1	1	0.341	0.04	4.18
8	2	6	40	6	1	0.357	0.046	4.17
9	3	1	70	10	1	0.346	0.02	3.57

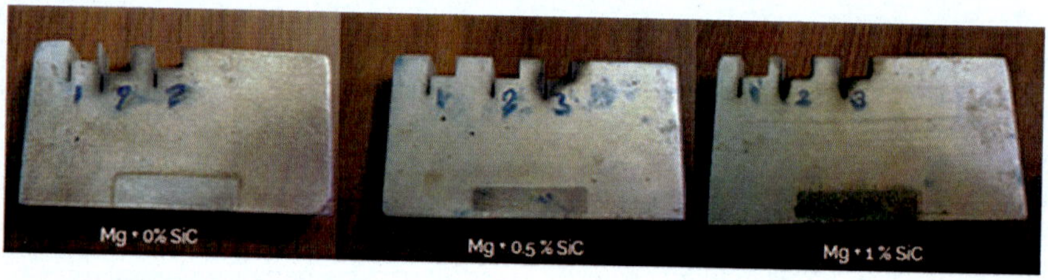

Fig. 1 WEDM processed workpieces

3 Results and Discussion

3.1 *Effect of WEDM on Surface Roughness*

The variation of the surface roughness values, with respect to various input parameters such as pulse on, voltage, and feed rate in terms different cuts and workpiece materials with different wt% of SiC is presented in Fig. 2. Among the other factors, wt% SiC nanoparticles is considered as the most dominating factor that affects the surface roughness [11]. It can be observed from Fig. 2 that the surface roughness values are decreased with increase in wt% of SiC nanoparticles initially which is mainly due to the improved brittleness and uniform distribution of nano-SiC particles in the composites that in turn enhance the surface finish [6], but further increase in the wt % of SiC nanoparticles increases the surface roughness due to the presence of hard SiC nanoparticles [12]. It is also noticed that the increase in pulse on time increases the discharge energy resulting in increased surface roughness while the increase in the voltage decreases the surface roughness as a result of reduced spark due to the increase in the gap between electrodes [13].

Table 3 Microhardness values at different depth profiles

Distance (μm)	Microhardness (HV)								
	Mg + 0% SiC			Mg + 0.5% SiC			Mg + 0.5% SiC		
	Cut 1	Cut 2	Cut 3	Cut 1	Cut 2	Cut 3	Cut 1	Cut 2	Cut 3
50	48.9	67.5	123.7	111.4	155.8	117.1	121.9	123.5	127.1
100	47	60.3	116.2	107.4	163.8	117.4	110.6	109.2	128.4
150	47.5	47.6	84.7	100.8	117.1	90.8	104.9	126.8	115.3
200	37.2	41.2	48.9	100.5	116.9	87.8	83.9	121.9	109.2

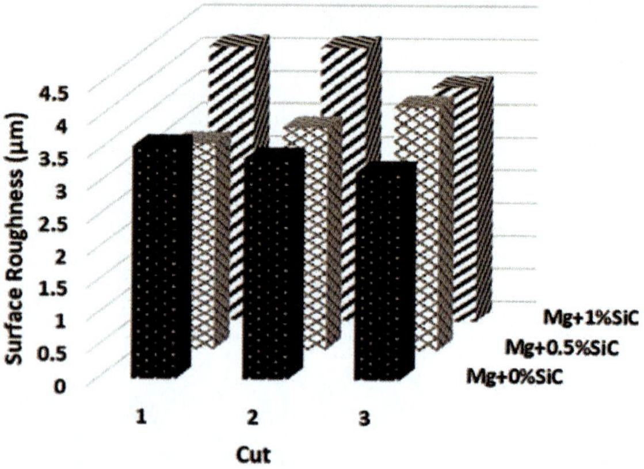

Fig. 2 Surface roughness variation with WEDM parameters and wt% of SiC

3.2 *Effect of WEDM on Microhardness*

Microhardness is considered as one of the important parameters in the surface integrity study. The effect of WEDM cutting on the microhardness of magnesium matrix nano-SiC reinforced composites with varying wt% of SiC nanoparticles is presented in Fig. 3a–c.

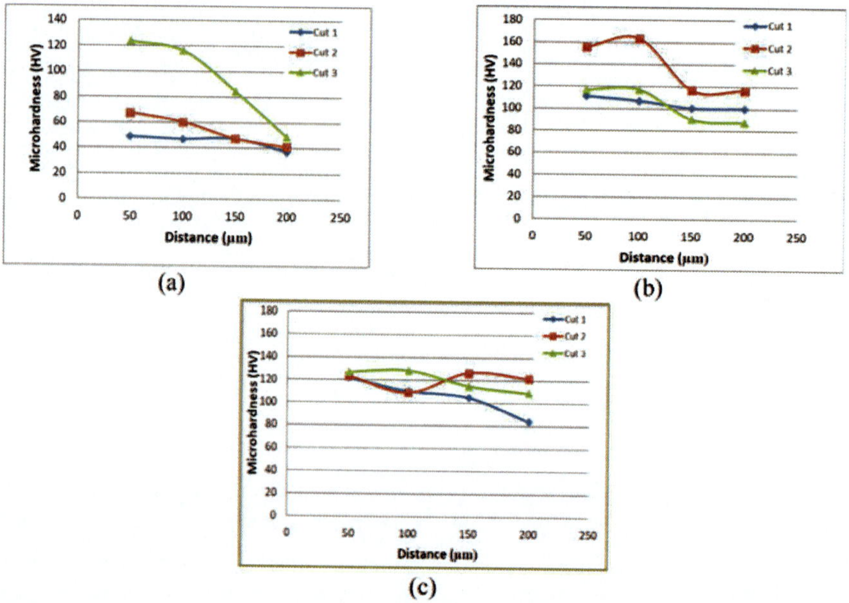

Fig. 3 **a** Microhardness of Mg + 0% SiC, **b** Microhardness of Mg + 0.5% SiC, **c** Microhardness of Mg + 1% SiC

It is observed from Fig. 3a–c that the microhardness decreases as the depth from the machined surface increases. The increased microhardness near the machined surface is mainly due to the heat generated at the cutting resulting in the work hardening of the deformed machined surface layer along the depth [14]. Pulse on time is one of the predominant factors that affect the microhardness [15]. Figure 3a–c indicate that the high level of pulse on time and voltage increase the hardness by forming hard surface as a result of high discharge energy. Microhardness profiles revealed that the subsurface deformation occurs up to a maximum of 200 μm below the machined surface and moving further away from the machined surface, and the hardness starts to decrease and attains the bulk hardness [6].

3.3 Subsurface Microstructure

The machined surface is melted by the sparks produced during WEDM process, and the molten metal is flushed away by the dielectric medium. The solidification of the remaining material form recast layer known as the white layer. During the WEDM process, heat affected zone (HAZ) is formed underneath the white layer owing to rapid heating and quenching. The white layer and HAZ encompass altered microstructure, microcracks, impurities, and other undesirable structures which seriously affect the surface integrity [16]. The subsurface microstructure of WEDMed workpieces with varying SiC nanoparticles wt% was observed using a scanning electron microscope (SEM) through the cross section of the machined surface and presented in Fig. 4a–c.

It can be seen from Fig. 4a that the thickness of white layer increases from 7.48 to 14 μm as the pulse on time increase from 1 to 10 μs since the discharge energy increases with the pulse on time. Higher pulse on time permits high temperature to infiltrate deeper into the subsurface, producing more molten material and eventually results in a thicker white layer [17]. The voltage has considerable influence on the white layer thickness after pulse on time [16]. But, the white layer thickness increases from 7.48 to 14 μm with the decrease in the voltage from 70 to 40v. The microcracks into the subsurface were not observed beyond the white layer.

4 Conclusions

The effects of WEDM on the surface integrity of magnesium nanocomposites with varying wt% nano-SiC reinforcement were investigated experimentally. Based on the results of the investigation, the following conclusions are drawn:

- The surface roughness values decrease with increase in wt% of SiC nanoparticles initially, but further increase in the wt % of SiC nanoparticles increases the surface roughness.

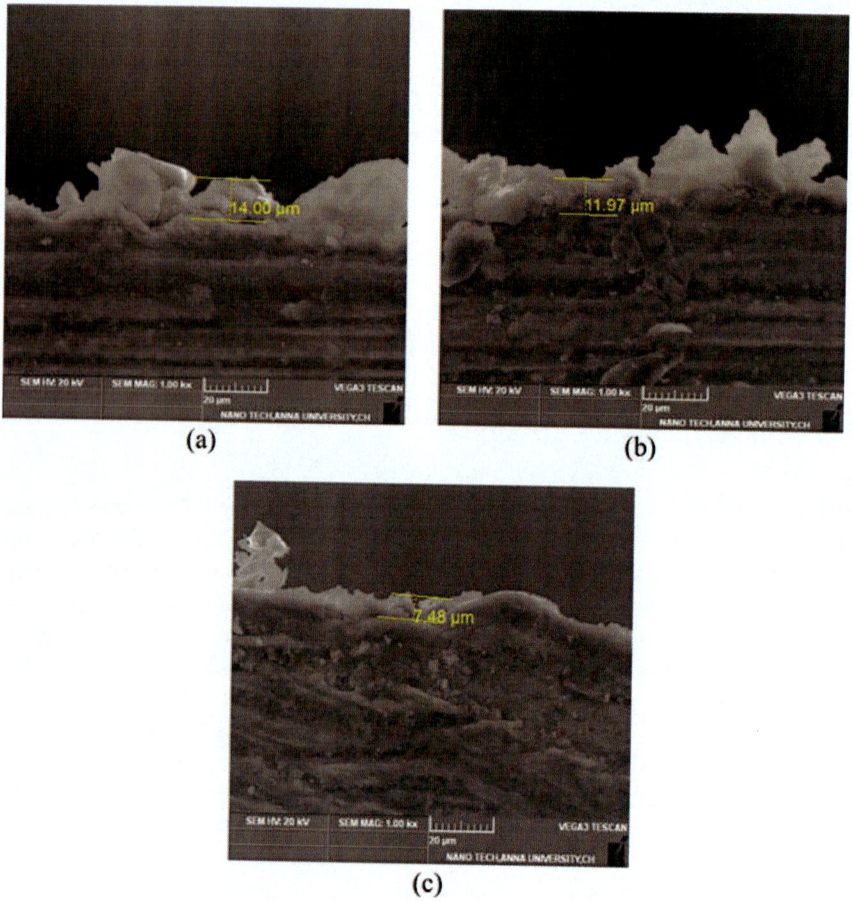

Fig. 4 **a** SEM at $T_{ON} = 10$ μs. **b** SEM at $T_{ON} = 6$ μs. **c** SEM at $T_{ON} = 1$ μs

- The increase in pulse on time increases the surface roughness while the increase in the voltage decreases it.
- The high level of pulse on time and voltage increase the hardness of the machined surface, and the subsurface deformation occurs up to a maximum of 200 μm below the machined surface.
- The white layer thickness increases from 7.48 to 14 μm as the pulse on time increases and the voltage decreases.

References

1. Gupta M, Wong WLE (2015) Magnesium-based nanocomposites: lightweight materials of the future. Mater Charact 105:30–46
2. Cao G, Konishi H, Li X (2008) Mechanical properties and microstructure of SiC-reinforced Mg-(2,4) Al-1Si nanocomposites fabricated by ultrasonic cavitation based solidification processing. Mater Sci Eng, A 486(1–2):357–362
3. Benedict GF (1987) Electric discharge machining (EDM)-Nontraditional manufacturing processes. Marcel Dekker, New York, Basel
4. Boothroyd G, Winston AK (1989) Non-conventional machining processes-Fundamentals of machining. Marcel Dekker, New York, Basel
5. Rajmohan T, Palanikumar K (2011) Optimization of machining parameters for multi-performance characteristics in drilling hybrid metal matrix composites. J Compos Mater 46(7):869–878
6. Vijayabhaskar S, Rajmohan T (2018) Experimental investigation and optimization of machining parameters in WEDM of nano-SiC particles reinforced magnesium matrix composites. Silicon. https://doi.org/10.1007/s12633-017-9676-0
7. Vijayabhaskar S, Rajmohan T, Pranay Sisir TV, Phani Abishek JVS, Sharukh Khan D, Mohankrishna Reddy R (2018) Review of WEDM studies on metal matrix composites. IOP Conf. Series: Materials Science and Engineering, vol 390, p 012051
8. Arikatla SP, Tamilmannan K, Krishnaiah A (2017) Surface integrity characteristics in wire electrical discharge machining of titanium alloy during main cut and trim cuts. Mater Today: Proc 4:1500–1509
9. Degarmo, E. P., Black, J.T., Kohser, R. A.: Materials and Processes in Manufacturing.9th ed. Wiley, (2003).
10. Rajmohan T, Prabhu R, Subbarao G, Palanikumar K (2012) Optimization of machining parameters in electrical discharge machining (EDM) of 304 stainless steel. Proc Eng 38:1030–1036
11. Ozana SL, Guleryuza F, Kasmanb S, İpeka R (2012) An investigation of WEDM process parameters on the surface roughness of Al/B4Cp metal matrix composites. AIP Conf Proc 1476: 317–320
12. Yan BH, Tsai HC, Huang FY, Lee LC (2005) Examination of wire electrical discharge machining of Al2O3p/6061Al composites. Int J Mach Tools Manuf 45:251–259
13. Aghdeab SH, Najm VN, Saleh AM (2018) Surface roughness evaluation in WEDM using Taguchi parameter design method. Eng Technol J 36:60–64
14. Rajmohan T, Palanikumar K, Davim JP (2012) Analysis of surface integrity in drilling metal matrix and hybrid metal matrix composites. J Mater Sci Technol 28(8):761–768
15. Khan NZ, Wahid MA, Singh S, Siddiquee AN, Khan ZA (2013) A study on micro hardness in wire electrical discharge machining based on taguchi method. Int J Mechan Prod Eng 1(1):10–15
16. Rao PS, Ramji K, Satyanarayana B (2018) Surface integrity of wire EDMed aluminum alloy: a comprehensive experimental investigation. J King Saud Univ Eng Sci 30(4):368–376
17. Li L, Guo YB, Wei XT, Li W (2013) Surface integrity characteristics in wire-EDM of Inconel 718 at different discharge energy. Procedia CIRP 6:220–225

Acoustic, Rheological and Optical Properties of Binary Mixtures of Aqueous Solutions of PEG 200—A Comparative Analysis

P. Dhivya, R. Padmanaban, A. Gayathri, and K. Venkatramanan

Abstract Prediction of density, viscosity, ultrasonic velocity and refractive indices of binary liquid mixtures is essential for the determination of composition of binary liquid mixtures. Ultrasonic velocity and refractive index measurements in combination with density, viscosity, boiling point, melting point and other analytical data are very useful industrially also for common substances which include oils, waxes, sugar syrups, etc. In the present study, it is planned to compute the density, viscosity, ultrasonic velocity and refractive index of binary liquid mixtures of polyethylene glycol (Molar mass: 200) (PEG 200) with water at different molar concentrations (0.0009, 0.0027, 0.0045, 0.0063, 0.0080, 0.0098, 0.0116, 0.0133, 0.0151 and 0.0168), at 303 K. The experimental values of viscosity, ultrasound velocity and refractive index are compared with various theoretical models. The most reliable method that matches with experimental method is identified by calculating average percentage error (APE).

Keywords Average percentage error · Polyethylene glycol · Ultrasonic velocity

1 Introduction

Ultrasonic study has become an interesting research tool in the field of polymers to determine the structure and molecular interactions in liquid systems. The study of the thermodynamic properties of binary mixtures is significant for many purposes. For analysing the properties and interactions taking place in polymer systems, knowledge of rheological properties is essential and viscometer is one of the efficient techniques for studying the properties. This is to obtain information on molecular features of the studied mixtures. The sign and magnitude of excess parameters investigate the interactions between the multi-components of a system [1–5]. Polyethylene glycol [PEG 200] has a lot of applications in pharmaceutical preparations, cosmetics, medical devices and industries. The molecular interaction properties of polyethylene glycol

P. Dhivya · R. Padmanaban · A. Gayathri · K. Venkatramanan (✉)
Faculty of Science, Sri Chandrasekharendra Saraswathi Viswa Mahavidyalaya, Kanchipuram, India
e-mail: Kv.scsvmv@gmail.com

T. Rajmohan et al. (eds.), *Advances in Materials and Manufacturing Engineering*, Springer Proceedings in Materials 7, https://doi.org/10.1007/978-981-15-6267-9_28

(PEG200) in water for various molar concentrations ranging from (0.0009, 0.0027, 0.0045, 0.0063, 0.0080, 0.0098, 0.0116, 0.0133, 0.0151 and 0.0168) at 303 K are reported in this paper. The present work aims to compare the experimentally determined ultrasonic velocity, viscosity and refractive index values with the computed values using different analytical models and empirical relations.

2 Materials and Methods

The material used in the present study is polyethylene glycol (PEG) 200 (spectra grade) which is obtained from Kavin Scientific Company, Chennai, India. The polymer solutions are prepared by dissolving the polymer in double distilled water in order to get required concentrations (0.0009, 0.0027, 0.0045, 0.0063, 0.0080, 0.0098, 0.0116, 0.0133, 0.0151 and 0.0168). All binary mixtures are prepared gravimetrically in air-tight bottles and adequate precautions have been taken to minimise evaporation loss. Ultrasonic velocity was measured by a single crystal variable path interferometer operating at a frequency of 2 MHz (VCT-70A model) (accuracy ±0.1 m/s). The viscosity of polymer solutions for above concentrations is determined using Brookfield viscometer (accuracy ±0.01 cP). The viscometer is connected to an electronically controlled thermostat having a thermal stability of ±0.1K for taking measurements at temperatures 303 K. The refractive index measurements are performed using Mittal make Abbe refractometer (accuracy ±0.001). In all property measurements, the temperature was controlled with in ±0.1 K using a constant temperature path by circulating water from the thermostat. In the present analysis, the binary mixtures were prepared for polyethylene glycol 200 in water for different mole fractions. Thermostat of high accuracy is used to maintain constant temperature throughout the experiment.

3 Results and Discussion

The experimental values of ultrasonic velocity, viscosity and refractive index for the binary mixtures of polyethylene glycol (PEG 200) in water and over the entire composition range (0.0009, 0.0027, 0.0045, 0.0063, 0.0080, 0.0098, 0.0116, 0.0133, 0.0151 and 0.0168) expressed in terms of mole fraction are listed in Tables.

3.1 Ultrasonic Velocity Deviation

Ultrasonic velocities of binary liquid mixtures of PEG 200 in water are measured at 303 K. The experimental values along with the values calculated mathematically using Nomoto's model, Van deal & Van geel, impedance relation and Rao's specific sound velocity method are presented in Table 1 using the standard relations [3].

In the present system, it is understood that the average percentage of deviation [2] is more in Rao's model and less in impedance model (Table 2). From this, it may be concluded that impedance relation is the most suited mathematical method for estimating the speed of ultrasound waves for the systems taken for study.

Table 1 Experimental and mathematical ultrasonic velocity of aqueous solutions of PEG 200

Mole fraction	Expt. values	Nomoto's model	Van deal & Van geel model	Impedance model	Rao's specific model
0.0009	1511.43	1394.93	1496.83	1503.53	1397.15
0.0027	1516.38	1396.08	1484.78	1503.67	1397.27
0.0045	1523.86	1397.19	1473.10	1503.81	1397.38
0.0063	1532.69	1398.27	1461.78	1503.95	1397.50
0.0080	1539.70	1399.31	1450.79	1504.10	1397.61
0.0098	1548.06	1400.31	1440.12	1504.24	1397.72
0.0116	1556.52	1401.29	1429.75	1504.38	1397.84
0.0133	1568.38	1402.23	1419.68	1504.52	1397.95
0.0151	1570.37	1403.15	1409.89	1504.65	1398.06
0.0168	1576.39	1404.04	1400.37	1504.79	1398.17

Table 2 Percentage of deviation of ultrasonic velocity of aqueous solutions of PEG 200

Mole fraction	Nomoto's model	Van deal & Van geel model	Impedance model	Rao's specific model
0.0009	7.7077	0.9662	0.5228	7.5609
0.0027	7.9335	2.0839	0.8381	7.8549
0.0045	8.3124	3.3312	1.3158	8.3001
0.0063	8.7703	4.6267	1.8749	8.8206
0.0080	9.1178	5.7745	2.3122	9.2282
0.0098	9.5440	6.9728	2.8309	9.7113
0.0116	9.9728	8.1439	3.3499	10.195
0.0133	10.594	9.4807	4.0719	10.865
0.0151	10.646	10.211	4.1849	10.975
0.0168	10.932	11.167	4.5416	11.301
APD	9.3534	6.2765	2.5843	9.4815

3.2 Viscosity Deviation

The viscosity of binary liquid mixtures based on zero adjustable parameter is computed using standard relations [5]. The theoretical values of viscosity for PEG 200 in water calculated by using various theoretical models (Kendall & Monroe, Arrhenius, Frenkel and Hind) are compared with experimental value (Table 3). From this study, it is observed that (Table 4) kendall and Monroe method is the most suited mathematical method for estimating relative viscosity for the systems taken for analysis.

Table 3 Experimental and mathematical viscosity of aqueous solutions of PEG 200

Mole fraction	Expt. values	Kendall & Monroe model	Arrhenius model	Frenkel model	Hind-Ubbelohde model
0.0009	0.88	1.0259	1.0100	1.0252	1.0506
0.0027	1.00	1.0381	1.0128	1.0359	1.1115
0.0045	1.04	1.0504	1.0156	1.0465	1.1722
0.0063	1.10	1.0627	1.0184	1.0572	1.2327
0.0080	1.13	1.0751	1.0212	1.0680	1.2930
0.0098	1.18	1.0876	1.0239	1.0788	1.3531
0.0116	1.26	1.1001	1.0267	1.0897	1.4121
0.0133	1.37	1.1126	1.0294	1.1006	1.4726
0.0151	1.39	1.1252	1.0322	1.1116	1.5321
0.0168	1.46	1.1378	1.0350	1.1227	1.5913

Table 4 Percentage of deviation of viscosity of aqueous solutions of PEG 200

Mole fraction	Kendall & Monroe model	Arrhenius model	Frenkel model	Hind-Ubbelohde model
0.0009	−16.5789	−14.7762	−16.508	−19.381
0.0027	−3.8125	−1.2816	−3.5878	−11.149
0.0045	−1.0009	2.3462	−0.6282	−12.713
0.0063	3.3868	7.4200	3.8869	−12.066
0.0080	4.8556	9.6320	5.4857	−14.427
0.0098	7.8329	13.2259	8.5733	−14.669
0.0116	12.6936	18.5152	13.5148	−12.140
0.0133	18.7883	24.8555	19.6607	−7.4905
0.0151	19.0511	25.7376	20.0260	−10.220
0.0168	22.0665	29.1086	23.1039	−8.9924
APD	11	14.69	11.50	−12.33

3.3 *Refractive Index Deviation*

Refractive index is calculated for PEG 200 in water for different molar concentration ranges (0.0009, 0.0027, 0.0045, 0.0063, 0.0080, 0.0098, 0.0116, 0.0133, 0.0151 and 0.0168) at 303 K. The experimental refractive index along with the values calculated mathematically using Newton method, Arago–Biot model, Gladstone–Dale equation and Eyring and John equation for the above systems is reported [2] (Tables 5 and 6).

From the above investigation, it is observed that the average percentage of deviation is more in Eyring and John model and less in Newton model. From this, it may be concluded that Newton relation is the best suitable mathematical method for estimating the refractive index of the liquid mixture in the systems taken for study.

Table 5 Experimental and mathematical refractive index of aqueous solutions of PEG 200

Mole fraction	Expt. values	Newton model	Arago–Biot model	Gladstone model	Eyring–John model
0.0009	1.335	1.3351	1.3350	1.3351	1.3351
0.0027	1.338	1.3373	1.3371	1.3371	1.3371
0.0045	1.340	1.3394	1.3392	1.3392	1.3390
0.0063	1.343	1.3415	1.3412	1.3411	1.3410
0.0080	1.346	1.3434	1.3431	1.3431	1.3427
0.0098	1.349	1.3453	1.3489	1.3449	1.3447
0.0116	1.350	1.3471	1.3467	1.3467	1.3464
0.0133	1.351	1.3489	1.3484	1.3484	1.3481
0.0151	1.353	1.3506	1.3500	1.3500	1.3497
0.0168	1.355	1.3523	1.3517	1.3517	1.3513

Table 6 Percentage of deviation of refractive index of aqueous solutions of PEG 200

Mole Fraction	Newton model	Arago–Biot Model	Gladstone model	Eyring–John Model
0.0009	−0.0094	−0.0056	−0.0057	−0.0040
0.0027	0.0512	0.0617	0.0617	0.0668
0.0045	0.0428	0.0597	0.0597	0.0678
0.0063	0.1143	0.1371	0.1371	0.1481
0.0080	0.1907	0.2190	0.2190	0.2327
0.0098	0.2718	0.3051	0.3051	0.3212
0.0116	0.2096	0.2476	0.2476	0.2659
0.0133	0.1518	0.1943	0.1943	0.2148
0.0151	0.1723	0.2188	0.2188	0.2413
0.0168	0.1967	0.2469	0.2469	0.2713
APD	0.1411	0.1696	0.1696	0.1834

4 Conclusion

In the present investigation, the experimental ultrasonic velocities, viscosities and refractive indices of binary mixtures for PEG 200 in water are determined as a function of concentration at 303 K. The experimental values are compared with the values obtained through various mathematical models and the most suited mathematical model is analysed.

Acknowledgements The authors thank SCSVMV University, Enathur, Kanchipuram, for providing facilities to carry out this research work.

References

1. Siddharthan N, Jayakumar S (2016) Theoretical evaluation of sound velocity, viscosity and density of binary liquid system. Int J Chem Sci 14(4):2981–2996
2. Padmanaban R, Venkatramanan K, Girivel K, Kasthuri A, Usharani A, Gayathri A, Vellaichamy R (2017) Mathematical and experimental analysis of ultrasound velocity and refractive index in binary mixtures of pharmaceutically important polymer-PEG 600. Springer Proc Phys 182:709–721
3. Suriya Shihab SK, Babu S, Sreehari Sastry S (2016) Experimental and theoretical studies of speed of sound and viscosity in some liquid mixtures of Methyl Benzoate: a comparative study. Rasayan J Chem 9(4):641–649
4. Anil Kumar K, Srinivasu Ch (2014) Experimental and theoretical evaluations-excess ultrasonic speeds in binary liquid mixture at different temperatures. Adv Appl Sci Res 5(1):224–230
5. Vikash Singh KK, Singh T, Singh VK (2008) Theoretical predictions of viscosity of binary and ternary liquid mixtures at 298.15 K. Asian J Chem 20(6):4173–4184

Comparative Analysis on Mechanical Properties of Luffa and Coconut Coir Fiber Reinforced Polyester Composites

T. V. Rajamurugan, A. Baskaran, S. Matheswaran, and S. Epriya Lavanya

Abstract The composite materials continuously alternate the conventional materials by possessing important properties. Due to demand of materials with specific characters, emerging of new materials is growing day by day. Luffa and coconut fiber (coir)reinforced polymer composites are more attractive due to their specific properties and biodegradability. In this work, important mechanical properties like tensile, impact and flexural strength are analyzed from the coir fiber and luffa fiber reinforced composites fabricated by compression molding method. These composites are compared on the basis strength, toughness, hardness and density. From the results obtained, the incorporation of luffa and coir fiber reinforced composites has proven to be a better alternate for glass fiber reinforced polymer composite.

Keywords Coir fiber · Luffa fiber · Polyester resin · NFRP

1 Introduction

Many modern technologies require material, with two or more material, which exhibits superior properties then the individuals. These materials are generally cheaper than synthetic reinforced composites. Natural fiber reinforced composites are green composites [1] which are used in automobile industries, since these composites exhibit good impact strength, flexural strength and low density. Several researches have been focussed in this direction. Synthetic fibers are replaced by natural fibers because most of the natural composite materials grows from the forest and agriculture [2, 3]. Coconut coir is a commonly available natural tree in India, and coir is an extract

T. V. Rajamurugan (✉) · A. Baskaran
Department of Mechanical Engineering, Government College of Engineering Srirangam, Tiruchirappalli, Tamil Nadu, India
e-mail: vrajamanu@gmail.com

A. Baskaran
e-mail: basau2006@gmail.com

S. Matheswaran · S. E. Lavanya
Government College of Engineering Srirangam, Tiruchirappalli, Tamil Nadu, India

T. Rajmohan et al. (eds.), *Advances in Materials and Manufacturing Engineering*, Springer Proceedings in Materials 7, https://doi.org/10.1007/978-981-15-6267-9_29

for it is used extensively for manufacturing of ropes, floor mats, thermal insulators and also as fuel due of its huge availability. The botanical name of as sponge gourd is Luffa cylindrica is also a natural resource, and its extract is used as a fiber reinforcement in polymer composite. Researchers are continuously working on natural fiber materials which are used for the manufacture of composite particle boards, thermal insulators and building materials by using that above natural fibers [4, 5]. The incorporation of natural fiber in polymers has been addressed adequately in many literatures. Hestiawan, Jamasri and Kusmono et al. predicted that the bonding characteristics between natural fibers and polymer resins are poor due to hydrophobic nature, and they concluded that the above problem can be rectified by chemical treatment [6, 7]. Sapuan et al. [8] conducted experiments on green composites and concentrated mainly on orientation, size and distribution of fibers. Venkatshwaran et al. [9] conducted one of the mechanical testing such as tensile test, and their banana and sisal fiber reinforced hybrid composites perform poor than that of hybrid composites due to the improper fabrication of the composites. Reddy et al. [10] studied the mechanical properties of coir fiber reinforced composites, and the results show natural fiber reinforced composite material is mainly influenced by fiber characteristics. Since most data in the literature cover only a specific loading fraction of fiber, this work was aimed to develop a novel class of hybrid composite materials covering a small range of fiber weight fraction and length, with the use of two different types of natural fiber that are short coir fiber and luffa mat and study their physical and mechanical behavior.

2 Experimentation

2.1 Preparation of Composite Specimen

Coir fiber has been procured from Agasteeswaram, Kanyakumari district, Tamilnadu, India, and luffa fiber obtained from local sources of Trichy. Polyester resin (P-502) and hardener (MEKP) supplied by Ciba Geigy India Ltd are used to fabricate the specimens. The bark and seed were removed from luffa vegetable carefully, and then, luffa fiber was cut carefully into mat like structure prepared with dimension 140 mm $\times$ 90 mm. Coconut coir fibers were first cut to the short fiber length. Composite mat of uniform thickness was prepared from luffa fiber and short coir fiber of particular short fiber length. Both fibers are reinforced with polyester resin. The polyester resin and hardener are mixed with correct proportion, i.e., the ratio of 10:1. The different hybrid composite sheet was prepared with luffa fiber, coir fiber and luffa + coir fiber as reinforcement and weight fraction. Proper releasing agent and mold release spray were used for smooth removal of the fabricated composite plate. The specimen is allowed to cure at atmospheric temperature for 36 h [11]. After curing, the composite was cut by the hack saw. By using compression molding technique, the three samples were fabricated. Table 1 shows the physical properties of the fibers.

Table 1 Physical properties of fibers

Physical properties	Luffa	Coconut coir
Density (gm/cc)	0.93 ± 0.10	1.40
Young's modulus	166 MPa	4–6 Gpa
Tensile strength in (MPa)	15.75	17.56
Shear strength in (MPa)	13.7	14.57
Flexural strength in (MPa)	34.63	44.89
Impact strength in (KJ/m^2)	280	384.99

2.2 *Mechanical Testing*

The prepared specimens were cut for different tests like tensile, flexural and impact test. The prepared specimens for different tests are shown in Fig. 1.

2.2.1 Tensile Test

The fabricated composites were cut as per ASTM standard D638. The tensile test is carried out by fixing the sample and applying the load until it fractures. The tensile test is conducted on the universal testing machine (UTM) [ASTMD 3039/100KN-±0.5%]. The three types of sample were fabricated, first sample with luffa fibers-NFRP, second sample with coir fiber-NFRP and third sample with luffa + coir-NFRP fibers. The prepared sample for tensile test is presented in Fig. 2. The experiments are carried out for three times, and the mean values are recorded [12].

(a) Luffa Fiber (b) Coir Fiber (c) Luffa+Coir Fiber

Fig. 1 **a–c** Fabricated specimen with different fibers. **a** Luffa Fiber. **b**. Coir Fiber. **c**. Luffa + Coir Fiber

Fig. 2 Tensile test specimen

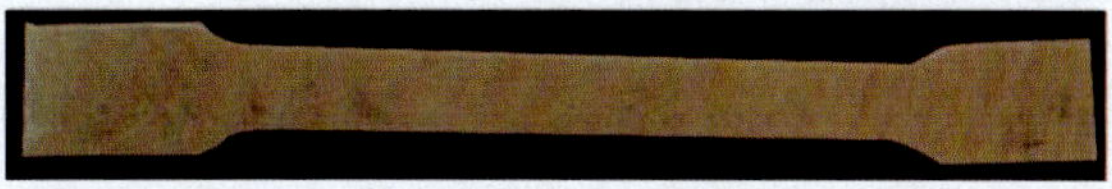

Fig. 3 Specimen before flexural test

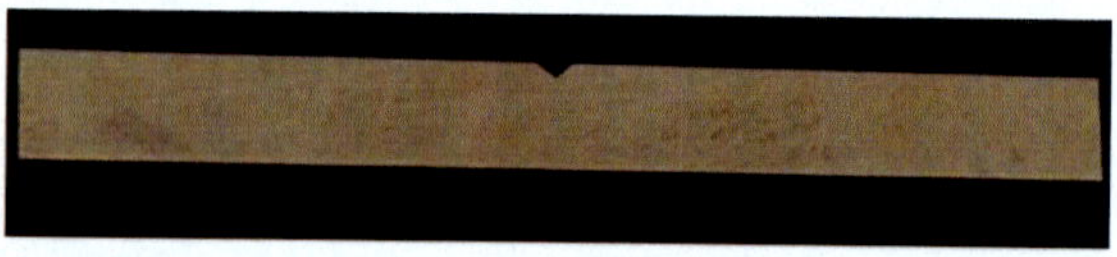

Fig. 4 Specimen before impact test

2.2.2 Flexural Test

Flexural test is conducted using a three point bend setup, and the distance between the two supports is 100 mm. The peak load carrying capacity of the composite laminates is noted. The flexural testing samples are fabricated as per the ASTM D790. The specimens are kept in the UTM machine, and force is applied upto it fractures and breaks, and the tests are repeated for three times, and the average value is taken. The sample used for carrying out the flexural test is presented in Fig. 3.

2.2.3 Impact Test

Impact tests are used for studying the toughness of the fabricated NFRP materials. The impact test sample is fabricated as per the ASTM-A370 standard. From the result obtained from the impact test, the energy required to fracture the material toughness and yield strength of the material can be calculated. The sample specimen used for impact testing is presented in Fig. 4.

3 Results and Discussions

In our study, green fibers are incorporated to the polyester resin to fabricate natural composite materials, and their influence on properties is analyzed. Three specimens were prepared NFRP (luffa), NFRP (coir), NFRP (luffa + coir) for testing.

3.1 *Tensile Properties*

Using the UTM, the three different sample were tested to obtain ultimate tensile strength. The specimen after tensile test is shown in Fig. 5. The results for the specimens are tabulated in Table 1. The maximum value (Peak load) is obtained for the combination of luffa + coir fiber as 1121.136 N which indicates luffa + coir reinforced polyester composites posses better results compared to that of other two composites [13].

3.2 Flexural properties.

The load bearing capacity of the samples is contradictory. The luffa fiber reinforced NFRP composites show an average of 50 N, whereas the coir fiber reinforced NFRP and luffa + coir fiber reinforced NFRP show 61 N and 54 N, respectively. The flexural tests results were shown in Table 2, and the result shows almost all the three composites posses same values. The fractured sample specimen after the test is shown in Fig. 6. The flexural test carried out for various samples along with its peak load, flexural strength and modulus are shown in Table 3.

Fig. 5 Tested sample

Table 2 Tensile test

Composition of Luffa & Coconut	S. no	Cross sectional area (mm^2)	Peak load (N)	Elongation in (mm)	UTS (MPa)
Coconut	1	75.000	922.022	0.870	12.292
	2	75.000	847.162	0.930	11.291
	3	75.000	875.984	1.590	11.684
Luffa	1	75.000	770.625	1.370	10.271
	2	75.000	532.967	1.370	7.102
	3	75.000	759.019	1.310	10.124
Luffa + Coconut	1	75.000	680.422	0.670	9.074
	2	75.000	1121.136	1.130	14.950
	3	75.000	760.520	1.170	10.144

Fig. 6 Tested sample

Table 3 Flexural test

Composition of Luffa & coconut fibers	S. no	Cross sectional area (mm^2)	Peak load in (N)	Flexural strength in (Mpa)	Flexural modulus in (Gpa)
Coconut	1	39.000	83.542	53.553	40.954
	2	39.000	52.572	33.700	2328.080
	3	39.000	48.030	30.788	2571.848
Luffa	1	39.000	46.960	30.103	49.145
	2	39.000	51.983	33.322	255.609
	3	39.000	54.907	35.197	2058.494
Luffa + Coconut	1	39.000	42.144	27.015	2025.641 lePara>
	2	39.000	82.296	52.754	−15.580
	3	39.000	38.416	24.626	1.781

3.2 *Impact Properties*

The impact results for all the three specimens were tabulated in Table 4. From the table, it is inferred that coir fiber reinforced composites exhibit better result compared to that of other two specimens. The specimen after conducting the impact test is shown in Fig. 7.

Table 4 Impact test

NFRP composites	S. no	IZOD impact value IN J fOR 3 mm thickness
Coir fiber reinforced composites	1	1.50
	2	1.20
	3	1.10
Luffa fiber reinforced composites	1	0.20
	2	0.25
	3	0.45
Luffa + coir fiber reinforced composites	1	1.00
	2	1.15
	3	0.75

Fig. 7 After impact test

4 Conclusions

In this investigation, three samples such as luffa fiber reinforced composites, coir reinforced composites and luffa + coirfiber reinforced polyester composites were fabricated, and their mechanical properties were evaluated. Based on experiments, the following points were concluded,

- The mechanical characteristics were observed for coir fiber NFRP, luffa fiber NFRP and a mixture of luffa + coirfiber reinforced polyester composites.
- Among the above, luffa + coirfiber reinforced polyester composites give best results for tensile, flexural and impact.
- A hybrid composite shows good mechanical properties.
- Proper care should be taken while using these composites.
- In future, further research may be concentrated on the traditional machining of the above composites.

References

1. Rajamurugan TV, Shanmugam K, Palanikumar K (2013) Analysis of delamination in drilling glass fiber reinforced polyester composites. Mater Des 45:80–87
2. Vinayagamoorthy R (2020) Mechanical performance of glass- and biofibre-reinforced hybrid composites. Glass Fibre Reinf Polym Compos, Adv Compos 12:1–16
3. Vinayagamoorthy R, Koteshwar TN, Madhav VV, Sai TK, Konda V (2019) Drilling associated parametric investigations on chemically treated natural fiber composite. Mater Today: Proc 16:277–283
4. Oboh IO, Aluyor EO (2009) Luffacylindrica-an emerging cash crop.. Afr J Agric Res 4(8):684–688
5. Vinayagamoorthy R, Konda V, Tonge P, Koteshwar TN, Premkumar M (2019) Surface roughness analysis and optimization during drilling on chemically treated natural fiber composite. Mater Today: Proc 16:567–573
6. Vinayagamoorthy R (2018) A review on the machining of fiber-reinforced polymeric laminates. J Reinf Plast Compos 37:49–59
7. Vazquez-Moreno JM, Sánchez-Hidalgo R, Sanz-Horcajo E, Viña J, Verdejo R, López-Manchado MA (2019) Preparation and mechanical properties of graphene/carbon fiber-reinforced hierarchical polymer composites. J Compos Sci 3(1):30–36
8. Sapuan SM, Harimiand M, Maleque MA (2003) Mechanical properties of epoxy/coconut shell filler particle composites. Arab J Sci Eng 28(2):171–182
9. Venkateshwaran N, Elayaperumal A, Sathiya GK (2012) Prediction of tensile properties of hybrid-natural fiber composites. Compos B Eng 43(2):793–796
10. Reddy TB (2013) Mechanical performance of green coconut fiber/HDPE composites. Int J Eng Res Appl 3(6):1262–1270
11. Vinayagamoorthy R, Manoj IV, Narendra Kumar G, Sai Chand I, Sai Charan Kumar GV, Suneel Kumar K (2018) A central composite design based fuzzy logic for optimization of drilling parameters on natural fiber reinforced composite. J Mechan Sci Technol 32:2011–2020
12. Vinayagamoorthy R, Influence of fibre pretreatments on characteristics of green fabric materials. Polym Polym Compos:096739112094346
13. Vinayagamoorthy R, Venkatakoteswararao G (2019) Synthesis and property analysis of green resin-based composites. J Thermoplast Compos Mater:089270571982878

Experimental Analysis on the Effect of Cu-ZSM5 on the Control of SI Engine Exhaust Emissions

A. Ananthu, P. Anu Nair, Anand K. Raj, Arjun K. Nair, K. G. Gokul, Arun Ajith, and K. S. Amruthunandu

Abstract The CO and HC emissions from S.I. engine are generally controlled by the catalytic converter based on costly noble metals. In our experiment, Cu-ZSM-5 cheaply available in India has been used as the catalyst, and it has been tested under wide range of engine exhaust condition to study the effect of temperature and air fuel ratio on CO and HC emissions. The maximum HC conversion efficiency of about 40% has been achieved at an A/F ratio of 14.87, temperature of 275 °C and space velocity of 37,500/h, while maximum CO conversion efficiency about 30% has been achieved at space velocity of 45,000/h and A/F of 14.87 and at an temperature of 275 °C. The catalyst was tested over a wide range of A/F ratio, and the peak performance was obtained at slightly leaner A/F ratios. For HC, it is about 15.25, and for CO, it is 15.5. The conversion efficiency is not encouraging as per emission standard norms. This is a preliminary approach; however, Cu-ZSM-5 can be processed with very low percentage of noble metals to achieve the conversion efficiency in the range of 80–90% to satisfy emission standard norms.

Keywords EGR · Catalytic converter · Emission

1 Introduction

The techniques for controlling the automotive exhaust emissions can be divided into two broad categories. One is combustion modification method which is adopted before or during the combustion process. They are spark retard, rich fuels/air metering, water injection, exhaust gas re-circulation, etc. Another is the exhaust after treatment, which is provided to the exhaust gases after combustion process is completed. This research work is completely based on the exhaust after treatment

A. Ananthu · A. K. Raj · A. K. Nair · K. G. Gokul · A. Ajith · K. S. Amruthunandu
Department of Mechanical Engineering, GISAT, Payyapady, Kottayam, Kerala 686516, India

P. Anu Nair (✉)
Department of Mechanical Engineering, Mar Baselios Christian College of Engineering & Technology, Kuttikkanam, Peermade, Kerala 685531, India
e-mail: anunairp@mbcpeermade.com

T. Rajmohan et al. (eds.), *Advances in Materials and Manufacturing Engineering*, Springer Proceedings in Materials 7, https://doi.org/10.1007/978-981-15-6267-9_30

method. Experiments were carried out to develop a three-way catalytic converter in which catalyst (Cu-ZSM-5) oxidizes CO and HC and at the same time can also reduce NO_x. Generally, costly noble metals are used as three-way catalyst. The current experimental work was performed using cheaper and easily available Cu-ZSM-5 as catalyst.

Several technical developments have taken place to combact the stringent emission standards world over that are ranging from new engine design for low emissions to fuels specification and advancement in three-way catalytic converter, oxidation catalytic converter [1–3]. The current emission standards would need to incorporate all the technologies together, and metal substrate type catalytic converters are in the market. Metallic substrates possess the technical properties necessary to conform with both current and future legislations, particularly with regard to durability, light off and space requirements. Due to the low heat capacity of the metal in relation to its weight, the amount of heat which the exhaust gas must transfer to the substrate is less in order to reach the light-off temperature [4–6]. The metallic catalytic converter can reach the operating temperature about twice as fact when compared to ceramic converter. This leads to significant decreases in emissions whenever the vehicle is started under cold starting condition. As the thermal conductivity of the metal substrate is higher, there is less thermal shock stress resulting in long life to the converter.

Generally, three ways of catalytic converter are employed in SI engines for reducing the CO, HC and NOx emission. Temperature is an important parameter for better chemical reactions [7]. However, catalyst aging tendencies are higher at very high temperatures [8, 9]. The problem of reduction of NOx emission using catalytic converter has attracted many people, and they have done lot of research on this problem.

2 Preparation of Cu-ZSM-5

ZSM-5 is a kind of synthetic zeolite developed by Mobil oil. ZSM-5 is an abbreviation for Zeolite Socony-Mobil-5 which is an alumino silicate zeolite with a high silica and low aluminum content. One of the most promising catalysts which reduce NO_x with hydrocarbons in oxidizing atmosphere is a copper (Cu) exchange form of the zeolite (Cu-ZSM-5). It is developed by ion-exchange process by which Na^+ ions in zeolite structure are replaced by copper (Cu) cat ions with the aim to change catalytic and adsorptive properties. High purity Cu-ZSM-5 crystals form into agglomerates having high physical strength and attrition resistance. Methods of forming the crystalline powder into agglomerates include the addition of inorganic binder, generally a clay, to the high purity zeolite powder in wet mixture. The blended clay zeolite mixture is extruded into cylindrical type pellets which are subsequently calcined in order to convert the clay to an amorphous binder of considerable mechanical strength.

3 Objective

(1) To develop catalytic converter based on non-noble metal, i.e., Cu-ZSM-5.
(2) To study the catalytic performance of catalyst Cu-ZSM-5 on the exhaust emissions of S.I. engine.

4 Experiment Setup

The schematic diagram of the experimental setup, used in this study, is shown in Fig. 1. The test engine (engine details are given on Appendix I) was coupled with a hydraulic dynamometer to measure the load of the engine. Consumption of air and gasoline by the engine was estimated by measuring air and gasoline flow rates separately. The flow through the catalyst bed was varied by regulating one valve. The volumetric flow rate of the gas through the catalyst bed was measured by a pressure calibrated orifice meter. Pressure difference across the catalyst bed was measured with the help of pressure gauges. Temperature of the exhaust gas was measured by a thermometer placed at a certain location of the catalytic converter. The concentration of CO and HC was measured with the help of a gas analyzer (Model NPM-CH1). The converter used during this experiment has a cylindrical hallow copper perforated casing which is covered by a cylindrical casing of mild steel. The two mild steel pipes were welded on either side of the converter.

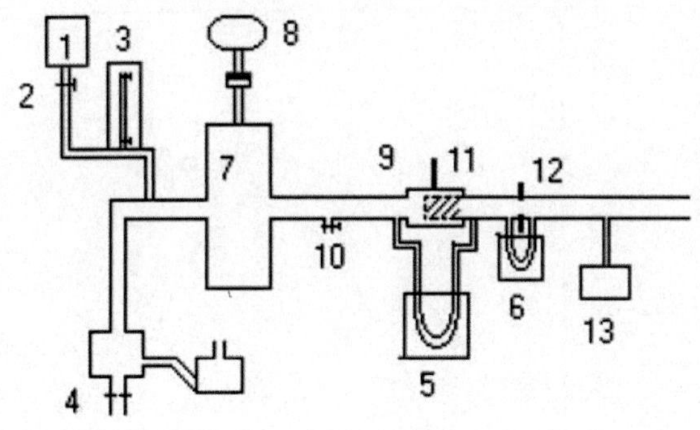

1. Fuel tank
2. Fuel control valve
3. Fuel measuring Burette
4. Air flow measuring orifice meter
5,6. Water manometer
7. Test engine
8. Dynamometer
9. Thermometer
10. Bypass Valve
11. Catalyst Bed
12. Gas flow measuring orifice meter
13. Gas analyzer

Fig. 1 Schematic diagram of the experimental setup

5 Results and Discussion

The results of the set of experiments conducted during the entire course of this work are discussed in this chapter. The results have been shown in the form of graphs which display CO and HC conversion efficiencies as a function of the catalyst temperature and A/F ratio.

5.1 *Effect of Temperature*

The performances have been achieved at different engine speeds, loads, A/F ratios and space velocity of exhaust gas over the catalyst bed. Effect of the catalyst temperature on the performance of Cu-ZSM-5 catalyst in reducing CO and HC has been shown in Figs. 2, 3, 4 and 5. The performances have been achieved at different engine speeds, loads, A/F ratios and space velocity of exhaust gas over the catalyst bed. The catalyst used here exhibits fairly good CO and HC conversion efficiencies at temperature ranges from 250 to 300 °C. The conversion efficiencies increase steadily with the increase in temperature till the peaks are reached and then tend to decrease with further rise in the catalyst temperature but generally at a slower rate than the rate of increase. Increase in reaction velocity is due to increase in specific reaction rate with increase in temperature. And decrease in reaction velocity is due to partial destruction of catalyst by coagulation of catalyst surface at higher temperature.

Fig. 2 Speed 1800 rpm, Load 6 kg

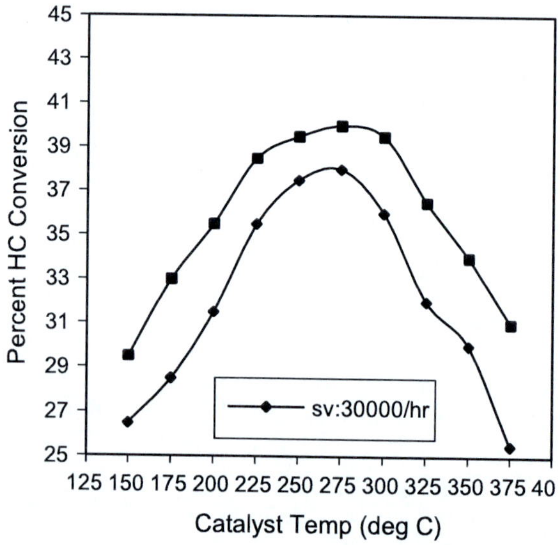

Fig. 3 Speed 1800 rpm, Load 6 kg

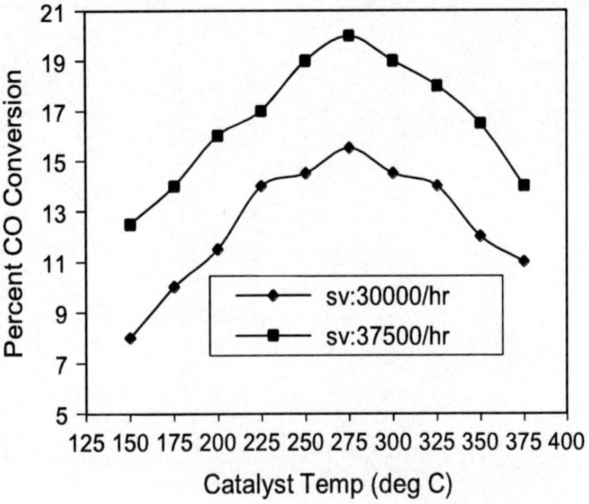

Fig. 4 Speed 1800 rpm, Load 8 kg

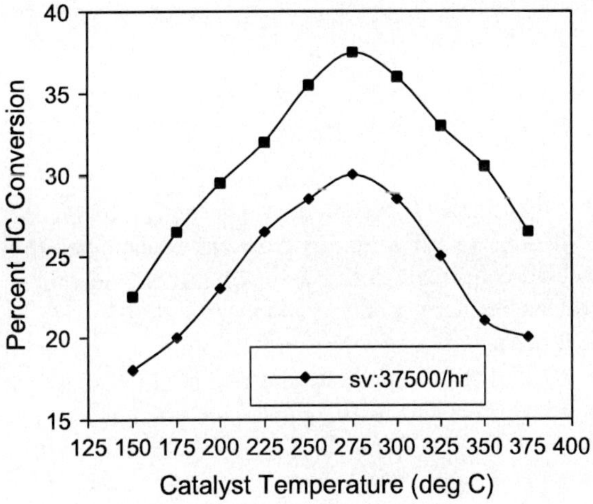

When the SVs are changed, then the temperature for the peak conversion efficiencies is found to be affected at the same A/F ratio. At higher SVs, peak CO and HC conversion efficiencies are observed to occur at higher temperatures.

Within the testing range of this catalyst, maximum HC conversion efficiency about 40% has been achieved at an A/F ratio of 14.87, temperature of 275 °C and space velocity of 37,500/h, while maximum CO conversion efficiency of about 30% has been achieved at space velocity of 45,000/h and A/F of 14.87 s and at a temperature of 275 °C Figs. 4, 5. The overall effective temperature range for HC conversion efficiency is 250–350 °C and that for CO conversion efficiency is 225–300 °C. It is found that the maximum conversion efficiency for HC is 10% higher than that for CO. It is also evident that CO and HC conversion efficiencies do not reach the

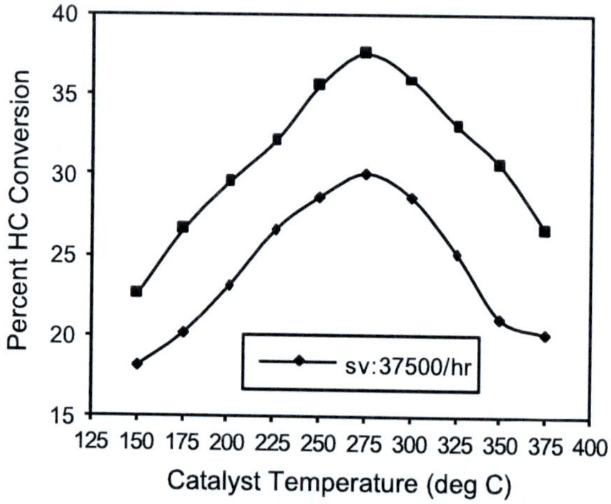

Fig. 5 Speed 1800 rpm, Load 8 kg

respective peaks at the same temperature; rather CO conversion reaches peaks at slightly higher temperatures than that for HC conversion.

5.2 *Effect of A/F Ratio*

It is found from this study that this catalyst maintains high performance through a. wide range of A/F ratio. In particular, the catalyst exhibits good performance for CO and HC conversion during lean A/F mixture operation. Peak CO and HC conversion efficiencies have occurred at an A/F ratio of 15.45, and performances close to the peak are maintained within an A/F ratio range of 14.87–16.23. But beyond A/F ratio of 16.23, both CO and HC conversion efficiencies fall sharply, and at an A/F ratio of 18.33, the CO and HC conversion efficiencies are obtained to be below 30% in some cases Figs. 6 and 7.

6 Conclusion

On the basis of the experimental results obtained using the Cu-ZSM-5 as the catalyst for controlling the exhaust emissions from a SI engine, some conclusions are drawn, and they are mentioned in this chapter. It is found that the conversion efficiencies of CO and HC are influenced primarily by two parameters. They are inlet concentrations of different component of gases, catalyst temperature and the gas flow rate through the catalyst bed. Although the results are obtained in a particular type of engine, they can be reproduced in other types of gasoline engines.

Major conclusion of this experimental investigations is as follows:

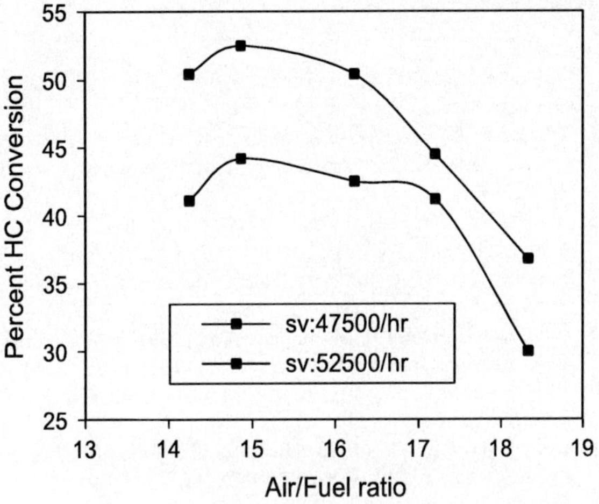

Fig. 6 Temperature at 350 °C

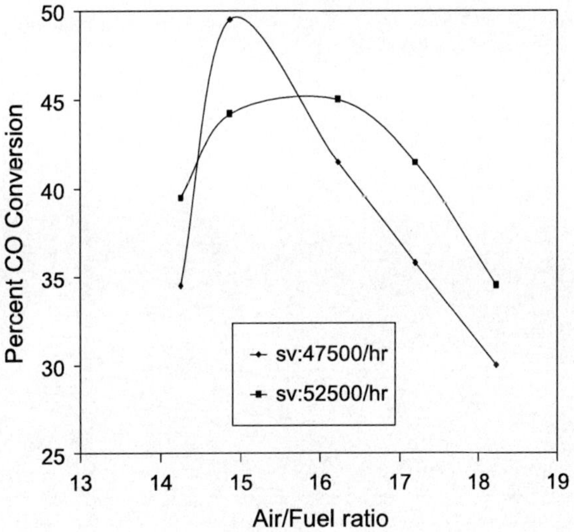

Fig. 7 Temperature at 350 °C

The CO and HC conversion efficiencies increase steadily with increase in temperature, reach the corresponding peaks and then start decreasing as the temperature is further increased. From this study, it has been concluded that the conversion efficiencies tend to increase with change in A/F ratio from rich to stoichiometric, maintain high performance upto a certain range of A/F ratio in the lean side and then tend to decrease for further lean mixtures.

References

1. Taylor KC (1983) Automotive catalytic converters. In: Catalysis science and technology. Springer, Berlin, vol 5, pp 120, 170
2. Hegedus LL, Summers JC, Schlatter JC, Baron K (1979) Poison-resistant catalysis for simultaneous control of hydrocarbons, carbon monoxide and nitrogen oxide emission. J. Catal 56:321–335
3. Yokota K, Muraki H, Fujitani Y (1985) Rh-free three way catalysis for automotive exhaust control. SAE, 850129 94:790–797
4. Summers JC, Silver RG (1990) Catalytic control issues associated with the use of reformulated gasoline. SAE 902072(99):756–567
5. Summers JC, Hiera JP, Williamson WB (1991) Noble metal uses reduction strategies for three-way emission control catalysis. SAE 911732(100):512–518
6. Williamson WB, Summers JC, Skowron JF (1988) Catalyst technologies for future automobiles. SAE 880103(97):41–51
7. Cooper BJ, Truex TJ (1985) Operational criteria affecting the design of thermally stable single-bed three-way catalysts. SAE 850123(94):782–789
8. Gottberg I, Rudquist JE, Backlund O, Wallman S, Maus W, Bruck R, Swars H (1991) New potential exhaust after-treatment technologies for 'clean car' legislation. SAE 910840(99):343–359
9. Harkonen MA, Talvite P (1992) Optimization of metallic TWC behavior and precious metal costs. SAE 920395(101):568–679

Optimization of SI Engine Cycle with Variable Composition and Specific Heat

Akhil Sukumaran, P. Anu Nair, S. Sourabh Gopal, P. S. Sabin, Jithu M. Suredran, and Yadhu P. Mohan

Abstract The combustion modelling and simulation of four-stroke gasoline engine has been discussed in the chapter. The combustion parameters viz., pressure, temperature, and energy are calculated in a step by step analysis by assuming a homogenous single zone within the spark ignition engine cylinder and applying the first law of thermodynamics during the closed part of the cycle. The first law of thermodynamics is written in a suitable form to facilitate the computational procedure involved in the iterative technique to determine the temperature at the end of the step. The variation in gas composition during combustion and expansion is considered. The variation of specific heats with temperature and composition is also considered. Only six gases are considered, i.e., oxygen, nitrogen, carbon dioxide, carbon monoxide, hydrogen, and water vapour, and during the heat release, the composition is varied as required for perfect combustion. The computer code also calculates indicated mean effective pressure, power of the engine, and thermal efficiency. The results show that this program can be quite useful for the design of internal combustion engine parts by considering the peak values of various combustion parameters obtained during the analysis.

Keywords Variable specific heat · SI engine · Composition

1 Introduction

Most of the researchers have developed several theoretical and empirical relations to enhance the spark-ignition engine performance. Engineers always try to work for the continuous improvement of engine performance. The performance of the SI engine depends on the interrelationship between speed, power developed, and SFC at

A. Sukumaran · S. Sourabh Gopal · P. S. Sabin · J. M. Suredran · Y. P. Mohan
Department of Mechanical Engineering, GISAT, Payyapady, Kottayam, Kerala 686516, India

P. Anu Nair (✉)
Department of Mechanical Engineering, Mar Baselios Christian College of Engineering & Technology, Kuttikkanam, Peermade, Kerala 685531, India
e-mail: anunairp@mbcpeermade.com

T. Rajmohan et al. (eds.), *Advances in Materials and Manufacturing Engineering*, Springer Proceedings in Materials 7, https://doi.org/10.1007/978-981-15-6267-9_31

each operating condition within the useful range of load and speed. The theoretical model takes more time to estimate the factors compared to the empirical methods. The mathematical model of the Otto cycle could be analyzed by using the computer program depending on real rights [1–4].

The sequence of events for the ideal constant volume cycle with a gas of constant composition and constant specific heats are two isentropic processes (Expansion and compression) and two constant volume processes (heat reception and heat addition). The present analysis with hydrocarbon-air mixture differs from the ideal cycle in a number of respects viz., the variation in gas composition and the variation of specific heats with temperature and composition. In the present work, a software package has been developed using Microsoft visual C++ computer language. The package is user-friendly which can be run by feeding the engine geometry details, fuel composition, and thermal properties of fuel as input to the package. The desired output will be in the form of plots such as volume, pressure, temperature versus volume step, pressure versus volume, and work done versus volume step.

2 Methodology

The assumptions made in the present analysis are: (a) the gas is homogenous and single phase; (b) the mixture is at thermally equilibrium state. The basic stoichiometric chemical equation for a hydrocarbon–oxygen reaction is given by

$$C_nH_m + \left[n + \frac{m}{4}\right]O_2 + 3.76 \times \left[n + \frac{m}{4}\right]N_2 \Rightarrow nCO_2 + \frac{m}{2}H_2O + 3.76 \times \left[n + \frac{m}{4}\right]N_2 \tag{1}$$

The above stoichiometric equation defines the correct mixture. To allow for mixtures different from the correct mixture, we introduce a parameter called equivalence Φ, which is defined as the ratio of the actual fuel/air ratio to the stoichiometric fuel/air ratio.

For w moles of fuel, the mixture composition will take the form:

$$w\left[C_nH_m + \left[n + \frac{m}{4}\right]\frac{1}{\varphi}O_2\right] + 3.76\left[n + \frac{m}{4}\right]\frac{1}{\varphi}N_2 \tag{2}$$

And the total moles of the mixture will be given by

$$M_A = w\left[1 + 4.76\left(n + \frac{m}{4}\right)\frac{1}{\phi}\right] \tag{3}$$

The magnitude of w is obtained from the ideal gas equation for the mixture:

$$P_A V_A = M R_{\text{mol}} T_A$$

First law of thermodynamics is used for the cycle calculations in the present analysis. For the combustion and expansion strokes, a second set of equations is used to represent thermal equilibrium for the chemical reactions. The derivation of these equations is explained in the following text. The analysis has been performed numerically which is based on the Newton–Raphson method. The cycle is examined step by step as discussed below.

2.1 Adiabatic Compression

Consider a small change in volume from V1 to V2. According to the first law of thermodynamics,

$$dQ - dW = dE$$

where Q, W and E are enthalpy, work done, and the internal energy, respectively. Since the process is adiabatic, $dQ = 0$, and thus

$$dE + dW = 0$$

For a change in pressure from P_1 to P_2, the work dW is approximately given by

$$dw = pdV = \left[\frac{P_1 + P_2}{2}\right][V_2 - V_1]$$

The change in internal energy, dE is given by

$$dE = E_2 - E_1$$

The internal energy of the mixture is a function of the composition of the cumulative gases present in the cylinder contents and gas temperature. The composition is considered to be constant during the compression stroke. For a mixture of gases having N species, the total internal energy is given by [1]

$$E(T) = R_{\text{mol}} \sum_{i=1}^{i=N} \left(\left(\sum_{j=1}^{j=5} U_{i,j} T^j \right) - T \right) \tag{4}$$

where w_i is number of moles of gas, i.

The change in internal energy is given by

$$E_2 - E_1 = E_R(T_2) - E_R(T_1) \tag{5}$$

For a change in volume from V_1 to V_2, the first law of thermodynamics can be written as $E_2 - E_1 + dW = 0$

Substituting for $(E_2 - E_1)$ and dW, we obtain

$$E_R(T_2) - .E_R(T_1) + \left[\frac{P_1 + P_2}{2}\right][V_2 - V_1] \tag{6}$$

To solve for T_2 and P_2, the Newton–Raphson method is used.

2.2 *Adiabatic Combustion at Constant Volume*

During the analysis, it is assumed that the products of combustion comprise of CO_2, CO, H_2O, H_2 O_2 and N_2.

The general chemical reaction for the combustion process will be of the form

$$\begin{aligned} &w\left[C_nH_m + \left[n + \frac{m}{4}\right]\frac{1}{\varphi}O_2\right] + 3.76\left[n + \frac{m}{4}\right]\frac{1}{\varphi}N_2 \\ &\rightarrow [a_1CO_2 + a_2CO + a_3H_2O + a_4H_2 + a_5O_2 + a_6N_2] \end{aligned} \tag{7}$$

To determine the number of moles, a1–a6, the dissociation reactions must be considered. In the present case, there are two such reactions:

$$CO + H_2 \rightarrow H_2 + CO_2$$
$$CO + {}^1\!/_2\, O_2 \rightarrow CO_2$$

The concentrations of CO_2, CO, H_2O, and H2 in the equilibrium mixture can be determined from their partial pressure-equilibrium constant relations. The equilibrium constants can be determined from the Gibbs functions for the nominal reactants and products, using Eq. (8).

$$\ln(K_p) = \sum\left(\frac{v_g(T)}{R_{mol}T}\right)_R - \sum\left(\frac{v_g(T)}{R_{mol}T}\right)_R - \frac{\Delta H_o}{R_{mol}T} \tag{8}$$

The total number of moles of the products per mol of fuel is given by

$$M_p = a_1 + a_2 + a_3 + a_4 + a_5 + a_6$$

To obtain the number of moles, a1–a6, a chemical mass balance is prepared for carbon, hydrogen, and oxygen, and is shown as

Carbon: $n = a_1 + a_2$.

Hydrogen: $m/2 = a_3 + a_4$.

Oxygen: $2\left[n + \frac{m}{4}\right]\frac{1}{\varphi} = 2a_1 + a_2 + a_3 + 2a_5$

The gas equation at state C is given by

$$P_c V_c = w\, R_{\text{mol}} T_c$$

$$\frac{P_c}{M_p} = \frac{w \times R_{\text{mol}} \times T_c}{V_c} \tag{9}$$

Combining the above equations, we obtain

$$f(a) = (a - B) - \frac{m}{2}\left[\frac{n - a}{n + Ca}\right] + \frac{2}{D}\left[\frac{a}{n - a}\right]^2 = 0 \tag{10}$$

where a = a_1,

$$\text{A} = \frac{P_c}{M_p}, B = \frac{2}{\varphi}\left[n + \frac{m}{4}\right] - \left[n + \frac{m}{4}\right], C = \left(\frac{1}{K_{p1}} - 1\right), D = AK_{P_2}^2$$

From the above expressions, the number of moles of all the species is calculated.

The temperature T_C is obtained from the first law of thermodynamics. The composition of combustion products is function of temperature. For adiabatic combustion, $dQ = 0$ and for constant volume combustion, $dW = 0$.

$$dE = E_c - E_B = 0$$

The internal energy E is defined as

$$E = E_o + E(T)$$

At the state B, the internal energy is

$$E_B = E_R = E_{OR} + E_R(T_B)$$

At the state C, the internal energy is

$$E_C = E_P = E_{OP} + E_P(T_C)$$

The internal energies in terms of the composition of the mixtures at B and C and also the specific internal energies are expressed using Eq. (5).

Finally, the change in the internal energy is given by

$$f(E) = [E_{OP} + E_P(T_c)] - [E_{OR} + E_R(T_B)] = 0 \tag{11}$$

The internal energies E_{OR}, $E_R(T_B)$ are fixed since they refer to the reactants at the temperature, T_B. The internal energies, E_{oP} and $E_p(T_c)$ will depend on the composition of the products which in turn depend on T_C. The internal energies are calculated using Eq. (4). Equation (11) can be solved numerically by Newton–Raphson method. To calculate the temperature T_C at the end of combustion, the mixture composition must be evaluated first and then the energy balance is performed using first law of thermodynamics. The equilibrium constant K_p can be calculated using the methods outlined already explained. The maximum temperature in the cycle corresponds to T_C which is dependent on the air/fuel ratio, initial temperature, and pressure at the commencement of combustion.

2.3 Adiabatic Expansion

Although the maximum temperature is reached during combustion, the chemical reactions continue to take place during the expansion stoke, and the reason being that the chemical species are in thermodynamic equilibrium. The composition of the gas mixture will, therefore, vary with pressure and temperature. To satisfy the thermodynamic equilibrium, the chemical reactions for water–gas and carbon dioxide, namely,

$$CO + H_2O \rightarrow H_2 + CO_2$$
$$CO + {}^1\!/_2\, O_2 \rightarrow CO_2$$

are continuously taking place. The method of analysis for the determination of the composition given for the combustion process is, therefore, the same as that for the expansion process. If a step from V_1 to V_2 is considered, then the pressure and temperature will change from P_1, T_1 to P_2, T_2.

These parameters are related by

$$P_1 V_1 = M_1 R_{\text{mol}} T_1 \text{and} P_2 V_2 = M_2 R_{\text{mol}} T_2$$

and the number of moles at 1 and 2 will be

$$M_1 = w[a_1 + a_2 + a_3 + a_4 + a_5 + a_6]T_2$$

$$M_2 = w[\mathrm{a}_1 + \mathrm{a}_2 + \mathrm{a}_3 + \mathrm{a}_4 + \mathrm{a}_5 + \mathrm{a}_6]T_2$$

Hence, if T_2 is known, number of moles for different species (a1–a6) can be determined. To obtain $T_{2,}$, the first law of thermodynamics is applied,

$$dQ - dW = dE.$$

For an adiabatic process, $dQ = 0$, and for a small step V_1 to $V_{2,}$ work done is approximately given by

$$dw = pdV = \left[\frac{P_1 + P_2}{2}\right][V_2 - V_1]$$

Now, $E = E_P = E_{op} + E_p(T)$.

The internal energies can be calculated as a polynomial function of temperature from Eq. (4). The number of moles a1–a6 is different at each temperature. At the initial temperature T_1, the internal energies E_{op1} and $E_p(T_1)$ are fixed. However, E_{op2} will not necessarily be equal to E_{op1} due to the change in composition of the mixture. The first law of thermodynamics then assumes the following form

$$f(E) = \left(E_{op_2} + E_p(T_2)\right) - \left(E_{op}1 + E_p(T_1)\right) + \left[\frac{P_1 + P_2}{2}\right][V_2 - V_1] = 0 \quad (12)$$

The above expression can be solved numerically by Newton–Raphson method.

It may be noted that the calculation involves the examination of T_2 and mixture composition followed by the accuracy check using the first law of thermodynamics. The important difference between the expression for adiabatic combustion at constant volume and adiabatic expansion is that in the later case, the work done by the combustion gases is added.

3 Results and Discussion

The results are discussed in respect of two cases. In the first case, the code has been executed for "KIRLOSKAR vertical 4-stroke, single acting, totally enclosed spark ignition enginem" considering engine specifications in the analysis. In the second case, the data has been taken from literature [5]. Figure 1 shows the total trapped volume versus the step volume; the plot gives the idea as to how the volume increment is taken for the analysis. It clearly shows that the volume increment is constant during the entire cycle. Figure 2 shows the variation of pressure with the cumulative volume inside the cylinder. There is a sudden rise in the pressure during constant volume process because of ideal cycle assumption. Figure 3 shows the variation of temperature with the cumulative volume inside the cylinder. Since the

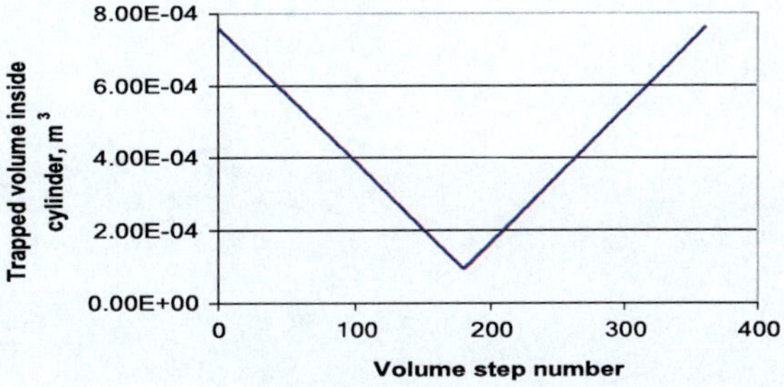

Fig. 1 Total tapped volume versus step volume

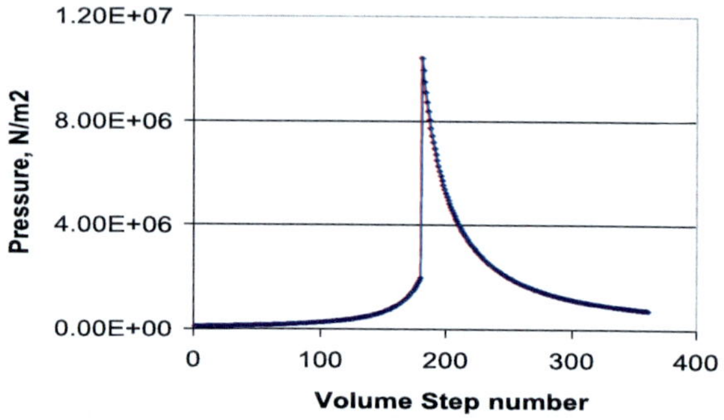

Fig. 2 Pressure versus step volume

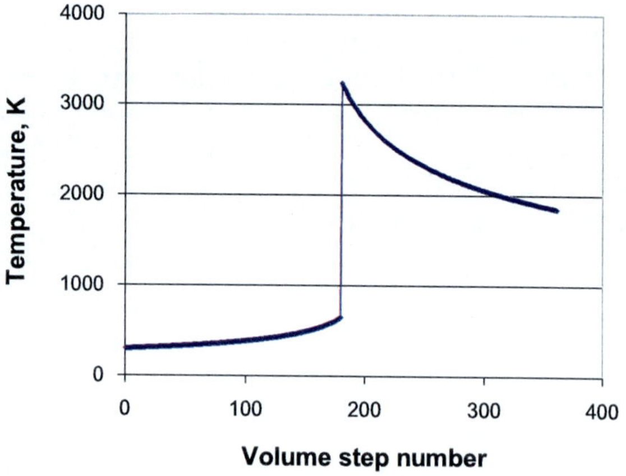

Fig. 3 Temperature versus step volume

ideal cycle assumption is taken in the analysis, there is a sudden increase in the temperature during constant volume process. Figure 4 shows the work output for

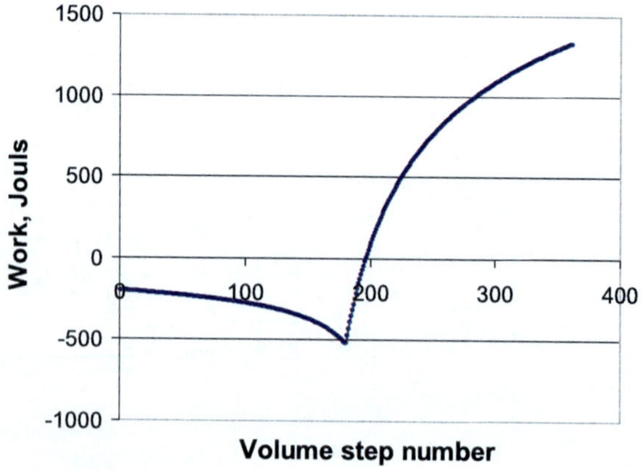

Fig. 4 Workdone versus step volume

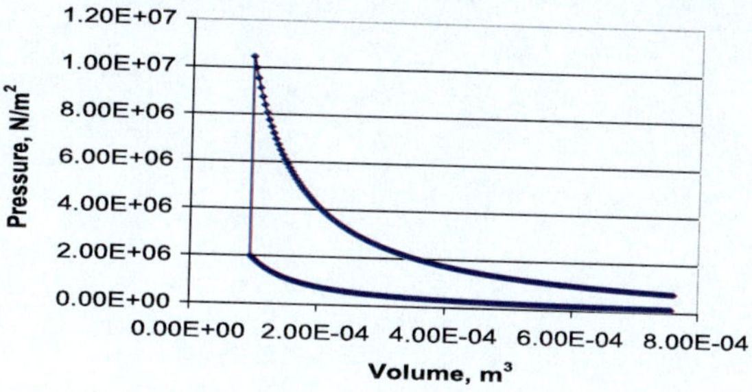

Fig. 5 PV diagram

every volume step. Integrating area under the curve gives total work done in a cycle. Since work is received during compression stroke, the curve is on the negative side.

4 Conclusion

In the present chapter, a computer code has been developed to analyze and simulate four-stroke spark ignition engine cycle. By considering the peak values in the analysis, the design of engine can be modified accordingly. This problem can be extended further by adopting pure real cycle which includes heat losses to the cylinder walls, heat loss to the exhaust, and multi-zone modelling of the combustion gases with chemical kinetics. The results are validated with the data pertaining to "vertical 4-stroke, single acting, totally enclosed spark ignition engine" and it is found that the results are in good harmony.

Reference

1. Öğüçlü Ö (1998) Thermodynamics model of the cycle of SI engine. Msc thesis, Graduate School of Natural and Applied Sciences of Dokuz Eylül University, Đzmir, pp 1–9
2. Bayraktar H, Durgun O (2003) Mathematical modeling of SI engine cycles. Energy Sources 25:439–445
3. Nişli S (1982) Development of a digital computer program for the power cycle and pollutant formation in spark ignited engines. Msc thesis, Metu Graduate School of Natural and Applied Sciences, Ankara, pp 1–10
4. Etiz U (1994) Simulation of the combustion process in A SI engine on a personal computer. MSc thesis, Metu Graduate School of Natural And App. Sciences, Ankara, pp 1–9
5. Baluswamy N (1978) A simple method of heat release analysis on experimental pressure crank-angle diagram by computer simulation. In: 5th national conference on IC engines and combustion, (1998), pp 135–142

Experimental Studies on the Effect of Mixed Metal Oxide $DeNo_x$ Catalyst on the Control of CI Engine Exhaust Emission

Jimin Reji, P. Anu Nair, Aravind B. Nair, S. Anas, Rahul Reji, and A. S. Akshay

Abstract Catalysts have come to occupy a prominent place in the reciprocating engine generated pollutants purifications. With the progressive tightening of the emission norms, there is a spurt in the research activity in this area. For the very survival of the piston engines in the new millennium, the engines have to adapt themselves to the challenges posed. As there is a vast population of single-cylinder diesel engine-powered irrigation pump sets in the rural areas of our country, it was thought fit to make these engines environmentally friendlier. As these engines are very rugged, by the fitment of fairly inexpensive exhaust gas treatment devices like the one discussed in this chapter, the emission levels of major pollutants like oxides of nitrogen and particulate matter can be significantly brought down. There is only a marginal increase in the fuel consumption rate due to the add-on device. The details of a mixed oxides catalyst-coated γ-alumina pellets filled converter system for the exhaust gases of a single cylinder, D.I. Diesel engine system are presented and discussed in this chapter. It is believed that such systems have a role to play in the maintenance of the ecosystem of rural India. Effective reduction of NO_x level will likely require exhaust after treatment devices such as catalytic NO_x converters. The effects of aluminium-doped manganese and copper dioxide catalysts on the NO_x emissions of the engine are presented and discussed in this chapter.

Keywords Egr,catalytic converter · $DeNO_X$ catalyst

1 Introduction

Internal combustion engines produce undesirable emissions during the combustion process. It is the dream of the researcher in the field of IC engines to develop fuel and

J. Reji · A. B. Nair · S. Anas · R. Reji · A. S. Akshay
Department of Mechanical Engineering, GISAT, Payyapady, Kottayam, Kerala 686516, India

P. Anu Nair (✉)
Department of Mechanical Engineering, Mar Baselios Christian College of Engineering & Technology (MBCCET), Kuttikkanam, Peermade, Kerala 685531, India
e-mail: anunairp@mbcpeermade.com

T. Rajmohan et al. (eds.), *Advances in Materials and Manufacturing Engineering*, Springer Proceedings in Materials 7, https://doi.org/10.1007/978-981-15-6267-9_32

engine with less quantity of harmful emissions (i.e., NO_x, HC and CO) generated, and these could be let into the surrounding without a major impact of the environment. The major causes of emission are dissociation of nitrogen, non-stoichiometric combustion, and impurities in fuel and air. The emission of the SI engine is worse than the CI engine.

Several technical developments have taken place to combact the stringent emission standards world over, that are ranging from new engine design for low emissions to fuels specification and advancement in three-way catalytic converter, oxidation catalytic converter, and EGR technology. The current emission standards would need to incorporate all the technologies together and metal substrate-type catalytic converters are in the market. Metallic substrates possess the technical properties necessary to conform with both current and future legislations, particularly with regard to durability, light off, and space requirements. Metal-supported automotive catalysts has also been reported by Hawker [1]. Due to the low heat capacity of the metal in relation to its weight, the amount of heat which the exhaust gas must transfer to the substrate is less in order to reach the light-off temperature. The metallic catalytic converter can reach the operating temperature about twice as fact when compared to ceramic converter. This leads to significant decrease in emissions whenever the vehicle is started under cold starting condition. As the thermal conductivity of the metal substrate is higher, there is less thermal shock stress resulting in long life to the converter. Nair et al. [2] conducted an experiment to study with the use of an exhaust gas recirculation (EGR) system on a given spark ignition engine for studying the performance and emission characteristics.

Generally, three ways of catalytic converter are employed in SI engines for reducing the CO, HC, and NOx emission. Temperature is an important parameter for better chemical reactions [3]. However, catalyst ageing tendencies are higher at very high temperatures [4]. The problem of reduction of NOx emission using catalytic converter has attracted many people, and they have done lot of research on this problem.

NO_x is one of the criteria pollutants as well as precursor to ozone formation. In addition to being an essential ingredient of photochemical smog and precursor to HNO_3, itself an ingredient of acid precipitation and fog, NO_x is the only important gaseous species in the atmosphere that absorbs visible light.

The mechanism of NO formation from atmospheric nitrogen has been given as below:

$$O + N_2 \rightarrow NO + N \tag{1}$$

$$N + O_2 \rightarrow NO + O \tag{2}$$

$$N + OH \rightarrow NO + H \tag{3}$$

Formation of NO_2 has been given as below:

$$NO + HO_2 \rightarrow NO_2 + OH \quad (4)$$

Subsequently, conversion of this NO_2 to NO occurs via

$$NO_2 + O \rightarrow NO + O_2 \quad (5)$$

2 Control of Oxides of Nitrogen

Many theoretical and experimental investigations have well established the fact that the concentration of NO_x in the piston engine exhaust is closely related to the peak cycle temperature and anything done to reduce this temperature will reduce the oxides of nitrogen.

The following are the proven two methods for NO_x control in the engine systems:

(i) Exhaust gas recirculation (EGR)
(ii) Catalytic converters.
(i) **Exhaust gas recirculation (EGR)**
A portion of the exhaust gas is conditioned and recirculated into the cylinder. This method reduces the peak combustion temperature, since the inert exhaust gas serves as a heat sink. This also reduces the quantity of oxygen available for combustion thus reducing temperature and thereby NO_x. It should be noted that most of the NO_x emissions occur during lean mixture limits when exhaust gas recirculation is least effective. The engine power loss is more when more than 15% exhaust gas is recirculated.
(ii) **Catalytic converter**

A catalytic converter system is one of the effective methods to reduce the pollutants from engine exhaust to very low levels. Oxides of nitrogen can be reduced in the presence of reductants like HC, CO, and H_2 over some suitable catalysts. One type of converter, namely pellet-type converter consists of small pellets or tablets made of an inert material such as ceramic alumina or they may be made of metals also which can provide catalyst support and thermal stability. The pellets are contained in a suitably designed enclosure typically made of metals. In a pellet-type converter, the exhaust gas coming from the engine flows up and down through the space between the pellets [5]; therefore, most of the gases comes into contact with the catalyst and gets converted. Therefore, the efficiency of a pellet-type catalytic converter is higher than that of other types of catalytic converters.

3 Role of Metal Oxide-Coated Catalysts and Construction of Catalytic Converter

In this chapter, the study of transition metal oxides of manganese and copper was tested for their activity toward reduction of NO into N_2 and O_2. The interest for the selection of these oxides stems from the fact that transition metals with varied valences have large propensity to attract, adsorb, activate, and finally decompose small molecules into their parent elements. More importantly, they have inherent redox properties which are necessary at this context of the study. The following general reaction scheme can be suggested to operate in the conversion of nitrogen oxides into nitrogen and oxygen.

$$MO + 2\,NO \rightarrow MO\,(NO)_2 \tag{6}$$

$$MO\,(NO)_2 \rightarrow MO + N_2 + O_2 \tag{7}$$

Where,
MO = MnO and CuO.

In this decomposition, the metal oxides are required to have high coordination and unsaturation in order to simultaneously adsorb two NO molecules in close proximity. Otherwise, the decomposition of the adsorbate nitric oxide into N_2 and O_2 would be difficult.

4 Experimental Study

The present experimental setup is for evaluating the performance of the catalytic converter. The tests were conducted on a typical 3.7 kW, single cylinder stationary CI engine. The loading of the engine was done by an electrical generator with resistance bank. The air to the engine was supplied through a surge tank. The air flow rate was measured by an orifice meter. The catalytic converter attachment to the engine exhaust pipe was made by a flanged coupling. The exhaust gas was made to pass through the converter axially. The catalytic converter was attached close to the engine exhaust outlet pipe. The manganese and copper oxides mixture-coated ceramic γ-alumina pellets were packed in a cylindrical container with both ends fixed with stainless steel wire mesh. This container was fitted inside the cylindrical catalytic converter unit by means of flanges. The manganese and copper oxide mixture coating on the ceramic alumina pellets was effected by the following procedure: A 5% aqueous solution of manganese acetate ($Mn(CH_3COO)_2)6H_2O$ and copper nitrate ($Cu(NO_3)_26H_2O$) each weighing 3.75 g was taken in a beaker. A mass of 145 g of the ceramic pellets was dropped in the solution and the contents shaken well and heated on a hot plate to get the dried metal oxide-coated pellets.

The engine was instrumented to measure fuel consumption, brake output, exhaust gas back pressure and emissions of NO_x, particulates, and smoke. While the NO_x level was directly measured in the system, the particulates were collected on filter chapter and weighed in physical balance. The tests were conducted with high speed diesel oil and without the catalytic NO_x converter in place before the trials with the converter were made. The readings were taken for the whole load range of the engine.

5 Results and Discussion

Figure 1 shows the NO_x variation trend. It could be seen that there is significant reduction in the NO_x level varying between 130 and 720 ppm with the converter attachment compared to non-converter operation emissions of 150 ppm to 1040 ppm, respectively, over the load range. The NO_x reduction peaks at overload amounting to 44.44%. Figure 2 indicates the NO_x conversion efficiency trend. The conversion efficiency peaks to about 42% at about 50% load on engine. The corresponding exhaust gas temperature was found to be 530 K. At very light loads, the conversion efficiency was rather low varying between 14 and 35% because the NO_x will not be adsorbed and activated by the catalysts and hence cannot be converted.

The brake-specific fuel consumption did not show any significant variation with the converter in place (Fig. 3), except a small increase at very light loads and close to full load. The increase can be expected due to back pressure build up because of the converter fitment. The brake thermal efficiency variation was not much except there was a maximum drop of 7.7% at full load (Fig. 4). The back pressure was more by about 17% maximum at around 65% load [6]. There was a marginal reduction in the specific particulate emissions of the order of 0.1–0.2 g/kWh over the load range.

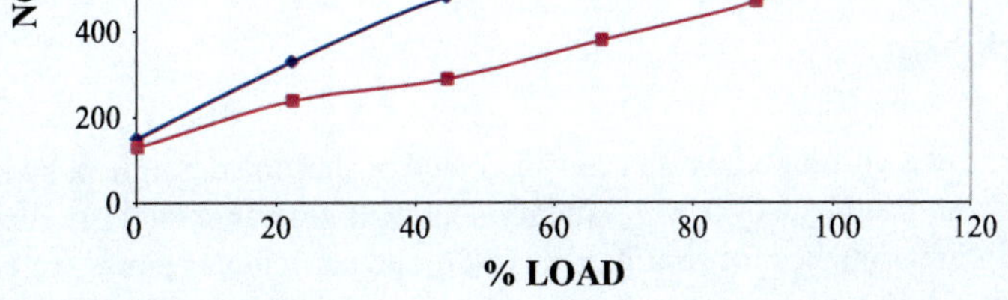

Fig. 1 NOx emission pattern

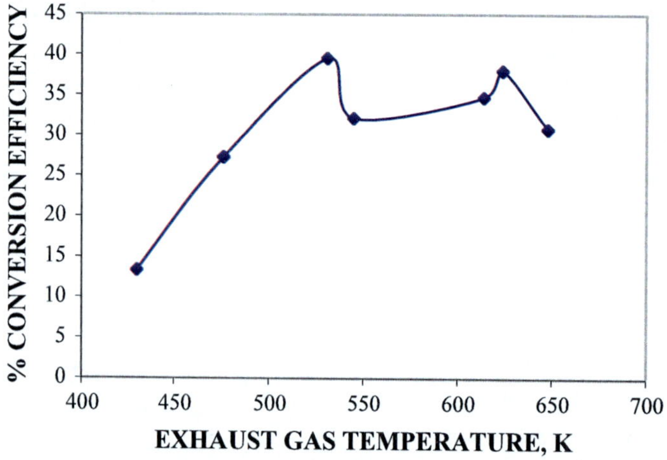

Fig. 2 NOx conversion efficiency pattern with exhaust gas temperature

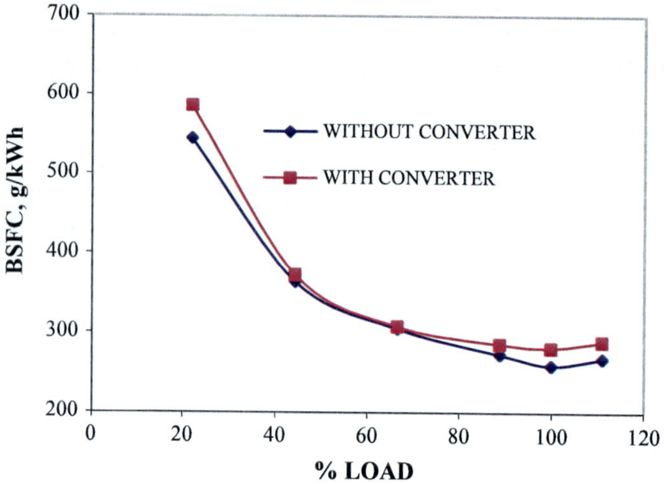

Fig. 3 Brake-specific fuel consumption pattern

6 Conclusion

An initial study with 'mixed oxide catalyst coated γ-alumina ceramic pellets' as the small diesel engine exhaust gas NO_x containment medium has been fairly successful. Further trials are proposed for simultaneous NO_x and particulates emission reduction with modified diesel fuelling as well. With the agricultural sector using small diesel engines in large numbers for purposes like irrigation water pumping, even a nominal 40% NO_x reduction is bound to make a lot of differences in the rural ecology.

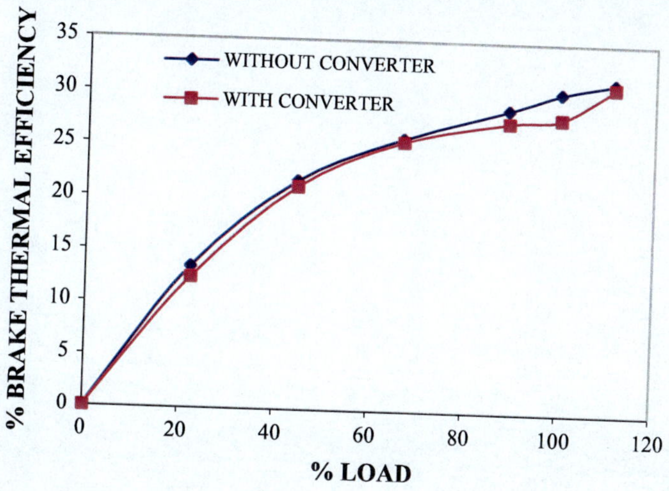

Fig. 4 Brake thermal efficiency characteristics

References

1. PN Hawker, C Jaffray, AJJ Walkins (1988) Metal supported automotive catalyst for use in Europe. SAE:8803173
2. Anu Nair P, Elias S, John V (2015) Performance and emission characteristics of exhaust gas recirculation system and ethanol operation in SI engine. Eur J Adv Eng Technol 2(8):82–86
3. Summers JC, Wilkanson WB, White JJ, Marmangali RE (1987) Performance of copper base metal catalyst in shychometric automotive exhaust. SAE 872132
4. Evans WDJ, Searles RA, Wilkins AJJ (1987) Catalytic exhaust emission control progress in the application of catalyst in Europe. Mech C:34318
5. Adomaitics JR, Smith JE, Achery DE (1980) Improved pelleted catalyst substrates for automotive emissions control. SAE paper 800084
6. Paul Day J, Gocha LS (1988) Impact of catalyst support design parameters on automotive emission. SAE paper 881590

Injection and Exhaust Gas Recirculation Strategies for Reducing Emissions of Cyclohexanol-Diesel Blends in CI Engine

S. Boopathi, J. Ravikumar, R. Devanathan, and S. A. Arokya Anicia

Abstract Cyclohexanol (C_6H_{13}–OH) is an advanced biofuel derived from lignocellulosic biomass that is suitable for compression ignition engine with several properties closer to fossil diesel. This study analyzes the emissions of a direct-injection (DI) diesel engine fueled with cyclohexanol/diesel blends containing 10% (CHX10), 20% (CHX 20), and 30% (CHX30) by volume and an investigate carried out naturally aspirated, exhaust gas recirculation (EGR) 0, 10 and 20% and varying the injection timings 19°, 21° and 23° CA bTDC. According to the experimental results, the CO emissions decreased in CHX20 blend at 21° CA bTDC compared to pure diesel, the HC emissions decreased in CHX30 blend at 21° CA bTDC compared to pure diesel, the NOx emissions lower in 21° CA bTDC at all test blends, and the smoke opacity lower in 23° CA bTDC at all test blends. This investigation concluded that the variation of injection timing and EGR had an extensive effect for reducing emissions characteristics of the diesel engine.

Keywords Cyclohexanol · Diesel engine · Emissions · High carbon alcohols · Advanced biofuel

1 Introduction

Exhausting oil resources and increasing demand of energy steered to a hunt for an alternative fuel which is both ecological and environmentally responsive [1]. Biodiesel is the alternative source and almost each country is preparing a policy on manufacturing and using of biodiesel in its transport division. Song et al. [2] used copper catalyst for the change of biomass-derived cyclohexanone to cyclohexanol. Long et al. [3] succeeded more than 97.74% of guaiacol change with 100% cyclohexanol selectivity in the occurrence of 20% Nickel/Magnesium-oxide catalyst. Yi et al. [4] achieved a lignin-derived mono-phenol to cyclohexanol over Pd/c-Al_2O_3 (Palladium/gamma-Alumina) compound with selectivity up to 98.6% with 100%

S. Boopathi (✉) · J. Ravikumar · R. Devanathan · S. A. A. Anicia
Department of Mechanical Engineering, Jeppiaar Institute of Technology, Kunnam, Sunguvarchatram, Sriperumbudur, Kancheepuram 631604, India
e-mail: boopshare@yahoo.co.in

T. Rajmohan et al. (eds.), *Advances in Materials and Manufacturing Engineering*, Springer Proceedings in Materials 7, https://doi.org/10.1007/978-981-15-6267-9_33

carbon balance by hydrogenation method. Literature study concluded that, many of them tested successfully in diesel engine new lignin derived bio-fuels with dissimilar molecular structures, provided that an optimal of alternative bio-fuels that can be favorably exploited for encouraging engine combustion, performance, and emission appearances. Currently, limited oxygenated aromatic and aliphatic compounds research in diesel engines. In the current study, 30% by volume of a cyclohexanol is chosen to be blended with diesel in a one-cylinder DI diesel engine that is mostly used in the Indian agricultural region [5]. This current work investigated diesel engine emission characteristics by fueled with cyclohexanol/diesel (up to 30% by volume) and under various engine modified operating conditions (EGR rates 0, 10, 20% and injection timing 19°, 21°, 23° CA b TDC).

2 Materials and Methods

2.1 *Test Fuels*

In this study, the mixing of test fuel is prepared by blends cyclohexanol/diesel mixing proportions of 10/70, 20/50, and 30/40 by volume and is stated as CHX10, CHX20, and CHX30, respectively. Cetane number not improved in the blends compared to base diesel. Blends properties tested by The ASTM test method at Polymeric Materials Research Laboratory in Alagappa College, Tamilnadu which is presented in Table 1.

Table 1 Properties of test fuel

Properties	Test method	ULSD	Cyclohexanol	CHX10	CHX20	CHX30
Oxygen (% by wt.)	– –	0	16.12	1.692	3.215	4.832
Calculated cetane index	ASTM D4737	53	–	49.29	46.51	43.71
Kinematic viscosity at 40 °C (mm^2/s)	ASTM D445	3.624	4.73	3.715	3.832	4.076
Low heating value (MJ/kg)	ASTM D240	43.44	38.382	43.104	42.708	42.089
Density (kg/m^3)	ASTMD4052	851	951	871	871	881
Flash point (°C)	ASTM D93	68	68	68	68	68

2.2 Test Engine and Facilities

Tests were conducted in a naturally aspirated, direct-injection, four-stroke, single-cylinder, constant-speed (1500 rpm) diesel engine [6] which is generally used for cultivated purposes to drive pump-set in India. Swinging field electrical dynamometer which contains a 5-kVA AC alternator (220 V, 1500 rpm) attached on bearings and on an unbending border for rocking field type loading is employed. The engine output power is determined by the reaction torque via a load cell-type strain gauge. HC, CO and NOx emissions measured by MRU Delta 1600 L exhaust gas analyzer and smoke opacity measured by AVL 439 opacimeter.

2.3 EGR Setup

EGR system is the effective method to reduce the formation of NOx. Here, in this work, the recycled gas is cooled to 35 °C. EGR valve controlled the EGR rate. The flow rate of exhaust gas is measured by orifice. EGR quantity was determined using Eq. 1.

$$\text{EGR\%} = \left[\frac{(CO_2)_{\text{intake}}}{(CO_2)_{\text{exhaust}}}\right] \times 100 \tag{1}$$

MRU gas analyzer is used to measure the amount of CO_2 in the exhaust. This is succeeded by regulating the control valve to differ the flow rate of the exhaust up until the preferred value of intake of CO_2 is reached. Analogous method was used before the EGR rates are determined in a same engine [5, 6].

2.4 Test Procedure

There are no modifications made on test engine. Experiments were performed under steady-state condition and at rated load of the engine, which corresponds to a brake mean effective pressure (bmep) of 5.3 bar. Emission characteristics of the test engine were recorded at nine operating conditions by progressively increasing the three cold EGR rates (i.e., 0, 10, and 20%).

3 Result and Discussion

3.1 Emission Analysis

CO emissions

Figure 1 shows the variations of CO emissions for all test fuels under natural condition with engine load. Due to high in-cylinder pressure that derived during such load, CO emissions remain low at high loads. With the addition of cyclohexanol, CO emission increases gradually but CO emissions decreased in CHX20 blend at 21° CA bTDC compared to pure diesel due to shortened of ignition delay, the degree of fuel–air mixing, and uniform burning could have enhanced.

HC Emissions

The variation of HC emissions with engine load under naturally aspirated, EGR, and varying injection timings all test fuels was shown in Fig. 2 with addition of cyclohexanol. HC increased and remains high for all blends. HC emissions decreased in CHX30 blend compared to pure diesel at 21° CA bTDC due to shortened of ignition delay, the degree of fuel–air mixing, and uniform burning could have enhanced.

NOx Emissions

The variation of NOx emissions for all test fuels as a function of engine under naturally aspirated, EGR, and varying injection timings in Fig. 3. NOx emission

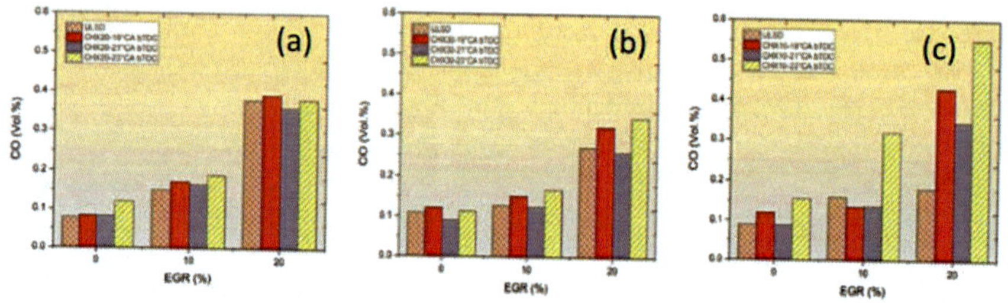

Fig. 1 Effect of EGR on smoke opacity at various injection timings for **a** CHX10, **b** CHX30 and **c** CHX30

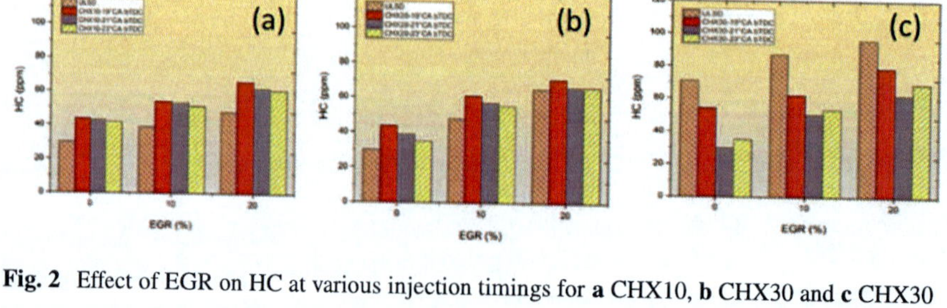

Fig. 2 Effect of EGR on HC at various injection timings for **a** CHX10, **b** CHX30 and **c** CHX30

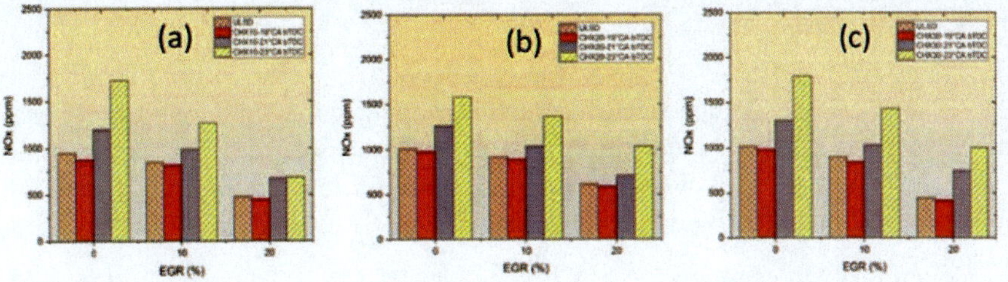

Fig. 3 Effect of EGR on NOx at various injection timings for **a** CHX10, **b** CHX30 and **c** CHX30

decreases with increase in cyclohexanol content in the blends at low/medium load. This is due to high latent heat vaporization and low energy content of cyclohexanol that cause low in-cylinder temperatures which decrease the formation of thermal norms. Even, opposite trends at high engine loads were shown by NOx behavior. Large quantity of fuel enters the combustion chamber to produce the same power output at high engine loads and there is a large influence of higher oxygen content at lower cetane number of cyclohexanol mixture over the cooling effect of cyclohexanol latent heat of vaporization. NO_X emissions lower in 21° CA bTDC at all test blends.

Smoke Opacity

The variations of smoke opacity for all test fuels under naturally aspirated, EGR, and varying injection timings with engine load were shown in Fig. 4. Lignin-derived cyclic oxygenates like benzyl alcohol, guaiacols, furans, anisole, cyclohexanone, cyclohexanane, cyclohexane ethanol, 2-phenyl ethanol, and γ-valerolactone show good soot suppression characters and cyclohexanol is not in option from the expected literature survey. With increasing cyclohexanol content in the mixture, smoke opacity was decreased. Increases oxygen availability even in rich fuel zones prohibiting the formation of soot due to the presence of fuel bound oxygen. Smoke opacity lower in 23° CA bTDC at all test blends.

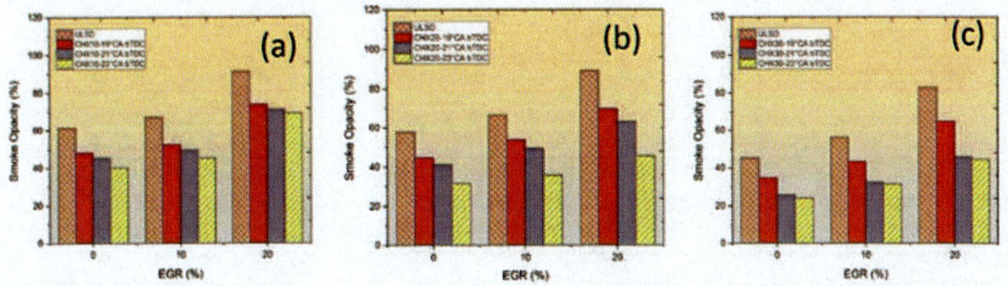

Fig. 4 Effect of EGR on smoke opacity at various injection timings for **a** CHX10, **b** CHX30, and **c** CHX30

4 Conclusion

This investigation concluded that the variation of injection timing and EGR had an extensive effect for reducing emissions characteristics of the diesel engine and the engine characteristics were summarized based on the experimental results in comparison with ULSD operation of the engine:

1. CO emissions decreased in CHX20 blend at 21° CA bTDC compared to pure diesel.
2. HC emissions decreased in CHX30 blend at 21° CA bTDC compared to pure diesel.
3. NOx emissions lower in 21° CA bTDC at all test blends.
4. Smoke opacity lower in 23° CA bTDC at all test blends.

References

1. Ziejewski M, Goettier H I(1986) Influence of vegetable oil based alternate fuels on residue deposits and components wear in a diesel engine. SAE technical paper, 860302
2. Song Z, Ren D, Wang T, Jin F, Jiang Q, Huo Z (2015) Highly selective hydro thermal production of cyclohexanol from biomass-derived cyclohexanone over Cupowder. Catal Today
3. Long J, Shu S, Wu Q, Yuan Z, Wang T, Xu Y (2015) Selective cyclohexanol production from the renewable lignin derived phenolic chemicals catalyzed by Ni/MgO. Energy Convers Manage 105:570–577
4. Yi J, Luo Y, He T, Jiang Z, Li J, Hu C (2016) High efficient hydrogenation of lignin derived mono-phenols to cyclohexanols over Pd/c-Al2O3 under mild conditions. Catalysts. 6(1):12
5. Rajesh Kumar B, Saravanan S (2016) Effects of iso-butanol/diesel and n-pentanol/diesel blends on performance and emissions of a DI diesel engine under premixed LTC (low temperature combustion) mode. Fuel 170:49–59
6. Ravikumar J, Saravanan S (2016) Performance and emission analysis on blends of diesel, restaurant yellow grease and n-pentanol in direct-injection diesel engine. Environ Sci Pollut Res 24(6):5381–5390

A Brief Review: Study on Mechanical Properties of Polycarbonate with Different Nanofiller Materials

Prudhvi Raj and Ravi Kumar

Abstract Recently, the use of polycarbonate has been studied intensively and many researchers started producing polycarbonate-based composites by introducing different carbon-based nanofillers. Significant improvement in mechanical properties of polycarbonate has been reported due to the inclusion of even small amount of nanofillers. Enhancement of properties through the inclusion of nanofillers into polycarbonate matrix will be useful for both application-based investigations and scientific studies. Current study provides some insights of different research papers based on polycarbonate composites. Present literature shows that inclusion of small quantity of nanofiller in to polycarbonates can significantly enhance various mechanical properties.

Keywords Polycarbonate · Composites · Nanofiller · Mechanical properties

Abbreviations

PC	Polycarbonate
ABS	Acrylonitrile-butadiene-styrene
PEI	Polyetherimide
EMA	Ethylene-methyl acrylate
PBT	Polybutylene terephthalate
EPC	Ethylene-propylene copolymer
MWCNT	Multiwalled carbon nanotube
SWCNT	Single wall carbon nanotube
GO	Graphene oxide
RGO	Reduced graphene oxide

P. Raj · R. Kumar (✉)
School of Mechanical Engineering, Lovely Professional University, Phagwara, Punjab 144411, India
e-mail: ravi.18747@lpu.co.in

T. Rajmohan et al. (eds.), *Advances in Materials and Manufacturing Engineering*, Springer Proceedings in Materials 7, https://doi.org/10.1007/978-981-15-6267-9_34

1 Introduction

1.1 Polycarbonate

Polycarbonate is a thermoplastic material which is strong and unbreakable. Polycarbonate contains polymers with carbonate groups and it is produced by reacting bisphenol A with phosgene. Polycarbonate possesses good mechanical, thermal, and electrical properties. Due to these properties, polycarbonates are used for industrial applications like window shades, CD's and DVD's, eye glasses, roof sheets, aeronautical, and military applications [1]. Polycarbonate has high impact strength, but it is less scratch resistance, scratches can be formed easily on applications like eye wears, automotive parts, so they are coated with scratch resistance coatings. The demand for production of polycarbonate is increasing all over the world due to its properties and applications [2]. Polycarbonate is transparent and lightweight compared to other thermoplastics. The properties of the polycarbonate can be further more increased by introduction of different nanoparticles in to polymer matrix [3]. The mechanical properties, e.g. tensile strength, impact resistance, and ultimate tensile strength can be enhanced by introducing nanofillers like graphene, graphene oxide, reduced graphene oxide, carbon nanotubes, and many other nanofillers. The improved properties can be useful for different applications [4].

2 Literature Review

Polycarbonate has good mechanical, electrical, and thermal properties. We can further more increase the properties of polycarbonate by incorporating polycarbonate matrix with different nanofillers. Nanofillers like CNT, graphene, carbon nanofillers, and other fillers can be reinforced with polymer matrix to improve mechanical properties [5]. Nanofiller like graphene can be used to improve mechanical properties of polycarbonate. Graphene is thinnest material available in the world. Graphene alone has good mechanical, electrical, and thermal properties. Incorporating of very small amount of graphene into polymer matrix increases mechanical, electrical, and thermal properties. Graphene is reinforced with polycarbonate matrix by using solution blending technique to prepare PC/G composite and its properties are investigated. By introducing graphene in to polymer matrix, the mechanical and electrical properties of the composites are boosted [6]. Carbon nanotubes are allotropes of carbon. Carbon nanotubes are of two types (1) SWCNT and (2) MWCNT. Single walled nanotubes have diameter less than 1 nanometer (nm), where as multiwalled nanotubes have diameters reaching up to 100 nm (nm). Both single walled and multiwalled nanotubes have high mechanical strength and electrical conductivity. Introducing small amount of single walled or multiwalled nanotubes in to polymer matrix increases the properties of that composite. Different types of MWCNT are reinforced with polycarbonate and investigated on mechanical, electrical, and glass transition behaviour of the composites. It is found that by reinforcing polycarbonate with MWCNT improves the mechanical properties but there is a slight decrease in the glass transition temperature of composites [7]. SWCNT are reinforced with polycarbonate to investigate on mechanical and electrical conductivity of the composite. The mechanical properties of the composites increased by addition of very less amount of SWCNT in polycarbonate matrix [8]. Nanofiller particles like ZnO, sic, TiO_2, ZrO_2, Al_2O_3 can be reinforced with polymer matrix to obtain better mechanical properties. These nanofillers are called nanooxides, and these fillers can be used as reinforcing agents in polymer matrix. Nanofiller like nanoclay can also use as reinforcing agent in polymer matrix, but nanoclay in some cases reduces the tensile strength of polymer composites and improves Young's modulus. Polycarbonate reinforced with TiO_2 and investigated on mechanical properties of composite. The mechanical properties like tensile strength and hardness of the nanocomposite can be increased by increasing TiO_2 content in the polymer matrix [9]. Polycarbonate is reinforced with glass fibre and its mechanical properties are investigated. Glass fibre with 10% and 20% taken as compositions in polycarbonate matrix and mechanical properties like tensile strength, flexural strength, and hardness is investigated. The results revealed that glass fibre with 10 wt.% has shown increase in tensile strength up to 253 N/m^2 compared to 247 N/m^2 of 20% glass fibre. Flexural strength has also increased to 160 N/m^2 compared to 130 N/m^2 of 20% glass fibre. From the results, it has been observed that increase in glass fibre content decreases the strength of the composite [10]. Polycarbonate reinforced with 3 wt.% CNT and the effect of recycling on the structure and properties of the composites are investigated. Polycarbonate with 3 wt.% CNT has Young's modulus of 6000 Mpa compared to neat PC of 2600 Mpa. Recycling of the composite up to 20 times is carried and the results were compared between neat PC and CNT/PC composites. After 20 recycles, it was found that the Young's modulus of CNT/PC is 5000 Mpa which is still higher than neat PC, and it is still suitable for many applications, whereas the tensile strength of the composite is decreased to 18 Mpa with less than neat PC of 51 Mpa., and the impact strength of the CNT/PC reduced to 60 J/m. From the results, it has been observed that recycling the composites affect the strength of the composites [11]. Polycarbonate was grafted on to functionalized graphene nanosheets, and properties are compared with polycarbonate graphene nanosheets of simple mixing. It has been observed that PC-g-MGNS of 3 wt.% has increased tensile strength up to 81 MPa and Young's modulus up to 2270 MPa which is 20.5 and 22.7% higher than that of conventional PC-s-GNS [12]. Polycarbonate reinforced TiO_2 nanocomposite films are prepared and its mechanical properties like tensile strength, hardness, and elastic modulus are investigated. The tensile strength of the composite increased up to 18% by addition of only 2 wt.% of TiO_2. The elastic modulus also increased by addition of only 0.8 wt.% of TiO_2 [13]. Polycarbonate is reinforced with multiwalled carbon nanotubes and its mechanical, and thermal properties are investigated and it was found that by addition of only 10 wt.% of multiwalled carbon nanotube in to polycarbonate matrix shows decrease up to 35% in tensile strength and 47% decrease in tensile strength of the composite.

Addition of 1–5 wt.% of MWCNT in polycarbonate shows good mechanical properties, but addition of 10 wt.% shows decrease in mechanical properties like tensile strength and bending strength (Table 1).

It has been observed from the above discussion that introducing small amount of nanofiller in polycarbonate matrix has improved mechanical properties like tensile strength, Young's modulus, impact resistance, and flexural strength of the composite.

3 Discussion and Conclusion

The present literature depicts study on improving mechanical properties of polycarbonate using different nanofillers. The nanofillers mainly used in this literature works are graphene, carbon nanotubes, clays, and nanooxides. By introduction of nanofillers in to polycarbonate matrix improved the properties like tensile strength, Young's modulus, and impact strength. Above study shows maximum tensile strength of 253 GPa when polycarbonate reinforced with glass fibre. Also, it depicts maximum Young's modulus of 6000 Mpa when polycarbonate reinforced with 3 wt.% CNT. Since PC shows different characteristics for different applications, there is a substantial room for improvement in various mechanical, optical as well as electrical properties.

Table 1 Mechanical properties of polycarbonate reinforced with different nanofillers

Matrix	Reinforcement	Tensile strength (MPa)	Young's modulus (MPa)	Impact (J/m)	References
PC/ABS	4 wt.% T-RGO	+135	–	–	[14]
PC	2 wt.% WO_3	+53.6	+1527	–	[15]
PC	3 wt.% ZrO_2	+59	–	+1240	[16]
PC	30 wt.% SGF	+107.4	+2966	−31.5	[17]
PC	2 wt.% Al_2O_3	+73	+3000	–	[18]
PC	7 wt.% GNP	+70	–	−20	[19]
PC	0.5 wt.% ZNO	+61.0	+1957	–	[20]
PC	Glass fibre	+138	+11.0	+218	[21]
PP/PC	1 wt.% GO	–	+2516	+40	[22]
PC/PLA	30 wt.% carbon fibre	+114.6	–	–	[23]
PC/PEI	Glass fibre	190.82	+ 6.25	−153.61	[24]
PC/EMA	10 phr Ir-GO	70	+90	–	[25]
PC/PBT	4 wt.% c-MWCNT	+65.9	+2116	–	[26]

(continued)

Table 1 (continued)

Matrix	Reinforcement	Tensile strength (MPa)	Young's modulus (MPa)	Impact (J/m)	References
PC/EPC	1 wt.% MWCNT	57.39	+1320	–	[27]
PC/ABS	10 phr MICA	−54.8	+1162	−582	[28]
PC	15 vol% SCF	+95	+2500	–	[29]
PC/ABS	3 wt.% GNP	+58.37	+2820		[30]
PC	0.5 phr organ clay	+65	+1600		[31]
PC	0.1 wt.% SWCNT	+26		+0.05	[32]
PC	5 vol% CNT/CU	467	111		[33]
PBT/PC	0.3 wt.% MWCNT	+85	+4.0	+72	[34]

References

1. Bagotia N, Choudhary V, Sharma DK (2018a) A review on the mechanical, electrical and EMI shielding properties of carbon nanotubes and graphene reinforced polycarbonate nanocomposites. Polym Adv Technol 29(6):1547–1567
2. Higgins BA, Brittain WJ (2005) Polycarbonate carbon nanofiber composites. Eur Polymer J 41(5):889–893
3. Krishnan AV, Stathis P, Permuth SF, Tokes L, Feldman D (1993) Bisphenol-A: an estrogenic substance is released from polycarbonate flasks during autoclaving. Endocrinology 132(6):2279–2286
4. Bagotia N, Choudhary V, Sharma DK (2017) Studies on toughened polycarbonate/multiwalled carbon nanotubes nanocomposites. Compos B Eng 124:101–110
5. Marquis DM, Guillaume É, Chivas-Joly C (2011) Properties of nanofillers in polymer, nanocomposites and polymers with analytical methods, Cuppoletti J (ed). ISBN: 978-953-307-352-1
6. Lago E, Toth PS, Pugliese G, Pellegrini V, Bonaccorso F (2016) Solution blending preparation of polycarbonate/graphene composite: boosting the mechanical and electrical properties. RSC Adv 6(100):97931–97940
7. Castillo FY, Socher R, Krause B, Headrick R, Grady BP, Prada-Silvy R, Pötschke P (2011) Electrical, mechanical, and glass transition behavior of polycarbonate-based nanocomposites with different multi-walled carbon nanotubes. Polymer 52(17):3835–3845
8. Hornbostel, B., Pötschke, P., Kotz, J., & Roth, S. (2006). Single-walled carbon nanotubes/polycarbonate composites: Basic electrical and mechanical properties. Physica Status Solidi (b) 243(13):3445–3451
9. Jahantigh F, Nazirzadeh M (2017) Synthesis and characterization of TiO_2 nanoparticles with polycarbonate and investigation of its mechanical properties. Int J Nanosci 16(05n06):1750012
10. Srimurugan R, Ramnath V, Ananthapadmanaban D, Ramanan N (2017) Glass fibre reinforced polycarbonate for automobile chassis application. JCHPS Special Issue 7
11. Zhang J, Panwar A, Bello D, Isaacs JA, Jozokos T, Mead J (2018) The effects of recycling on the structure and properties of carbon nanotube-filled polycarbonate. Polym Eng Sci 58(8):1278–1284

12. Yoon SH, Jung HT (2017) Grafting polycarbonate onto graphene nanosheets: synthesis and characterization of high performance polycarbonate–graphene nanocomposites for ESD/EMI applications. RSC Adv 7(73):45902–45910
13. Shahbazi N, Momeni A (2016) Mechanical properties of polycarbonate-TiO2 nanocomposite film. Indian J Pure Appl Phys (IJPAP) 54(4):241–250
14. Li Y, Wang A, Meng L, Jiang N (2018) Preparation of graphene and its application in polycarbonate/acrylonitrile-butadiene-styrene composites. J Polym Eng 38(4):399–407
15. Sangpraserdsuk T, Phiriyawirut M, Ngaotrakanwiwat P, Wootthikanokkhan J (2017) Mechanical, optical, and photochromic properties of polycarbonate composites reinforced with nano-tungsten trioxide particles. J Reinf Plast Compos 36(16):1168–1182
16. Rostamiyan Y, Ferasat A (2017) High-speed impact and mechanical strength of ZrO2/polycarbonate nanocomposite. Int J Damage Mech 26(7):989–1002
17. Dayananda Jawali N, Siddaramaiah S, Lee JH (2008) Polycarbonate/short glass fiber reinforced composites—physico-mechanical, morphological and FEM analysis. J Reinf Plast Compos 27(3):313–319
18. Hakimelahi HR, Hu L, Rupp BB, Coleman MR (2010) Synthesis and characterization of transparent alumina reinforced polycarbonate nanocomposite. Polymer 51(12):2494–2502
19. Oyarzabal A, Cristiano-Tassi A, Laredo E, Newman D, Bello A, Etxeberría A, Müller AJ et al (2017) Dielectric, mechanical and transport properties of bisphenol A polycarbonate/graphene nanocomposites prepared by melt blending. J Appl Poly Sci 134(13)
20. Carrion FJ, Sanes J, Bermúdez MD (2007) Influence of ZnO nanoparticle filler on the properties and wear resistance of polycarbonate. Wear 262(11–12):1504–1510
21. Pisanova E (2011) Polycarbonate composites. In: Wiley Encyclopedia of composites, pp 1–9
22. Tiwari SK, Oraon R, De Adhikari A, Nayak GC (2017) A thermomechanical study on selective dispersion and different loading of graphene oxide in polypropylene/polycarbonate blends. J Appl Polym Sci 134(28):45062
23. Hazer S, Coban M, Aytac A (2018) A study on carbon fiber reinforced poly (lactic acid)/polycarbonate composites. J Appl Polym Sci 135(48):46881
24. Singh JC, Srivastava M, Singh RK, Yadaw SB (2018) Mechanical properties thermoplastic laminates of polycarbonate-polyetherimide blend with glass fibers. J Textile Sci Eng 8:366
25. Bagotia N, Choudhary V, Sharma DK (2018b) Superior electrical, mechanical and electromagnetic interference shielding properties of polycarbonate/ethylene-methyl acrylate-in situ reduced graphene oxide nanocomposites. J Mater Sci 53(23):16047–16061
26. Hu Y, Song S, Lv X, Sun S (2018) Enhanced properties of PBT/PC blends with the addition of carboxyl-functionalized multiwalled carbon nanotube. 폴리머 42(2):206–214
27. Taraghi I, Fereidoon A, Paszkiewicz S, Roslaniec Z (2017) Electrically conductive polycarbonate/ethylene-propylene copolymer/multi-walled carbon nanotubes nanocomposites with improved mechanical properties. J Appl Poly Sci 134(14)
28. Fahmi Asyadi M, Jawaid AH, Wahit MU (2013) Mechanical properties of mica-filled polycarbonate/poly (acrylonitrile-butadiene-styrene) composites. Poly Plast Technol Eng 52(7):727–736
29. Sharma R, Kar KK, Das MK, Gupta GK, Kumar S (2017) Short carbon fiber-reinforced polycarbonate composites. In: Composite materials. Springer, Berlin, Heidelberg, pp 199–221
30. Pour RH, Hassan A, Soheilmoghaddam M, Bidsorkhi HC (2016) Mechanical, thermal, and morphological properties of graphene reinforced polycarbonate/acrylonitrile butadiene styrene nanocomposites. Polym Compos 37(6):1633–1640
31. Suin S, Maiti S, Shrivastava NK, Khatua BB (2014) Mechanically improved and optically transparent polycarbonate/clay nanocomposites using phosphonium modified organoclay. Mater Des 1980–2015(54):553–563
32. Sharma A, Tripathi B, Vijay YK (2010) Dramatic improvement in properties of magnetically aligned CNT/polymer nanocomposites. J Membr Sci 361(1–2):89–95
33. Lim BK, Mo CB, Nam DH, Hong SH (2010) Mechanical and electrical properties of carbon nanotube/Cu nanocomposites by molecular-level mixing and controlled oxidation process. J Nanosci Nanotechnol 10(1):78–84

34. Rejisha CP, Soundararajan S, Sivapatham N, Palanivelu K (2014) Effect of MWCNT on thermal, mechanical, and morphological properties of polybutylene terephthalate/polycarbonate blends. J Poly

Determination of Impact and Hardness Properties of Neem-Kenaf Fiber Reinforced Polymer Composites

B. Vijaya Ramnath, S. Rajesh, C. Elanchezhian, and G. Pon Senthil Kumar

Abstract Hybrid natural fiber composites can be used to meet the present technological development and innovation needs. These natural fibers are environmentally friendly and have a better economical value. This chapter discusses the fabrication and study on mechanical characteristics of hybrid composites which consist of neem, kenaf and GFRP. Here, the composite was fabricated by hand layup technique. The arrangements of fibers are the combinations of neem and kenaf fibers placed alternatively to form the hybrid composites. For better adhesion between fibers, the matrix namely Epoxy resin LY556 and HY951 hardener is used in the ratio of 10:1. Glass fiber composites are enclosed on both the sides in order to improve the surface finish and also to avoid damages occurs during loading condition. The volumetric fraction is one-third of the total volume occupied by neem and kenaf fiber. Test result shows that hybrid natural composite has excellent impact and hardness properties.

Keywords Kenaf fiber · Neem fiber · Hand layup method · Mechanical testing

1 Introduction

Composite materials are made up of two or more constituents with different physical and chemical properties to give different properties than the individual constituents. These composite materials are more advantageous than individual constituents or conventional materials due to their increased hardness, strength, high thermal and wear resistance and suits very well for various engineering applications. Navaranjan and Neitzert [1] experimented that, neem fibers and kenaf fibers are cellulose-based plant fibers which are good capable of acting as a reinforcement with glass fiber polymer. Resin and hardener are used as a medium for fabrication of natural fiber

B. Vijaya Ramnath (✉) · S. Rajesh · C. Elanchezhian · G. Pon Senthil Kumar
Department of Mechanical Engineering, Sri Sairam Engineering College, West Tambaram, Chennai 600 044, India
e-mail: vijayaramnathb@gmail.com

S. Rajesh
e-mail: rajeshsri02@gmail.com

T. Rajmohan et al. (eds.), *Advances in Materials and Manufacturing Engineering*, Springer Proceedings in Materials 7, https://doi.org/10.1007/978-981-15-6267-9_35

composites. Vijaya Ramnath et al. [2, 3] inferred that the main concept of combining two fibers is to compensate the poor property of one fiber by the other and making it as superior when combined together. The surface modification of natural fibers is done by alkali treatment to improve the adhesion between fiber matrix interfaces. But, the mechanical properties of the natural fiber are affected by the hydrophilic nature which is the major disadvantage of the natural fiber. Srinivasan et al. [4] studied that modulus of elasticity is better for dry fibers, and wet fibers showed good adhesion between matrix and fiber. The flexural strength, impact strength and tensile strength for the polymeric composites are found to be better when the kenaf fibers are alkalized. Vijayakumar et al. [5] studied that the neem tree (Azadirachta Indica) is a major source of fibers for various marine applications. The strength of the neem fiber depends on many factors such as harvesting time, growing conditions of the plant and extraction methods which influences the chemical composition and structure of the fiber. Natural fibers which are extracted manually have 20% higher strength than mechanically extracted ones. Vijaya Ramnath et al. [6, 7] investigated that neem fiber reinforced polyester composites are prepared by extracting neem tree fibers and incorporating them in polyester resin matrix to study the tensile strength and impact strength of the resultant composites. Subasinghe et al. [8] studied that in natural fiber composite (NFC), there has been a rapid growth in R&D and innovation sectors. This interest is created because of the superiority in the advantages, which includes low environmental impact, low cost and support wide range of applications when compared to other fibers such as synthetic fiber composites. Vijaya Ramnath et al. [9, 10] inferred that, aeronautical and defense application sectors demand precision engineering, weight saving, finite tolerances, simplifies production and operations uses natural fiber reinforced composites, whereas synthetic fiber composites are limited to these applications because of high cost of material and fabrication methods. Structural members are strengthened by fiber reinforcement polymers even after they have been severely damaged because of its low density and high stiffness. This is because of the increased toughness, impact resistance and the energy absorption provided by the fiber to the composite. The modulus of elasticity is directly proportional to the volume of the fiber, and the ultimate strength is inversely proportional to the volume of the fiber in the composite. The usage of natural fibers in curved pipes has resulted in reduction of cost and weight of about 20% and 23%, respectively. Pickering et al. [11] investigated that, Africa is the origin of kenaf crop (Hibiscus cannabinus). It is widely cultivated around the world for use as fiber, paper or biofuel. It acts as a binderless thermal insulator. Subasinghe et al. [8] inferred that, kenaf fiber because of its rapid growth can be used as reinforcement in natural fiber composites. Developments of kenaf-based industries have been encouraged by the government. Kenaf degrades earlier because of the presence of lignin and hemicellulose. After 300 °C, the char content formed from lignin present in the kenaf fibers causes a slow rate of degradation. The kenaf fibers have high lignin content so the thermal resistance can be increased by the incorporations of lingo cellulosic kenaf fibers into a polypropylene matrix. Subasinghe et al. [12] studied that, the photosynthesis rate is three times than that of the usual plants. It can filter carbon dioxide at 14 times its own weight which is higher than others. Tharazi et al. [13] investigated that, the unidirectional long kenaf fiber which is reinforced with polylactic acid were fabricated into biodegradable composites by hot pressing method. To optimize the tensile strength and to determine the significance of the factors influencing it, analysis of variance (ANOVA) and response surface methodology (RSM) were used. For optimum tensile strength, the combination of hot pressing parameters was 200 °C temperature, 3 MPa pressure and heating time at 8 min. Confirmation test run yields error which is less than 7% verified the validity of the model. Tharazi et al. [13] studied that kenaf fibers are more sustainable that concrete reinforced. The properties are studied based on nano-indentation principle and effect of water absorbing fibers are noted. The result shows that there is not much significant difference in the structure but it has some variations based on relative volume fractions. Vijayakumar et al. [5] studied the performance of kenaf fibers treated with nitrogen and phosphorous from the soil and it shows the effect of carbon dioxide with the presence of air at significantly high rate. Due to this, there is an enormous consumption in fuel and controlled emission in the automotive vehicles. Zadeh et al. (2014) investigated the absorption of sound for kenaf fibers under various conditions. The result reveals that it has good mechanical properties as well as sound frequencies which vary from 0.5 to 500 Hz. Lima et al. [14] investigated the performance of kenaf fiber which is used to produce low cost carbon using acidic chemical activating agent. The result shows that it has got high percentage of carbon and it produces very less ash which leads to good material for production of porous carbon. Bharath et al. [15] studied that, for cleaning oil spills, kenaf core has been evaluated as an absorbent material by a technical report TR-2101-ENV of the Naval Facilities Engineering service center. Navaranjan and Neitzert [1] experimented that, the ability of a material to withstand applied stress or shock load is called impact resistance. The factors influence the impact strength are strength of the material, elastic modulus of the fiber material, orientation and length of the fibers, interfacial bond strength of fiber matrix, and finally, the method of impact testing. Several applications include interior and exterior components of automobiles, aircraft, constructions and building materials which are used as engineering materials of high impact strength. Vijaya Ramnath et al. [16, 17] inferred that, the low viscous liquids are more absorbed by natural composite materials than high viscous liquids. Infinite material permutation including fiber and resin types, architecture quantities, production methods used and interfaces influences the impact behavior of composites. Rajesh et al. [18, 19] experimentally investigated the mechanical characteristics of various natural fiber composites and result shows that there is a significant improvement in the mechanical properties. Also, scanning electron microscopy is done to observe the internal structure of the composite specimen. Bajuria et al. [20] investigated flexural and compressive properties of kenaf with the combination of silica nano-particles in the epoxy medium by using vacuum infusion process and found the above properties are mostly significant.

Fig. 1 Neem leaves and stem

2 Materials and Methods

2.1 Neem Fiber

Neem leaves and stem as shown in Fig. 1 is one of the natural fibers which meet a wide range of applications as composites reinforced with polymer matrix, resin and suitable hardener. The strength of the neem fiber depends on many factors such as harvesting time methods of extraction of neem fibers.

2.2 Kenaf Fiber

It is certainly one of the important plants cultivated for natural fibers globally, next to cotton, which is endemic to ancient Africa. Kenaf fibers as shown in Fig. 2 are reinforced with polymer matrix to form fiber reinforced polymeric composites which perfectly improve the features of the polymers.

Fig. 2 Kenaf leaves and fibers

Fig. 3 GFRP mat

2.3 *Glass Fiber Reinforced Polymer*

Figure 3 shows the glass fiber reinforced polymer mat. It is the most commonly used polymer which is made up of many strands of silica glass fiber. It is extremely strong and robust material. The strength to weight ratio of glass fiber is high and can be easily manufactured by extruding and molding processes. It has very good bulk and weight properties when compared to metals. The erosive wear rate of the composite is increased by the glass polymer.

2.4 *Epoxy Resin and Hardener*

HY951 hardener has good mechanical and chemical properties. LY556 epoxy resin has good bonding strength. The resin-hardener severs the purpose of binder between various layers of fiber. The mixing ratio between resin and hardener is 10.1. LY556 epoxy resin and HY951 hardener are shown in Fig. 4 give the best binding property in room temperature of 28–30 °C.

3 Fabrication Procedure

The fabrication of this proposed composite laminate is made by hand layup technique. Initially, the fibers undergo retting process followed by the drying action under sunlight for about 24 h. After drying the fibers, they are well combed and straightened followed by cutting them at equal lengths as straight flakes required forming the laminate. Epoxy resin (LY556) and the HY951 hardener are employed in the process. The bottom GFRP sheet is placed on a plane surface and rollers are used to remove the major and minor sprinkles on the sheet thoroughly. Then, the neem fibers are placed on the sheet in a stipulated setting time after which resin-hardener mixture applied. The layers of neem fibers and kenaf fibers are arranged between the GFRP laminate in perpendicular and parallel order and taken as Sample 1 and

Fig. 4 Resin and hardener

Table 1 Arrangement of fibers

Sample 1 (perpendicular orientation)	Sample 2 (parallel orientation)
GRFP LAMINATE	GRFP LAMINATE
NEEM FIBER	NEEM FIBER
KENAF FIBER	KENAF FIBER
GRFP LAMINATE	GRFP LAMINATE

Sample 2, respectively. The arrangements of these composite samples are shown in Table 1. The fabricated samples are made to dry for 12–14 h under application of 15 kg weight to bind fibers together firmly.

4 Testing of Composites

4.1 Impact Test

In this test, the amount of energy absorbed by the specimen during the breakage is the impact strength of the material. A pendulum is setup to drop on specimen to fracture it from an angle of 135°. Charpy impact test is employed in this case. The specimen is prepared as per ASTM: D256 standards, which are shown in Fig. 5. Composite specimen before and after the impact test is shown in Fig. 6.

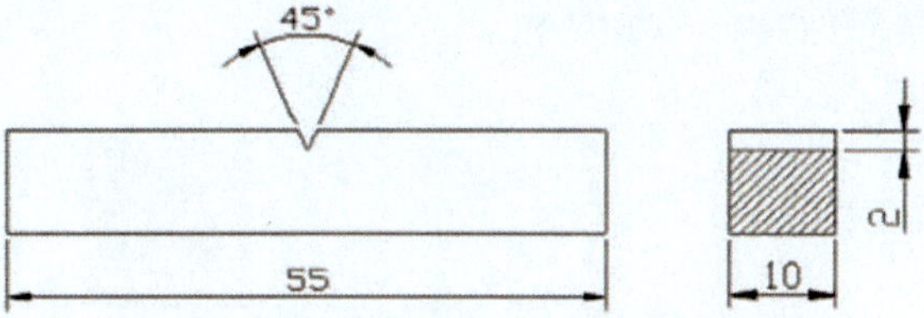

Fig. 5 ASTM standard for impact testing

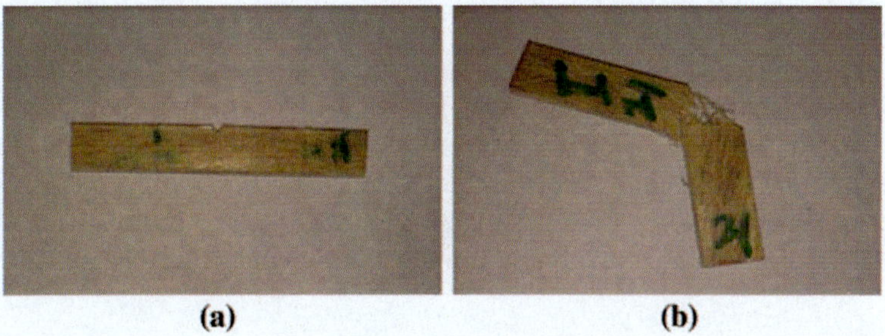

Fig. 6 **a** Specimen before impact. **b** Specimen after impact test

4.2 Hardness Test

The hardness of the composites was measured using a Rockwell hardness testing machine according to ASTM D785-98. The hardness value of an composite corresponds to the amount of indentation occurred when the steel ball is made to indent on the composite specimen.

5 Result and Discussion

5.1 Results of Impact test

The impact test results obtained from the Charpy impact test is shown in Table 2. The

Table 2 Results of impact test

S. No.	Sample	Energy absorbed (J)	Average value (J)
1	1	3.8	3.6
		3.5	
		3.5	
2	2	4.8	4.6
		4.5	
		4.5	

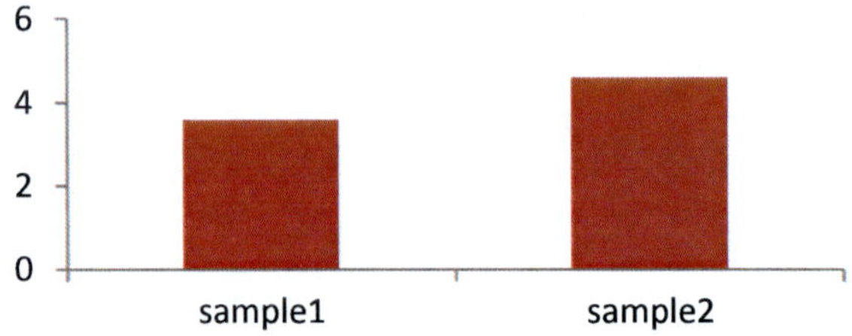

Fig. 7 Results of impact test (J)

impact strength values corresponding to three trials of each sample are noted. The average energy absorbed is calculated in joules and noted. The values are plotted as graph is shown in Fig. 7. The total energy absorbed is 4.6 J for Sample 2 due to the presence of kenaf and neem fiber in the composite laminate at 0° orientations which resists the fibers to break at maximum load. This orientation increases the contact area between the laminate which in turn increases the contact area against impact load.

5.2 *Results of Hardness Test*

The experimental results of the value of indentation obtained during the hardness test carried out using Rockwell hardness machine are furnished in Table 3. Three trials are carried out for each sample and the hardness number is noted as shown in Table 3. The maximum indentation occurs for Sample 2 with average 64.6 Rockwell number. The results are plotted as graphics as shown in Fig. 4. Sample 2 has high hardness as the fibers are arranged in parallel orientation, which resist the indentation. This is the result of high bonding between fibers and increased contact area due to parallel orientation. The graph shows the result of Hardness test for the two samples (Perpendicular and Parallel orientation respectively) (Fig. 8).

Table 3 Results for hardness test

S. No.	Sample	Rockwell hardness number (RHN)	Average value
1	1	56	58.6
		60	
		60	
2	2	64	64.6
		68	
		62	

Fig. 8 Result of hardness test (RHN)

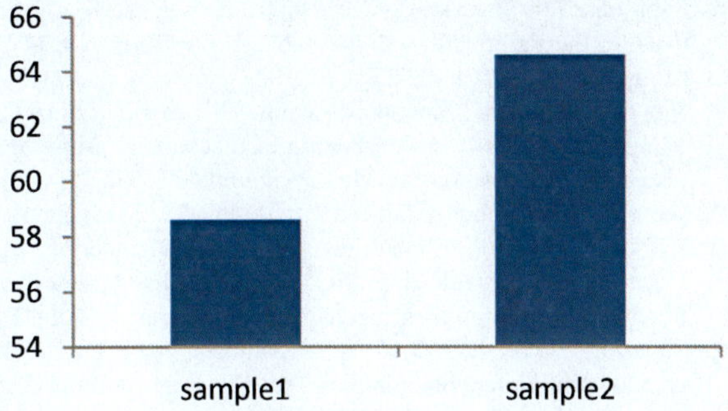

6 Conclusion

Based on impact and hardness experimental results, the following conclusions have been made. For impact, the maximum amount of energy absorbed is recorded as 4.6 J. This is due to the presence of kenaf and neem fibers which resist the force to minimum, thereby reduce the damage of the composite specimen. For hardness test, the value of indentation is 64.6 Rockwell hardness number. The indentation on the specimen is restricted due to the presence of strong adhesion between the fibers and the matrix which does not allow the penetration due to load into the composite specimen. Hence, due to high value of impact and hardness properties, this hybrid composite can be implemented in various engineering applications where high impact is in demand.

References

1. Navaranjan N, Neitzert T (2017) Impact strength of natural fibre composites measured by different test methods a review. In: MATEC web of conferences, vol 109
2. Vijaya Ramnath B, Elanchezhian C, Nirmal PV, Prem Kumar G, Santhosh Kumar V, Karthick S, Rajesh S, Suresh K (2014) Experimental investigation mechanical behavior of jute-flax based glass fiber reinforced composite. Fibers Poly 15:1251–1262
3. Rajesh S, Vijaya Ramnath B (2018) Investigation of hardness and impact property of a kevlar composite. Int J Appl Eng Res 10:6215–6218
4. Srinivasan VS, RajendraBoopathy S, Vijaya Ramnath B (2015) Investigation of flexural property of kenaf-flax hybrid composite. Appl Mech Mat 10
5. Vijayakumar T, Ramana KV, Balamurali KV, Shahjahan P (2012) Tensile & impact behaviour of neem fiber-polyester composites. Gold Res Thoughts 2
6. Vijaya Ramnath B, Rajesh S, Elanchezhian C, Elanchezhian C, Santosh Shankar A, Pithchai Pandian S, Vickneshwaran S, SundarRajan R (2016) Investigation on mechanical behaviour of twisted natural fiber hybrid composite fabricated by vacuum assisted compression molding technique. Fibers Poly 17:80–87
7. Sakthivel M, Vijayakumar S, Vijaya Ramnath B (2018) Investigation on mechanical and thermal properties of stainless-steel wire mesh-glass fibre reinforced polymer composite. Silicon 10:1–9

8. Subasinghe A, Somashekar AA, Bhattacharyya D (2018) Effects of wool fibre and other additives on the flammability and mechanical performance of polypropylene/kenaf composites. Comp Part B 136:168–176
9. Vijaya Ramnath B, Manickavasagam VM, Elanchezhian C, Vinodh Krishna C, Karthik S, Saravanan K (2014) Determination of mechanical properties of intra-layer abaca–jute–glass fiber reinforced composite. Mat Design 60:643–652
10. Ramnath BV, Arvind R, Dinesh I, Prasadh MH (2018) Review on natural fiber composites. In: IOP conference series: materials science and engineering, vol 390
11. Pickering KL, AruanEfendy MG, Le TM (2016) A review of recent developments in natural fibre composites and their mechanical performance. Comp Part A Appl Sci Manuf 83:98–112
12. Subasinghe ADL, Das R, Bhattacharyya D (2015) Fiber dispersion during compounding/injection molding of PP/kenaf composites flammability and mechanical properties. Mat Design 86:500–507
13. Tharazi AB, Sulonga N, Muhamada CHC, Harona D, Tholibona NF, Ismaila MKF, Radzia M, Razaka Z (2017) Optimization of hot press parameters on tensile strength for unidirectional long kenaf fiber reinforced polylactic-acid composite. Proc Eng 184:478–485
14. Lima ZY, Putraa A, Nora MJM, Yaakobb MY (2018) Sound absorption performance of natural kenaf fibres. Appl Acoust 130:107–114
15. Bharath VRR, Ramnath BV, Manoharan N (2015) Kenaf fibre reinforced composites: a review. ARPN J Eng Appl Sci 10:5483–5485
16. Vijaya Ramnath B, Junaid Kokan S, Niranjan Raja R, Sathyanarayanan R, Elanchezhian C, Rajendra Prasad A, Manickavasagam VM (2013) Evaluation of mechanical properties of abaca–jute–glass fibre reinforced epoxy composite. Mat Design 51:357–366
17. Sharavanan S, Vijayaramnath B, Elanchezhian C, Rajesh S (2018) Experimental investigation of tensile and impact behaviour of hemp flax hybrid composite. Int J Mech Prod Eng Res Devel (IJMPERD) 8:549–556
18. Manickavasagam VM, Ramnath BV, Elanchezhian C, Aravinthan V (2018) Natural fibre composites—a review. In: IOP conference series: materials science and engineering, vol 390
19. Rajesh S, Vijaya Ramnath B, Elanchezhian C, Abhijith M, Dinesh Riju R, KathirKishan K (2018) Investigation of tensile behavior of kevlar composite. Mat Today Proc 5:1156–1161
20. Bajuria F, Mazlana N, Ishaka MR, Imatomi J (2016) Flexural and compressive properties of hybrid kenaf/silica nanoparticles in epoxy composite. Proc Chem 19: 955–960

Study on Wear and Corrosion Behaviour of Aluminium Hybrid MMC

B. Vijaya Ramnath, E. Naveen, S. Abhishek Subramanian, R. Rakesh, and S. Sharun Krishnan

Abstract Nowadays, composite materials replace conventional material by virtue of their properties like high strength to weight ratio, improved wear resistance and corrosion resistance. In this work, the aluminium metal matrix composites were fabricated with help of stir casting process. Here, five samples of composites were fabricated with aluminium as matrix material and fly ash, bamboo ash and copper as reinforcements with different proportions, increasing from 1 to 5%. Wear and corrosion behaviour of fabricated composites were studied using sliding wear test and salt spray corrosion test. Morphological analysis was also done to study the internal structure of corrosion and wear tested samples.

Keywords Aluminium metal matrix composite · Fly ash · Bamboo ash · Copper

1 Introduction

Currently, manufacturing industries require materials which possess high corrosion and wear resistance. At the same time, materials with high strength to weight ratio characteristics are also necessary as they reduce the weight of the component and increase the strength. For this purpose, aluminium metal matrix composites (AMMC) have been seen to be the most promising material to be used. AMMC have been seen to be used in many industry applications such as in automotive industries to make engine parts, wheel hub and brake parts to ultimately increase the efficiency and in aerospace industries to reduce the weight of the body. Most frequently, aluminium base metals are reinforced with silicon carbide and alumina which is commonly used in most of the industries for giving high strength to low weight ratio and high hardness. Bodunrin et al. [1] studied the effect of reinforcement of aluminium metal matrix composites with agrowaste, industrial waste and ceramic materials. Fly ash as by-products from industries have been used to improve the wear resistance and machinability which is

B. Vijaya Ramnath (✉) · E. Naveen · S. Abhishek Subramanian · R. Rakesh · S. Sharun Krishnan
Department of Mechanical Engineering, Sri Sairam Engineering College, West Tambaram, Chennai 600044, India
e-mail: vijayaramnath.mech@sairam.edu.in

T. Rajmohan et al. (eds.), *Advances in Materials and Manufacturing Engineering*, Springer Proceedings in Materials 7, https://doi.org/10.1007/978-981-15-6267-9_36

due to the solid lubricating effect it has and the microstructural properties. It has influenced the uniform distribution of the particles in metal matrix. The ceramic materials when reinforced with the aluminium base metal like graphite, CNT has lead to the increase in the load bearing capacity and pinning effect of the composite. Liu et al. [2] studied the effect of friction stir processing (FSP) of Al base metal with magnesium, lanthanum and nickel as the reinforcement. They concluded that an increase in the hardness of the surface and the maximum tensile strength of 410 Mpa is obtained and the corrosion results show that the composites have a lower current corrosion density and high passivation current density which leads to increased corrosion resistance effect. Aribo et al. [3] conducted an experiment on the wear and corrosion analysis of Al 6063 hybrid composites with snail-shell-ash (SSA) and SiC as the reinforcements with varying compositions. They observed that an increase in the Vicker hardness number (VHN) with a maximum increase of 50VHN and the corrosion resistance is found to be maximum with the lowest current corrosion density both in 7.5%SiC and 7.5%SSA composition composite. Bienia et al. (2003) have studied the microstructure and corrosion behaviour of aluminium fly ash composites was produced by squeeze casting technology. They found that uniform structural homogeneity with low porosity levels and a high interfacial bonding between the particles. The pitting corrosion was also found to be enhanced in comparison with the unreinforced matrix and the properties of the oxide film forming on the surface of the composites were also determined using this result. Alaneme et al. [4–7] investigated the corrosion and tribological behaviour of aluminium hybrid composites with bamboo leaf ash and alumina as the reinforcements in chloride medium. The samples were prepared with different compositions with 2, 3 and 4 wt% alumina and 10% bamboo leaf ash was prepared using double stir casting technique. The result shows an increase in corrosion resistance with the addition of BLA particles and the coefficient of friction and the wear rate was found to be comparable in the 2% and 3% BLA composites showing slightly superior wear resistance. Selvam et al. [8, 9] have studied the effect on aluminium 6061 hybrid composites with reinforcing materials of SiC and fly ash. They concluded that the tensile strength and the microhardness level of the composite increased with increase in wt. percentage of the SiC particles. Praveen Kumar et al. [10] studied the effect of addition of bamboo leaf ash to aluminium base metal and analysed the microstructure and mechanical properties where the composites were prepared by stir casting method. The BLA of 2, 4, 6% compositions were produced and the results were appreciable. They found that the hardness of the composites increased with the increase in BLA content. The hardness was maximum in 4% BLA composition composite. The SEM analysis micrographs showed good interfacial bond between BLA particles and the matrix. Also, the highest yield and tensile strengths were observed at 4% BLA particle. Chethan et al. [11] conducted a mechanical characterization of aluminium 6061 with bamboo ash charcoal as the reinforcing material. The bamboo charcoal of 2, 4, 6% by weight was used to prepare the composite using stir casting method. The composites prepared were normalized at 550 °C for 6 h and different cooling techniques were used to cool it. They concluded that the charcoal is uniformly distributed, as seen in the SEM results. As the charcoal content increased, there was an increase in the hardness of the composite and

6%BLA composition of the furnace cooling sample had the highest hardness of 112 VHN, which is actually more compared to many other alloys. They found that the tensile strength reduced with the increase in the bamboo ash content and also the density of the composites slightly varied with the increase in the weight composition of bamboo charcoal.

2 Innovation in the Work

In this work, the effect of addition of bamboo ash, fly ash and copper as the reinforcing particles with aluminium 6061 as the base metal is investigated. The reinforcing materials were chosen considering each of its own unique properties. The presence of fly ash helps in improving the machinability of the workpiece, it also reduces the thermal conductivity as its melting point is more than about 1000 °C. It is also chemically inert. The bamboo ash helps in improving the fracture toughness, corrosion resistance and copper has good thermal conductivity and corrosion resistance and it alloys easily when combined with base metals. Considering the above factors, the reinforcements were chosen and the composites with different compositions were prepared and tested. The fabrication of these composites was carried out using stir casting method which is one of the most prominent methods to manufacture aluminium composites. Also, comparatively, it is a low-cost method and gives flexibility to produce any shape and size. In this work, the wear and corrosion behaviour of aluminium metal matrix composites reinforced with bamboo ash, fly ash and copper is evaluated.

3 Materials and Methods

3.1 Materials

The materials used in this work include pure commercially available aluminium 6061 which is the matrix material and the reinforcements include finely powdered fly ash, bamboo ash and copper taken in the preheated condition to relieve the water content present in them to get a sound and good composite. For bamboo ash, the bamboo was fired, dried and crushed into powder form so that it can be used effectively by completely fusing with the base metal. The aluminium 6061 is taken in the form of a pipe of some diameter and is cut according to mould size provided with the friction stir casting equipment. The different compositions of samples fabricated using friction stir casting methods are shown in Table 1. Aluminium 6061 is a precipitation-hardened aluminium alloy containing magnesium and silicon as its major alloying elements. Fly ash is a by-product from burning pulverized coal in electric power generating plants. During combustion, mineral impurities in the coal (clay, feldspar,

Table 1 Composition of composites

Sample number	% of aluminium	% of copper	% of fly ash	% of bamboo ash
1	97	1	1	1
2	94	2	2	2
3	91	3	3	3
4	88	4	4	4
5	85	5	5	5

quartz and shale) fuse in suspension and float out of the combustion chamber with the exhaust gases. It is a soft, malleable and ductile metal with very high thermal and electrical conductivity. Copper is used as a conductor of heat and electricity, as a building material, and as a constituent of various metal alloys.

3.2 *Production of Composite*

There are different types of casting methods available to fabricate metal matrix composite. Its advantages lie in its simplicity, flexibility and mainly applicable to large quantity production with cost advantage. In this work, aluminium metal matrix composites reinforced with bamboo ash, fly ash and copper at different composition percentages were casted using a bottom pouring stir casting machine. A special designed furnace in which pouring of melt into the mould happens from its bottom through a remote-control switch. This type of furnace does not require the user to lift and pour the melt into the mould. This machine can be used to produce casting where base metal of that casting should have the melting point less than 1000 °C (Fig. 1).

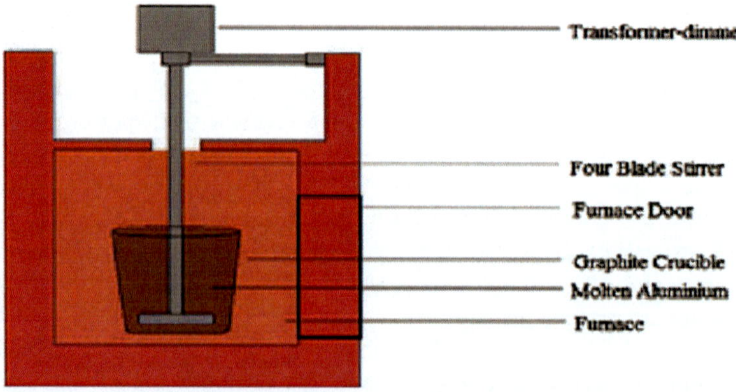

Fig. 1 Basic stir casting equipment setup

Table 2 Wear test specifications

Serial number	Test condition	Requirements	Actual
1	Chamber temperature (°C)	35 ± 2	34.2–34.7
2	pH of solution	6.5–7.2	6.8–7.0
3	Air pressure (psi)	12–18	15
4	Concentration of sodium chloride (%)	5 ± 1	5.1–5.2
5	Collection of solution per hour (ml)	1–2	1.1–1.4
6	Test hours (h)	72	72

4 Wear Test

In this work, wear test is performed to find the wear rate of the fabricated composites. In the abrasive wear test, a hard-rough surface slides over a softer surface. ASTM international defines it as the loss of material due to hard particles forced against a solid surface. There are a number of factors which influence abrasive wear and hence the manner of material removal. Abrasive wear can be measured as loss of mass by the Taber abrasion test according to ISO 9352 or ASTM D 4060. Table 2 shows the wear test specifications used in this work.

5 Corrosion Test

The salt spray test is one of the standardized corrosion test used to analyse the corrosion resistance of materials. Salt spray testing is an fast paced corrosion test that produces a corrosive attack to samples in order to evaluate the suitability of the coating to use as a protective finish. The specification of corrosion test apparatus is shown in Table 3.

Table 3 Corrosion test specification

Serial number	Contents	Value
1	Cylinder size	⌀ 150 mm and 500 mm length
2	Material of coarser abrasive sheet	60 and 80 grade
3	Equivalent revolution	84 times
4	Rotational frequency	40 ± 1 rpm
5	Load applied	1, 2, 3, 4 and 5 kg

6 Results and Discussion

6.1 Result of Wear Test

The results of the wear test have been tabulated by finding the wear rate using the relation given below. This method of finding the wear rate using initial weight and final weight is known as gravimetric analysis. The wear tested samples are shown in Fig. 2.

$$\text{WEARRATE} = \frac{\text{I.W} - \text{F.W}}{\rho \times T}$$

where, I.W is the initial weight of the composite in grams, F.W is the final weight of the composite in grams, ρ is the theoretical density in gm/cm^3, T is the time of sliding in mins, theoretical density is calculated as follows:

$$\text{TheoreticalDensity}\,[\rho] = \frac{\rho Al \times \rho Cu \times \rho FA \times \rho BA}{[\rho Al \times \rho Cu \times \rho FA \times M.F(BA)] + [\rho Cu \times \rho FA \times \rho BA \times M.F(Al)] + [\rho FA \times \rho BA \times \rho Al \times M.F(Cu)] + [\rho BA \times \rho Al \times \rho Cu \times M.F(FA)]}$$

where, M.F is the mass fraction for the corresponding reinforcement, which is found as follows:

$$\text{M.F} = \frac{\text{Massofonesubstance}}{\text{MassofTotalMixture}}$$

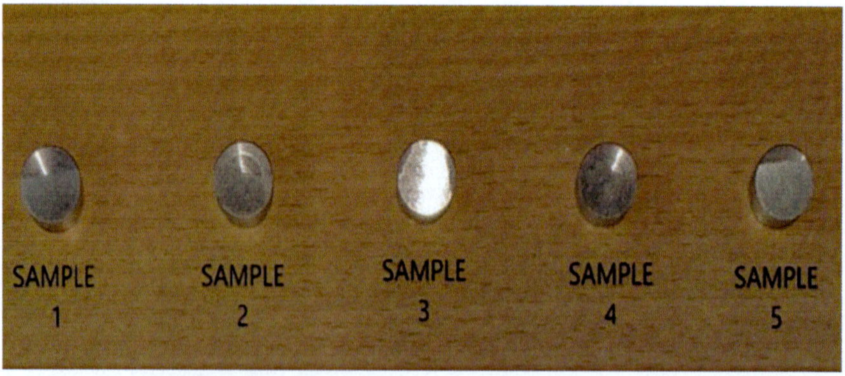

Fig. 2 Wear tested samples

ρ_{Al} is the density of Al = 2.7 gm/cm^3, ρ_{Cu} is the density of copper = 8.96 gm/cm^3, ρ_{FA} is the density of fly ash = 0.86 gm/cm^3, ρ_{BA} is the density of bamboo ash = 1.4 gm/cm^3.

6.1.1 Model Calculation for Theoretical Density

For sample 1,

Mass Fraction for Al $= \frac{0.97}{1} = 0.97$, Mass Fraction for FA $= \frac{0.01}{1} = 0.01$, Mass Fraction for BA $= \frac{0.01}{1} = 0.01$, Theoretical Density $[\rho] =$

$$\frac{2.7 \times 8.96 \times 0.86 \times 1.4}{[(2.7 \times 8.96 \times 0.86 \times 0.01)] + [(8.96 \times 0.86 \times 1.4 \times 0.97)] + [(0.86 \times 1.4 \times 2.7 \times 0.01)] + [(1.4 \times 2.7 \times 8.96 \times 0.01)]}$$

Theoretical Density for sample 1 $[\rho] = 2.63$ gm/cm^3.

Similarly, for other samples, the theoretical density was calculated and is shown in Table 4.

The wear rate for 60 grit size composite for all loading conditions is furnished in Table 4. From Fig. 3, it is evident that the sample 1 undergoes maximum wear under all the loading conditions compared to the other samples applied under the same

Table 4 Wear rate for 60 grit size

Sample	Actual density (gm/cm^3)	Theoretical density (gm/cm^3)	10 N	20 N	30 N	40 N	50 N
S1	2.34	2.63	1.073	1.311	1.657	1.727	2.05
S2	2.23	2.57	1.061	1.246	1.564	1.618	1.873
S3	2.14	2.52	1.031	1.223	1.439	1.545	1.772
S4	2.04	2.46	0.93	1.185	1.338	1.53	1.613
S5	1.95	2.41	0.853	1.01	1.258	1.363	1.5

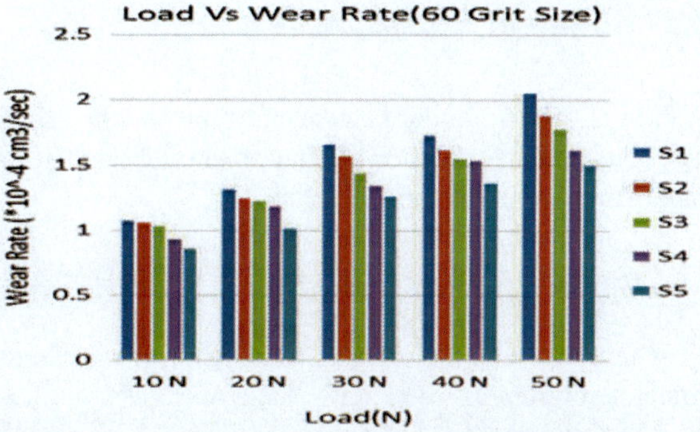

Fig. 3 Load versus wear rate (for 60 grit size)

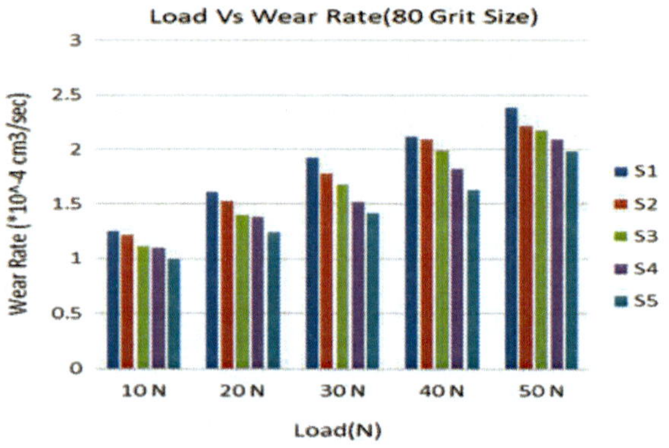

Fig. 4 Load versus wear rate (for 80 grit size)

Table 5 Wear test results (for 80 grit size)

Sample	Actual density (gm/cm^3)	Theoretical density (gm/cm^3)	10 N	20 N	30 N	40 N	50 N
S1	2.34	2.63	1.251	1.614	1.927	2.12	2.39
S2	2.23	2.57	1.221	1.532	1.778	2.091	2.218
S3	2.14	2.52	1.116	1.401	1.681	1.991	2.175
S4	2.04	2.46	1.102	1.382	1.523	1.82	2.092
S5	1.95	2.41	1.001	1.244	1.417	1.627	1.981

load. It noted that, sample 5 has the maximum wear resistant property as compared with other samples (Fig. 4).

Similarly, for grit size 80, the wear test has been performed for all loading conditions as furnished in Table 5. From the wear results, it can be seen that the wear rate for sample 5 which has 5%wt composition of reinforcements has a minimum wear rate which means, it gives more resistance to wear compared to other compositions.

6.2 *Result of Corrosion Test*

ASTM B117-16 standard has been followed when finding the corrosion test values. The results of the corrosion test have been tabulated in Table 6 by finding the corrosion rate using the relation given below:

$$CR = (87.6 \times W)/D \times A \times T$$

where CR = Corrosion Rate(mm/year), W = Weight Loss (mg), D = Density of Specimen (g/cm^3).

A = Exposing Area of Specimen (cm^2), T = Time of Exposure (h).

The theoretical density of the specimen is found by using the following relation:

Table 6 Result of corrosion test

Sample number	Weight loss (mg)	Corrosion rate (mm/year)
1	1300	2.979748
2	1100	2.645695
3	500	1.253165
4	300	0.788757
5	200	0.550107

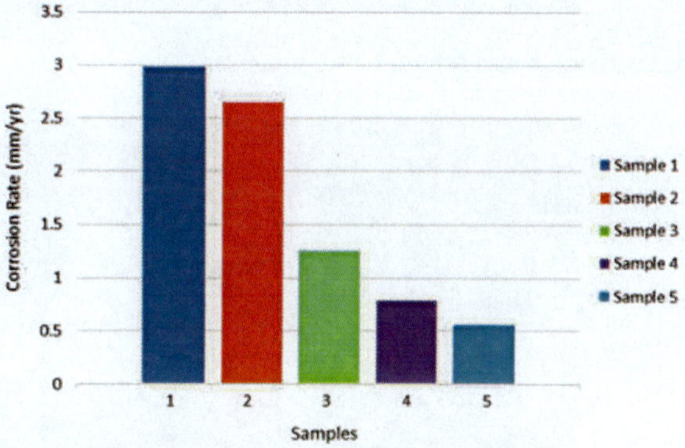

Fig. 5 Result of corrosion rate

Fig. 6 Fabricated and corrosion tested sample 1

Theoretical Density (D) = 100/[(*% of X material/Density of X material*) + (*% of Y material/Density of Y material*) + …].

From Fig. 5, it is evident that the corrosion rate of the samples decreases as the reinforcements are added in proportions the corrosion resistance increases as the reinforcements are added.

6.2.1 Corrosion Tested Samples

See Figs. 6, 7, 8, 9 and 10.

Fig. 7 Fabricated and corrosion tested sample 2

Fig. 8 Fabricated and corrosion tested sample 3

Fig. 9 Fabricated and corrosion tested sample 4

Fig. 10 Fabricated and corrosion tested sample 5

7 Conclusion

From the results obtained from the above tests, the following conclusions can be made:

The wear rate for grit size of 60 under different load conditions show that for sample 1 the wear rate is found to be high and contrastingly for sample 5 it is appreciably the lowest. For the grit size of 80, the results found are the same and sample 5 has the maximum wear resistant property. On the other hand, the corrosion rate results on the other hand show that sample 5 showed the least corrosion rate among other samples and the corrosion rate has been decreasing as the reinforcements have been added. As the composition of the reinforcements tend to increase from 1 to 5%, the wear rate is found to be decreasing. This shows that the reinforcements added tend to increase the wear resistant property of the sample. When considering the corrosion behaviour, the sample with 4% wt concentrations of different reinforcements has lead to poor results in the corrosion test. While the sample with 5%wt concentrations of different reinforcements produced good results.

References

1. Bodunrin MO, Alaneme KK, Chown LH (2015) Aluminium matrix hybrid composites: a review of reinforcement philosophies; mechanical, corrosion and tribological characteristics. J Mater Res Technol 4(4):434–445
2. Peng Liu A, Shi Q, Zhang Y (2013) Microstructural evaluation and corrosion properties of aluminium matrix surface composite adding Al-based amorphous fabricated by friction stir processing. Comp Part B 52:137–143
3. Aribo S, Fakorede A, Ige O, Olubambi P (2017) Erosion-corrosion behaviour of aluminum alloy 6063 hybrid composite. Wear. 376–377. 608–614

4. Alaneme KK, Olubambi PA, Afolabi AS, Bodurin MO (2014a) Corrosion and tribological studies of bamboo leaf ash and Al reinforced Al-Mg-Si alloy matrix hybrid composites in chloride medium. Int J Electrochem Sci 9:5663–5674
5. Alaneme KK, Ademilua BO, Bodunrin MO (2013) Mechanical properties and corrosion behaviour of aluminium hybrid composites reinforced with silicon carbide and bamboo leaf ash. Tribol Ind 35:25–35
6. Bodunrina MO, Alanemea KK, Chownb LH (2015) Aluminium matrix hybrid composites: a review of reinforcement philosophies; mechanical, corrosion and tribological characteristics. J Materrestecnol 4(4):434–445
7. Alaneme KK (2011) Corrosion behaviour of heat-treated Al-6063/SiCp composites immersed in 5 wt% NaCl solution. Leonardo J Sci 18(18):55–64
8. David Raja Selvam J, Robinson Smart DS, Dinaharan I (2013) Synthesis and characterization of Al6061-fly ash p-SiCp composites by stir casting and compocasting methods. Energy Proc 34:637–646
9. David Raja Selvama J, Robinson Smart DS, Dinaharan I (2013) Microstructure and some mechanical properties of fly ash particulate reinforced AA6061 aluminum alloy composites prepared by compocasting. Mat Design 49:28–34
10. Praveen Kumar B, Birru AK (2017) Microstructure and mechanical properties of aluminium metal matrix composites with addition of bamboo leaf ash by stir casting method. Trans Nonferrous Met Soc China 27:2555−2572
11. Chethan KN, Pai A, Keni LG, Singhal A, Sinha S (2018) Fabrication and tribological response of aluminium 6061 hybrid composite reinforced with bamboo char and boron carbide micro-fillers. In: IOP conference series: materials science and engineering, vol 310, p 012028

Effect of Pineapple/Coconut Sheath Fiber Reinforced with Polyester Resin Matrix on Mechanical and Microstructure Properties of Hybrid Polyester Composite

L. Natrayan and M. S. Santhosh

Abstract Natural fiber-reinforced composites are gaining more attention among industries and researchers globally for its greater availability and properties like elasticity, malleability, and fatigue strength. The proposed research focuses on tensile, impact, and flexural properties estimation of randomly aligned hybrid composites made up of pineapple/coconut sheath fiber reinforced with polyester (PCSFP) matrix. The hybrid PCFSP composites were fabricated by book press compression moulding with a various volume fraction of reinforcement and matrix materials. Fracture surface analysis was done using scanning electron microscope for predicting the impact response of the hybrid laminates. The targeted output of hybrid PCSF-reinforced polyester composites is suitable for commercial products like thermoset shelters, containers as well as light-weight automobile body bonnets, headliners, and trunk trims.

Keywords Compression moulding · Pineapple fiber · Tensile test · Flexural test · Impact test · SEM · Coconut sheath fiber · Polyester matrix

1 Introduction

The concern in legitimate usage of polymer matrix composites made with natural fibers is growing exponentially mainly due to their biodegradability, strength-to-weight ratio, wear and tear resistance, lesser toxicity along with conventional properties. The inclusion of natural fibers in thermo and thermoset polymers are desirable for various customary applications in engineering industries like automobile, marine, and air force [1]. The composite technology focuses on the usage of natural fibers in automobile and aerospace parts since the 1950s [2]. Natural fiber composites are considered as a better replacement for traditionally used metals like aluminum and

L. Natrayan (✉)
School of Mechanical Engineering, VIT, Chennai 600127, India
e-mail: natrayanphd@gmail.com

M. S. Santhosh
Selvam Composite Materials Research Laboratory, Namakkal 637003, India

T. Rajmohan et al. (eds.), *Advances in Materials and Manufacturing Engineering*, Springer Proceedings in Materials 7, https://doi.org/10.1007/978-981-15-6267-9_37

copper [3]. Polymer composites reinforced with mudar, sisal, banana leaf, Prosopis juliflora, pineapple, and coir fibers are desired by material scientists all over the world. Chiefly, jute fibers (Geotextile material) are used as raw material to fabricate ropes, sacks, cords, and bags as they are less toxic, eco-friendly, and recyclable [4]. The invention of synthetic fibers like glass has now replaced the jute market [5]. Presently, the composites made with natural and synthetic fiber reinforcement are used as a substitute for timers and other wood products. Ranganathan et al. [6] fabricated the polypropylene composites reinforced with coir kenaf, jute, sisal, and hemp fibers by film stacking compression mould setup and compared the various mechanical properties for optimum hybrid configuration. Kim et al. [7] fabricated cassava and pineapple floor-reinforced biocomposites with enhanced thermal and mechanical properties. PLA bio composites possess better thermal stability. Suresh Kumar et al. [8] reported that characterized row and treated coconut sheath fiber composites possess better mechanical and thermal properties than raw coconut fiber composites. Nur [9] treated pineapple leaf/poly lactic acid composites and stated that effective fiber–matrix adhesion and fiber roughness can be obtained by alkaline treatment. From the above literature, it is observed that few authors reported pineapple leaf fibers and most of the researches have been done with epoxy and polypropylenes and still, there is an option to fabricate pineapple/coconut sheath hybrid laminates with polyester matrix. The present research has fabricated the pineapple/coconut sheath hybrid laminates with different wt% of polyester matrix and has studied the tensile, impact, SEM, and flexural behavior of PCSFP hybrid composites.

2 Experimental Method

2.1 Materials

The PCSF reinforcements and coconut sheath were collected from the farms around Coimbatore, Tamilnadu, India, and they were separated as fibers. Figure 1 shows the separation process of pinapple leaves and coconut sheath. Commercially available unsaturated polyester resin matrix is well known for its applications like bulk and the sheet moulding compounds were supplied by Covai seenu pvt ltd, Coimbatore, India. Cobalt napthenate and MEKP, Methyl ethyl ketone peroxide were used as an accelerator and catalyst, respectively. The polymer matrix and catalyst accelerator were mixed in the ratio of 67:33.

2.2 Hybrid Laminate Fabrication

The targeted hybrid configurations were prepared using book press-type compression molding process [10]. The pretreated (cleaned and dried) PCSFs were prepared for

Fig. 1 Pineapple leaves fiber plant and coconut sheath

10 mm filaments and the suitability examination of coconut sheath and pineapple fiber was done before fabricating the composites. Initially, a known weight of 10 mm length pineapple fibers was randomly spread on the flat surface followed by coconut fiber and finally, pineapple fibers were spread over coconut fibers [11]. The whole configuration was in the sandwich structure of Pineapple fiber-coconut sheath-pineapple fiber. The fabricated sheets were kept in the temperature of 55 °C under vacuum pressure. Finally, sheets were compressed and cut into the size of 300 mm × 300 mm × 3 mm using the water jet machining process [11]. The composite laminates with different weight fractions of 60%, 65%, and 70% were fabricated. Figures 2 and 3 represent step by step fabrication process and fabricated hybrid laminate.

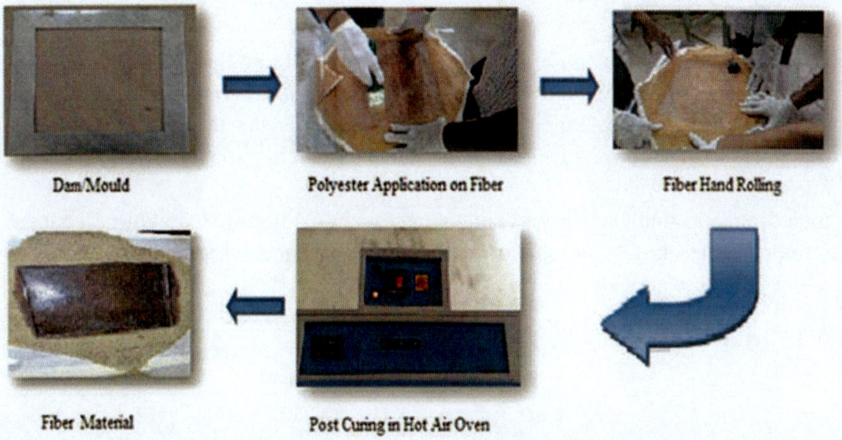

Fig. 2 Fabrication process

Fig. 3 Fabricated composite laminate

2.3 *Mechanical Tests*

The required size of specimens for tensile, impact, and flexural testing were cut as per ASTM D 638-03, ASTM D 256-88, and ASTM D 790-86, respectively, using the water jet machining process [12]. Shimadzu universal testing machine of 50 KN capacity with 0.5 mm/min cross-head speed consisting for tensile, flextural, and impact test. Minimum of three tests for each configuration of test samples were conducted in the ambient temperature—21 °C and 60% of relative humidity. Finally, the average values were reported.

3 Results and Discussion

3.1 *Tensile Behavior*

Table 1 represents the details of density, tensile strength, and elastic modulus [13] for comparison and it is evident that pineapple fibers have better mechanical properties than coconut fibers. Pineapple fibers have less density than other traditional natural fibers which is one of the attractive parameters for manufacturing light-weight composites.

Including a particular fiber percentage excess than a nominal weight percentage may improve the elastic modulus or behavior of laminates which proves the break

Table 1 Properties of natural fibers [13]

Fibers	Density (g/cm)	Tensile strength (Mpa)	Elastic modulus (Gpa)
Kenaf	1.45	930	53
Pineapple	0.8–1.6	400–627	14.5
Jute	1.3	393–773	26.5
Bamboo	0.6–1.1	140–1150	11–17
Coconut	1.15	131–175	4–6

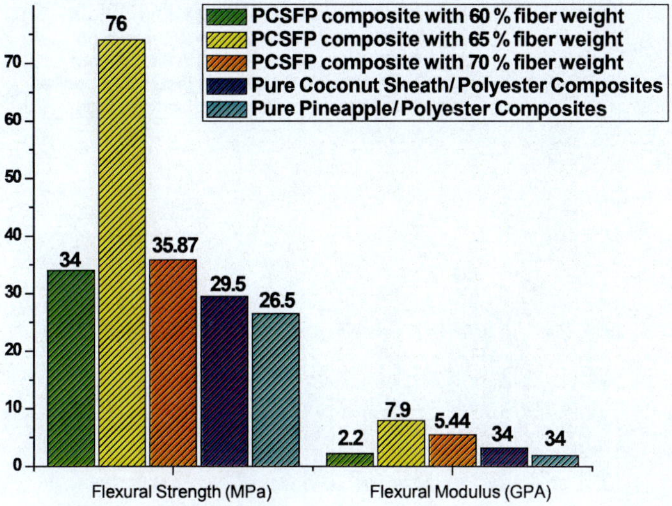

Fig. 4 Tensile properties of PCSFP composites at various fiber wt%

elongation values of composite structures [14]. It is also noted that stress level increased for fiber weight percentage for given strain levels for increase in total fiber volume fraction percentage. The relationship between fiber weight percentage and tensile strength of short PCSFP composites and pure coconut sheath/pineapple/polyester composites is demonstrated in Fig. 4. The tensile strength of PCSFP composites increased by increasing the fiber weight up to 65%. The tensile strength decreased when the fiber weight percentage was increased above 65%. The tensile results indicate that 65% of fiber weight showed better results compared to fiber wt% of PCSFP composites.

The optimum fiber wt% composite showed high mechanical properties since the load transfered from fiber to matrix were high. The maximum elastic modulus, elongation %, and tensile strength of short PCSFP composites were 0.7 GPa, 5.03%, and 11.70 MPa, respectively.

3.2 Flexural Behavior

Figure 5 shows the flexural properties on different fiber wt% of PCSFP composites with 10 mm fiber length. Elastic modulus and flexural strength of PCSFP composites increased by increasing the fiber weight up to 70%. Elastic modulus and flexural strength decreased by increasing the fiber weight percentage above 65%. The maximum elastic modulus and flexural strength of short PCSFP composites were 7.9 GPA, 76.44 MPa, respectively. From the flexural results, fiber weight 65% showed better results compared to fiber wt% of PCSFP composites.

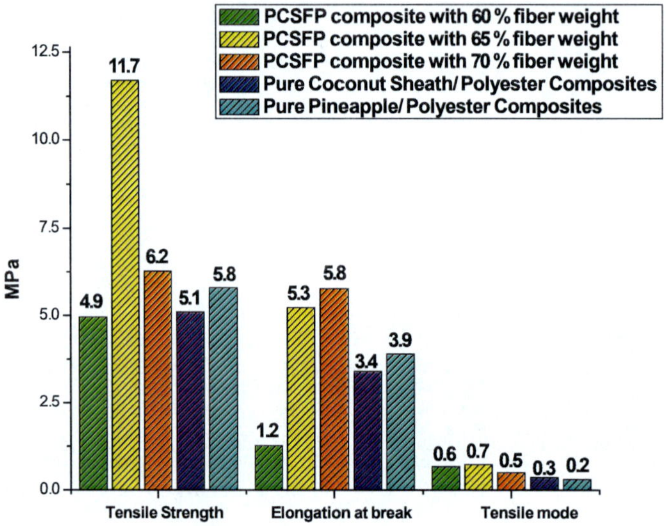

Fig. 5 Flexural properties of PCSFP composites at various fiber wt%

3.3 *Impact Behavior*

Figure 6 shows the impact strength on different fiber wt% of PCSFP composites with 10 mm fiber length. The impact strength of PCSFP composites increased by increasing the weight fraction of fibers. The impact strength decreased by increasing the PCSF weight from 65 to 70%. From the flexural results, fiber weight 65% showed

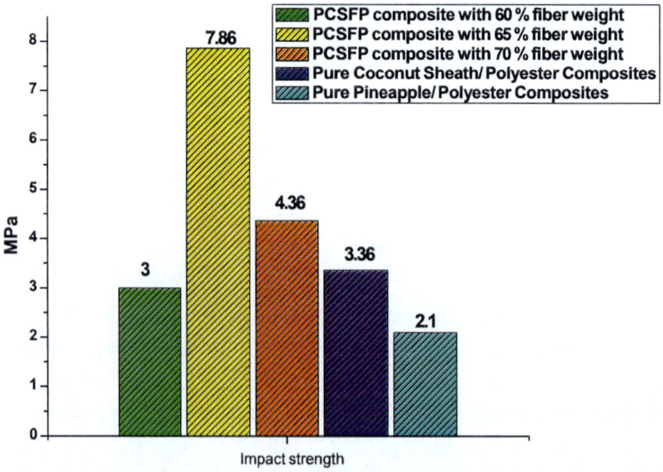

Fig. 6 Impact properties of PCSFP composites at a different fiber weight percentage

Fig. 7 a, b Tensile specimen—SEM micrograph of PCSFP laminates

better results compared to fiber wt% of PCSFP composites. The optimum impact strength of PCSFP composites was 7.86 kJ/m^2.

3.4 *Scanning Electron Microscope (SEM)*

The specimens were placed under the SEM machine to study the SEM structure of the PCSFP composites [15]. Figure 7a, b represents micrographs of the fractured surface and the interfacial connection among matrix material and fiber. Figure 7a shows the physical microstructure of coconut and pineapple fibers that are altered due to alkaline pretreatment. Elimination of fiber impurities by alkaline treatment results in better bonding between fiber and the matrix.

Besides, the tensile specimen SEM image of the PCSFP laminate shown in Fig. 7b indicates that fiber pullout, fiber breakage, and voids are present in the fractured samples. It occurs due to the moderate wettability of fiber and matrix [16]. The investigation of fracture surfaces suggests that the wettability of the PCSFP has to be improvised to obtain good mechanical properties.

4 Conclusion

The following conclusions were drawn from the experimental results.

- The impact, tensile, and flexural properties of arbitrarily aligned short PCSFP composites mainly depend on fiber volume fraction percentage.
- The results indicate that 65 wt% of fiber shows better results compared to 60 wt %, and 70% volume fractions.
- The maximum elastic modulus, elongation %, and tensile strength of short PCSFP composites were 0.7 GPa, 5.03%, and 11.70 MPa, respectively.

- The maximum elastic modulus and flexural strength of short PCSFP composites were 7.9 GPA, 76.44 MPa, respectively.
- The optimum impact strength of PCSFP composites was 7.86 kJ/m^2.
- Microstructure analysis discloses that general defects like delamination and fiber pullouts occur at the high load conditions and these defects greatly spoil the flexibility of hybrid laminates.

References

1. Dean A, Reinoso J, Sahraee S, Rolfes R (2016) An invariant-based anisotropic material model for short fiber-reinforced thermoplastics: coupled thermo-plastic formulation. Compos A Appl Sci Manuf 90:186–199. https://doi.org/10.1016/j.compositesa.2016.06.015
2. Hemanth RD, Senthil Kumar M, Gopinath A, Natrayan L (2018) Evaluation of mechanical properties of E-Glass and coconut fiber reinforced with polyester and epoxy resin matrices. Int J Mech Prod Eng Res Devel 7(5):13–20
3. Gagné M, Therriault D (2014) Lightning strike protection of composites. Prog Aerosp Sci 64:1–16. https://doi.org/10.1016/j.paerosci.2013.07.002
4. Manousaki A, Jancheva M, Grigorakis S, Makris DP (2016) Extraction of antioxidant phenolics from agri-food waste biomass using a newly designed glycerol-based natural low-transition temperature mixture: a comparison with conventional eco-friendly solvents. Recycling 1(1):194–204. https://doi.org/10.3390/recycling1010194
5. Santhosh MS, Sasikumar R, Natrayan L, Senthil Kumar M, Elango V, Vanmathi M (2018) Investigation of mechanical and electrical properties of Kevlar/E-glass and Basalt/E-glass reinforced hybrid composites. Int J Mech Prod Eng Res Devel 8(3):591–598
6. Ranganathan N, Oksman K, Nayak SK, Sain M (2016) Structure property relation of hybrid biocomposites based on jute, viscose and polypropylene: the effect of the fibre content and the length on the fracture toughness and the fatigue properties. Compos A Appl Sci Manuf 83:169–175. https://doi.org/10.1016/j.compositesa.2015.10.037
7. Kim KW, Lee BH, Kim HJ et al (2012) Thermal and mechanical properties of cassava and pineapple flours-filled PLA bio-composites. J Therm Anal Calorim 108:1131–1139. https://doi.org/10.1007/s10973-011-1350-y
8. Suresh Kumar SM, Duraibabu D, Subramanian K (2014) Studies on mechanical, thermal and dynamic mechanical properties of untreated (raw) and treated coconut sheath fiber reinforced epoxy composites. Mater Des 59:63–69. https://doi.org/10.1016/j.matdes.2014.02.013
9. Ramli SNR, Md SHS, Fadzullah ZM (2016) the effect of alkaline treatment and fiber length on pineapple leaf fiber reinforced poly lactic acid biocomposites. Jurnal Teknologi. 79:111–115. https://doi.org/10.11113/jt.v79.11293
10. Sreekumar PA, Thomas Selvin P, Jean Marc S, Kuruvilla J, Unnikrishnan G, Sabu T (2009) Effect of fiber surface modification on the mechanical and water absorption characteristics of sisal/polyester composites fabricated by resin transfer molding. Comp Part A 40(11):1777–1784. https://doi.org/10.1016/j.compositesa.2009.08.013
11. Natrayan L, Sivaprakash V, Santhosh M S (2018) Mechanical microstructure and wear behavior of the material AA6061 reinforced SiC with different leaf ashes using advanced stir casting method. Int J Eng Adv Technol 8: 366–371
12. Tarawneh MA, Sahrim HJ, Ahmad Rozaidi Rasid SY, Bahri YS, Seyong Ehnoum AR, Ka KZ, Seng LY (2011) Mechanical properties of thermoplastic natural rubber (TPNR) reinforced with different types of carbon nanotube. Sains Malaysiana 40(7):725–728

13. Natrayan L, Senthil Kumar M, Palani Kumar K (2018) Optimization of squeeze cast process parameters on mechanical properties of Al_2O_3/SiC reinforced hybrid metal matrix composites using taguchi technique. Mater Res Express 5:066516. https://doi.org/10.1088/2053-1591/aac873
14. Varga CS, Miskolczi N, Bartha L, Lipoczi G (2011) Improving the mechanical properties of glass-fibre-reinforced polyester composites by modification of fibre surface. Mater Des 31:185–193. https://doi.org/10.1016/j.matdes.2009.06.034
15. Natrayan L, Santhosh M S, Mohanraj R, Hariharan R. (2019) Mechanical and tribological behaviour of Al_2O_3 & SiC reinforced aluminium composites fabricated via powder metallurgy. IOP Conf Ser Mater Sci Eng 561(1): 012038. https://doi.org/10.1088/1757-899X/561/1/012038
16. Natrayan L et al (2017) An experimental investigation on mechanical behaviour of SiCp reinforced Al 6061 MMC using squeeze casting process. Int J Mech Prod Eng Res Devel 7(6):663–668

Chemical Treatment, Influence of Fiber Content, and Optimization of Hybrid Natural Fiber-Reinforced Composites

G. Venkatesha Prasanna, Tirunagari Jayadeep, and Nikhitha Poornabhodha

Abstract The principle purpose of this substance treatment was to shrink the water assimilation property of biofibrils and furthermore to improve the congeniality with resin. Right now, biocomposites were created by mixing 90% epoxy with 10% vinyl ester grid and fortifying the biofibrils in to the resin mix. The mechanical presentation of the composites was likewise affected by fiber amount and compound treatment of the fibrils. Exterior of the fibrils was altered by soluble base treatment prompting higher crystallinity of fibrils. The examination of tensile, compressive, and flexural properties of Bagasse–Luffa fibrils fortified biocomposites was completed for 10, 20, 30, 40, and 50% of surface changed composites and other untreated biocomposites. 40% fiber amount benzene diazonium chloride treated composite showed higher and ideal condition for the previously mentioned mechanical properties than the, 5% NaoH treated, 10% NaoH treated and 10, 20, 30, and 50% fiber amount exterior treated and untreated biocomposites.

Keywords Bagasse–Luffa fibers · Surface treatment · Mechanical tests

1 Introduction

Aside from the favorable circumstances utilizing artificial fibers in diverse applications, these man-made fibers have genuine impediments as far as their biodegradability, beginning preparing costs, recyclability, vitality utilization, machine scraped area, wear and well-being perils, and so forth. Be that as it may, the unfavorable ecological effects by man made fibers, like resurgence method produces a huge natural burden as artificial fibers are not decayed no problem at all. So as to beat this emergency, need of normal/inexhaustible fiber-based composites is requesting to make the world 'green' fibers [1, 2]. Biofibrils have been utilized as an effective material to engineered strands just as support for polymer composite materials. Biofibrils fundamentally comprise of cellulose, hemicelluloses, gelatin, lignin, and waxes

G. Venkatesha Prasanna (✉) · T. Jayadeep · N. Poornabhodha
Department of Mechanical Engineering, CVR College of Engineering, Hyderabad 501510, India
e-mail: gvpcvrmech@gmail.com

T. Rajmohan et al. (eds.), *Advances in Materials and Manufacturing Engineering*, Springer Proceedings in Materials 7, https://doi.org/10.1007/978-981-15-6267-9_38

[3, 4]. The wet retention property possessed by the common filaments constrains the utilization of these strands as potential support in the creation of composites. Polymer lattice is hydrophobic in nature [5, 6]. This may cause fiber and framework inconsistent and brings about feeble interfacial holding between the fiber and the network. The primary intension of this concoction treatment is to decrease the water assimilation property of the biofibrils and furthermore to improve the similarity with polymer grid. The polymer mix have numerous focal points interms of the items for explicit end use applications and upgrading matrix mix successful use, execution, and properties [5–8].

Epoxy matrix have generally utilized due to their unrivaled thermomechanical properties and brilliant processability. Be that as it may, the utilization of thermoset materials is frequently constrained due to low toughness property. Further mixing procedure can be utilized productively to defeat the sub-par properties of both the parts. Miscible polymer mixes produce a recently improved material from less unrivaled individual parts [9]. Thusly, when the mass measures of epoxies are utilized, typically it will mix with toughening operators [8, 9]. Vinyl ester a toughening material which can be utilized alongside epoxy materials. Vinyl ester is the extra result of an epoxy pitch with unsaturated carboxylic acid, for example, acrylic or methacrylic tart. Because of the adaptability present in the structure of epoxy, it is utilized as one of the mix materials in the current work. The hardness of the epoxy tar is expanded by mixing it with adaptable polymers and elastomers [10–12]. Henceforth, an appropriate polymeric material was expected to improve the impact resistance, stress–strain properties by preserving stiffness and warm soundness of the epoxy. To accomplish this, advancement between cross-connected polymer system of thermoset-thermoset mixes was examined on account of their expanded mechanical properties. Vinyl ester is the generally utilized thermosetting material on account of their less expense and great blend of properties like: chemical confrontation, mechanical, and thermal properties. The hydroxyl bunches upgrade the extremity of vinyl ester and encourage grip and fibrils wetting properties, which drove the wetting of the filaments in resins mix. For this reason, a mix of epoxy/vinyl ester (90/10% w/w) polymers was readied [5, 13, 14]. Composite examples with the 10, 20, 30, 40, and 50% substance of untreated, soluble base, and benzene diazonium chloride-treated Bagasse–Luffa strands with mix of epoxy/vinylester were manufactured.

2 Experimental

2.1 Materials

The two resins namely: 1. Epoxy araldite LY 556 as resin, hardener HY 951, 2. Vinyl ester.

The two fibers namely: 1. Luffa fiber, 2. Bagasse fiber were used to fabricate the composites.

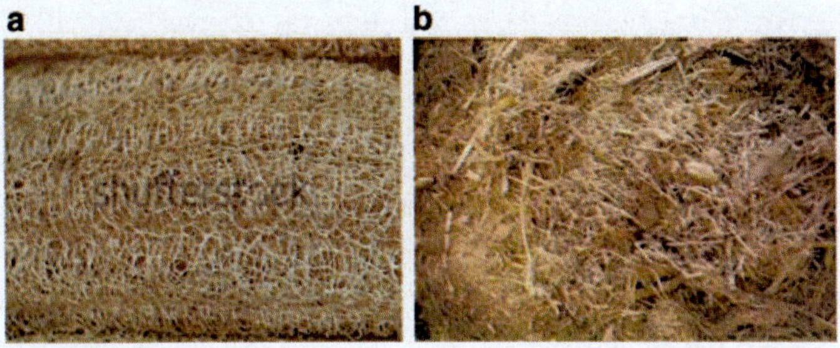

Fig. 1 **a** Luffa fiber, **b** Bagasse fiber

2.2 Surface Modification of Natural Fibers

Because of the nearness of hydrophilic hydroxyl bunches in the bo fibrils, scarcely any issues happen along the interface during the fortification of common fiber into the matrix. The hydrophilic behavior of biofibers thwarts the successful holding of filaments with resin. Aside from this, pectin, grease, oil, and waxy substances spread the responsive practical gatherings of the fiber and go about as an obstruction to interlock with the framework. To improve the viability of interfacial glue holding, surface of fiber should be altered with various chemical actions.

2.3 Chemical (Alkali) Treatment

So as to improve the viability of interfacial attachment/holding, fiber surface should be altered with various substance medications, responsive added substances, and coupling operators. Synthetic medications produce progressively responsive gatherings on the outside of the fiber and this encourages effective coupling with the lattice. Surface adjustment of strands by alkali treatment with different concentrations of NaOH: 5 and 10 wt% was done in this research. After the alkali treatment, the filaments were inundated in refined water for 24 h to expel the remaining NaOH. Further washing was completed with refined water containing a limited quantity of acetic acid. Strands were then dried out in a broiler at 70° C for 3 h.

2.4 Preparation of Benzene Diazonium Chloride

8 cm^3 of concentrated HCl was added to a boiling tube containing 3 cm^3 of phenyl amine (aniline) and 10 cm^3 of water, the mix was vibrated until the amine get dissolved, and then the solution was cooled to 5 °C by cooling it in an ice bath. Following that, a solution of sodium nitrite (3 g in 8 cm^3 of water), pre-cooled to 5

°C was added. The temperature of the combination was retained below 10 °C during the addition of sodium nitrite [15].

2.5 *Benzene Diazonium Treatment of Bagasse–Luffa Fibers*

Bagasse–Luffa fibers were sliced to a length of 10 mm, washed with cleansed water, and were then dry out in an oven at 70 °C for 24 h. The desiccated fibers were inundated in a 6% NaOH solution taken in a 2.0 L glass beaker for 10 min at about 5 ° C. A latestly prepared diazo solution was then transferred gradually into the above mixture with invariable stirring. Fibers were then taken out, washed with cleanser arrangement followed by refined water, lastly dried in an outdoors for 48 h [8, 15].

2.6 *Fabrication of Blended Hybrid Biocomposites*

The mould cavity was coated with a thin layer of hard wax for simple discharge and take out of fabricated composite specimens. At that point, after the wax was restored, a meager layer of fluid arrangement of poly vinyl alcohol (PVA) was applied. Biofiber hybrid composites were created utilizing the hand lay-up procedure. At that point, the mixture biostrands of the untreated and treated Bagasse–Luffa filaments were fortified to the grid mix to get the required biocomposites for the investigation of performance & and mechanical properties. At that point, air bubbles were evacuated cautiously by utilizing the roller with delicate rolling. With the end goal of complete fix, composite specimens were post relieved at temperature of 800 °C for 2 h by keeping the examples in hot broiler. After the total relieving, untreated and treated examples of biocomposite examples were tried.

3 Results and Discussion

3.1 *Tensile Strength Properties*

Specimens for the tensile property test were prepared with 150 mm × 15 × 3 mm and were evaluated as per ASTM standards. On account of hydrophilic behavior of biol fibers, i.e., water assimilation, contrariness was seen between biofilaments and matrix leads to lacking wettability, which brings about mediocre behavior of composites. To beat this, surface alterations of fibers were carried out and which reduce its hydrophilic tendency and thus improve compatibility with the matrix. The influence of fiber surface modification by the chemical treatment, % alkali treatment, and fiber quantity on the tensile property of biocomposites was indicated in

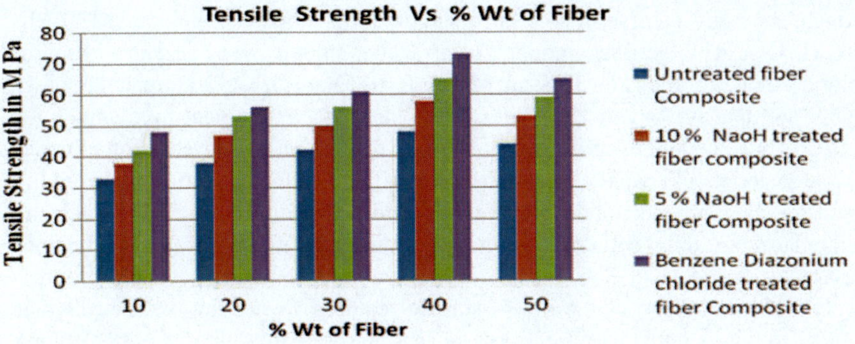

Fig. 2 Tensile strength of untreated, alkali-treated, and benzene diazonium chloride-treated composites

Fig. 2. In case of untreated bio iber composites, the repellency between hydrophobic matrix and hydrophilic natural fibers which would antagonistically influences the interfacial holding between fiber surface and grid and which brings about lessening of the fortification limit of the fiber because of the inappropriate stress move from the resin blend to strands. Among the different surface adjustments performed, it was likewise seen that 40% fiber amount benzene diazonium chloride-treated composite displayed unrivaled and ideal condition for ductile property than the, 5% NaoH treated, 10% NaoH treated and 10, 20, 30, and 50% fiber amount chemically treated and untreated biocomposites [4, 15]. The most elevated tensile strength was watched for 40% fiber stacking because of the improved fiber circulation in a lattice material which upgraded the morphological properties. Further in benzene diazonium treatment, the fiber surface turns out to be increasingly uniform because of the end of smaller scale voids and subsequently the pressure move limit between a definitive cells improves [12, 15]. Nonetheless, obvious and thus the stress transfer capacity between the ultimate cells improves [12, 15]. Nonetheless, obvious higher rigidity was seen in alkali-treated Bagasse–Luffa composites and noteworthy increment was seen in the tensile strengths of benzene diazonium chloride-treated Bagasse–Luffa fiber composites. Benzene diazonium chloride treatment decreases the hydrophilic idea of Bagasse–Luffa strands contrasted with antacid treatment because of the coupling of the hydroxyl gathering of fiber with benzene diazonium chloride. Chemical treatment increases the aspect ratio and increases surface roughness [1, 11]. In this manner upgrades the powerful fiber surface region for good attachment with the grid which prompts increment in the interfacial grip between the Bagasse–Luffa strands and the polymer network material. The examination unmistakably proposes that the chemical reaction formed a strong interface with fiber cell walls and coupling response will happen between benzene diazonium chloride and cellulose of fiber bringing about the development of a diazocellulose compound, which represents the watched huge increment in tensile properties of composites. The ideal condition was a result of ideal fiber amount and concoction (benzene diazonium chloride) treatment, which advanced the better interfacial cement holding between strands framework mix brought about sufficient resulted in adequate stress transfer and performance [3, 4, 9]. Yet, where as at higher fiber amount, i.e., 50% fiber amount even benzene diazonium chloride-treated composite, 5% NaoH treated, 10% NaoH treated, untreated biofiber composite displayed mediocre values [16]. The surface treatment by methods for both benzene diazonium chloride, NaoH, gives the route for the evacuation of undesirable materials like waxes, oils, debasements, hemicellulose, and lignin from the characteristic filaments notwithstanding lessening of dampness ingestion yields the higher level of (alpha) cellulose in common strands and furthermore cleans the fiber surface [8, 10, 15, 17]. Benzene diazonium chloride treatment advances progressively powerful region of the fiber's surface required for wetting by the polymer contrasted with soluble base treatment so as to have great holding [12, 18, 19] and results in improved mechanical properties [6, 20, 21].

3.2 *Flexural Properties*

The specimens with measurements 150 mm × 15 mm × 3 mm were created and tried for flexural properties according to ASTM D 5943-96 determinations. The variety in the estimations of flexural quality as an element of fiber amount, with and without synthetic treatment was appeared in Fig. 3. After the examination, it was seen that the flexural property of the 40% fiber amount, benzene diazonium chloride-treated biofibers composite indicated higher and predominant incentive than the 40% fiber amount, 5% NaoH treated, 10% NaoH treated, 10, 20, 30, and 50% fiber amount untreated and treated composites [3, 10, 15]. Because of more alkali concentration at (10% NaoH) than the (5% NaoH), abundance delignification of the fiber happens, which brought about debilitating or harming of the fiber surface that brought about diminished flexural quality, showed by Fig. 3 [22, 23]. High fiber-framework similarity, great fiber–network connection and holding was watched for 5% NaoH treated biofibers composite than 10% NaoH-treated biocomposites [17, 18, 23]. Better and

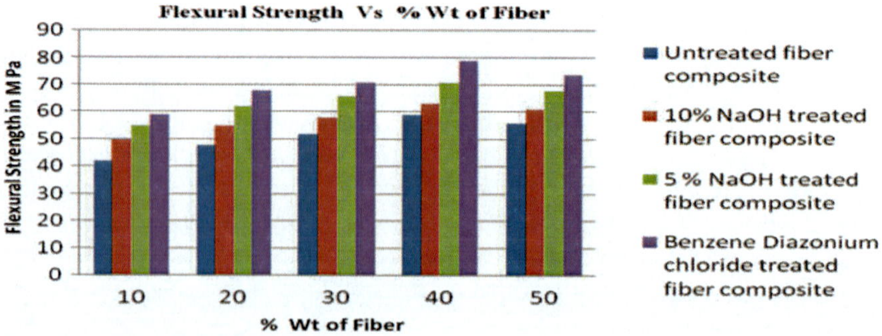

Fig. 3 Indicates the change of flexural property of unprocessed and chemically treated composites with the variation percentage of fibers content

significant improvement of fiber conveyance in a grid, aspect ratio & and stress shift capability was watched in benzene diazonium chloride-treated biofibril composites than in alkali-treated biofibril composites and untreated composites [24–26].

3.3 Compressive Strength

The specimens were fabricated & tested for compressive strength properties as per ASTM D specifications. The alteration in the values of compressive strength property with compressive quality property regarding fiber amount with and without surface change was appeared in Fig. 4. It was likewise seen that benzene diazonium chloride-treated biocomposites displayed higher and ideal condition for compressive property than the soluble base-treated and untreated biocomposites [3, 4, 10, 15, 27]. The ideal condition was a direct result of ideal fiber stacking, and concoction treatment, which advanced the better interfacial holding between filaments framework mix brought about satisfactory resulted in adequate stress transfer and performance [8, 11, 12, 16]. Past works uncover that fiber surface change by concoction treatment decreases the fiber's distance across and expands surface harshness [28, 29]. In any case, decline in compressive property of the composite was seen at 50% fiber amount. For 50% fiber content, untreated & treated composites (high fiber load), decline in the compressive property shows poor fiber and network interfacial holding and small-scale break development at the interface prompts inadequate stress transfer [30–32]. Consequently, most noteworthy compressive quality was seen at 40% fiber stacking [21, 26]. Moreover, benzene diazonium treatment occupies the voids spaces of fiber and adjusts lopsided fiber, more as thought about soluble base-treated composites [33–35].

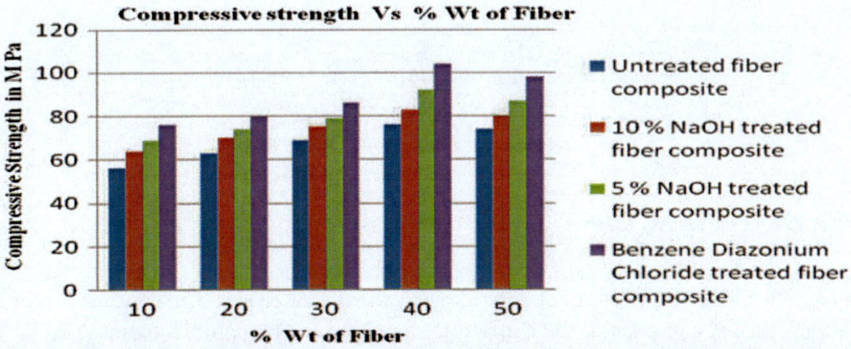

Fig. 4 Compressive strength of untreated, alkali-treated, and benzene diazonium chloride-treated composites

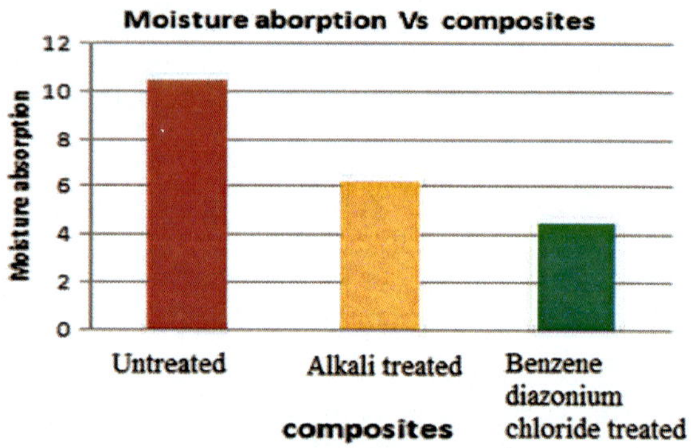

Fig. 5 Moisture content absorption of untreated, alkali-treated, and benzene diazonium chloride-treated composites

3.4 *Moisture Content Test*

Figure 5 demonstrates the rate retention of dampness amount by natural fiber composite, alkali-treated, and benzene diazonium chloride-treated composites [5, 21, 32]. The biofibril composites instigated from the hydrophilic nature of the fiber reveal hydrophilic properties [4, 11, 15]. But the alkali sensitive hydroxyl groups are present.

Alkali sensitive hydroxyl groups present among the molecules are broken down, which then react with water molecules (H-OH) and move out from the fiber structure [20, 25, 31]. Because of which, hydrophilic hydroxyl groups were diminished and increases the fibers dampness opposition property [3, 9, 16]. Due to the chemicals treatment, the fiber exterior becomes more uniform which leads to the elimination of micro voids and increases the effective fiber surface area for good bonding with the resin. The rate ingestion of dampness amount by natural fiber composite is more when contrasted with alkali-treated and Benzene diazonium chloride-indulged composites [8, 29].

$$\%\text{water absorption} = 100 \times (W_f - W_i)/W_i$$

3.5 *Dielectric Strength*

To learn the dielectric strength of unprocessed, alkali-treated and benzene diazonium chloride-treated composites, the composite samples were prepared as maintained by the ASTM D-149 standards. The composites having measurements of 120 mm × 120 mm × 3 mm were armored with fibers in one direction along 120 mm length. The dielectric break down voltage was instituted at five points for every specimen, and normal worth was considered for investigation. The test was done at 50 Hz recurrence and room temperature. Advanced micrometer of 0.001 mm least tally was utilized

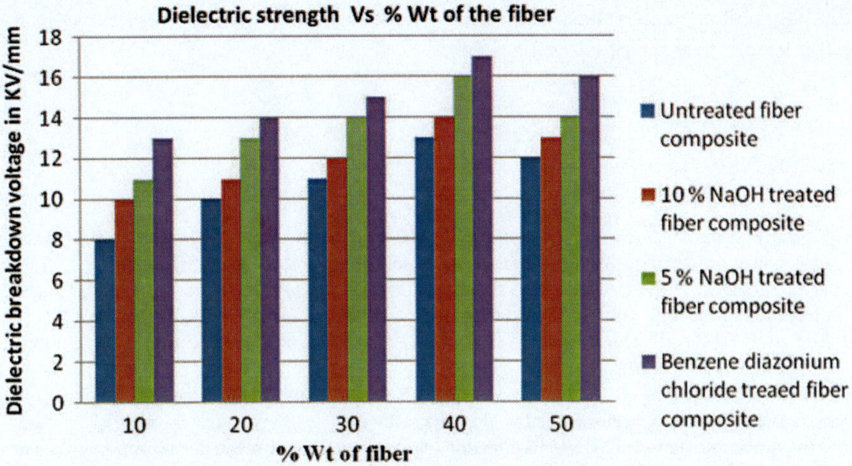

Fig. 6 Reveals the dielectric strength of untreated and chemical-treated composites

to discover the thickness of the composites at separate point [36]. It was likewise calculable to see that the dielectric quality of mixture fiber composites increments with increment in amount of fiber from 10–40% weight of fiber yet diminishes at 50% fiber amount composite [14, 20, 21, 33]. At 50% fiber quantity composite, decrease in dielectric strength was due to poor interfacial bonding between the fibers and matrix blend [4, 10, 12, 15].

4 Conclusion

The work did shows that the concoction treatment impressively improves the fiber quality, fiber–network bond, and thus the exhibition of the common fiber composites. The tensile, compressive, and flexural property were seen to have enhanced for both alkali and benzene diazonium chloride treatment of fiber composites. The outcome confirms that the tensile, compressive, and flexural of the composites expanded with increment in fiber stacking up to 40% and past 40% indicated an abatement in ductile, compressive, and flexural quality. Generally, overabundance measure of fiber rests onto each other as opposed to being blended in with 50% fiber loads, which additionally prompts decreased stress transfer ability. The rate retention of dampness amount by natural fiber composite was more when contrasted with soluble base-treated and benzene diazonium chloride-treated composites. Yet, after the chemical action or surface alteration of filaments, it was discovered that upgrade in the properties of the composites with the evacuation of contaminations like hemi-cellulose, oils, waxes, dampness, and lignin from the surfaces of the strands, and yields harsh surface geology, which improves the viewpoint proportion of strands and expands the strengthening effectiveness of the fiber with the improvement of stress move from the framework to stack bearing filaments.

References

1. Kaith BS, Singha AS, Sanjeev K, Susheel K (2008) Merceriza-tion of flax fibre improves the mechanical properties of fibre-reinforced composites. Int J Polym Mater 57(1):54–72
2. Tiwari S, Bijwe J, Panier S (2011) Tribological studies on Polyetherimide composites based on carbon fabric with optimized oxidation treatment. Wear 271:2252–2260
3. Edeerozey AM, Akil HM, Azhar AB, Ariffin MZ (2007) Chemical modification of kenaf fibres. Mater Lett 61:2023–2025
4. Venkatesha Prasanna G, Jha NK (2019) Optimisation and mechanical testing of hybrid bio composites. Elsevier, Materials today, pp 1395–1401
5. Venkatesha Prasanna G (2018) Modification and optimization of mechanical properties of bio fibers blended composites. Int J Eng Technol 7(2.33):799–801
6. Murali Mohan Rao K, Rao M (2007) Extraction and tensile properties of natural fibers: Vakka, date and bamboo. Compos Struct 77:288–295
7. Venkatesha Prasanna G, Venkata Subbaiah K, Varada Rajulu A (2012) Chemical resistance, impact, flexural, compressive properties and optimization of fibers of natural fibers reinforced blend composite. Schol J Eng Res 1(6):85–89
8. Punyamurth R, Sampathkumar D, Bennehalli B, Srinivasa PB, Vekateshappa C (2014) Surface modification of abaca fiber by benzene diazonium chloride treatment and its influence on tensile properties of abaca fiber reinforced polypropylene composites. Ciencia Technologia dos Materials 26(2):142–149
9. Gowdagiri Venkatesha P, Mahurkar J, Arun P, Sangawar N (2019) Effect of hybridization, Influence of surface modification on the properties and optimization of composites. Springer, The smart innovation, systems and Technologies, ISSN 2190-3018
10. Venkatesha Prasanna G, Kumar Jha N, Harish C (2018) Chemical treatment and analysis of mechanical properties of hybrid biocomposites. Int J Mech Prod Eng Res Devel, Special Issue:189–193. ISSN (P):2249-6890; ISSN (E): 2249-8001
11. Venkatesha Prasanna G, Shivananda HK, Venkata Subbaiah K (2011) Flexural & impact properties of banana-palmyra fibers reinforced epoxy-unsaturated polyster blended composites. Int J Mech Eng 4(1)
12. Raghavendra Rao H, Ashok Kumar M, Ramachandra Reddy G (2011) Hybrid composites: effect of fibers on mechanical properties. Int J Macromol Sci 1(1):9–14
13. Ray D, Sarkar BK, Rana AK, Bose NR (2001) The mechanical properties of vinylester resin matrix composite. Compos A Appl Sci Manuf 32:119–127
14. Venkatesha Prasanna G, Kethamukkala K, Akhil Kumar K (2019) Surface modification, characterization & optimization of hybrid bio composites. Springer, The smart innovation, systems and Techhnologies, ISSN 2190–3018
15. Alamgi K, Islam MRM, Bledzki AK (2010) Bio Res 5:1618
16. Bessadok A, Marais S, Roudesli S, Lixon C, Metayor M (2008) Influence of chemical modifications on water sorption and mechanical properties of Agave fibre. Compos Part A 39:29–45
17. Yu T, Ren J, Yuan S, Li H, Li Y (2010) Effect of fibre surface-treatments on the properties of poly (lactic acid)/ramie compo-sites. In: Composites part a: applied science and manufacturing, vol 41, pp 499–505
18. Kalia S, Kaith BS, Kaur I (2009) Pretreatments of natural fibres and their application as reinforcing material in polymer compo-sites—a review. Polym Eng Sci 49:1253–1272
19. Mohanty AK, Misra M, Drzal LT (2001) Surface modifications of natural fibers and performance of resulting bio-composites. Compos Interf 8(5):313–343

20. Qianqian S, Xianhua C (2008) Effect of rare earths surface treatment on tribological properties of CFs reinforced PTFE composite under oil lubricated condition. J Rare Earths. 26(4):584–592
21. Wambua P, Ivens J, Verpoest I (2003) Natural fibers: can they replace glass in fiber reinforced plastics. Compos Sci Technol 63:1259–1264
22. Seena J, Koshy P, Thomas S (2005) The role of interfacial interactions on the mechanical properties of banana fibre reinforced phenol formaldehyde composites. Compos Interf 12:581–600
23. Zhang XR, Zhao P, Pei XQ, Wang QH, Jia Q (2007) Flexural strength and tribological properties of rare earth treated short CFs/polyimide composites. Ex Poly Lett 1(10):667–672
24. Gowdagiri Prasanna V, Venkata subbaiah K (2013) Modification of tensile, compressive properties & chemical resistance of hybrid biocomposites. Int J Nanomater Biostruct 3(1):9–12. ISSN 2277-3851
25. Rana AK, Mandal, Bandyopadhyay AS (2003) Short jute fiber reinforced polypropylene composites: effect of compatibiliser, impact modifiber and fiber loading. Compos Sci Technol 63:801–806
26. Murali Mohan Rao K, Ratna Prasad AV, Ranga Babu MNV, Mohan Rao K, Gupta AVSSKS (2007) Tensile properties of elephant grass fiber reinforced polyester composite. Mater Sci 42:2666–2672
27. Myrtha K, Holia O, Anung S (2007) Physical and mechanical properties of natural fibers filled polypropylene composites and its recycle. J Biol Sci 7:393–396
28. Zhang XR, Pei XQ, Wang QH (2007) The effect of fiber oxidation on the friction and wear behaviors of short-cut CFs/polyimide composites. Ex Poly Lett 1(5):318–325
29. Sreekala MS, Thomas S (2003) Effect of fiber surface modification on water-sorption characteristics of oil palm fibres. Compos Sci Technol 63:861–869
30. Venkatesha Prasanna G, Venkata Subbaiah K (2013) Hardness, tensile properties and morphology of blend hybrid biocomposites. Sch J Eng Res 2(1):1–29
31. Venkatesha Prasanna G, Venkata Subbaiah K (2013) Modification, flexural, impact, compressive properties & chemical resistance of natural fibers reinforced blend composites. Malay Polym J 8(1):38–44
32. Venkatesha Prasanna G, Jayadeep T, Reddy KM (2019) Influence of fiber length & surface treatment on the mechanical properties of hybrid composites, Springer, The smart innovation, systems and Technologies, ISSN 2190-3018
33. Su F, Zhang Z, Wang K, Jiang W, Liu W (2005) Tribological and mechanical properties of the composites made of carbon fabrics modified with various methods, 36:1601–1607
34. Zhiwei X, Yudong H, Yuanjun S, Chunhua Z, Li L (2007) Surface characteristics of rare earth treated CFs and interfacial properties of composites. J Rare Earths 25:462–468
35. Akin DE, Gamble GR, Morrison WH, Rigsby LL (1996) Chemical and structure analysis of fiber and core tissues from flax. J Sci Food Agric 72:155–165
36. Morrison RT, Denmi Z, Boyd RN (1989) Organic Chemistry. Prentice-Hall, London

Intelligent Automatic Guided Vehicle for Smart Manufacturing Industry

S. Chandramohan and M. Senthilkumaran

Abstract In the vision of "Smart Industry," the industrial process and control, automation, monitoring and network connectivity are extended to advancements in technology from very simple to the point where anything can connect through Internet of things (IoT). Simplifying, Industry 4.0 is the next industrial revolution, which exists throughout the automatic guided vehicles (AGV), software defined networks (SDN), robotics, IoT and cloud computing. In this paper, integrating the industrial robots with automatic guided vehicle is proposed in order to make autonomous decision-making capabilities. Subsequently, the possible solutions for industrial wireless networks through IIoT are discussed. Finally, this chapter demonstrates the industrial robot equipped with active RFID for effective logistics environment in a real factory scenario.

Keywords Industrial robot · AGV and active RFID

1 Introduction

Smart Industry is a concept of utilizing the manufacturing process by making use of advanced technology. In this regard the entire process can be facilitated using wireless sensor networks (WSN), Software Define Networks with virtualization, adaptive decision-making models and cloud based data analytics. In the context of Industry 4.0, smart manufacturing system is considered to be the future manufacturing system that is secured by adopting latest models and technologies to commute the traditional system into a intelligent manufacturing industry [1].

S. Chandramohan (✉) · M. Senthilkumaran
Department of ECE, SCSVMV, Kanchipuram, India
e-mail: chandramohan@kanchiuniv.ac.in

M. Senthilkumaran
e-mail: msenthilkumaran@kanchiuniv.ac.in

T. Rajmohan et al. (eds.), *Advances in Materials and Manufacturing Engineering*, Springer Proceedings in Materials 7, https://doi.org/10.1007/978-981-15-6267-9_39

Table 1 Characteristics of IMI

S.No.	Key components	Description
1	Customization	Flexibility and control for certain products
2	Scalability	Usage of system objects with respect to expansion or shrinkage
3	Inerrability	Quick integration of new components and technologies
4	Mobility	Local, unrestricted mobility of objects within a system
5	Convertibility	Small conversion times between various manufacturing statuses
6	Interconnectivity	Permit different connections inside and outside of a system

2 Automated Guided Vehicle

More than a decade, automated guided vehicles (AGVs) have been operating impressively in medium- and large-scale industries. These vehicles have capability to adapt in the process to the automation [2]. The biggest consumer of AGVs is the automotive production industries like warehousing and distribution, textiles, etc. In manufacturing industry, logistic systems play a vital role towards efficient delivery. Smart AGV is popularly used and can be considered as the suitable option for effective utilization of the multiple products delivery. In a dynamic environment and situation like decision-making inside a factory, it needs guidance, which is highly complex in a large-scale industry. Therefore, this paper provides a possible solution to use industrial robots for the guidance and make AGV more flexible in a challenging environment [3]. An optimized production is the concept of intelligent manufacturing industry (IMI) by utilizing the technological advancement. In the realization of the above process, industrial Internet of things (IIoT), virtual management (VM) and cloud computing play as vital role. Also the human contribution in an IMI can be minimized (Table 1).

3 Industrial Automation

A robot is a re-programmable and multi-functional operator, designed to shift materials, objects, products and tools through programmed motions. It may vary accordingly with the task, which also acquire relevant information and move intelligently in response. For example, technique like application-layer automation reflects the activities of a rules-based process with no compromise for the existing IT architecture. In the industrial automation, process can expedite procurement, supply chain management and human resources or business processes that require swing around access to multiple existing systems [4, 5].

Fig. 1 Autonomous robots

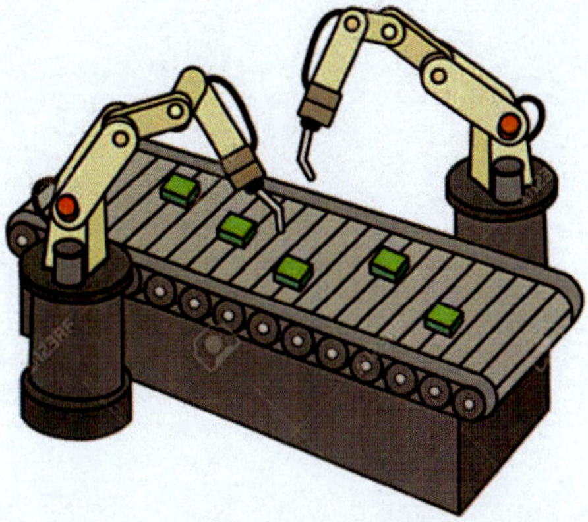

3.1 Robot-Based Process Automation

The objective of robot-based process automation (RBPA) is to replace repetitive task performed by humans. It does not need the development of logic, nor does it needs direct access. Also, the interaction with the existing IoT architecture with no complex system integration required [6].

RBPA is the recent technique used to configure software or a "Smart Robot" to interpret existing applications for computing data and communicating with other digital systems. RBPA "Intelligent Robot" is revolutionizing the way we think in the business perspective [7, 8]. Finally, it provides good improvements in accuracy, cycle time and increased productivity by removing repetitive task. Hence, we introduce a new direction for robot automation, as shown in Fig. 1.

3.2 Intelligent AGV and RBPA

Automated guided vehicles have been utilized effectively for product and material handling and mainly for transportation inside a manufacturing industry. Despite this may be the ideal scenario, the reality is that some tasks demand humans and the shop floor is a mix of automated that change by situations [9, 10]. To handle these complex situation, such as a change in layout for different environments, the AGV has to adopt intelligently. In our proposed work, an RFID enabled intelligent AGV with an arm and gripper attached, as shown in Fig. 2. It is used to pick the objects and place according to robot fixed laid-out path. RFID is used for guidance, motion control and object identification. It transmits information via RF waves of a specific frequency [11].

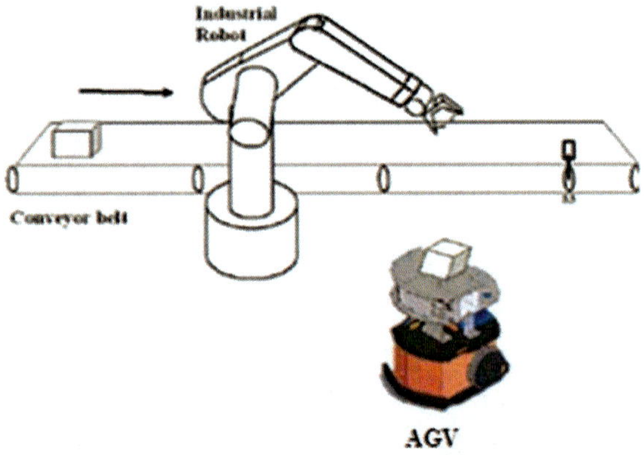

Fig. 2 Integration of AGV and RBPA

Hence, this chapter focuses on the technique for designing a smart AGV system reconfigurable in challenging environments [12, 13]. Since RBPA provides technological solutions to various types of businesses, operating models that provide automation will enhance quality and efficiency.

4 Internet of Things (IoT)

"Internet of Things, Data and Service" is recommended by Germany Trade and Invest in the appearance of the fourth industrial revolution called "Industry 4.0" or "Smart Factory" that will transform the traditional manufacturing production. "Industry 4.0" is a technological revolution from traditional manufacturing systems to cyber-physical systems (CPS) [14]. In the vision of the Internet of things (IoT), network connectivity is extended to electronic devices from very simple to the point where anything can connect to the Internet.

4.1 Industrial Internet of Things (IIoT)

This CPS interacts between the virtual versus real world and intelligent "smart" objects are a new aspect on production processes. Cyber-physical systems represent the next revolutionary step based on existing embedded systems. IIoT found with the advantages of optimized production, real-time monitoring and autonomous decision-making towards feature of predictive and proactive maintenance. Also, the IoT combined with the Internet of Services will enable Industry 4.0. Therefore, these systems promise to revolutionize our interaction with the physical world just as the Internet has transformed communication and personal interaction.

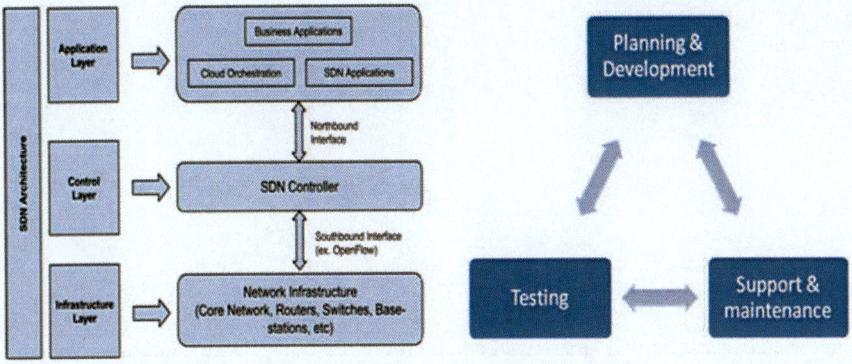

Fig. 3 SDN open flow architecture with RBPA methodology

In a larger-scale smart manufacturing company, Industrial IoT is a key element which brings together modern cloud computing, trend of automation and AI to create intelligent, self-optimizing industrial equipment and facilities. Thus, the design principles on smart product utilize the latest sensor technology to enhance various equipment with remote monitoring capabilities and its applications extend through data-driven automation practices in the industrial sector of economy. Finally, the IIoT deliberates priority for manufacturing companies as it empowers them to provide more value to their customers.

5 Design and Implementation

To validate our work, we implement the AGV to interact the entire system using this architecture (Fig. 3). In order to implement IIoT in the context of Industry 4.0, all kinds of devices should interact effectively. Under the architecture of SD-IIoT [15], we propose IIoT as an evolution of robot to machine (R2M) and machine to machine (M2M) by introducing intelligent automatic guided vehicle with industrial robotics.

6 Results and Discussion

We first review the correlations of related devices such as AGV, RBPA and cyber-physical systems (CPS). SD-IIoT has been multiple incarnations, such as sensor networks, M2M and industrial robot. While the correlations among M2M, WSNs, CPS and robot are still having some limitations. The proposed work improves the utilization rate by integrating the intelligent AGV and RBPA. In the material handling process at conveyer belt, our intelligent AGV loads parallel along with the fixed industrial robots. Therefore, our proposed work utilizes the resources effectively. Figure 4

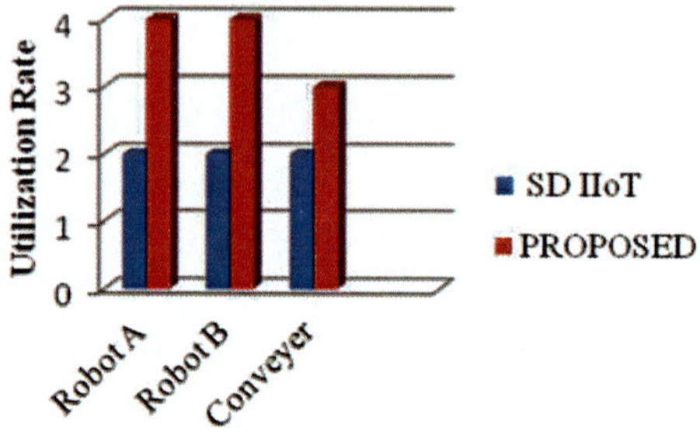

Fig. 4 Material management

shows the utilization rate is increased when compared with SD-IIoT architecture. In the mean while, the total energy consumption is reduced significantly.

7 Conclusion

In the context of Smart Industry, intelligent AGV, RBPA and cloud resources nurture the production of smart factories. The work designs a structure of the RBPA-AGV model for shop floor. To enhance the intelligence level of these systems, which feature new concepts through the integration of SD-IIoT, AGV and industrial robots systems are crucial [16]. In this chapter, we introduce an intelligent AGV-robot-based smart control model for material handling to improve the efficiency of equipment utilization rate and reduce the energy.

References

1. Wang S, Wan J, Zhang D, Li D, Zhang C (2016) Towards the smart factory "a self-organized multiagent system assisted with big data based feedback and coordination Elsevier computer networks," Comput Netw 101:158–168
2. Keliang Zhou R, Lu S, Wan J, Ya H, Liu*n Q (2013) Enabling cyber physical systems with M2M technologies. Ad Hoc Ubiquitous Comput 13(3/4)
3. Mehami J, Zhong RY, Mauludin N (2018) Automated guided vehicles for manufacturing in the context of smart industry 4.0" 2351-9789, Elsevier under responsibility of the scientific committee of the North American manufacturing research conference
4. Kim S, Jung K, Lee I, Song H, Kim J (2012) Vision guidance system for AGV using ANFS. In: International conference on intelligent robotics and applications, pp 377–385
5. Nielsen A (2017) A methodology for implementation of mobile robot in adaptive manufacturing environments. Intell Manuf 28(5):1171–1188
6. Luo Y, Ying D, Pace P, Li W, Fortino G (2018) Workshop networks integration using mobile intelligence in smart factories, 0163-6804/18/$25.00 © 2018 IEEE. In: IEEE communications magazine
14. Chengliang L, Lloret J, Wan J, Tang S, Hua Q (2018) IOT journal of cloud robotics for material handling in cognitive industrial internet of things. IEEE Internet Things J 5(4)
8. Harashima F, Lee S, Lee KC, Lee H (2002) The Integration of mobile vehicles for automated material handling using Profibus and IEEE networks. IEEE Trans Ind Electron 49(3):693–701

9. Zhong K, Newman ST (2017) The intelligent manufacturing in the context of industry 4.0: a review. Elsevier Eng 3:616–630
10. Xu C, Zhong RY, Lu SP, Wang LH (2017) A RFID-enabled positioning system in automated guided vehicle for smart factories. J Manuf Syst 44:179–190
11. Zhong SL, Xu C (2016) An active-RFID tag enabled locating approach with multipath effect elimination in AGV. IEEE Trans Autom 13(3)
12. Bryan N, David S, Alonzo K Infrastructure free AGV based on computer vision. In: IEEE robotics & automation magazine
13. Zhong RY, Wang H, Xu (2017) IoT enabled real time machine status monitoring approach for cloud manufacturing.Procedia CIRP 63:709–714
14. Ahmad J, Chen B, Imran M, Li D, Liu C (2018) Towards dynamic resource management for IoT based manufacturing. IEEE Commun 0163-6804/18/2018
15. Li D, Shenglong T, Wan J, Shu Z, Wang S, Athanasios (2016) Software defined IIoT (SD-IIoT) in the context of industry 4.0. IEEE Sens J 16(20)
16. Lee DY, DiCesare F (1994) Integrated scheduling of flexible manufacturing systems employing automated guided vehicles. IEEE Trans Ind Electron 41(6):602–610

Sensitivity Analysis for Welding Parameters in Dissimilar Friction Stir Welded Joints Using Response Surface Methodology

K. Palani, C. Elanchezhian, B. Vijaya Ramnath, and A. Adinarayanan

Abstract This investigation aimed to study the influence of the critical friction stir welding (FSW) parameters involved on the tensile properties, viz. tensile strength and tensile elongation of the dissimilar AA8011-H24 aluminium alloys and Ti3Al2.5 titanium alloy joints. The process parameters, viz. rotational speed, axial feed, and welding speed, with five various tool pin profiles have used the FSW experiments on the responses, such as tensile strength and tensile elongation. The mathematical models are developed from the responses using the four factors, five-level central composite design of response surface methodology. The effectiveness of the FSW process parameters for the tensile properties and changes in welding parameters affects the tensile strength and tensile elongation using the sensitivity analysis.

Keywords Dissimilar friction stir welding · Response surface methodology · Sensitivity analysis

1 Introduction

The potential applications of joining of Ti alloys due to its high strength to weight ratio, excellent specific strength, and better corrosion resistance with Al alloys due to its reduction in weight density, and low cost of the structures which play the critical challenge to reduce fuel cost and improve the performance in aviation, automobile, shipbuilding, and nuclear industries [1]. The critical parts and the rest of the structures are made of Ti alloys and Al alloys in the aerospace and marine industries [2].

K. Palani (✉)
Department of Mechanical Engineering, Sri Chandrasekharendra Saraswathi Viswa Maha Vidyalaya, Enathur, Tamil Nadu 631561, India
e-mail: palani.k@kanchiuniv.ac.in

C. Elanchezhian · B. Vijaya Ramnath
Department of Mechanical Engineering, Sri Sairam Engineering College, Chennai, Tamil Nadu 600044, India

A. Adinarayanan
Department of Mechanical Engineering, AMET University, Chennai, Tamil Nadu 603112, India

T. Rajmohan et al. (eds.), *Advances in Materials and Manufacturing Engineering*, Springer Proceedings in Materials 7, https://doi.org/10.1007/978-981-15-6267-9_40

However, it is very difficult to fabricate the sound welds due to the presence of brittle inter-metallic compounds (IMCs) in the Ti/Al welded joints. The IMCs deteriorates the strength properties due to the large difference in the metallurgical, chemical, and physical properties of Ti alloy and Al alloy of the joints in conventional fusion welding processes [3]. The quality of the welded joints is decided by the selection of suitable process parameters on the strength properties of the welded joints. Several approaches are made earlier to fabricate the dissimilar friction stir welded joints. The effect of pin profiles and rotational speed on the microstructure and tensile strength of dissimilar AA8011-H24 and AA6061-T6 friction stir welded aluminium alloys are discussed [4]. They studied that the performance on the tensile strength is examined in the weldment based on the different rotational speeds and the tools used in the FSW process.

The optimum process parameters are determined using the statistical technique, response surface methodology to enhance the strength properties of the welded joints [5–7]. The sensitivity analysis is implemented to find the optimum effective process parameters which are involved the quality of the friction stir welded AA6061-T6 aluminium alloy joints [8]. The dissimilar FSW joints with the sensitivity of the process parameters are not discussed from the current researchers. In this investigation, the brief discussion is made on the sensitivity analysis of welding parameters in dissimilar friction stir welded AA8011-H24 aluminium alloys and Ti3Al2.5 V titanium alloys joints using response surface methodology.

2 Materials and Methods

In this present work, the 6 mm thick rolled plates of Ti3Al2.5 V titanium alloys and AA 8011-H24 aluminium alloys were procured from Kataria steel distributors in Mumbai with the dimensions of 100 mm $\times$ 100 mm to fabricate the friction stir welded (FSW) butt joints [9]. The AA8011-H24 aluminium alloys have the major alloying elements of 0.74% Fe, 0.56% Si, 0.13% Cu and remaining Al with 140 Mpa tensile strength, while the Ti3Al2.5 V titanium alloys have 0.3% Fe, 3% Al, 2.5% V and remaining Ti with 620 MPa tensile strength. The five different tool profiled tools, viz. straight cylindrical, tapered threaded cylindrical, straight hexagon, tapered hexagon, straight threaded cylindrical, are used with the dimensions of 6 mm pin diameter, 18 mm shoulder diameter and 5.7 mm pin length and are made of M2 tool steel with Cobalt coating. The experiments are conducted based on the four factors, five levels of central composite design with the rotational speeds (250, 300, 350, 400 and 450 RPM), axial feeds (1, 2, 3, 4 and 5 mm) and welding speeds (25, 30, 35, 40 and 45 mm/min) to examine the tensile properties of the joints. The tensile samples and hardness samples are prepared as per the standard dimensions.

3 Results and Discussions

The design matrix which contains the different combinations of the input process parameters, viz. rotational speed, tool pin profile, axial feed and welding speed, and corresponding responses are listed in Table 1. After conducting the experiments based on the design matrix, the mathematical models of tensile strength and tensile elongation are developed for the dissimilar alloy joints of FSW process with the input process parameters of rotational speed, tool pin profile, axial feed, and welding speed are represented in Eqs. 1 and 2, respectively, using the Design Expert 11 software.

$$\begin{aligned} TS\,(\text{MPa}) = & -245.271 + (1.158 * Ns) + (23.433 * Tp) + (22.40 * Fa) + (7.80 * Sw) \\ & - (0.03 * Ns * Tp) - (0.01 * Ns * Fa) - (0.003*) Ns * Sw) \\ & - (0.15 * Tp * Sw) - (0.239 * Fa * Sw) - (0.0013 * Ns^2) \\ & - (1.456 * Tp^2) - (1.568 * Fa^2) - (0.082 * Sw^2) \end{aligned} \tag{1}$$

$$\begin{aligned} TE\,(\%) = & -16.044 + (0.057 * Ns) + (1.314 * Tp) - (0.416 * Fa) + (0.489 * Sw) \\ & + (0.0017 * Ns * Tp) - (0.0004 * Ns * Sw) - (0.044 * Tp * Sw) \\ & + (0.016 * Fa * Sw) - (0.00007 * Ns^2) - (0.0497 * Tp^2) \\ & - (0.0597 * Fa^2) - (0.0037 * Sw^2) \end{aligned} \tag{2}$$

In this investigation, the effectiveness of FSW parameters is examined based on the responses using the sensitivity analysis. It is used to identify the most influencing input FSW parameters on the tensile properties, such as tensile strength and tensile elongation of the dissimilar AA8011-H24 aluminium alloys and Ti3Al2.5V titanium alloys from the mathematical models of the dissimilar joints. The sensitivity equations are attained by partial differentiating of the tensile strength and tensile elongation with respect to five different tool pin profiles and the other input parameters such as welding speed and axial feed are kept constant are represented in Eqs. 3 and 4, respectively.

$$\partial TS/\partial Ns = +0.45 - (1.47 * Tp) - (0.52 * Fa) - (0.69 * Sw) - (6.67 * Ns) \tag{3}$$

$$\partial TE/\partial Ns = +0.13 + (0.084 * Tp) + (0.026 * Fa) - (0.098 * Sw) - (0.34 * Ns) \tag{4}$$

The above equations represent the sensitivity of rotational speed with respect to the various tool pin profiles on the responses, viz. tensile strength and tensile elongation of the butt joints, respectively.

The sensitivity of tensile strength and tensile elongation of the rotational speeds with five various profiled tools and the other parameters, viz. welding speed and axial feed, kept constant are given in Table 2. This investigation shows the effectiveness of the responses which is attained by the small changes in applied parameters in the FSW process. The positive sensitivity values indicate that there is an increment in the

Table 1 Design matrix with inputs parameters and experimental responses

Expt	Rotational speed Ns (RPM)	Tool pin profile (Tp)	Axial feed Fa (mm)	Welding speed Sw (mm/min)	Tensile strength TS (Mpa)	Tensile elongation TE (%)
1	300	2	2	30	149.9	3.82
2	400	2	2	30	156.2	4.15
3	300	4	2	30	152.9	4.36
4	400	4	2	30	153.2	4.88
5	300	2	4	30	155.4	3.73
6	400	2	4	30	159.7	3.98
7	300	4	4	30	158.5	4.35
8	400	4	4	30	156.9	4.95
9	300	2	2	40	154.2	4.42
10	400	2	2	40	157.8	4.22
11	300	4	2	40	154.2	4.08
12	400	4	2	40	151.9	4.14
13	300	2	4	40	155.1	4.62
14	400	2	4	40	156.4	4.48
15	300	4	4	40	155.1	4.28
16	400	4	4	40	150.7	4.62
17	250	3	3	35	149.1	3.74
18	450	3	3	35	150.8	4.32
19	350	1	3	35	158.8	4.17
20	350	5	3	35	156.1	4.78
21	350	3	1	35	154.8	4.36
22	350	3	5	35	159.2	4.51
23	350	3	3	25	155.9	4.22
24	350	3	3	45	154.2	4.39
25	350	3	3	35	163.3	4.65
26	350	3	3	35	163.4	4.73
27	350	3	3	35	163.2	4.71
28	350	3	3	35	163.2	4.65
29	350	3	3	35	163.2	4.68
30	350	3	3	35	163.3	4.58
31	350	3	3	35	163.3	4.56

Table 2 Sensitivity analysis on tensile strength and tensile elongation

Tool pin profile	Constant welding speed: 40 mm/min and constant axial feed: 3 mm			Sensitivity of tensile strength (Mpa)	
	Rotational speed (RPM)	Tensile strength (Mpa)	Tensile elongation (%)	$\partial TS/\partial N_s$	$\partial TE/\partial N_s$
Straight cylindrical	250	139.120	3.744	16.040	0.544
	300	151.810	4.112	9.370	0.204
	350	157.840	4.144	2.700	−0.136
	400	157.210	3.840	−3.970	−0.476
	450	149.910	3.200	−10.640	−0.814
Threaded tapered cylindrical	250	144.990	3.867	14.570	0.628
	300	156.210	4.319	7.900	0.288
	350	160.770	4.435	1.230	−0.052
	400	158.670	4.215	−5.440	−0.392
	450	149.900	3.659	−12.110	−0.732
Straight hexagon	250	147.940	3.890	13.100	0.712
	300	157.690	4.426	6.430	0.372
	350	160.784	4.626	−0.240	0.032
	400	157.210	4.490	−6.910	−0.308
	450	146.980	4.018	−13.580	−0.648
Tapered hexagon	250	147.980	3.813	11.630	0.796
	300	156.270	4.433	4.960	0.456
	350	157.890	4.717	−1.710	0.116
	400	152.850	4.665	−8.380	−0.224
	450	141.150	4.277	−15.050	−0.564
Straight threaded cylindrical	250	145.110	3.636	10.160	0.880
	300	151.930	4.340	3.490	0.540
	350	152.080	4.708	−3.180	0.200
	400	145.570	4.740	−9.850	−0.140
	450	132.400	4.436	−16.520	−0.480

responses, viz. tensile strength and tensile elongation by small changes in parameter, while the negative sensitivity values show that the decrement in the responses [10].

3.1 Sensitivity Analysis on the Tensile Strength

The sensitivity analysis is carried out on the tensile strength with the input parameter rotational speed for five different friction stir welding tools represented in Fig. 1. From Fig. 1, it can be seen that the straight threaded cylindrical tool pin has the

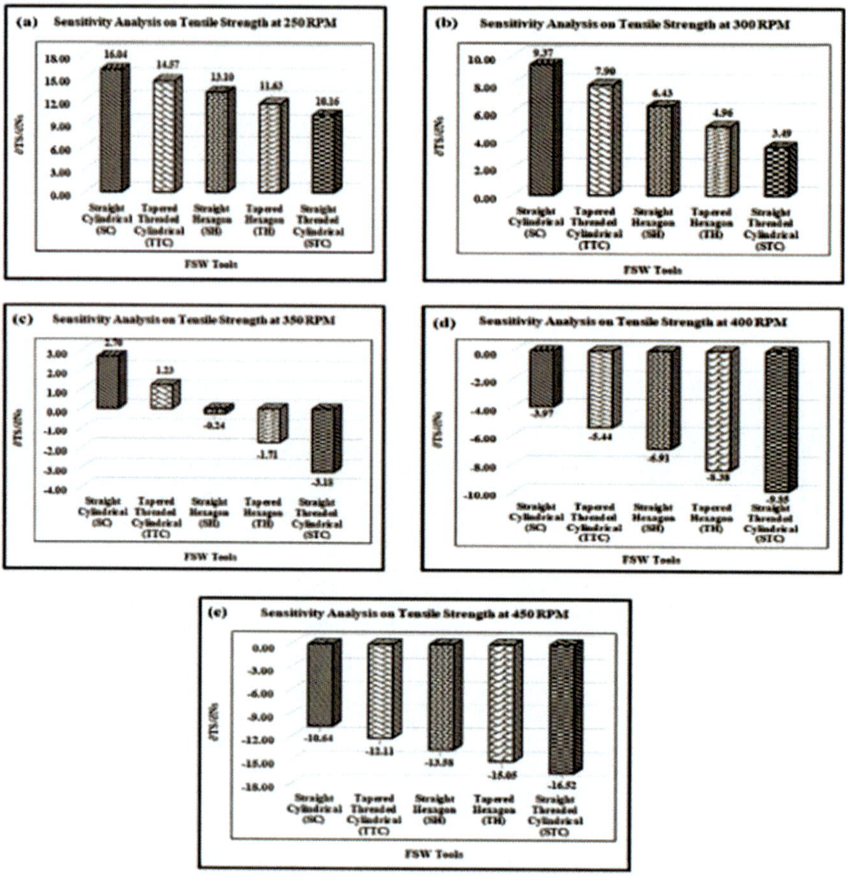

Fig. 1 Sensitivity analysis on the tensile strength of the dissimilar FSW joints

minimum positive sensitivity values of the tensile strength at the rotational speeds of 250 RPM and 300 RPM, while the negative sensitive values are obtained at the rotational speeds of 400 RPM and 450 RPM, respectively. At the same time, the straight cylindrical tool pin gives the superior sensitivity values of the tensile strength at the rotational speed of 250 RPM, whereas the minimum sensitivity is observed at the 450 RPM rotational speed [11].

The minimum variations are observed on the tensile strength of the dissimilar FSW joints using the Straight hexagon tool pin at the 350 RPM rotational speed. In this regard, the other tool pin profiles, straight cylindrical and tapered threaded cylindrical tool pins, provide the positive sensitivity values, whereas the tapered hexagon and straight threaded cylindrical tool pin provided negative sensitivity results on the tensile strength. This is due the variations in frictional heat developed in the weld nugget zone and material flow, which changes the tensile strength of the butt joints [12].

3.2 Sensitivity Analysis on the Tensile Elongation

The sensitivity analysis applied on the tensile elongation with the input parameter rotational speed for five different friction stir welding tools is represented in Fig. 2. It can be seen that the straight cylindrical tool pin gives the minimum positive sensitivity values on tensile elongation at the rotational speeds of 250 RPM and 300 RPM, whereas the negative sensitivity values of tensile elongation are obtained in the rotational speed range of 350 RPM–450 RPM respectively. At the meanwhile, the straight threaded cylindrical tool pin delivers the minimum positive sensitivity value of the tensile elongation at the rotational speed of 350 RPM, whereas the negative sensitivity results are obtained for the tensile elongation for all the friction stir welding tools at the rotational speeds of 400–450 RPM, respectively. The straight

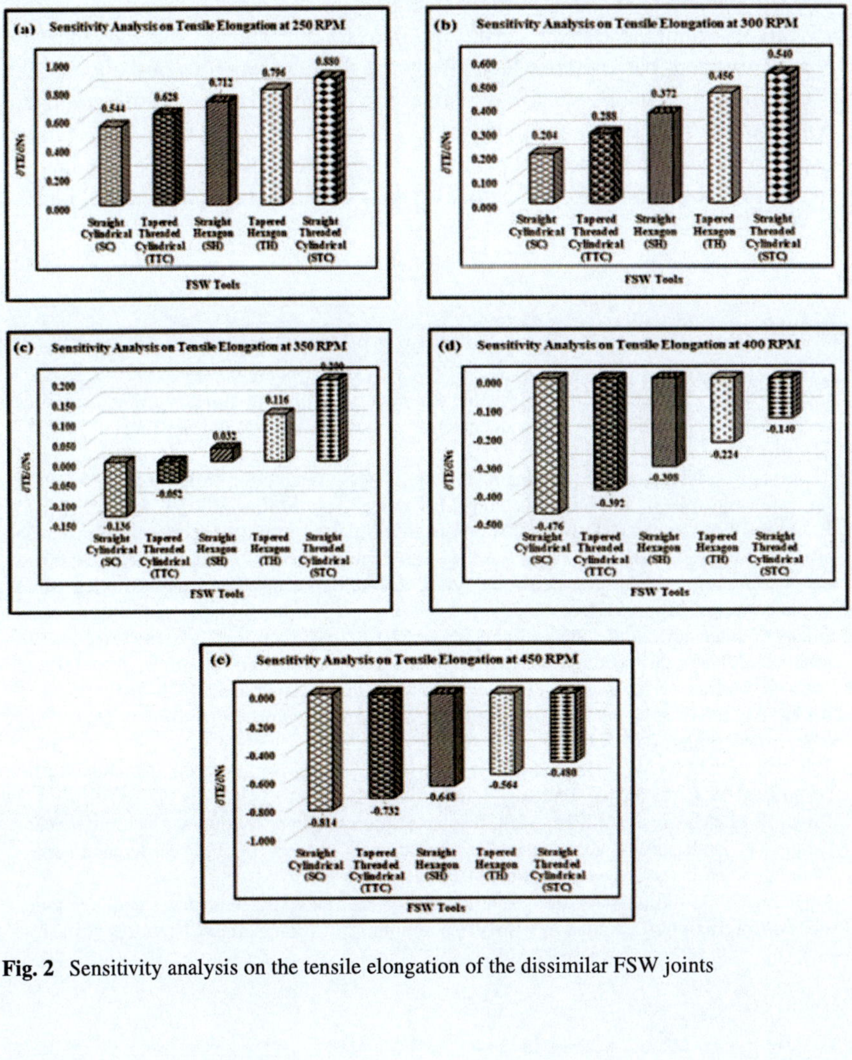

Fig. 2 Sensitivity analysis on the tensile elongation of the dissimilar FSW joints

hexagon tool pin provides the minimum fluctuations on the tensile elongation due to the ductile behaviour of the dissimilar welded joints at the weld nugget zone [13].

4 Conclusions

In this investigation, the mathematical relationship between the input parameters and the responses, viz. tensile strength and tensile elongation, has studied on the dissimilar FSW joints and the significant conclusions were listed as follows:

1. The developed mathematical models are employed to examine the relationship between the tensile properties and the input process parameters of the FSW process using response surface methodology.
2. The effectiveness of the welding parameters in the developed mathematical models is examined by the sensitivity analysis and studied the influence of changes in input parameters which affects the tensile properties of the joints.
3. The minimum negative sensitivity values are obtained on the tensile strength, while minimum positive sensitivity values are produced for the tensile elongation using the straight hexagon tool pin at the 350 RPM rotational speeds compared to other combinations of the rotational speeds with different profiled tool pins.

References

1. Charit I, Mishra RS (2018) Effect of friction stir processed microstructure on tensile properties of an Al-Zn-Mg-Sc alloy upon subsequent aging heat treatment. J Mater Sci Technol 34:214–218
2. Thomas WM, Nicholas ED, Needham JC, Murch MG, Temple-Smith P, Dawes CJ (1991) Improvements relating to friction welding. G.B. Patent Applications. 9,125,978.8
3. Palani K, Elanchezhian C, Raju M (2018) Sensitivity analysis of process parameters on tensile properties in plasma arc welding of AA8011-H24 aluminium alloys and Ti3Al2.5V titanium alloys using response surface methodology. In: IOP conference series: materials science and engineering, vol 390, p 012042
4. Palani K, Elanchezhian C, Vijaya Ramnath B, Bhaskar GB, Naveen E (2018) Effect of pin profile and rotational speed on microstructure and tensile strength of dissimilar AA8011 and AA6061-T6 friction stir welded aluminum alloys. Mater Today Proc 5:24515–24524
5. An H, Green DE, Johrendt J (2010) Multi-objective optimization and sensitivity analysis of tube hydroforming. Int J Adv Manuf Technol 50(1–4):67–84
6. Tian X, Liu Y, Deng W (2015) Sensitivity analysis for process parameters influencing electric arc cutting. Int J Adv Manuf Technol 78(1–4):481–492
7. Palani K, Elanchezhian C (2018) Multi response optimization of friction stir welding process parameters in dissimilar alloys using grey relational analysis. In: IOP conference series: materials science and engineering, vol 390, p 012061
8. Rajakumar S, Muralidharan C, Balasubramanian V (2011) Response surfaces and sensitivity analysis for friction stir welded AA6061-T6 aluminium alloy joints. Int J Manuf Res 6(3):215–235

9. Palani K, Elanchezhian C (2015) Multi response optimization of process parameters on AA8011 friction stir welded aluminium alloys using RSM based GRA coupled with DEA. Appl Mech Mater 813:446–450
10. Lua SK, Secgin A (2007) Sensitivity analysis of submerged arc welding process parameters. J Mater Process Technol 202:500–507
11. Karaoglu S, Seçgin A (2008) Sensitivity analysis of submerged arc welding. Int J Mater Process Technol 202:500–507
12. Palani PK, Murugan N (2006) Sensitivity analysis for process parameters in cladding of stainless steel by flux cored arc welding. Int J Manuf Process 8:90–100
13. Karthikeyan R, Balasubramanian V (2012) Optimization and sensitivity analysis of friction stir spot-welding process parameters for joining AA 6061 aluminum alloy. Int J Manuf Res 7(3):257–272

Experimental Investigation of Welding Parameters on Microhardness and Tensile Behaviour of Similar and Dissimilar Submerged FSWed Aluminium Alloy Joints

K. Palani, C. Elanchezhian, B. Vijaya Ramnath, D. Charan, Ch. Divya Prakash, R. Kathiravan, and O. Narendra Reddy

Abstract This investigation focused to study the effects of submerged friction stir welding (SFSW) process parameters on the microhardness and tensile strength of the AA8011–H24 aluminium alloy and AA5083-H321 aluminium alloy joints. The nine different experiments are carried out based on the SFSW process parameters, viz., rotational speed, welding speed and tool tilt angle with hexagonal tool pin profile under the three different mediums, namely distilled water, brine solution and salt water. The microhardness and tensile tests are conducted on the welded samples to identify the strength of the welded joints at weld stir zone. It is concluded that the superior tensile strength of 69.67 Mpa is examined for dissimilar AA5083–H321 with AA8011–H24 alloy joint than the other joints and also, the superior microhardness is found for the similar AA5083–H321 aluminium alloy under distilled water compared to other fabricated joints.

Keywords Aluminium alloy · Microhardness · Submerged friction · Stir welding · Tensile strength

1 Introduction

The potential applications of joining of different grades of aluminium alloys due to its high strength to weight ratio, better corrosion resistance and low cost, which play the major role in fuel consumption with cost reduction and it enhances the performance in the automobile, marine and automobile industries [1]. Due to the presence of defects in the welded structured components using the traditional fusion welding

K. Palani (✉) · D. Charan · Ch. Divya Prakash · R. Kathiravan · O. Narendra Reddy
Department of Mechanical Engineering, Sri Chandrasekharendra Saraswathi Viswa Mahavidyalaya University, Enathur, Kanchipuram, Tamil Nadu 631561, India
e-mail: palani.k@kanchiuniv.ac.in

C. Elanchezhian · B. Vijaya Ramnath
Department of Mechanical Engineering, Sri Sairam Engineering College, West Tambaram, Chennai, Tamil Nadu 600044, India

T. Rajmohan et al. (eds.), *Advances in Materials and Manufacturing Engineering*, Springer Proceedings in Materials 7, https://doi.org/10.1007/978-981-15-6267-9_41

processes, the friction stir welding is mainly focused on the industrial sectors [2, 3]. In marine applications, the green innovative solid-state welding technique, such as submerged friction stir welding process, which has the combinations of extrusion and forging process under the submerged medium with non-consumable rotating tool [4]. The newly fabricated non-consumable tool, which is used to fabricate the butt joint configuration under the submerged medium [5]. The investigation is done on the effects of the parameters, viz., tool pin profile, rotational speed and welding speed on the mechanical properties of the AA6063 aluminium alloy using the SFSW process [6, 7]. The effect of pin profiles and rotational speed on the microstructure and tensile strength of dissimilar AA8011–H24 and AA6061-T6 friction stir welded aluminium alloys [8]. The study is made on the performance of the tensile strength in the weldment based on the different rotational speeds and the tools used in the FSW process. The effect of welding speed on the microstructures and mechanical properties of the underwater friction stir welded AA2219 aluminium alloy joints are discussed. The superior strength is achieved at the lower welding speed, which has the defect-free welded joints at the constant rotational speed of the tool [9, 10]. The limited discussions have made by the current researchers using the submerged friction stir welding (SFSW) process on the different grades of aluminium alloys. In this investigation, the study is carried on the effects of the FSW parameters using the non-consumable D2 steel tool under a different submerged medium.

2 Materials and Methods

In this investigation, the AA5083–H321 and AA8011–H24 aluminium alloys are procured from Kataria steel distributors in Mumbai with the dimensions of 100 mm × 100 mm to fabricate the SFSW welded joints using the conventional friction stir welding machine [3, 11]. The mechanical properties and chemical composition of the AA5083–H321 and AA8011–H24 aluminium alloys are represented in Table 1. The tank is designed and it is prepared with a dimension of 600 mm × 500 mm × 200 mm for the SFSW of dissimilar joints.

Besides, the specially designed and fabricated non-consumable FSW tool, which is made of D2 steel with the dimensions of shoulder diameter of 18 mm, pin diameter

Table 1 Mechanical properties and chemical composition of the dissimilar alloys

Aluminium alloy	Mechanical properties		Chemical composition							
	Tensile strength (Mpa)	Elongation (%)	Cu	Fe	Si	Mn	Mg	Ti	Zn	Al
AA8011–H24	103	25	0.13	0.7	0.5	0.5	0.3	0.002	0.1	Rem
AA5083–H321	307	17	0.1	0.4	0.4	0.5	4.5	0.15	0.25	Rem

Table 2 Various combinations of the aluminium alloys for SFSW process

S. No.	Combinations of aluminium alloys	Submerged medium
1	AA8011–AA5083	Distilled water
2	AA8011–AA5083	Salt water
3	AA8011–AA5083	Brine solution
4	AA8011–AA5083	Distilled water
5	AA8011–AA5083	Salt water
6	AA8011–AA5083	Brine solution
7	AA8011–AA5083	Distilled water
8	AA8011–AA5083	Salt water
9	AA8011–AA5083	Brine solution

of 6 mm and a pin length of 5.7 mm and the side length of 2 mm is selected for the experiments.

The three different submerged mediums, namely distilled water, salt water and brine solution are used to fabricate the SFSW aluminium alloy joints and the nine different combinations of the aluminium alloys with the submerged medium are listed in Table 2. The tensile specimens are sliced from the perpendicular direction of the welded joints using the wire-cut EDM process based on ASTM E8M-09 guidelines using SHANTHA Universal Testing Machine [8]. Besides, the hardness test is carried out using Vickers microhardness tester (Shimadzu, HMV 2 T E) with a load 2 kg with 15 s holding time-based on ASTM E92 guidelines. The hardness values are measured with the 0.5 mm indentation on the weld centre.

3 Results and Discussion

3.1 Microhardness of SFSW Joints Under the Distilled Water

The submerged friction stir welding is carried out on the AA5083–H321 and AA8011–H24 aluminium alloys under the distilled water and it is depicted in Fig. 1. The inverted '*W*' shaped microhardness profiles are examined for the fabricated joints. The superior microhardness is found at the weld stir zone for the friction stir welded AA5083–H321 under the distilled water than the other fabricated joints [8]. Besides, due to the presence of intermetallic compounds in the weld SZ, 35.71 and 66.33% of the hardness are examined for the dissimilar joints under distilled water, which is more than the hardness of both base materials.

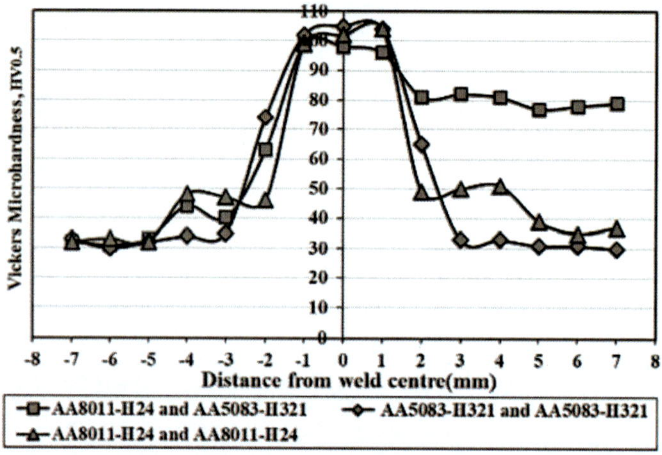

Fig. 1 Vickers microhardness profiles of SFSW joints under the distilled water

3.2 *Microhardness of SFSW Joints Under the Salt Water*

The submerged FSW is carried out on the AA5083–H321 and AA8011–H24 aluminium alloy joints under the salt water and it is depicted in Fig. 2. The inverted 'W'-shaped microhardness profiles are examined for the fabricated joints. The lower microhardness is examined at the stir zone for the AA5083–H321 aluminium alloy joints than the AA8011–H24 aluminium alloy joints. Also, it is more than the hardness of the dissimilar AA5083–H321 and AA8011–H24 aluminium alloy joints, due to the presence of the intermetallic compounds in the weld SZ.

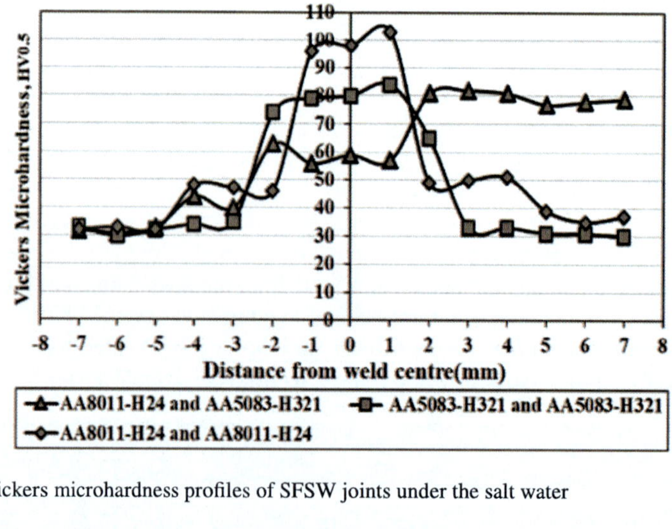

Fig. 2 Vickers microhardness profiles of SFSW joints under the salt water

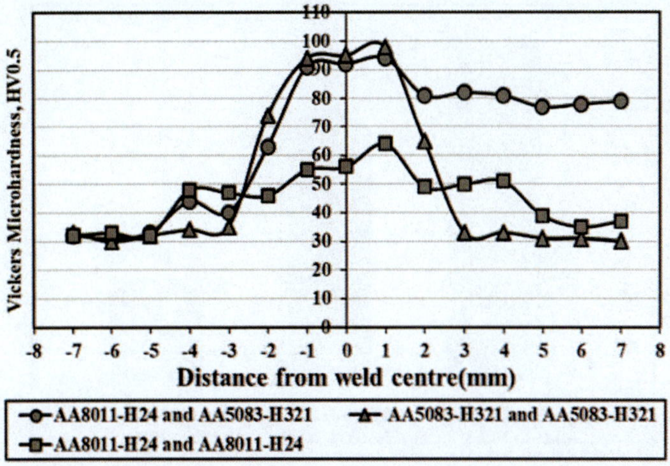

Fig. 3 Vickers microhardness profiles of SFSW joints under the brine solution

3.3 Microhardness of SFSW Joints Under the Brine Solution

The submerged friction stir welding is carried out on the AA5083–H321 and AA8011–H24 aluminium alloy joints under the brine solution and it is depicted in Fig. 3. The inverted '*W*'-shaped profile of microhardness is achieved for the SFSW AA5083–H321 and AA8011–H24 aluminium alloy joints. The higher microhardness appears at the SZ for the AA5083–H321 aluminium alloy joints than the AA8011–H24 aluminium alloy joints and also, these are almost equal to the hardness appeared in the dissimilar joints. Besides, the 53.19% and 65.9% hardness of the dissimilar submerged friction stir welded AA8011–H24 and AA5083–H321 aluminium alloy joints are examined at the weld SZ, which is higher than the TMAZ and BMZ, respectively.

3.4 Tensile Strength of SFSW Joints Under the Distilled Water

The tensile strength of the submerged friction stir welding is carried out on the AA5083–H321 and AA8011–H24 aluminium alloy joints under the distilled water and it is depicted in Fig. 4. The superior tensile strength of 69.67 MPa is observed in the dissimilar joints at the process parameters of 1750 rpm rotational speed and 40 mm/min welding speed when it is compared to other fabricated joints. It is due to the presence of intermetallic compounds appeared in the welded region, which initiates the increase in strength properties by the sufficient frictional heat with dynamic recrystallization of the materials [10, 11].

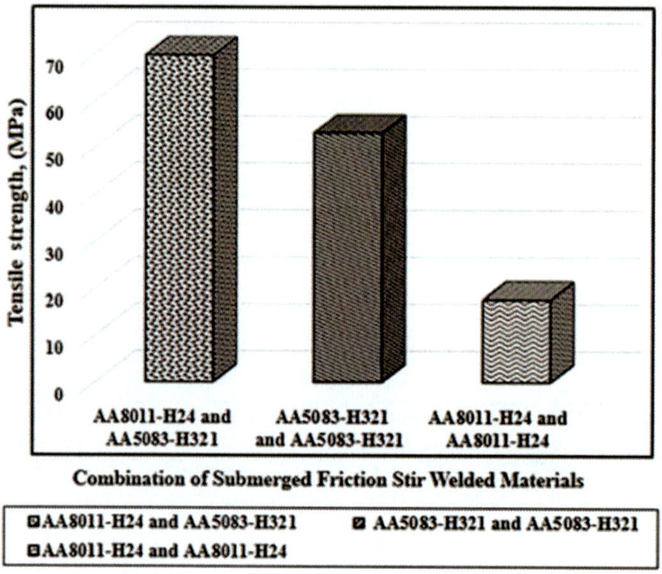

Fig. 4 Effect of the tensile strength of SFSW joints under the distilled water

The lower tensile strength of 17.67 MPa is observed in the AA8011–H24 aluminium alloy joints, because of the formation of defects, which leads to the inferior strength properties in the welded region.

3.5 *Tensile Strength of SFSW Joints Under the Salt Water*

The tensile strength of the submerged friction stir welding is carried out on the AA5083–H321 and AA8011–H24 aluminium alloy joints under the salt water and it is depicted in Fig. 5. The superior tensile strength of 27.15 MPa is observed in the dissimilar joints at the process parameters of 1750 rpm rotational speed and 40 mm/min welding speed when it is compared to other fabricated joints. It is due to the presence of brittle intermetallic compounds appeared in the weld SZ, which initiates the inferior strength properties by the lack of frictional heat with dynamic recrystallization of the materials [12].

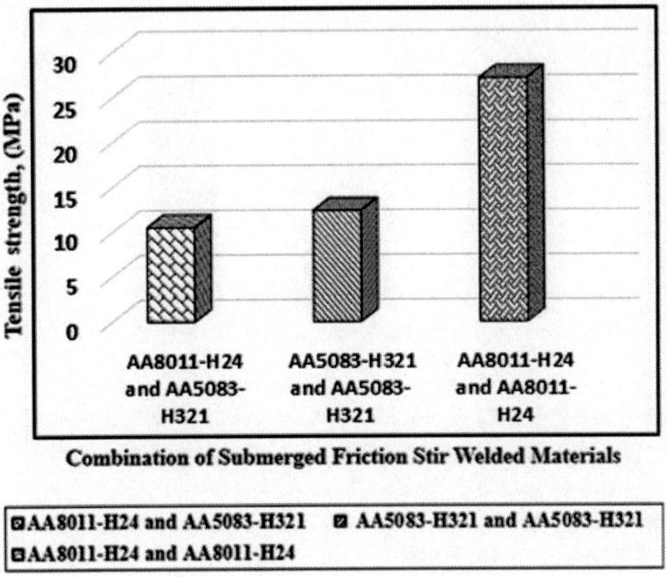

Fig. 5 Effect of the tensile strength of SFSW joints under the salt water

3.6 Tensile Strength of SFSW Joints Under the Brine Solution

The tensile strength of the submerged friction stir welding is carried out on the AA5083–H321 and AA8011–H24 aluminium alloy joints under the brine solution and it is depicted in Fig. 6. The superior tensile strength of 54.9 MPa is observed in the dissimilar joints at the process parameters of 1750 rpm rotational speed and 40 mm/min welding speed when it is compared to other fabricated joints.

It is due to the presence of brittle intermetallic compounds appeared in the weld SZ, which initiates the inferior strength properties by the lack of frictional heat with dynamic recrystallization of the materials [12–14]. The lower tensile strength of 13.84 MPa is observed in the SFSW AA5083–H321 aluminium alloy joints, by the presence of defects in the welded region.

4 Conclusions

In this investigation, the effect of welding parameters on Vickers microhardness and tensile strength of submerged friction stir welding AA8011–H24 and AA5083–H321 aluminium alloy joints are investigated and the following inferences are discussed.

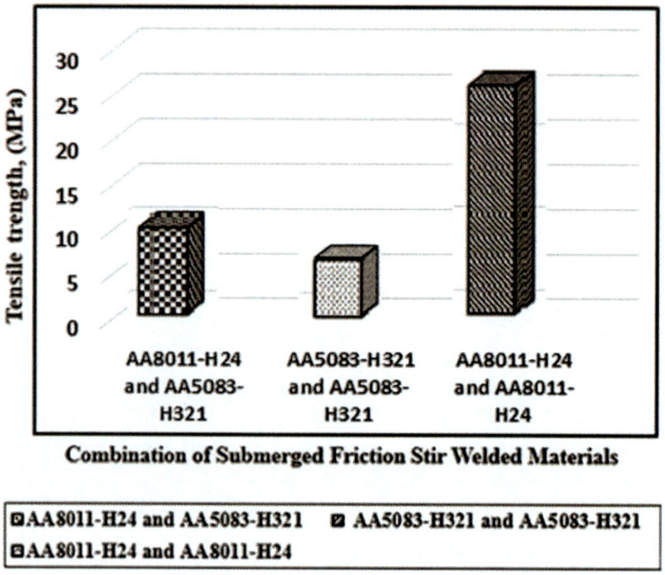

Fig. 6 Effect of the tensile strength of SFSW joints under the brine solution

1. The submerged friction stir welding process is carried out on the AA8011–H24 and AA5083–H321 aluminium alloy joints under different submerged mediums, namely distilled water, salt water and brine solution.
2. The superior microhardness is appeared at the TMAZ for dissimilar AA5083–H321 aluminium alloy joints under the medium of distilled water than the other fabricated joints.
3. The superior tensile strength of 69.67 MPa observed for the dissimilar AA5083–H321 and AA8011–H24 aluminium alloy joints under the distilled water than the other combinations of joints.

References

1. Thomas WM, Nicholas ED, Needham JC, Murch MG, Temple-Smith P, Dawes CJ (1991) Improvements relating to friction welding. G.B. Patent Applications. 9,125,978.8
2. Shanavas JS, Dhas ER, Murugan N (2018) Weldability of marine grade AA5052 aluminium alloy by underwater friction stir welding. Int J Adv Manuf Technol 95(9–12):4535–4546
3. Palani K, Elanchezhian C, Raju M (2018) Sensitivity analysis of process parameters on tensile properties in plasma arc welding of AA8011-H24 aluminium alloys and Ti-3Al-2.5V titanium alloys using response surface methodology. IOP Conf Ser Mater Sci Eng 390:012042
4. Palani K, Elanchezhian C (2018) Multi response Optimization of Friction stir welding process parameters in dissimilar alloys using Grey relational analysis. IOP Conf Ser Mater Sci Eng 390:012061

5. Babu KT, Muthukumaran S, Bharat Kumar CH (2018) A study on grain size, mechanical properties and first mode of metal transfer in underwater friction stir welded AA5052-O. Key Eng Mater 775:466–472
6. Zhao Y, Wang Q, Yan HK (2014) Microstructure and mechanical properties of spray formed 7055 aluminum alloy by underwater friction stir welding. Mater Des 56:725–730
7. Mofid MA, Abdollah-Zadeh A, Ghaini FM (2012) Gur H (2012) Submerged friction-stir welding (SFSW) underwater and under liquid nitrogen: an improved method to join alloys to Mg alloys. Metall Mater Trans A 43(13):5106
8. Palani K, Elanchezhian C, Vijaya Ramnath B, Bhaskar GB, Naveen E (2018) Effect of pin profile and rotational speed on microstructure and tensile strength of dissimilar AA8011 and AA6061-T6 friction stir welded aluminium alloys. Mater Today Proc 5:24515–24524
9. Bahemmat P, Haghpanahi M, Besharati Givi MK, Seighalani KR (2012) Study on dissimilar friction stir butt welding of AA7075-Oand AA2024-T4 considering the manufacturing limitation. Int J Adv Manuf Technol 59(9):939–953
10. Rathinasuriyan C, Senthil KV, S. (2017) Experimental investigation of weld characteristics on submerged friction stir welded 6061–T6 aluminium alloy. J Mech Sci Technol 31(8):3925–3933
11. Palani K, Elanchezhian C (2015) Multi response optimization of process parameters on AA8011 friction stir welded aluminium alloys using RSM based GRA coupled with DEA. Appl Mech Mater 813:446–450
12. Heidarzadeh A, Khodaverdizadeh H, Khodaverdizadeh A, Ebrahim M (2012) Tensile behavior of friction stir welded AA 6061–T4 aluminum alloy joints. Mater Des 37:1206–1212
13. Palani PK, Murugan N (2006) Sensitivity analysis for process parameters in cladding of stainless steel by flux cored arc welding. Int J Manuf Process 8:90–100
14. Palanivel R, Mathews PK, Murugan N, Dinaharan I (2012) Effect of tool rotational speed and pin profile on microstructure and tensile strength of dissimilar friction stir welded AA5083-H111 and AA6351-T6 aluminium alloys. Mater Des 40:7–16

Study on the Influence of Nanosized Silica Reinforcement in Microrubber Blended Epoxy Carbon Composite Laminate Subjected Under Dynamic Mechanical Analysis

R. Ramesh, C. Senthamaraikannan, Niranjan Suresh, and B. Lokesh

Abstract The woven fabric carbon/epoxy composite is one of the commercially well-recognized composite materials for structural applications like aerospace systems, automobiles and also in ship building industries. In the present investigation, dynamic mechanical behavior of woven fabric carbon epoxy composite laminate fabricated, by blending micron sized rubber and nanosized silica particles of varying weight fraction in epoxy matrix was studied. Silica in nanosize was added in two different weight fractions say 6 and 11% by weight in 9 wt% rubber particle blended epoxy matrix. The composite plates were fabricated by hand layup method with 40% of fiber volume fraction. The storage modulus, loss modulus and tan delta values were evaluated by conducting dynamic mechanical analysis for all the specimens as per the ASTM standards. Results from the DMA tests under temperature sweep for the constant frequency range of 1 Hz reveals that there is significant increase in the tested parameters like storage modulus and loss modulus of the hybrid composite.

Keywords Carbon composite laminate · Microrubber · Nanosilica · Damping ratio · Dynamic analysis · Storage modulus · Loss modulus · Glass transition temperature

R. Ramesh (✉) · C. Senthamaraikannan · N. Suresh · B. Lokesh
Department of Mechanical Engineering, Sri Venkateswara College of Engineering, Pennalur 602117, India
e-mail: rameshraja@svce.ac.in

C. Senthamaraikannan
e-mail: senthamarai@svce.ac.in

N. Suresh
e-mail: niranjansuresh1998@gmail.com

B. Lokesh
e-mail: lokeshbaskaran5@gmail.com

T. Rajmohan et al. (eds.), *Advances in Materials and Manufacturing Engineering*, Springer Proceedings in Materials 7, https://doi.org/10.1007/978-981-15-6267-9_42

1 Introduction

Woven fabric carbon epoxy composite was mostly used in structural applications due to their superior specific strength and specific stiffness ratio [1]. Laminated composite plates are used to fabricate aircraft structures in the aerospace industry, vehicle parts in the automotive industry, and in ship building industry. Costa et al. [2] studied the formation of rubber particles in epoxy matrix by carboxyl terminated process and further they extended their study to analyze rubber-modified carbon/epoxy laminates for aerospace field applications and concluded that they are most suitable for structural components of aerospace applications. Johnsen et al. [3] proved that the brittleness of the epoxy can be reduced by adding reactive liquid rubber such as carboxyl terminated butadiene acrylonitrile copolymer, CTBN, and thereby improving their toughness. Recent review on multi-functional composite materials reveals that tensile modulus; tensile strength and fracture toughness of epoxy considerably decreases with the addition of CTBN particles [4]. Senthamaraikannan et al. [5] investigated CTBN rubber-modified carbon composite laminates under free vibration and concluded that the addition of CTBN rubber particles significantly enhances damping performance of carbon epoxy laminate. An optimum content of the reactive liquid rubber microparticles is arrived as 9% by wt, for significant enhancement in passive damping with moderate mechanical properties. Nanosilica reinforced composites are gaining importance due to their capability for improving static/dynamic modulus, microhardness and fracture toughness of composites. Manjunatha et al. [6] demonstrated that the addition of nanosilica along with microrubber particles in epoxy matrix significantly raises the properties of the neat epoxy resin. There is limited experimental work in studying the material damping performance of microrubber and nanosilica modified woven carbon epoxy composites. Experimentally, dynamic mechanical analysis (DMA) is very useful in studying the dynamic modulus and damping parameter of polymer over a range of temperature [7]. This paper presents our recent study on dynamic mechanical properties of carbon fiber/epoxy laminates by adding microsized rubber and nanosized particles in epoxy matrix by dynamic mechanical analysis. The epoxy was modified by adding different weight fractions of nanosilica with 9 wt% rubber particles and DMA tests were conducted on modified carbon epoxy laminate.

2 Fabrication of Carbon Epoxy Composite Plates for Evaluating DMA Characteristics

To study the viscoelastic behavior of microrubber and nanosilica added carbon epoxy composite laminate, twill weave carbon fabric was considered. Epoxy blended with 40% by wt microrubber particles of 5 μm in size and nanosilica particles of 25 nm average size was purchased from Evonik® Industries AG, Germany. Further, purchased epoxy was diluted into different weight fractions by blending and mixing

Table 1 Composition of particles in hybrid epoxy formulations and its corresponding material codes are tabulated

Material codes are given for various composition	Weight (%)		
	Epoxy	Rubber	Silica
CE	100	–	–
M9	91	9	
M9 S6	85	9	6
M9 S11	80	9	11

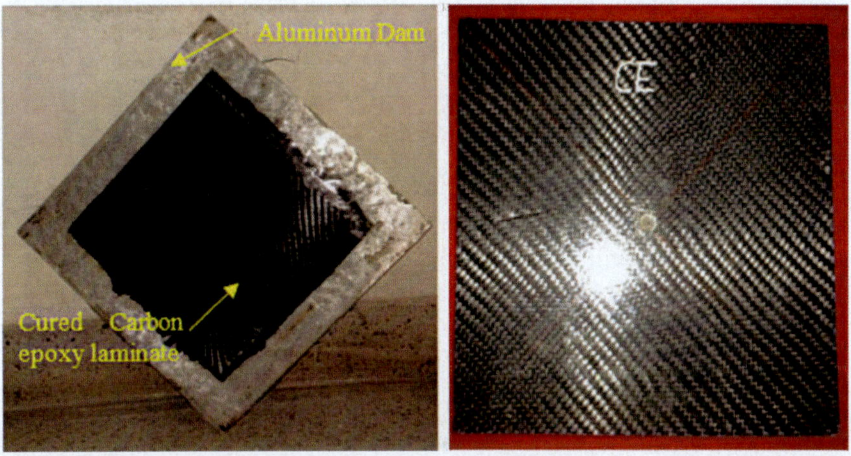

Fig. 1 Fabricated carbon epoxy composite

with the same type of virgin epoxy resin purchased from the same company for considered hybrid epoxy composition as shown in Table 1.

Mixing, blending and stirring procedures were adopted as per the previous research study [8] supported by probe sonicator for 30 min at 60 °C. The volume fraction of fiber is maintained at 40% by weight as preferred in earlier researches [5, 9]. Laminates were prepared by hand layup method using a precisely sized dam and cured under 350 psi using a compression molding machine for 24 hrs. After curing for predetermined hours, laminates were trimmed and shown in Fig. 1.

3 Dynamic Mechanical Analysis of Hybrid Carbon Epoxy Composites

Dynamic mechanical analysis was conducted using a dynamic mechanical analyzer (DMA Q800 V20.6 Build 24), as per the ASTM Standard D4065 [10]. The samples with dimensions 35 mm long × 15 mm wide × 3 mm thick were tested from 20 to 180 °C at a heating rate of 10 °C/min and at frequency of 1.0 Hz. Three samples were used for conducting the experiments in three-point bending mode and results were discussed.

Storage Modulus

The variation of dynamic elastic modulus as a function of temperature for the considered carbon epoxy laminates is shown in Figs. 2 and 3. The storage modulus curves of the modified and unmodified composites present three zones, a glassy region, an abrupt modulus decreasing region and the high temperature rubbery region. Storage modulus was increased by adding the nanosilica particles in the epoxy matrix. Storage modulus at the glassy region was shown in Table 2. It is observed that storage modulus of all composites increases with the addition of 6 wt% and 11 wt % of silica with 9 wt% rubber particle modified epoxy, by 35% and 49%, respectively. Samples are subjected to tension–compression deformation mode as similar to three-point bending mode at flexural test. Storage modulus will be increased when higher amount of stress transfer take place between fiber matrix interactions. This is due to effect of deformation of micro and nanoparticles inside the epoxy matrix. More loads can be taken up by the particles if their weight content in the epoxy matrix increases. This can be further described by considering the deformation of epoxy matrix in two phases. In the first phase, the constituent particles will deform within the matrix; in the second phase, this deformation will behave as bulk deformation of matrix including the fiber. As there is increase in micro/nanoconstituents' content in each layer, this increases the rate of bulk deformation by providing more local individual particle deformation. Thereby increases the amount of elastic energy storage in the laminates. Distribution of particles in the matrix causes more amount of energy storage because of added flexibility to the epoxy matrix. Thus, the deformation of the rubber/silica particle is responsible for enhanced stress transfer. Additional deformation of adding rubber particle is supported by previous studies [5].

Sprenger et al. [11] reported that adding epoxidized rubber particle in the matrix, decreases the tensile and flexural modulus due to its additional deformation effect, thereby increasing the plasticizing effect in the matrix. Kinloch [12] also confirms that the addition of rubber particle in the epoxy matrix enhances more energy absorbing capacity due to flexibility of rubber particles. Han et al. [13] describe that the inclusion of nanofiller is mobile enough to form the temporary bonds that allow it to

Table 2 Effect of micro/nanoparticles on dynamic mechanical properties of carbon epoxy composites

Parameters	Units	CE	M9	M9 S6	M9 S11
Storage modulus	MPa	13,116	16,720	22,583	24,976
Loss modulus	MPa	1930	2487	3074	4497
tan delta	–	0.41	0.49	0.55	0.77
Glass transition temperature	°C	101.3	114.8	83.5	80.7

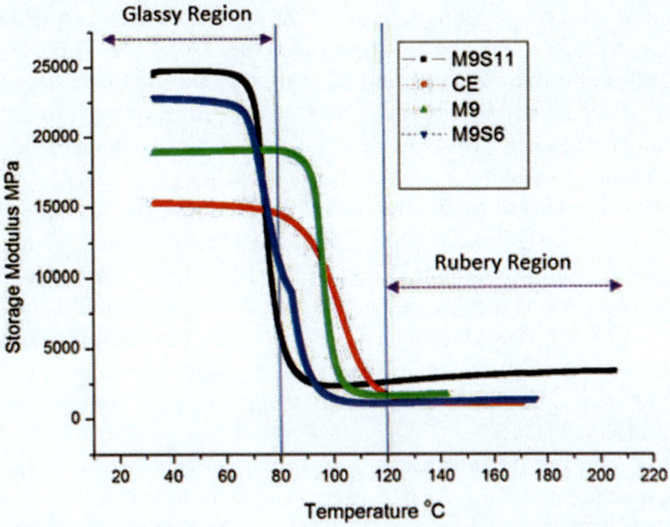

Fig. 2 Effect of adding silica particles on storage modulus of rubber-modified carbon epoxy composites

dissipate energy and thus rubbery matrices benefit the most from the addition of nanofillers. This phenomenon enhances the storage modulus of the nanosilica added rubber-modified carbon epoxy composite. There is increase in storage modulus due to increase in nanosilica fillers.

During transition state due to temperature rise, sudden collapse of bonding causes abrupt decrease in storage modulus. Addition of rubber particle influences more de-bonding in the epoxy network; it is clearly visible that M9 specimens show abrupt decrease in storage modulus value. In rubbery state, as further temperature increases more loosening of epoxy matrix causes increased mobility within the matrix and loses their close packing arrangement, thus, decreases storage modulus for increasing temperature. Also, the addition of nanosilica particle in this state influences by increasing trend in the storage modulus after 120 °C.

4 Loss Modulus

The effect of temperature on loss modulus of carbon epoxy composite by the addition of rubber/silica particles is shown in Fig. 4. As discussed previously, laminates are subjected to cyclic load of 1 Hz under three-point bending mode. The loss modulus, E'', is proportional to the lost or dissipated energy per cycle. At low temperatures up to 60 °C, viscous behavior of rubber particles influences the loss modulus of the composite by increasing its viscoelastic property for the addition of rubber in epoxy matrix. Further, the nanoconstituents increase the surface area and promotes more amount of energy dissipating capacity of the composites. In room temperature, loss modulus of M9S11 laminate was better than other considered concentrations. Dissipation of energy at low temperature was very effective for this weight fraction. In transition region, the transition peak was found to become stronger for increasing silica particle concentration. At this temperature, nanoparticles exhibit pseudo-plastic deformation, which further increases the energy dissipating capacity of composites.

Further, the comparison of loss modulus curves shows that the loss modulus peak or glass transition temperature Tg shifts toward lower temperature for M9S11 laminate due to the inclusion of more silica particles in the epoxy matrix. Francis et al. [14] explains that there is increase in epoxy particle mobility with the inclusion of filler particles. This is further confirmed by comparing the broadening and narrowing curve regions. Broadened curve can be seen for unmodified carbon epoxy composite. Curve become narrow for higher concentration silica particle modified composite M9S11 than M9S6 (Fig. 3).

From Table 2, the loss modulus for the M9S11 composite is found to be the highest which is an indication of the improved energy dissipation, which could contribute to their better energy dissipating properties due to the above-said reasons. There is a maximum increase of 36% in energy dissipating capability of carbon epoxy composites due to the addition of rubber particles in the epoxy matrix. Similar trend was observed by Baller et al. [15], when incorporating nanosilica filler in virgin epoxy matrix. In this study, Baller et al. explain that the non-dissipative nature of the nanosilica filler in epoxy matrix causes the very small interfacial losses and that the matrix/filler interfacial bonding is accordingly good. This increased interfacial adhesion leads to an overall increase in the modulus under DMA test.

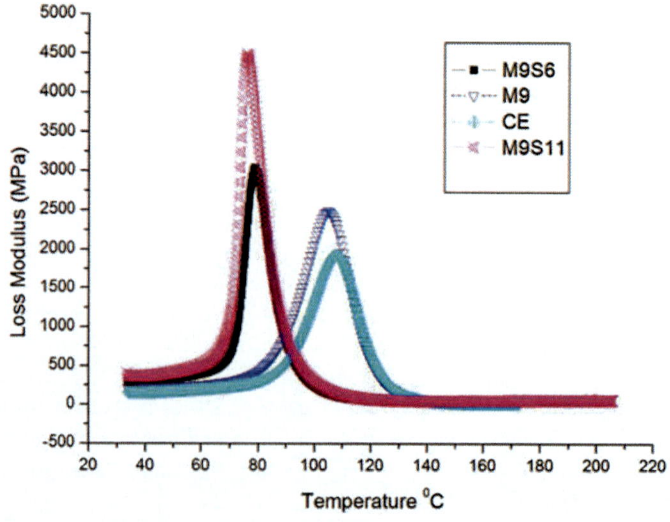

Fig. 3 Effect of nanosilica particles on loss modulus of carbon epoxy composite

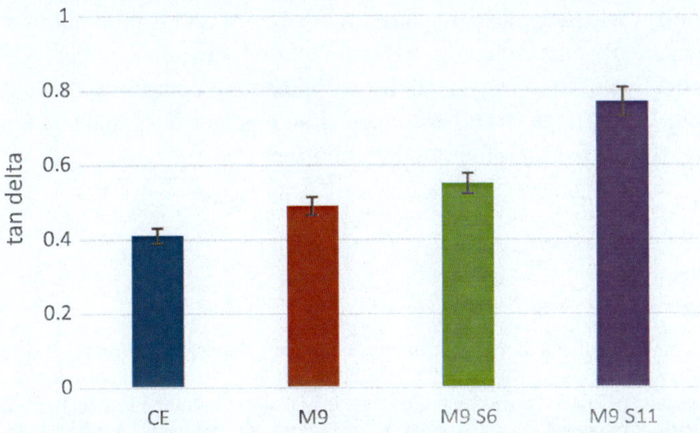

Fig. 4 Variation in damping factor (tan δ) values of carbon epoxy composites with addition of rubber and silica particles

5 Damping Factor

The magnitude of the damping factor is an indicative of the material damping present in polymer system and it gives balance between the elastic and viscous phase in a polymeric material. Damping factor at glass transition temperature was extracted for each considered specimen and shown in Fig. 4 shows that damping factor increases with increase in nanosilica concentration in 9 wt% rubber added carbon epoxy composite and reaches maximum value for M9S11 composite. Chen et al. [16, 17] demonstrated that the thermal and rheological effect of adding nanosilica in epoxy matrix, and the results show the nanosilica particles induce an exclusion zone that lowers the reaction temperature and can also lower Tg of the nanocomposite at high-filler loadings.

6 Conclusions

Dynamic mechanical behavior of woven fabric carbon epoxy laminate was conducted with modified epoxy, blended with microrubber and nanosized silica particles as per standards. Dynamic mechanical test results show that the addition of microrubber particle in epoxy matrix in carbon laminate improves the material damping capacity of the composite. Further, the addition of nanosilica aides the dynamic mechanical properties under temperature sweep. Study indicated the nanosilica particles induce an exclusion zone in epoxy matrix which in turn increases the storage modulus. Non-dissipative nature of the nanosilica causes filler, higher amount of energy storage and release is possible for composite due to increased deformation of matrix for same loading conditions. At lower temperature, energy absorbing capacity is better for increased weight content; thus, M9S11 wt% laminates have better storage modulus and loss modulus values. Carbon epoxy laminate with 9 wt% rubber and 11 wt% silica concentration show better glass transition temperature and loss modulus values at room temperature. Results indicate that the nanoparticles in microrubber added epoxy matrix are effective reinforcing agents for carbon epoxy composite. Because of the improved properties and the excellent dispersion, this appears capable combination for use in the manufacturing of tailored composites.

References

1. Long AC (2005) Design and manufacture of textile composites. CRC Press
2. Costaa ML, Rezendea MC, Paivab JMF, Botelhoc EC (2006) Structural carbon/epoxy prepregs properties comparison by thermal and rheological analyses. Polymer-Plast Technol Eng 45(10)
3. Johnsen BB, Kinloch AJ, Mohammed RD, Taylor AC, Sprenger S (2007) Toughening mechanisms of nanoparticle-modified epoxy polymers. Polymer 48(2):530–541
4. Gibson RF (2010) A review of recent research on mechanics of multifunctional composite materials and structures. Composite 92(12):2793–2810
5. Senthamaraikannan C, Sarathkumar SK, Ramesh R (2014) Experimental investigation on modal response of woven fabric carbon composite plate reinforced with particles of micro rubber blended epoxy matrix under free vibration condition. Adv Mater Res 984–985: 273–279
6. Manjunatha CM, Taylor AC, Kinloch AJ, Sprenger S (2009) The cyclic-fatigue behavior of an epoxy polymer modified with micron-rubber and nano-silica particles. J Mater Sci 44(2009):4487–4490
7. Pothan LA, Thomas S, Groeninckx G (2006) The role of fibre/matrix interactions on the dynamicmechanical properties of chemically modified banana fibre/polyester composites. Compos Part A 37:1260–1269
8. Manjunatha CM, Sprenger S, Taylor AC, Kinloch AJ (2010) The tensile fatigue behavior of a glass-fiber reinforced plastic composite using a hybrid-toughened epoxy matrix. J Compos Mater 44:2095–2109
9. Rahmani H, Heydar Mahmoudi Najafi S, Ashor A (2015) Mechanical performance of epoxy/carbon fiber laminated composites. J Reinf Plast Compos 33(8): 733–740
10. ASTM D4065-01 (2004) Standard practice for plastics: dynamic mechanical properties: determination and report of procedures. American Society of Testing and Materials, West Conshohocken. PA, USA
11. Sprenger S, Kothmann MH, Altstaedt V (2014) Carbon fiber-reinforced composites using an epoxy resin matrix modified with reactive liquid rubber and silica nanoparticles. Compos Sci Technol 105:86–95
12. Kinloch AJ (2003) Toughening epoxy adhesives to meet today's challenges. Mater Res Soc Bull 28:445–448
13. Han JT, Cho K (2006) Nanoparticle-induced enhancement in fracture toughness of highly loaded epoxy composites over a wide temperature range. J Mater Sci 41:4239–4245
14. Francis B, Rao L, Ramaswamy R, Jose S, Thomas S, Raju KVSN Morphology, viscoelastic properties, and mechanical behavior of epoxy resin modified with hydroxyl-terminated poly(ether ether ketone) oligomer with pendent tert-butyl groups. Polym Eng Sci 45(12): 1548–2634
15. Baller J, Becker N, Ziehmer M, Thomassey M, Zielinski B, Muller U et al (2009) Interactions between silica nanoparticles and a epoxy resin before and during network formation. J Polym 50–14:4895–4905
16. Chen C, Justice RS, Schaefer DW, Baur JW (2008) Highly dispersed nanosilica–epoxy resins with enhanced mechanical properties. Polymer 49:3805–3815
17. Senthamaraikannan C, Ramesh R Study of free and forced vibration characteristics of micro rubber and nano silica reinforced woven fabrichybrid carbon composite beams for structural applications. Ph.D thesis, Anna University, Chennai, India

Moth-Flame Optimization Algorithm for Improving the Surface Roughness on FDM Processed Parts

A. Tamilarasan, D. Rajamani, P. Pranay, P. Manohar, A. Venkata Akhil, and B. Thirupathi Reddy

Abstract This paper presents an optimal method for determining the best FDM processing parameters by minimizing the surface roughness using a nature-inspired moth-flame optimization algorithm. The experiment design is formulated using the Box–Behnken design. The quadratic model derived from ANOVA results shows that the model was significant and satisfactory. Moth-flame optimization algorithm is effectively implemented and found to be an excellent optimization tool for fast convergence and global optima location.

Keywords FDM process · Surface roughness · Moth-flame algorithm · Optimization

1 Introduction

Additive manufacturing technology is an innovative development technique derived directly from the CAD data set to produce layer-by-layer components [1]. In the additive production processes including the FDM process are required for the supply of high quality components. The higher demand for FDM in the manufacturing sector of industrial components such as medical devices, telecommunications, electronics and aerospace needs ever-increasing dimensional precision rates [2]. Since the mechanism is complex, the resulting properties of constructed components are difficult to model and evaluate [3]. Consequently, researchers have suggested several conceptual models developed via experimental data analysis. The group method for predictive processing and differential evolution used for the optimization of FDM process parameters was introduced by Rayegani and Onwubolu [3]. The obtained

A. Tamilarasan (✉) · P. Pranay · P. Manohar · A. Venkata Akhil · B. Thirupathi Reddy
Department of Mechanical Engineering, Sri Chandrasekharendra Saraswathi Viswa Mahavidyalaya, Enathur, Kanchipuram 631561, Tamil Nadu, India
e-mail: tamilrj2010@gmail.com

D. Rajamani
Centre for Autonomous System Research, Department of Mechanical Engineering, Vel Tech Rangarajan Dr. Sagunthala R&D Institute of Science and Technology, Chennai, Tamil Nadu 600062, India

T. Rajmohan et al. (eds.), *Advances in Materials and Manufacturing Engineering*, Springer Proceedings in Materials 7, https://doi.org/10.1007/978-981-15-6267-9_43

results are highly promising and hence the approach proposed for practical use in the design and manufacturing of components using additive processing technologies. Sood et al. [4] tried to find an optimal parametric combination which could simultaneously improve the ABS P400's tensile, flexural and impact strengths. Rao and Rai [5] suggested TLBO and non-TLBO sorting algorithms for the optimization of FDM method parameters. The TLBO algorithm has been found to be superior to the GA and QPSO algorithms. Many researchers have made significant efforts to enhance FDM processed components in terms of surface roughness and different strengths. This paper therefore proposes to find optimal combinations of ABS material FDM parameters to improve surface roughness with a recent moth-flame optimizing algorithm. The experiments were based on the response surface methodology (RSM)-based Box–Behnken method (BBD) was used. Moreover, the RSM provides a statistical model for surface roughness. The proposed flame-moth optimization algorithm is used for the lowest surface roughness in order to ensure the best possible parameters environment for the FDM process parameters.

2 Experimental Planning

2.1 RSM-Based Box–Behnken Design

RSM is an experimental methodology that is mostly used to evaluate and solve problems by examining relations between predictors and responses to identify variables that significantly influence the quality and properties of the product [6]. Box–Behnken design (BBD) investigates the relationships of multiple variables of description with one or more variables of response. The key BBD range should be limited to a situation where drastic responses are not expected [6]. This present analysis has chosen and assigned four FDM parameters as X_1, X_2, X_3 and X_4, respectively. The low, middle and high levels of each variable were labeled as -1, 0, and $+1$, respectively, as shown in Table 1. The variables were coded by the following equation as:

$$x_i = \frac{X_i - X_0}{\Delta X}, \quad i = 1, 2, 3 \tag{1}$$

Table 1 Process parameters and levels

Parameter (symbol)	Unit	-1	0	$+1$
Layer thickness (X_1)	mm	0.13	0.23	0.33
Part orientation (X_2)	Degree	0	45	90
Raster width (X_3)	mm	0.2	0.35	0.5
Raster angle (X_4)	Degree	0	30	60

where X_i and x_i are the actual and coded values, respectively. X_0 is the value of X_i at the center point of the domain and ΔX is the increment of X_i corresponding to one unit of X_i. For the optimization of process parameters, the statistical models of the experimental data generate the secondary polynomial equation. The following quadratic (second order) polynomial model explains the behavior of the system:

$$Y = \beta_0 + \sum_{i=1}^{k} \beta_i x_i + \sum_{i=1}^{k} \beta_{ii} x_i^2 + \sum_{i<j} \beta_{ij} x_i x_j \tag{2}$$

where Y is the response and x_i are the values of the ith FDM process parameters; β_0 is the model constant; β_i represents the linear coefficient; β_{ii} denotes the quadratic coefficient; β_{ij} is the interaction coefficient; k is the number of the factors or variables; and ε is the statistical experimental error [6].

2.2 *Experimental Procedure*

Firstly, the components are modeled on and exported to CATIA V5 as an STL format. STL file is being imported to FDM applications (Insight) software. The material used for the production of the component is ABS P400. The temperature of the printer's nozzle during the experiment was set at a temperature of 808 °C, a fill rate of 100% and no support. All experiments were performed in conjunction with the BBD design experimental program. The parts shall be made by means of the FDM Fortus 400MC system machine Stratasys and manufactured specimens as shown in Fig. 1.

Surface roughness was calculated using a Mitutoyo SJ-210 profilometer sampling duration of 2.5 mm (Fig. 2). The overall surface roughness measurement is taken and displayed in Table 2.

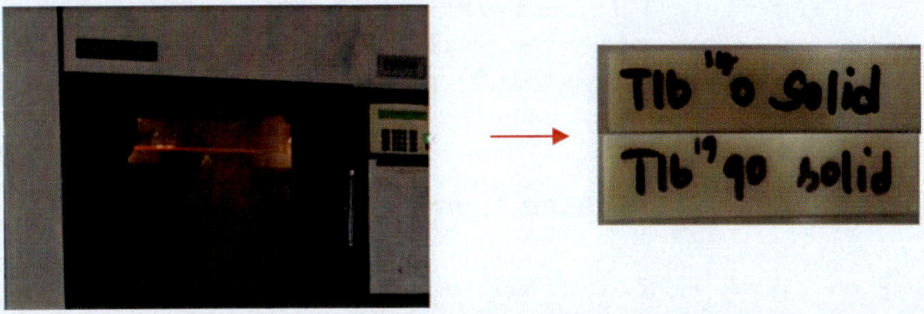

Fig. 1 FDM processing zone and fabricated specimens

Fig. 2 Surface roughness measurement

3 Statistical Modeling of Experimental Work

The use of a BBD matrix, which fits very accurately with the second-order response, results in coefficients of regression. The commercial Design Expert 7 software is used to obtain these coefficients. Using these coefficients, the final model is established and the polynomial second-order model for surface roughness was obtained in Eq. (3)

$$\begin{aligned} R_a = & -2.97155 + 49.94286X_1 + 0.016206X_2 \\ & + 2.09247X_3 + 0.045127X_4 \\ & - 0.065264X_1X_4 - 113.07928X_1^2 \\ & - 1.34493E - 004X_2^2 - 3.50958X_3^2 \\ & - 2.70090E - 004X_4^2 (R^2 = 0.9752;\ R^2_{\text{Adj}} = 0.9634;\ R^2_{\text{Pred}} = 0.94) \end{aligned} \quad (3)$$

As can be seen in Eq. 3, the determination coefficient (R^2) is 97.52%, ensuring high regression functions validity. The predicted R-squared of 0.94 is in reasonable agreement with the adjusted R-squared of 0.9634.

4 Moth-Flame Optimization Algorithm

Mirjalili [7] developed the moth-flame optimization algorithm and it works based on the navigation method of moths through night by maintaining a fixed angle with respect to the moon. Moths are very similar to butterflies as a group of insects. One of the moths most interesting behaviors is their unique approach to navigation. They fly through a fixed angle with respect to the moon to travel long distances in a straight path [7]. This effective approach is referred to as transverse orientation. The

Table 2 BBD design matrix with results

Exp. Run	X_1 (mm)	X_2 (Degree)	X_3 (mm)	X_4 (Degree)	R_a (μm)
1	0.229	45	0.381	30	3.948
2	0.229	0	0.381	60	3.608
3	0.229	45	0.559	60	3.821
4	0.229	45	0.381	30	3.967
5	0.330	45	0.381	0	1.995
6	0.229	0	0.381	0	2.717
7	0.229	90	0.381	60	4.075
8	0.330	45	0.203	30	2.222
9	0.229	90	0.559	30	3.481
10	0.127	45	0.381	0	2.198
11	0.229	45	0.381	30	3.906
12	0.127	90	0.381	30	3.057
13	0.229	45	0.559	0	3.226
14	0.330	90	0.381	30	2.420
15	0.330	0	0.381	30	1.868
16	0.127	45	0.559	30	2.972
17	0.229	90	0.381	0	3.099
18	0.127	0	0.381	30	2.759
19	0.229	0	0.559	30	3.396
20	0.229	45	0.381	30	3.712
21	0.229	45	0.381	30	4.189
22	0.229	0	0.203	30	3.538
23	0.330	45	0.381	60	2.547
24	0.127	45	0.203	30	3.141
25	0.127	45	0.381	60	3.545
26	0.229	45	0.203	0	3.354
27	0.330	45	0.559	30	2.123
28	0.229	45	0.203	60	4.036
29	0.229	90	0.203	30	3.969

transverse orientation's effectiveness strongly depends on the light source's distance. For example, when the source of light is close to the moth, the moth begins to fly around the light in a spiral path. Eventually, this spiral fly path converges the moth to the light [7]. The moths are regarded as the candidate solutions in the MFO algorithm and their position is considered as a vector of variables of decision. Therefore, each moth can fly freely in the problem's solution space. The flowchart for MFO algorithm as given in Fig. 3.

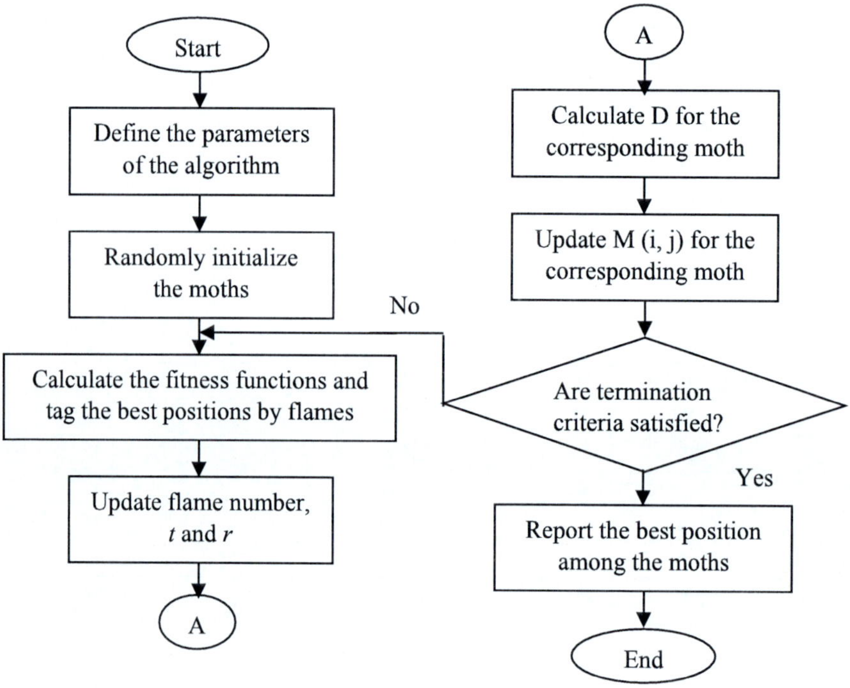

Fig. 3 Flowchart for MFO algorithm

5 Implementation of MFO Algorithm

In this study, MFO algorithm was used to calculate optimum FDM process parameters for producing the lowest surface roughness value. Equation (3) derived by response surface method is taken to minimize the fitness function with suitable lower and upper bounds. The optimal parameters are obtained as shown in Table 3 by applying MFO algorithm. The obtained convergence curve is depicted in Fig. 4. It is observed that best fitness and range decreases with increased number of iterations. It is observed that higher layer thickness, raster width and lower part orientation, raster angle values result in lowest surface roughness is noted.

Table 3 Confirmation experiments and their comparison with the results

Process parameters				Avg. surface roughness		
X_1 (mm)	X_2 (Degree)	X_3 (mm)	X_4 (Degree)	Pred	Actual	Error (%)
0.329	4.38	0.463	4.12	1.590	1.624	0.020936

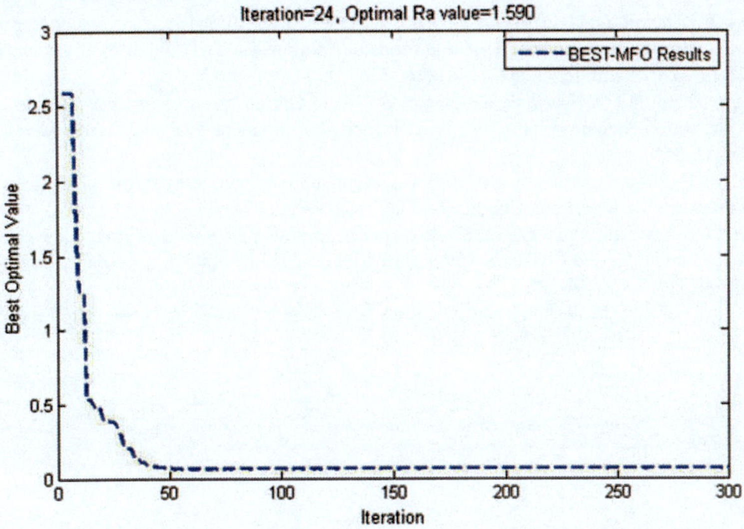

Fig. 4 Convergence plot during MFO optimization

6 Conclusions

The conclusions of this present study were drawn and as follows:

- RSM-based BBD has been used to determine the surface roughness attained by various FDM process parameters.
- The mathematical model was developed to describe the relationship between the independent process parameters and surface roughness at 95% confidence intervals.
- A nature-inspired MFO algorithm has produced an optimal combination of process parameters to optimize the surface roughness of the FDM processed specimens.

References

1. Mohamed O, Masood S, Bhowmik J (2017) A parametric investigation of the friction performance of PC-ABS parts processed by FDM additive manufacturing process. Polym Adv Technol 28(12):1911–1918
2. Rajamani D, Balasubramanian E, Arunkumar P, Silambarasan M, Bhuvaneshwaran G (2018) Experimental investigations and parametric optimization of process parameters on shrinkage characteristics of selective inhibition sintered high density polyethylene parts. Exp Tech 42(6):631–644

3. Rayegani F, Onwubolu G (2014) Fused deposition modelling (FDM) process parameter prediction and optimization using group method for data handling (GMDH) and differential evolution (DE). Int J Adv Manuf Technol 73(1–4):509–519
4. Sood A, Chaturvedi V, Datta S, Mahapatra S (2011) Optimization of process parameters in fused deposition modeling using weighted principal component analysis. J Adv Manuf Sys 10(02):241–259
5. Rao R, Rai D (2016) Optimization of fused deposition modeling process using teaching-learning-based optimization algorithm. Eng Sci Technol Int J 19(1):587–603
6. Tamilarasan A, Marimuthu K (2014) Multi-response optimisation of hard milling process parameters based on integrated Box-Behnken design with desirability function approach. Int J Mach Mach Mater 15(3–4):300–320
7. Mirjalili S (2015) Moth-flame optimization algorithm: a novel nature-inspired heuristic paradigm. Knowledge-Based Syst 89:228–249

Multi-response Optimization of AWJ Process Parameters in Cut Quality Characteristics of Hastelloy C-276

A. Tamilarasan, S. Arumugam, D. Rajamani, P. Changareddy, E. Balasubramanian, and P. Pranay

Abstract In this study, the AWJ cutting of Hastelloy C-276 has been performed to investigate the influence of four process parameters such as water jet pressure, jet traverse speed, abrasive flow rate and standoff distance on surface roughness and kerf taper angle. Based on the experiments carried out, two models have been developed using Box–Behnken design. Response surface analysis on cutting characteristics of AWJ was made based on the developed models. The desirability function approach was applied to acquire the compromised solutions between the multiple responses. The errors between experimental and predicted values at the optimal combination of parameter settings lie within ±2% which was found through confirmation test.

Keywords Abrasive water jet · Surface roughness · Kerf taper angle · Desirability function

1 Introduction

The abrasive water jet cutting technique contributes significantly to the development of product as a non-traditional method. The water jet was used to machine different materials in pure form or mixed with abrasives. The AWJ cutting process was environmentally friendly with good machining capabilities for handling materials such as steel, Inconel, granite and titanium [1]. The AWJ process's complex and stochastic nature makes it difficult to obtain an analytical model based on the process's physics. Therefore, the optimization of AWJ process parameters is necessary for obtaining the desirable cut quality characteristics of different materials.

A. Tamilarasan (✉) · S. Arumugam · P. Changareddy · P. Pranay
Department of Mechanical Engineering, Sri Chandrasekharendra Saraswathi Viswa Mahavidyalaya, Enathur, Kanchipuram, Tamil Nadu 631561, India
e-mail: tamilrj2010@gmail.com

D. Rajamani · E. Balasubramanian
Centre for Autonomous System Research, Department of Mechanical Engineering, Vel Tech Rangarajan Dr. Sagunthala R&D Institute of Science and Technology, Chennai, Tamil Nadu 600062, India

T. Rajmohan et al. (eds.), *Advances in Materials and Manufacturing Engineering*, Springer Proceedings in Materials 7, https://doi.org/10.1007/978-981-15-6267-9_44

Uthayakumar [2] addressed on the abrasive water jet processing of based superalloys using the three different parameters such as water jet pressure, jet nozzle traverse speed and standoff distance. It is observed that the most influential factor in surface morphology and surface quality is the water jet pressure. Therefore, in cutting Hastelloy C-276, the present work aims to simultaneously minimize surface roughness and kerf taper angle. The experiments are conducted in AWJ process using RSM-BBD methodology. Then, mathematical models are developed at 95% confidence intervals. Further, the desirability function-based multi-response optimization is used to optimize the AWJ cutting process parameters.

2 Experimental Procedure

In this study, nickel superalloy based on Hastelloy C-276 was used as the target material and this material is prepared in the dimensions of $150 \times 30 \times 5\ \text{mm}^3$. The box-behnken design, a RSM approach has been utilized to design and perform the experiments within the levels of selected process parameters, such as water pressure of 2000, 3000 and 4000 bar, jet traverse speed of 150, 200 and 250 mm/min, abrasive flow rate of 300, 400 and 500 g/min, and stand-off distance of 1, 2 and 3 mm, respectively. Figure 1 shows the experimental setup used for this study. The experiments were carried out in a water jet Germany manufacturing model S 3015 injection type AWJ machining center which can generate a maximum pressure of 620 MPa and 10 m/in traverse speed with a water discharge of 3.2 L per min.

The surface roughness (R_a) and kerf taper angle (K_t) were the output responses considered in this study. The surface roughness is measured by Mitutoyo making surface tester type SJ-201P. The tool makers microscope (Model 98—SCHERR TUMICO) was used to measure the top and bottom kerf widths at 10× magnification. The angle of the kerf taper (K_t) was calculated using the formula:

Fig. 1 Cutting zone of AWJ process

$$K_t(\text{deg}) = \frac{(W_t - W_b) \times 180}{2\pi t} \quad (1)$$

where W_t is the top kerf width, W_b is the bottom kerf width and t is the thickness of the material. For each cut, three measurements were taken and averaged value taken for best reading in order to avoid the statistical error. Table 1 lists the BBD design matrix with measured responses.

Table 1 BBD design matrix with measured responses

Exp. No.	X_1 (bar)	X_2 (mm/min)	X_3 (g/min)	X_4 (mm)	R_a (μm)	K_t (°)
1	4000	200	400	3	2	1.37
2	3000	150	500	2	2.68	2.32
3	3000	200	500	1	2.54	2.12
4	3000	200	400	2	3.21	1.42
5	3000	250	400	1	2.44	1.15
6	4000	200	500	2	1.67	2.25
7	3000	200	300	1	2.64	1.3
8	3000	200	400	2	3.21	1.44
9	3000	200	400	2	3.11	1.4
10	3000	200	400	2	3.21	1.41
11	2000	200	300	2	2.47	1.39
12	3000	250	400	3	3.21	1.25
13	3000	150	400	3	2.84	1.46
14	4000	200	300	2	1.91	1.15
15	3000	200	500	3	3.01	2.29
16	3000	150	300	2	2.64	1.36
17	3000	200	400	2	3.23	1.4
18	3000	150	400	1	2.34	1.32
19	3000	200	300	3	3.28	1.25
20	4000	200	400	1	1.57	1.21
21	4000	250	400	2	1.91	1.14
22	2000	200	500	2	2.34	2.28
23	3000	250	300	2	3.13	1.18
24	3000	250	500	2	2.74	2.15
25	2000	200	400	1	1.91	1.37
26	4000	150	400	2	1.47	1.2
27	2000	150	400	2	2.17	1.46
28	2000	200	400	3	2.79	1.45
29	2000	250	400	2	2.41	1.22

3 Results and Discussion

3.1 Mathematical Model for R_a and K_t

The fit summary recommended that the developed surface roughness and kerf taper angle quadratic model is analyzed and be statistically significant at 95% confidence interval. ANOVA Table 2 presents the results of quadratic models. The highly significant models are indicated by the F-test with a very low probability value ($p < 0.0001$). The developed models are examined by the determination coefficient ($R^2 = 0.995$ for Ra, $R^2 = 0.995$ for K_t), which meant that the sample variation of more than 99.5% is attributed to the variables and that each model could not explain only 0.5% of the total variance. Using an Adjusted-R^2 is to assess the suitability and fitness of the model [3, 4].

The adjusted-R^2 value corrects the sample size R^2 value and the model number of terms. The adjusted-R^2 value (0.992 for both R_a and K_t) is also high, promoting a high correlation between the observed values and the predicted values. The predicted R^2 of 0.987 (i.e., for R_a) and 0.9799 (i.e., for K_t) is in reasonable agreement with the Adjusted R^2 values. Furthermore, in this model, the value of adequate precision (AP), which compares the range of the predicted value at the design point with the average predictive error, is well above 4. Overall, statistical analysis showed that the experimental values fit well with the predicted values and that the polynomial model's accuracy and general availability were adequate for further optimization work.

$$R_a = -13.66927 + 5.36346E - 003X_1 + 0.048620X_2 \\ + 0.014139X_3 + 1.16228X_4 + 1.00500E$$

Table 2 ANOVA results-model validation

Source	SS	df	MS	F	p	R^2	Adj R^2	Pre R^2	AP
Surface roughness									
Model	8.45	12	0.7	321.9	<0.0001	0.995	0.992	0.987	56.33
Total	8.49	28							
Residual	0.035	16	0.021						
Lack of fit	0.026	12	0.022	1.019	0.5461				
Pure error	0.0863	4	0.021						
Kerf taper angle									
Model	4.31	11	0.39	315.8	<0.0001	0.995	0.992	0.9799	53.23
Total	4.33	28							
Residual	0.021	17	0.001						
Lack of fit	0.02	13	0.001	0.015	0.0723				
Pure error	0.0128	4	0.000						

$$
\begin{aligned}
&- 006X_1X_2 - 1.14000E - 004X_1X_4 \\
&- 2.09000E - 005X_2X_3 + 1.33500E \\
&- 003X_2X_4 - 9.38925E - 007X_1^2 \\
&- 1.07820E - 004X_2^2 - 1.35800E - 005X_3^2 - 0.19493X_4^2
\end{aligned} \tag{2}
$$

$$
\begin{aligned}
\text{KTA} = &-13.66927 + 5.36346E - 003X_1 + 0.048620X_2 \\
&+ 0.014139X_3 + 1.16228X_4 + 1.00500E \\
&- 006X_1X_2 - 1.14000E - 004X_1X_4 \\
&- 2.09000E - 005X_2X_3 + 1.33500E - 003X_2X_4 \\
&- 9.38925E - 007X_1^2 - 1.07820E - 004X_2^2 \\
&- 1.35800E - 005X_3^2 - 0.19493X_4^2
\end{aligned} \tag{3}
$$

3.2 Effect of Process Parameters on $\mathbf{R}_a$

Figure 2a, b shows the estimated response surface for R_a in relation to the process parameters of water jet pressure, jet traverse speed and abrasive flow rate. These figures indicate all the process variables significantly affected the surface roughness. In Fig. 2a, higher value of water jet pressure (4000 bar) and lower value of jet transverse speed (150 mm/min) result in a lower value of surface roughness (1.4 μm), while the lowest values of jet transverse speed and abrasive flow rate lead minimal surface roughness (2.63 μm) as illustrated in Fig. 2b.

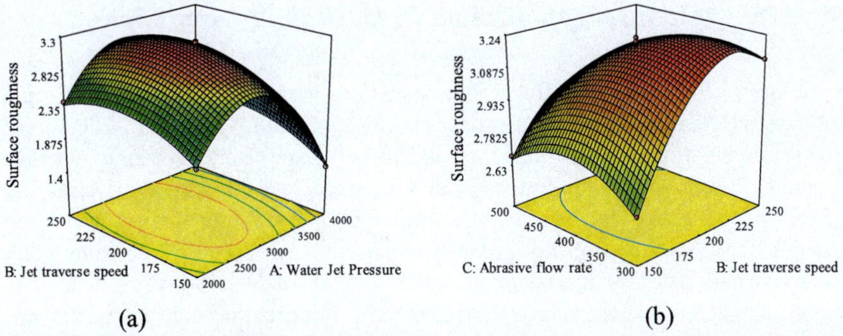

Fig. 2 **a**, **b** shows the effect of parameters on R_a

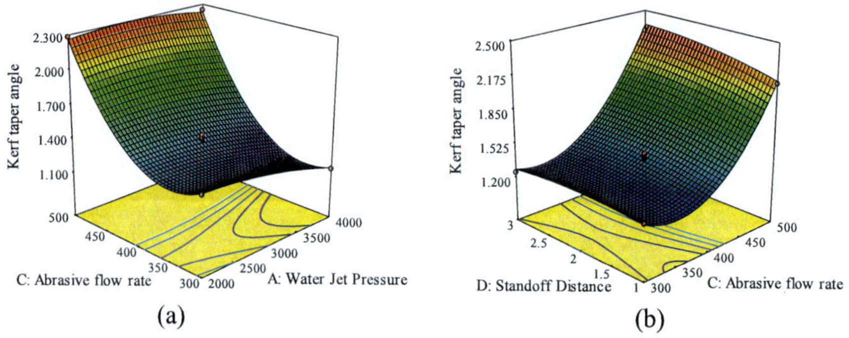

Fig. 3 **a**, **b** shows the effect of parameters on K_t

3.3 Effect of Process Parameters on K_t

Based on a developed model of K_t through the response surface methodology, Fig. 3a, b presents the effect of abrasive flow rate on kerf taper angle at various water jet pressure and standoff distances. From Fig. 3a, it is found that the kerf taper angle increased as the abrasive flow rate increased from 300 to 500 g/min for all values of water jet pressure. This scenario may be due to the fact that the outer rim of the diverged jet still has enough energy to cut material and due to the diverged jet energy [4]. Figure 3b shows that, with increased abrasive flow rate from 300 to 500 g/min, the kerf taper angle is found to increase; this is due the widening of the lower part of the kerf by the jet decreases as the standoff distance, which increases the higher value of K_t [5].

3.4 Multi-response Optimization by Desirability Approach

By the need of requirements for each response in the set, a combination of input variables optimizes a set of two responses in this desirability approach. The aim of present investigation is to optimize the AWJ process parameters in order to enhance the quality indices such as minimized R_a and K_t. To achieve this goal, the desirability function approach has been used. As seen in Fig. 4 by a numerical optimization procedure, the ramp desirability graph reached optimum points. Each ramp's dot denotes the parameter setting reflection, and the amount of desirability is indicated by the dot's height. Using the optimized parameters, a confirmation study was performed after optimizing each parameter level and with its corresponding response value. The value of R_a and K_t is obtained from the confirmation study which is closely linked to the data obtained from the optimization of desirability using Box–Behnken design. The triplicate confirmation experiments have been conducted, and the average values are reported in Table 3. The results of the confirmation experiment show that

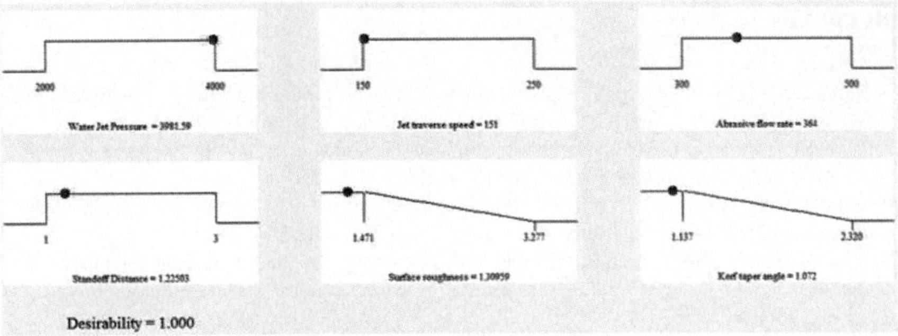

Fig. 4 Desirability ramp for numerical optimization

Table 3 Confirmation experiment results

R_a (μm)			K_t (°)		
Pred	Actual	Error (%)	Pred	Actual	Error (%)
1.309	1.41	1.1	1.072	1.087	−0.01399

the differences between the predicted values and the experimental values are below ±2%. This proximity of values therefore confirms the excellent reproducibility of the experimental conclusions.

4 Conclusions

Based on the detailed experimental study carried out on AWJ cutting of Hastelloy C-276, the following points can be concluded.

- The experimental work was performed using RSM-based BBD methodology.
- The empirical mathematical models showed a good fitting of the experimental data. And, the multi-response optimization using desirability approach was utilized for the search of simultaneous optimal values of the surface roughness and kerf taper angle.
- The simultaneous maximum values of 1.309 μm and 1.072° for surface roughness and kerf taper angle, respectively, corresponded to the optimal process parameters of water jet pressure(3981.59 bar), jet traverse speed(151 mm/min), abrasive flow rate (364 g/min) and standoff distance(1.22 mm).

References

1. Tamilarasan A et al (2018) Application of crow search algorithm for the optimization of abrasive water jet cutting process parameters. IOP Conf Ser-Mat Sci Eng 390(012034):1–11
2. Uthayakumar M et al (2015) Machinability of nickel-based superalloy by abrasive water jet machining. Mater Manuf Process 31(13):1733–1739
3. Tamilarasan A, Rajamani D (2017) Multi-response optimization of Nd:YAG laser cutting parameters of Ti-6Al-4V superalloy sheet. J Mech Sci Technol 31:813–821
4. Santhanakumar M, Adalarasan R, Rajmohan M (2015) Experimental modelling and analysis in abrasive waterjet cutting of ceramic tiles using grey-based response surface methodology. Arab J Sci Eng 40(11):3299–3311
5. Naresh Babu M, Muthukrishnan N (2014) Investigation on surface roughness in abrasive water-jet machining by the response surface method. Mater Manuf Processes 29(11–12):1422–1428

Application of Water Cycle Algorithm for Optimizing the PAC Process Parameters in Cutting Ti–6Al–4V Alloy

A. Tamilarasan, T. Rajmohan, S. Arumugam, A. Arunpremnath, K. Mohan, and P. Manohar

Abstract Plasma arc cutting (PAC) is a thermal cutting process using a heavy plasma gas jet to melt and separate metal. In this work, RSM-based Box–Behnken designed experiments were conducted taking into account four process parameters, namely cutting speed, current flow rate, torch height and gas pressure, to achieve maximum metal removal rate(MRR) in cutting of Ti–6Al–4V alloy. The experimental design, ANOVA, regression analysis and quadratic model have been developed using RSM according to BBD design methodology. A new evolutionary algorithm, i.e., the water cycle algorithm, a technique based on the continuous movement of water in nature, is used to determine the optimum parametric combinations of PAC process. The predicted optimum values were validated through the confirmation experiment, and the estimated relative error is 2.688%.

Keywords Plasma arc cutting · Box–Behnken design · Metal removal rate · Water cycle algorithm · Optimization

1 Introduction

Plasma arc cutting is a method used to cut metals of various thicknesses. The compressed air (nitrogen, argon, etc.) is used to produce a plasma jet, and the metals are cut using a high-temperature plasma arc and plasma jet mechanical erosion [1]. This process is faster and cheaper than other processes of cutting such as water jet and laser cutting. Torch geometry, plasma gas properties, flow rate, cutting speed, velocity, current, distance between torch and workpiece are very important process parameters which have a significant effect on cutting quality and production rate. For the smooth operation of the plasma arc cutting process, process parameters need to be optimized. Adalarasan et al. [2] implemented the effective GT-RSM technique for predicting the optimal setting of PAC cutting parameters for AISI304L stainless steel. Naik and Maity [3] implemented the DRSM technique to predict the optimum

A. Tamilarasan (✉) · T. Rajmohan · S. Arumugam · A. Arunpremnath · K. Mohan · P. Manohar
Department of Mechanical Engineering, Sri Chandrasekharendra Saraswathi Viswa Mahavidyalaya, Enathur, Kanchipuram, Tamil Nadu 631561, India
e-mail: tamilrj2010@gmail.com

T. Rajmohan et al. (eds.), *Advances in Materials and Manufacturing Engineering*, Springer Proceedings in Materials 7, https://doi.org/10.1007/978-981-15-6267-9_45

solution of input parameters for plasma arc cutting process. Maity and Bagal [4] applied RSM coupled with gray relational analysis and principal component analysis to optimize plasma arc cutting processes with multi-objective criteria. The literature has demonstrated an interesting potential through the use of different optimization approaches. Therefore, the present describes the application of the water cycle algorithm for optimizing the PAC process parameters in cutting Ti–6Al–4V alloy using the selected process parameters. The necessary experiments are conducted based on Box–Behnken design and are used for fitting of mathematical model.

2 Experimental Plan and Work

Response surface methodology (RSM)-based Box–Behnken design is employed to describe the liaison between plasma arc cutting parameters and their responses. A second-order polynomial response to study the influence of PAC process can be developed as given by the equation

$$y = \beta_0 + \sum_{i=1}^{k} \beta_i x_i + \sum_{i=1}^{k} \beta_{ii} x_i^2 + \sum_{i} \sum_{j} \beta_{ij} x_i x_j + \varepsilon \tag{1}$$

where β_0 is constant, β_i, β_{ii} and β_{ij} are the coefficients of linear, quadratic and cross-products terms, respectively. In this study, the PAC performance is evaluated in terms of metal removal rate (MRR). The selected process variables were equally spaced, and the low, middle and high levels of each variable were coded as -1, 0 and $+1$, respectively, as given in Table 1.

The EDX analysis of Ti alloy in Fig. 1a shows the presence of Ti (higher content), Al and V elements in the titanium alloy material. Experiments are carried out on the CNC plasma arc cutting (EPP-450 Model), which is supplied by ESAB. Air is used to cut gas, and special electrodes made of water-cooled copper are used with metal inserts (hafnium). A schematic diagram of plasma arc cutting process is shown in Fig. 1b. The 50-mm-length cuts were made with the use of compressed air as plasma gas on 3-mm-thick titanium alloy plate. A nozzle's outlet diameter is 1.2 mm and 6 bar plasma gas pressure. To obtain the amount of material removal rate (MRR), the weight of the workpieces (specimens) must be measured before and after the cutting

Table 1 Process parameters and levels

Levels	X_1: cutting speed (mm/min)	X_2: current flow rate (Ampere)	X_3: torch height (mm)	X_4: gas pressure (Bar)
−1	550	70	2	5
0	650	80	3	7
1	750	90	4	9

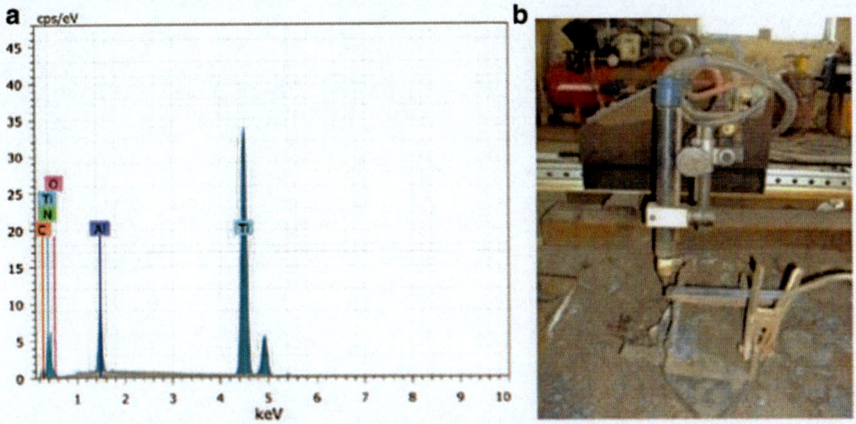

Fig. 1 **a** EDX of analysis of the titanium alloy, **b** experimental setup

process. The MRR can be defined as the volume of material removed divided by the machining time. MRR is defined by:

$$\text{MRR} = \text{WRW}/T\,(\text{g/min}) \tag{2}$$

where WRW: workpiece removal weight (g), T: cutting time (s). A Box–Behnken experimental design with 29 runs of MRR values is tabulated in Table 2.

3 Model Development and Checking the Adequacy Based on RSM.

Based on the calculated values of the regression coefficients using design expert software, a polynomial regression model equation is proposed as follows, fitting 97.7% of the data variation (actual values):

$$\begin{aligned}\text{MRR} = {} & +18.63846 - 0.0266X_1 - 0.47611X_2 + 8.36737X_3 \\ & + 0.086176X_4 - 5.49858\text{E} - 003X_1X_3 - 2.68944\text{E} \\ & - 003X_1X_4 - 0.043015X_2X_3 + 0.023942X_2X_4 \\ & + 0.16001X_3X_4 + 4.56965\text{E} - 005X_1^2 + 2.67827\text{E} \\ & - 003X_2^2 - 0.47955X_3^2 \\ & - 0.058038X_4^2 (R^2 = 0.977)\end{aligned} \tag{3}$$

The significant terms associated with the positive sign coefficient indicated a synergistic effect, and an antagonistic effect was indicated by the negative sign

Table 2 BBD design matrix

Exp. Run	Process parameters				Response
	Cutting speed (mm/min)	Current flow rate (Ampere)	Torch height (mm)	Gas pressure (Bar)	MRR (g/s)
1	550	70	3	7	2.985
2	750	70	3	7	2.177
3	550	90	3	7	2.557
4	750	90	3	7	2.053
5	650	80	2	5	2.049
6	650	80	4	5	0.460
7	650	80	2	9	0.913
8	650	80	4	9	0.603
9	550	80	3	5	1.644
10	750	80	3	5	2.385
11	550	80	3	9	2.512
12	750	80	3	9	1.101
13	650	70	2	7	1.446
14	650	90	2	7	2.136
15	650	70	4	7	1.673
16	650	90	4	7	0.642
17	550	80	2	7	1.919
18	750	80	2	7	2.500
19	550	80	4	7	2.171
20	750	80	4	7	0.552
21	650	70	3	5	2.493
22	650	90	3	5	1.440
23	650	70	3	9	1.290
24	650	90	3	9	2.152
25	650	80	3	7	1.685
26	650	80	3	7	1.746
27	650	80	3	7	1.855
28	650	80	3	7	1.634
29	650	80	3	7	1.765

coefficient [5, 6]. The analysis of variance has been done to check the adequacy of the model. The calculated R^2 value 0.977 shows good relationships between the response's experimental and predicted values. The results in the relationships between the independent variable and the response could be well explained.

4 Water Cycle Algorithm

Initially, the water cycle algorithm (WCA) was introduced by Eskandar et al. [7]. It is a modern algorithm that imitates the essence of the water cycle in order to solve optimization problems. The WCA is developed on powerful mathematical principles, which allow the algorithms to solve optimization problems efficiently. The WCA is focused on rivers and streams to the sea and was developed from observing the water cycle method. Suppose any rain or precipitation [7] phenomena are present. After the precipitation cycle, the population is generated along with the original design variables (i.e., population of streams). The best individual (i.e., the best stream) is chosen as the sea with the lowest cost function [7]. Therefore, a number of good streams are chosen as rivers (i.e., cost feature values similar to the latest best record), and the remaining streams are drawn into the rivers and the sea. The pseudo-code is given as follows [7]: set the parameters of WCA such as N_{pop}, N_{sr} and maximum number of iterations.

```
for i=1:Npop
    Create a stream
    Calculate the objective function value of the stream
end for
sort the streams from best to worst based on their objective function value
Sea ← the first stream
Rivers ← nsr - 1
Stream ← npop - nsr
Determine the intensity of flow for rivers and sea
i=0;
While   i < maximum number of iterations
i=i+1;
        Update the position of streams
        Stream_objective=objective function value of the new stream
    for     each stream
              if    stream_objective < river_objective
                    River_position= the new stream
              if      stream_objective < sea_objective
                           Sea_position= the new stream
              end if
            end if
            if river_objective < sea_objective
                      Sea_position =River_position
            end if
end for
```

```
    for    each river
Update the position of rivers
river_objective =objective function value of the new river
        if river_objective < sea_objective
                Sea_position =River_position
        end if
    end for
      for Rivers and streams
d=calculate the distance between each river or stream and the sea
        if  d <d max
            raining process (for both rivers and streams)
        end if
      end for
      Decrease the max d
end while
```

5 Implementation of WCA Algorithm in PAC Process

The problem of optimization of this study is indicated as maximizing the MRR criteria, subject to PAC cutting constraints. Therefore, the restricted problem of optimization with WCA is given by Eq. 3 taken as the maximization objective function.

Taking into account the lowest level (Level −1) and highest level (Level +1) of the experiment design specified for PAC cutting parameters as shown in Table 1, the lowest and highest limitation range for optimization constraints is as follows: 550 mm/min < cutting speed < 750 mm/min; 70 A < current flow rate < 90 A; 2 mm < torch height < 4 mm; 5 bar < gas pressure < 9 bar. The water cycle algorithm-based optimization process is performed using MATLAB environment to find the optimal MRR solution. The initial parameters for the WCA were selected as 50, 8 and 1E-05, for N_{pop}, N_{sr} and d_{max}, respectively. The convergence curve obtained from the algorithm is shown in Fig. 2. There are 200 runs performed in the WCA algorithm in the current work. The WCA achieves a good fitness function as can be seen in Fig. 2; in iteration 90, and after that, there is no significant improvement. Table 3 shows the results of WCA compared with experimental data. The relative error percentage is achieved that 2.688%.

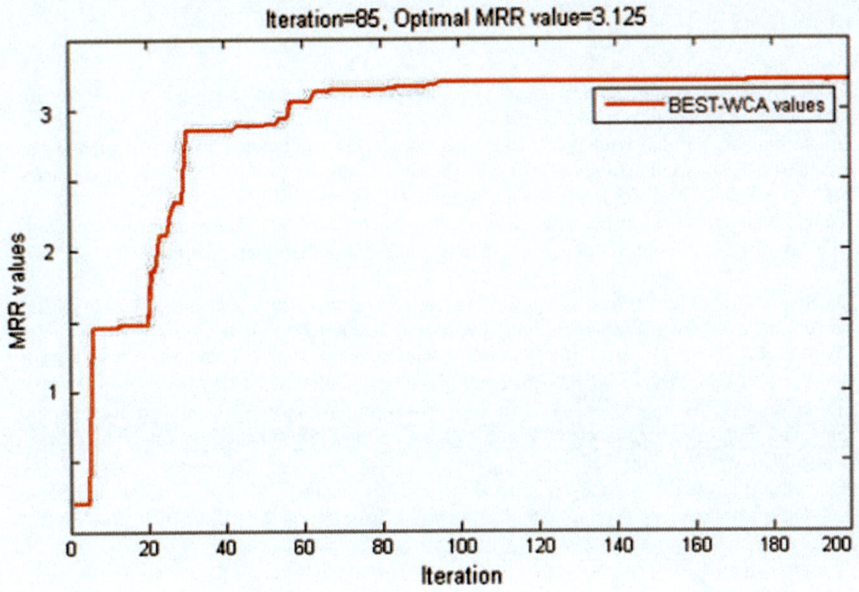

Fig. 2 Convergence plot for WCA algorithm

Table 3 Validation test results

Cutting speed (mm/min)	Current flow rate (Ampere)	Torch height (mm)	Gas pressure (Bar)	MRR (g/min)		
				Predicted	Experiment	Error %
739.62	71.47	2.6	5.28	3.125	3.041	2.688

6 Conclusions

In this work, an application of RSM based water cycle algorithm is utilized for maximizing the MRR in PAC cutting Ti–6Al–4V alloy and optimized parameters are validated through confirmation experiments. The main conclusions can be summarized as follows.

- The empirical relation established has the potential to MRR in PAC cutting of Ti–6Al–4V alloy within the ranges of process parameters studied.
- A MATLAB program was developed for WCO optimization algorithm to optimize the parameters.
- The relative error between the experimental and predicted value was 2.688%, which indicated that the proposed method is feasible.

References

1. Çelik Y (2013) Investigating the effects of cutting parameters on materials cut in CNC plasma. Mater Manuf Process 28(10):1053–1060
2. Adalarasan R, Santhanakumar M, Rajmohan M (2015) Application of grey taguchi-based response surface methodology (GT-RSM) for optimizing the plasma arc cutting parameters of 304L stainless steel. Int J Adv Manuf Technol 78(5–8):1161–1170
3. Naik D, Maity K (2018) Application of desirability function based response surface methodology (DRSM) for investigating the plasma arc cutting process of sailhard steel. World J Eng 15(4):505–512
4. Maity K, Bagal D (2014) Effect of process parameters on cut quality of stainless steel of plasma arc cutting using hybrid approach. Int J Adv Manuf Technol 78(1–4):161–175
5. Tamilarasan A, Rajamani D (2017) Multi-response optimization of Nd:YAG laser cutting parameters of Ti-6Al-4V superalloy sheet. J Mech Sci Technol 31:813–821
6. Tamilarasan A, Marimuthu K (2014) Multi-response optimisation of hard milling process parameters based on integrated box-behnken design with desirability function approach. Int J Mach Mach Mater 15(3–4):300–320
7. Eskandar et al (2012) Water cycle algorithm—a novel metaheuristic optimization method for solving constrained engineering optimization problems. Comput Struct 110–111:151–166

Butterfly Optimization Algorithm for Optimization of Roller Burnishing Process Parameters

A. Tamilarasan, S. Arumugam, D. Rajamani, S. Vijayabhaskar, R. Balakumar, and B. Thirupathi Reddy

Abstract In this work, an attempt is made to optimize the roller burnishing process parameters using a novel butterfly optimization algorithm. A total of 17 experiments were conducted according to Box–Behnken design considering three process parameters, namely burnishing speed, burnishing feed, burnishing depth to achieve higher surface hardness value. Single-objective optimization is performed using a nature-inspired butterfly optimization algorithm in which objective function is developed using RSM to select the optimal setting of burnishing process parameters. A comparison of the experimental and optimization results for surface hardness shows that the percentage error lies $\pm 2\%$.

Keywords Roller burnishing · Butterfly optimization algorithm · Surface hardness

1 Introduction

The burnishing process, which is an attractive finishing process that can improve metal surface finishing, surface hardness and dimensional accuracy, has been given considerable attention in the recent past. The burnishing process also reduces surface defects and alters the microstructure of conventional and unconventional machined surfaces. Related to the burnishing process research, great efforts have been made by many researchers [1–5]. Nguyen and Le [2] optimized the interior roller burnishing process for improving surface quality using archive-based micro-genetic algorithm. Seemikeri et al. [3] focused on surface roughness, microhardness, surface integrity and fatigue life aspects of AISI 1045 working material using full factorial experiment design. Khabeery and Axir [4] investigated the use of roller burning to improve the

A. Tamilarasan (✉) · S. Arumugam · S. Vijayabhaskar · R. Balakumar · B. Thirupathi Reddy
Department of Mechanical Engineering, Sri Chandrasekharendra Saraswathi Viswa Mahavidyalaya, Enathur, Kanchipuram, Tamil Nadu 631561, India
e-mail: tamilrj2010@gmail.com

D. Rajamani
Centre for Autonomous System Research, Department of Mechanical Engineering, Vel Tech Rangarajan Dr. Sagunthala R&D Institute of Science and Technology, Chennai, Tamil Nadu 600062, India

T. Rajmohan et al. (eds.), *Advances in Materials and Manufacturing Engineering*, Springer Proceedings in Materials 7, https://doi.org/10.1007/978-981-15-6267-9_46

surface integrity of 6061 aluminum alloy using a vertical milling machine. Tamilarasan et al. [5] utilized the lion optimization algorithm for improving the surface quality of roller burnished surfaces. Evaluating the results of optimal solutions in the literature, there are many studies on the operating parameters affecting the burnishing process parameters. Therefore, the present paper aims at the optimization of various process parameters for finding the best surface hardness value using new type nature-inspired butterfly optimization algorithm.

2 Experimental Framework

2.1 RSM-Based Box–Behnken Design

RSM is a widely accepted experimental statistical technique [6]. The RSM approach conducts statistically designed experiments, then evaluates the coefficients in a mathematical model and predicts the response and examines the model's adequacy [7]. It is very useful to model and predict the interest reaction affected by a number of input variables in order to optimize the responses. Among different designs, the BBD method was selected for the experimental design because it provides a reasonable balance between predictability and run economy [6]. The modeling of the desired response to several independent input variables can be obtained through the use of experiment design and the application of regression analysis. The response surface can be expressed as follows if all variables are assumed to be measurable and quadratic model of Y_u can be written as follows

$$Y_u = b_o + \sum_{i=1}^{k} b_i X_i + \sum_{j>i} b_{ij} X_i X_j + \sum_{i=1}^{k} b_{ii} X_i^2 \pm \varepsilon \quad (1)$$

The coefficient b_o is the free term, the coefficients b_i are the linear terms, the coefficients b_{ij} are the interaction terms, and the coefficients b_{ii} are the quadratic terms.

2.2 Experiment Procedure

The material used in this study is brass and delivered in the form of bars with diameter of 30 mm.These bars have been cut in length to 100 mm. As shown in Fig. 1, the roller burning operation was performed by clamping it on a lathe machine's tool post. The machine's three-jaw chuck and tailstock center clamped the workpiece. During burnishing, no coolant was used. The variable spindle speed of the lathe machine is up to 5000 rpm with a maximum 25 kW power. Burnishing experiments

Fig. 1 Experimental zone of roller burnishing process

were performed with roller burnishing tools on external surfaces. The using BBD method, three burnishing parameters were selected to perform the experiments. The burnishing parameters examined include: (1) burnishing speed (range: 30–150 rpm), (2) burnishing feed rate (range: 40–60 mm/rev) and (3) burnishing depth (range: 5–10 mm). Using the Mitutoyo surface hardness measuring device, surface hardness values were measured. For each sample, the surface hardness values are measured three times and their arithmetic mean value has been considered for investigation.

3 Regression Analysis for the BBD Design Matrix

Table 1 shows the arrangements and outcomes of the17 experiments conducted on the basis of BBD design. Based on the proposed second-order polynomial model, the mathematical model is developed from coefficients obtained using the Design–Expert software, which expressed the relationship between surface hardness using the selected burnishing process parameters (burnishing speed, burnishing feed rate and burnishing depth). The model equation for surface hardness in terms of actual factors is given by

$$\begin{aligned} \mathrm{SH} = & -19.47156 + 6.10417E - 003X_1 + 0.54687X_2 \\ & + 1.43950X_3 + 3.95833E - 004X_1X_2 \\ & - 1.31667E - 003X_1X_3 - 7.70000E - 003X_2X_3 \\ & - 5.31250E - 005X_1^2 - 4.56250E - 003X_2^2 \\ & - 0.0282X_3^2 (R^2 = 0.993;\ R^2_{\mathrm{Adj}} = 0.984) \end{aligned} \tag{2}$$

The mathematical regression model's statistical results have been calculated. The predicted values of the responses will ideally match the corresponding experimental results if the value of R^2 is equal to one. Equation (2) indicating the R^2 value was 0.993. The model's fit degree was high enough to account for 99.3 percentage surface hardness values. The $R^2{}_{\mathrm{Adj}}$ value (i.e., 0.984) is also high, promoting a high correlation between the observed values and the predicted values. The actual versus predicted

Table 1 BBD design matrix

Exp. Run	Burnishing speed (rpm)	Burnishing feed rate (mm/rev)	Burnishing depth (mm)	Surface hardness (Hv)
1	30	40	7.5	2.25
2	150	40	7.5	2.49
3	30	60	7.5	3.34
4	150	60	7.5	4.53
5	30	50	5	1.5
6	150	50	5	2.73
7	30	50	10	4.53
8	150	50	10	4.97
9	90	40	5	1.14
10	90	60	5	2.7
11	90	40	10	4.02
12	90	60	10	4.81
13	90	50	7.5	3.86
14	90	50	7.5	3.92
15	90	50	7.5	3.63
16	90	50	7.5	3.8
17	90	50	7.5	3.79

values graph depicted in Fig. 2 clearly shows that the predictions made by the mathematical model are in good agreement with the experimental values. Hence, the regression analysis is substantiating the model developed.

Fig. 2 Actual versus predicted values

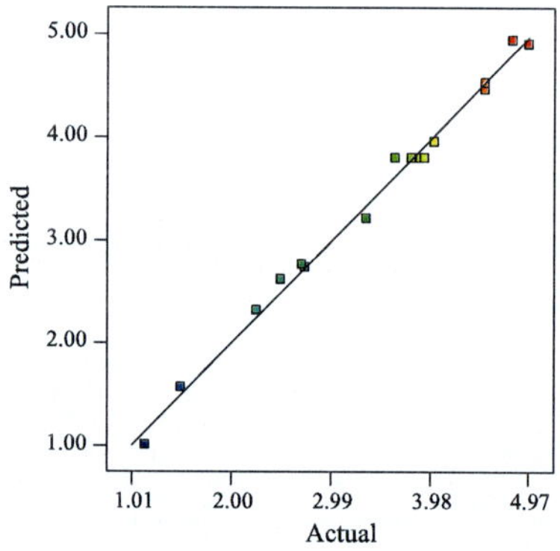

4 Butterfly Optimization Algorithm

In the past, various researchers have gained a lot of attention from nature-inspired metaheuristic algorithms. Butterfly optimization algorithm (BOA) is a part of the same class of algorithms for optimization [8]. It is essentially inspired by the behavior of butterflies foraging food [8]. In BOA, these butterflies are used as search agents to optimize the process variables. Butterflies use their sense of smell, vision, taste, touch and hearing to find a partner for food and mating. These senses also help to migrate from one place to another, to escape from predator and to lay eggs in suitable places. Among all these senses, smell is the most important sense that helps butterflies find food, usually nectar, even far away. Based on different observations behavior of the butterflies, the pseudo-code of butterfly optimization algorithm is developed and given in 'Algorithm 1' [8].

Algorithm 1: Butterfly Optimization Algorithm

```
Objective function f(x) , x= (x1, x2,...xdim) , dim = no. of dimensions
Generate initial population of n butterflies xi = (i=1,2,...., n)
Stimulus Intensity Ii at xi is determined by f(xi)
Define sensor modality c, power exponent a and switch probability p
While stopping criteria not met do
        for each butterfly bf in population do
                Calculate fragrance for bf
        end for
        Find the best bf
        for each butterfly bf in population do
                Generate a random number r from [0,1]
                    if r < p then
                        Move towards best butterfly / solution
                else
                        Move randomly in the search space
                end if
        end for
        Update the value of a
end while
Output the best solution found.
```

5 Optimization Through BOA Algorithm

The computer code for the BOA algorithm is developed in MATLAB environment as selected for the easy convergence of the optimization problem using the regression Eq. (2). The main control parameters are sensor modality (c), power exponent (a) and switch probability (p). On the basis of trail runs, these control parameters are selected and the results obtained with these selected parameters are close to the optimum results for the given problem. Figure 3 shows the evolution of the generations. The figure shows that the optimal fitness value after 40 iterations is 5.09 Hv based on the BOA algorithm. The corresponding optimized combinations of parameters are shown in Table 2, respectively.

Following the selection of optimal process parameters, experiments were conducted to check the corresponding induced surface hardness under the optimal combination of process parameters. The obtained percentage of error between the values predicted and the values tested in the experimental work. The calculated error is found to be very small, confirming the excellent reproducibility of the experimental conditions.

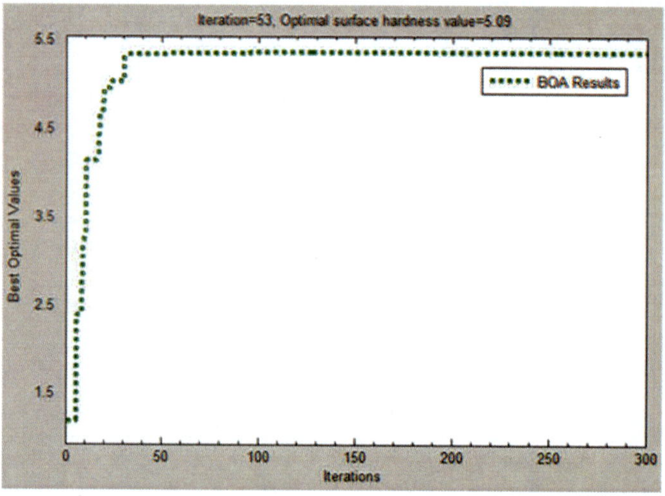

Fig. 3 Fitness curve with BOA algorithm

Table 2 Confirmation test results

Burnishing speed (rpm)	Burnishing feed rate (mm/rev)	Burnishing depth (mm)	Surface hardness (Hv)		
			Predicted	Experiment	Error %
137.2	53.61	9.98	5.09	5.056	0.6679

6 Conclusions

The main conclusions observed in this study are as follows:

- A Box–Behnken design effectively used for the experiment.
- The mathematical model developed using the RSM-BBD methodology described the relationship between the independent process parameters and surface hardness at 95% confidence intervals.
- A nature-inspired BOA algorithm has produced an optimal combination of process parameters to optimize the surface hardness of the roller burnished specimens.
- In the optimization, when the fragrance is 30, the BOAs search the global optimum with reasonable convergence speed.
- The best possible optimum conditions of this burnishing process are the following: 137.2 rpm of burnishing speed, 53.61 mm per rev of burnishing feed rate and 9.98 mm of burnishing depth to yield the highest value of surface hardness which is 5.09 Hv.

References

1. Łętocha A, Miller T, Kalisz J (2017) Optimization of measurement and analysis parameters of burnishing surfaces. Mechanik 11:1030–1034
2. Nguyen T, Le X (2018) Optimization of interior roller burnishing process for improving surface quality. Mater Manuf Process 33(11):1233–1241
3. Seemikeri C, Brahmankar P, Mahagaonkar S (2008) Investigations on surface integrity of AISI 1045 using LPB tool. Tribo Int 41(8):724–734
4. El-Khabeery M, El-Axir M (2001) Experimental techniques for studying the effects of milling roller-burnishing parameters on surface integrity. Int J Mach Tools Manuf 41(12):1705–1719
5. Tamilarasan et al (2018) Optimization of roller burnishing process parameters using lion optimization algorithm. IOP Conf Ser-Mat Sci Eng 390(012063):1–9
6. Tamilarasan A, Rajamani D (2017) Multi-response optimization of Nd:YAG laser cutting parameters of Ti-6Al-4V superalloy sheet. J Mech Sci Technol 31:813–821
7. Tamilarasan A, Marimuthu K (2014) Multi-response optimisation of hard milling process parameters based on integrated Box-Behnken design with desirability function approach. Int J Mach Mach Mater 15(3–4):300–320
8. Arora S, Singh S (2018) Butterfly optimization algorithm: a novel approach for global optimization. Soft Comput 23(3):715–734

Multi-objective Optimization of WEDM Process Parameters Using NSGA-II Algorithm

A. Tamilarasan, G. Sriram, S. Arumugam, D. Vijayan, D. Rajamani, and A. Venkata Akhil

Abstract In this paper, NSGA-II algorithm is applied to look for the variety of optimum solutions in WEDM process problem. Four parameters, namely wire feed, pulse-on time and pulse-off time, and two machining criteria, i.e., kerf width and material removal rate, are, respectively, considered as input variables and responses. The obtained results showed that the proposed NSGA-II algorithm is an effective and appropriate strategy for optimizing the process of WEDM machining process parameters.

Keywords WEDM · Kerf width · MRR · NSGA-II algorithm · Optimization

1 Introduction

Wire electrical discharge machining (WEDM) is an essential machining technique to produce complicated cut outs through hard to machine metals without the use of high-cost grinding or expensive formed tools [1].The most important performance measurements are material removal rate (MRR), machined surface roughness (SR) and kerf, i.e., cutting width(KW). Significant research has been carried out in the field of WEDM process by various researchers. Faisal and Kumar [1] implemented a multi-objective optimization for the EDM process of EN31 steel for MRR and roughness parameter by means of PSO and BBO techniques. Mohanty et al. [2] optimized the EDM process parameters using utility concept and QPSO algorithm. The demonstrated results showed the good convergence and computational effort of QPSO algorithm. Ma et al. [3] proposed an integrated optimization model for WEDM using Gaussian process regression and wolf pack algorithm approach in

A. Tamilarasan (✉) · G. Sriram · S. Arumugam · D. Vijayan · A. Venkata Akhil
Department of Mechanical Engineering, Sri Chandrasekharendra Saraswathi Viswa Mahavidyalaya, Enathur, Kanchipuram, Tamil Nadu 631561, India
e-mail: tamilrj2010@gmail.com

D. Rajamani
Centre for Autonomous System Research, Department of Mechanical Engineering, Vel Tech Rangarajan Dr, Sagunthala R&D Institute of Science and Technology, Chennai, Tamil Nadu 600062, India

T. Rajmohan et al. (eds.), *Advances in Materials and Manufacturing Engineering*, Springer Proceedings in Materials 7, https://doi.org/10.1007/978-981-15-6267-9_47

machining of SiCp/Al composite. Critical review of the past studies suggests that a significant amount of work is dedicated to technological improvement and statistical and numerical modeling to improve and analyze WEDM processes. In view of this, the current study focuses on multi-objective optimization of WEDM process parameters using NSGA-II algorithm for multiple performance characteristics viz. kerf width and MRR which are functions of selected process variables.

2 Experimental Methodology

Response surface methodology is an effective design of experiment approach for statistical evaluation of the relationship between independent variables and the responses [4–6]. In this work, central composite design (CCD) is selected and consists of 2*n* factorial runs (coded to the usual $\pm$ notation) with 2*n* axial runs ($\pm\alpha$, 0, 0, ..., 0), (0, $\pm\alpha$, 0, 0, ..., 0), ..., (0, 0, ..., $\pm\alpha$) and accentor runs (six replicates, 0, 0, 0, ..., 0). The main objective of this work is to identify the mathematical models describing the response dependence on the three parameters of viz., wire feed (X_1), pulse-on time (X_2) and pulse-off time (X_3). An empirical model has been developed to correlate the response to WEDM process and is based on a quadratic second-order model as

$$Y = \beta_0 + \sum_{i=1}^{k} \beta_i x_i + \sum_{i=1}^{k} \beta_{ii} x_i^2 + \sum_{i<j} \beta_{ij} x_i x_j \tag{1}$$

where Y is the response, x_i (1, 2, ..., k) are the independent k quantitative process variables, β_0 is a constant and β_i, β_{ii} and β_{ij} are the coefficients of linear, quadratic and interaction terms. The ranges of the three factors (i.e., wire feed: 4–8 mm/min; pulse-on time: 100–125 s; pulse-off time: 50–60 s) can be defined based on the previous experience and engineering analysis. A WEDM was selected as a machine tool by Sodick (Model: VL600Q), India (Fig. 1). High-strength low-alloy, 100 mm $\times$ 350 mm $\times$ 10 mm workpiece material, is used. This steel finds its wide application in the manufacture of high load bearing parts of cars due to its high strength-to-weight ratio, components such as steering, chassis parts, door intrusions and suspension parts. The kerf width was measured using the microscope of the Carl Zeiss tool maker (100). The kerf width value is the average of three measurements per cut piece length. First of all, the initial weight of the workpiece was taken before the WEDM operation was performed. The MRR is calculated as

$$\text{MRR} = \text{KW} \times T \times V_c \tag{2}$$

where KW is the kerf width, T the thickness of workpiece (10 mm) and V_c the cutting speed. The final weight for the workpiece was then compared to the initial weight and the difference between the two weights that yielded the MRR. Experiments have

Fig. 1 Schematic of WEDM process

been repeated twice to minimize the experimental error caused by noise factors .The experimental results of measured responses are given in Table 1.

3 Mathematical Modeling

The mathematical models are produced on the basis of the proposed second-order polynomial model using the Design–Expert software, which expressed the relationship between kerf width and MRR through the parameters of WEDM process. The final model equations for both responses are determined by the actual factors given by Eq. (3)–(4). The model demonstrates that the experimental data were accurately described, and that, the correlation between the three variables was successful. Furthermore, The R^2, adjusted R^2 values are 98.28% and 98.23%, respectively, for KW model, while the R^2, adjusted R^2 are 99.81% and 99.67%, respectively, for MRR model.

$$\begin{aligned}\text{KW} = &-1898.5335 + 44.9143X_1 + 11.3961X_2 + 42.0062X_3 \\ &- 0.11980X_1X_2 - 0.06184X_2X_3 - 2.0431X_1^2 - 0.0301862X_2^2 \\ &- 0.29275X_3^2 (R^2 = 0.9828;\ R^2_{\text{Adj}} = 0.9823)\end{aligned} \tag{3}$$

$$\begin{aligned}\text{MRR} = &+2852.14455 - 31.2237X_1 - 24.49425X_2 \\ &- 49.36818X_3 + 0.10423X_1X_2 + 0.15144X_2X_3 + 1.34445X_1^2 \\ &+ 0.065519X_2^2 + 0.3106X_3^2 (R^2 = 0.9981;\ R^2_{\text{Adj}} = 0.9967)\end{aligned} \tag{4}$$

Table 1 CCD design matrix with responses

Exp. Run	Process parameters			Responses	
	X_1: wire feed	X_2: pulse on time	X_3: pulse off time	KW (micron)	MRR (mm^3/min)
1	8	125	60	165.71	60.037
2	4	100	60	134.55	42.828
3	4	125	60	144.45	16.097
4	6	112.5	55	160.48	9.033
5	6	112.5	63.4	163.25	61.367
6	6	133.52	55	153.05	42.565
7	4	125	50	126.94	55.010
8	4	100	50	100.34	46.909
9	6	112.5	55	156.34	46.157
10	2.63	112.5	55	109.28	21.951
11	8	100	50	134.45	64.576
12	6	112.5	46.59	108.14	31.022
13	8	125	50	147.83	26.693
14	6	112.5	55	156.45	54.920
15	6	112.5	55	156.37	18.494
16	6	91.47	55	133.06	18.757
17	8	100	60	166.55	18.795
18	6	112.5	55	161.74	18.547
19	6	112.5	55	161.54	17.345
20	9.36	112.5	55	157.28	19.291

4 NSGA-II Algorithm

Srinivas and Deb [7] proposed a modified version of fast elitist multi-objective optimization algorithm of NSGA-II. The algorithm generates a set of evenly distributed solutions using non-dominated sorting and a crowded-comparison approach. The step-by-step procedure of algorithm is as follows:

1. Set the initial run parameters for the algorithm, viz. population size (N), maximum number of generations (g_{max}), crossover probability (P_c), mutation probability (P_m; generation, $g = 0$).
2. Randomly create an initial population P_g of size N with a good coverage of the search space and thereby have a diverse gene pool with potential to explore as much of the search space as possible.
3. Evaluate the objective values and rank the population using the concept of domination. Each solution is assigned a fitness (or rank) equal to its non-domination level (1 is the best level, 2 is the next best level and so on).

4. Perform the crowding sort procedure and include the most widely solutions by using crowding distance value.
5. The child population Q_g is produced from the parent population P_g using binary selection, recombination and mutation operators.
6. Then, the two populations are combined together to produce R_g (=P_g U Q_g), which is of size 2 N.
7. After this, the population R_g undergoes non-dominated sorting to achieve a global non-domination check.
8. The new population $P_g + 1$ is filled based on the ranking of the non-dominated fronts.
9. Since the combined population is twice the size of the population size N, all the fronts are not allowed to be used. Therefore, a crowding distance sorting is performed in descending order and the population is filled. Thus, for this new population $P_g + 1$, the whole process is repeated.
10. Update the number of generations, $g = g + 1$.
11. Repeat steps 3–10 until a stopping criterion is met.

5 NSGA-II Algorithm-Based Optimization of Parameters

Based on the empirical relationships obtained, NSGA-II algorithm is implemented in the MATLAB environment. The problem has two conflicting goals: minimizing the kerf width and maximizing the material removal rate. Maximization of MRR has been converted into minimization by taking it in reverse and only takes the problem of minimization. The range of lower and upper limits is set by LB = [4 100 50]; UB = [8 125 60]. The control parameters are adjusted to achieve the best solutions. Finally, following several simulations, the tuned parameters were selected to obtain the best optimal Pareto solutions. The parameters are: population size = 200, crossover probability = 0.9, generation number = 1000. Table 2 shows a set of 25 solutions with appropriate WEDMed conditions. The choice of important response (KW or MRR or combined) within the experimental range is very important because the process parameters selected are determined to improve the efficiency of production and quality of the cut specimens. The process engineer can select any of the optimal cutting conditions solutions for the optimal Pareto solution set, depending on the requirements. Since, it is observed from the Table 2, none of the solutions in the non-dominated set are better than any other solution in the set. A little consideration has shown that it is more desirable to boost productivity at higher MRR. The appropriate solutions are 1, 8 and 15, while the kerf width values will be compromised. In same way, the solutions of 3, 5 and 19 were noted for obtaining the minimum kerf width. And, MRR values are compromised. Similarly, with one response at a time, the other responses are compromised. The remaining solutions are treated with minimal kerf width during the WEDM process to obtain balanced higher MRR values.

Table 2 Optimal combination of parameters

Sol. Set	X_1	X_2	X_3	KW	MRR
1	4.001	100.11	50.101	97.622	61.509
2	4.0039	119.34	50.005	120.98	20.199
3	5.1852	123.6	50.005	136.46	10.624
4	4.0301	119.59	50.014	121.56	19.776
5	6.8196	123.68	50.002	145.57	7.0254
6	4.5468	123.45	50.017	129.95	14.025
7	5.6196	123.63	50.04	140.13	8.9793
8	4.0043	105.87	50.004	107.17	43.749
9	4.0202	112.6	50.009	115.71	28.865
10	4.0043	106.12	50.005	107.53	43.101
11	4.0314	121.26	50.026	122.56	18.562
12	4.9642	123.46	50.029	134.48	11.703
13	4.0049	107.13	50.007	108.95	40.528
14	4.7251	123.51	50.033	132.04	12.967
15	4.0049	102.74	50.008	102.42	52.625
16	4.0101	105.07	50.003	106.09	45.84
17	4.009	109.32	50.007	111.85	35.397
18	4.0063	105.26	50.009	106.35	45.337
19	6.1859	122.69	50.011	143.26	7.6983
20	4.0034	110.48	50.007	113.14	33.007

Finally, the first three sets of WEDMed solutions were selected to verify the predicted KW and MRR values through experimental work. The optimal parameters and experiment are presented in Table 3. Between the predicted optimum values and the optimum values obtained from the experiment, there is a good agreement based on Table 3.

Table 3 Validation of predicted optimal results with experiment

Sol. Set	X_1	X_2	X_3	KW	MRR
				Predicted/experiment	Predicted/experiment
1	4.001	100.11	50.101	97.622/97.01	61.509/60.245
2	4.0039	119.34	50.005	120.98/119.78	20.199/19.874
3	5.1852	123.6	50.005	136.46/136.12	10.624/9.232

6 Conclusions

This article presented a RSM and NSGA-II application to select the optimum combination values of WEDM process parameters affecting both kerf width and MRR values in cutting of HSLA steel. Based on that, the following conclusions have been drawn from the present work:

- Mathematical models for kerf width and MRR were developed and statistically validated using RSM through the understanding of four process variables.
- A multi-objective optimization approach based on NSGA-II is proposed to obtain Pareto-optimal solutions.
- Experiments have verified the optimal parametric setting obtained by NSGA-II algorithm and the proposed hybrid approach can be effectively utilized for optimization of WEDM process parameters.

References

1. Faisal N, Kumar K (2018) Optimization of machine process parameters in EDM for EN 31 using evolutionary optimization techniques. Technologies 6(2):1–16
2. Mohanty C, Mahapatra S, Singh M (2017) An intelligent approach to optimize the EDM process parameters using utility concept and QPSO algorithm. Eng Sci Technol Int J 20(2):552–562
3. Ma et al (2018) Integrated optimization model in wire electric discharge machining using Gaussian process regression and wolf pack algorithm approach while machining SiCp/Al composite. Adv Mech Eng 10(9):1–17
4. Tamilarasan A, Marimuthu K (2014) Multi-response optimisation of hard milling process parameters based on integrated Box-Behnken design with desirability function approach. Int J Mach Mach Mater 15(3–4):300–320
5. Tamilarasan A, Rajamani D (2017) Multi-response optimization of Nd:YAG laser cutting parameters of Ti-6Al-4V superalloy sheet. J Mech Sci Technol 31:813–821
6. Tamilarasan A, Rajamani D, Esakki B (2018). Parametric optimisation in Nd-YAG laser cutting of thin Ti-6Al-4V super alloy sheet using evolutionary algorithms. Int J Mat and Prod Tech 57(1/2/3): 71–91
7. Srinivas N, Deb K (1994) Multiobjective optimization using nondominated sorting in genetic algorithms. Evolutionary Comp 2(3):221–248

Water Quality Index and Correlation Study of Temple Ponds in Kanchipuram, Tamil Nadu, India

P. Meenakshi and G. Sriram

Abstract This research work aims to accept the quality of water in temple ponds in Kanchipuram. The water quality parameters such as pH, DO, total dissolved solids, chlorides, sulphates, iron and biological parameter like total coliforms were tested for temple pond water. The results were evaluated and compared with ISI-IS: 2296–1982 and BIS standards. The results show according to Class C standards the pond water was medium and good quality so the water nature is drinking water with usual treatment next by disinfection. Presence of dissolved oxygen according to Class C standards majorly affects the water quality. Dharmalingeswarar temple pond water is majorly affected and it having very poor quality. Correlation and regression values were calculated by using SPSS software. Analysis result shows TDS having good positive correlation between chloride ($r = 0.989$) and sulphate ($r = 0.970$). Chloride shows good positive correlation with SO_4 ($r = 0.988$). Results obtained from this study shows that the analysed temple pond water was fit for surface water.

Keywords WQI · Pond water · Correlation · Regression

1 Introduction

The water near in surface sources ponds [1], lakes, rivers, hydraulic structure dams, etc., which is only use by peoples for engineering [2], houses and farming purposes. Some pond water has heavy metals also like sulphur [3], iron, etc. Temple pond water having seasonal changes [4] due to the environment this causes the change of quality of water [5]. In this temple pond, [6] also one of the structure to store water and it is used for increasing the groundwater level. Kancheepuram is situated in south-west direction at a distance of 76 km from Chennai. Kanchipuram [7] has

P. Meenakshi (✉) · G. Sriram
Sri Chandrasekharendra Saraswathi Viswa Mahavidyalaya (Deemed to be University), Enathur, Kanchipuram, Tamil Nadu 631561, India
e-mail: meenathavamani82@gmail.com

G. Sriram
e-mail: drg.sriram@gmail.com

T. Rajmohan et al. (eds.), *Advances in Materials and Manufacturing Engineering*, Springer Proceedings in Materials 7, https://doi.org/10.1007/978-981-15-6267-9_48

a divine river Vegavathi, and a tributary of the river Pallar. Kancheepuram covers an area of 36.14 km^2 and has an elevation of about 83 m (275 ft.) above sea level. The term water quality index [8] give a symbol of how fitting the water for human use. The index [9] transforms a many quantity of variables data into a solo number, which represent the water worth stage. The strength and direction of the linear relationships between pairs of variables can be measured by correlation coefficients [10, 11]. To study the relationship [12] between one dependent and one independent [13] variable *regression* is used [14]. The study [15, 16] was conducted on March 2017, mainly related to knowing the nature of the temple pond water and how it is related to drinking water nature. Kanchipuram is highly holiest city and vigorously developing in industry side also, so to maintain the city holy and save water in ponds, the analysis temple pond water is at present very important. Renovation of ponds is important topic at present scenario.

2 Materials and Methods

The present survey was undertaken to estimate the worth of temple pond water in Kanchipuram. Seventeen stations were identified for sampling in Kanchipuram. The location and level of water in temple pond is shown in Table 1.

The study was conducted in the period of March 2017. The samples were collected in pre-cleaned polyethene bottles, immediately brought to the laboratory and analysed for various physicochemical parameters and total coliform count. The results were verified with ISI-IS: 2296–1982 standards. As per the standards the water classified into Class A to Class E. The details are given below.

(Class—A) = Water used for drinking without usual action but after disinfection.
(Class—B) = Bathing.
(Class—C) = Water used for drinking with usual treatment but after disinfection.
(Class—D) = Culture of fish and wildlife circulation.
(Class—E) = Irrigation, industrial waste removal.

2.1 Water Quality Index—(WQI)

WQI means worth of water in terms of digit. This number will give the quality of water. It also indicates the relative weight of each parameter. The computation of WQI was done using weighed arithmetic index method. There is number of parameter in the water sample result q_n represents the quality of nth parameter in the water sample with reference to standard permissible value.

$$\text{Quality rating } (q_n) = 100(\text{observed value} - \text{ideal value})/(\text{Standard value} - \text{ideal value}).$$

Table 1 Temple name with level of water

Pond No.	Name of the temple	Level of pond water
P1	Lakshmi narayana temple	Low
P2	Kamakshi amman temple	Medium
P3	Kailasanathar temple	Medium
P4	Kasi viswanathar temple	High
P5	Astabhujakoram temple	Low
P6	Puniya kotteswarar temple	Low
P7	Kusala kottam	Low
P8	Kachabeshwarar temple	Medium
P9	Varadharaja perumal temple (front side)	Medium
P10	Varadharaja perumal temple (back side)	Low
P11	Sonnavannam Saitha perumal temple	Low
P12	Dharmalingeswarar temple	Medium
P13	Ekambareswarar temple	Medium
P14	Santhaleeswarar temple	Low
P15	Vilakoli perumal temple	Low
P16	Kailasanathar temple	High
P17	Thamarai kulam	High

Ideal value = 0 except in pH, dissolved oxygen parameters.

Calculation of quality rating for pH and DO.

Quality rating (q_n) for pH = 100 (observed value for pH – 7)/(8.5 – 7) and

Quality rating (q_n) for DO = 100 (observed value for DO – 14.60)/(15.00 – 14.60).

Unit weight (W_n) = proportionality constant (k)/standard acceptable value given for nth parameter (s_n).

WQI value is calculated by the following equation.

$$\text{WQI} = \sum (w_n q_n) / \sum w_n \tag{1}$$

Water quality index and category value of water quality as shown in Table 2.

Table 2 Water quality index (WQI) and category value

S. No.	Category value	WQI
1	Excellent nature of water	(0–25)
2	Good nature of water	(26–50)
3	Poor nature of water	(51–75)
4	Very poor nature of water	(76–100)
5	Unfit for drinking	(>100)

3 Results and Discussion

Analysis of pond water is in the period of March 2017 shows the results in Table 3. All the results were compared with Class C standards of ISI-IS: 2296–1982.

3.1 PH

The pH value [16] indicates the measure of the hydrogen ion application in water. The pH value depends on the impurity enter into the water. According to Class C

Table 3 Observed values of parameters for temple pond water

Pond No.	pH	DO (mg/l)	TDS (mg/l)	CL (mg/l)	SO_4 (mg/l)	Fe (mg/l)	TCC (MPN)
P1	7.44	6.7	230	38	25	0.54	1800
P2	8.42	6.2	382	104	49	0.67	500
P3	7.47	7.3	161	23	10	0.72	500
P4	7.42	7.1	618	171	94	0.82	880
P5	7.4	7.4	573	159	87	0.68	1800
P6	7.47	7.9	305	61	24	0.53	500
P7	7.59	7.3	306	63	29	0.46	500
P8	7.25	7.6	159	38	16	0.37	820
P9	7.85	6.9	165	37	19	0.42	500
P10	7.48	7.9	424	94	39	0.54	2300
P11	7.94	6.7	932	244	118	0.74	1200
P12	9.07	6.2	1326	403	187	0.82	1440
P13	7.45	6.8	132	37	14	0.65	500
P14	7.76	6.8	676	165	72	0.65	800
P15	7.85	6.7	835	202	83	0.58	1200
P16	7.96	6.8	574	136	67	0.46	620
P17	7.4	7.2	425	96	32	0.67	1800

standards, the maximum limit is 8.5. Measured pH value ranging between 7.25 and 9.07. Test results show maximum temple ponds having a pH value less than 8.5. Dharmalingeswarar temple (P12) only having greater pH value 9.07. The high pH value indicates that the water is alkaline. The pH is considered as a derived water contaminant whose impact is considered a visual alarm.

3.2 Dissolved Oxygen

It means the stage of free, non-compound oxygen here in water. The quality of water [15] is majorly depends on DO because it indicates the amount of living organisms present in water. It enters the water through the air slowly or mixed quickly by aeration process. Bacteria and fungi need dissolved oxygen for decomposition and this process is main for nutrient recycling. Dissolved oxygen value varies between 6.2 and 7.9 mg/l. In the test result, all pond water having DO value greater than standard value 4. It majorly affects the water quality index.

3.3 Total Dissolved Solids

TDS indicates the presence of mineral content [4] of water. It refers to the presence of salts like sodium, chloride, calcium, magnesium, potassium, sulphates and bicarbonates. Presence of any one mineral greater than the limit causes the health risk. Commonly high TDS causes diarrhoea, corrosiveness, scaling and less effectiveness of detergents. Total dissolved solids value ranges between 161 and 1326 mg/l. The result shows all measured values with in the limit so TDS not affect the water quality in all temple ponds.

3.4 Chloride

Chloride is a salt naturally present in water, it can easily dissolve in water and transport into another place through groundwater. Chloride salt is important for functioning kidney, the nervous system and nutrition. No health effects created by chloride alone. If it combines with sodium and exceeds the limit can cause heart disease or kidney disease. Chloride value ranges between 21 and 403 mg/l. According to test results chloride is the minimum value comparing with standards for all 17 temple pond water.

3.5 Sulphate

One of the natural materials is sulphate. It contains sulphur [3] and oxygen. It majorly present in soil. Combustion of fossil fuels releases sulphur oxides to the atmosphere, it converted into sulphate. Decomposition of plant and animal may release sulphate into the water. Sulphate individually non-toxic. The combination of magnesium or sodium sulphate may cause diarrhoea and dehydration. Sulphate value varies between 10 and 187 mg/l. Sulphate is with in the limit for all 17 temple water.

3.6 Iron

Fe is the natural material present in soil. Iron is important for energy metabolism. Deficiency of Fe causes anaemia. It present in natural food also. It is the main a component of haemoglobin, it transports oxygen and carbon dioxide in the blood. Less iron causes toxicity; more value damage blood vessels, liver and kidneys. Iron value varies between 0.37 and 0.82 mg/l. Iron value is within the limit for all sampling points so it does not affect the quality of water.

3.7 Total Coliform Count

Total coliforms means a group of bacteria [16] present in water. Disease-causing bacteria, viruses in drinking water varied from area to area. Commonly it causes vomiting and diarrhoea. Major value affects the lungs, skin, eyes, nervous system, kidneys or liver. Total coliform count varies between 500 and 2300 MPN. These values are with in the limit according to surface water standards.

The above result shows total dissolved solids, chloride, sulphate, iron, total coliform count is within the standard limit according to Class C ISI-IS: 2296-1982 standards.

3.8 Water Quality Index

Calculation of element weight of each parameter is indicated in Table 4.

The WQI valued of each temple pond water is shown in Table 5.

WQI [2] shows the quality of water, most of the pond water in poor quality up to 13 temples, two temple pond water good quality, one temple pond water very poor quality. The overall water quality index status for 17 temple pond water is shown in (see Fig. 1).

Table 4 Unit weight of observed parameters

Parameter	ISI-IS: 2296–1982 Class C standards	Unit weight (W_i)
pH	8.5	0.29975
DO	4	0.63697
TDS	1500	0.00169
Cl	600	0.00424
SO_4	400	0.00637
Fe	50	0.05095
TCC	5000	0.00051
		1.00051

Water quality index is majorly affected due to the excessive presence of dissolved oxygen. Next, the variation of PH value is also next affects the water quality index. Dharmalingeswarar temple pond water having very high pH value, TDS, chloride, sulphate and iron value also high compared to other temple pond water so it is under very poor quality.

3.9 *Correlation and Regression Analysis*

There is a good positive correlation [12] between TDS with Cl ($r = 0.989$), with SO_4 ($r = 0.970$), with pH ($r = 0.689$) and negative relationship with DO ($r = -0.473$). Chloride gives good positive correlation with sulphate ($r = 0.988$), with pH ($r = 0.729$) and negative relationship with dissolved oxygen ($r = -0.502$). Similarly, good positive correlation between Fe with Cl ($r = 0.599$), with TDS ($r = 0.599$) with SO_4 ($r = 0.617$). Dissolved oxygen shows negative correlation with TDS ($r = -0.473$), chloride ($r = -0.502$), sulphate ($r = -0.503$), iron ($r = -0.390$).

Correlation analysis results are shown in Table 5.

Regression equation of observed parameters is shown in Table 6.

Linear regression between TDS with SO_4, Cl is shown in (see Fig. 2). Linear regression between Fe with SO_4, Cl is shown in (see Fig. 3). Linear regression between pH with SO_4, Cl is shown in (see Fig. 4).

4 Conclusion

The present study shows that the quality of water in 17 number of temples taken in the month of March 2017. As per the Class C standards for surface water quality given by ISI-IS: 2296–1982 only two number of temple ponds shown the good quality and the remaining 14 numbers of temples came under poor quality. The higher value of

Table 5 Water quality index for temple pond water

Pond No.	P1	P2	P3	P4	P5	P6	P7	P8	P9	P10	P11	P12	P13	P14	P15	P16	P17
WQI	56.4	74.6	53.1	53.8	51.6	49.3	54.9	47.8	61.4	49.6	64.4	85.2	55.9	60.9	62.9	63.9	52.7
status of water quality	Poor	Poor	Poor	Poor	Poor	Poor	Poor	Good	Poor	Good	Poor	V. Poor	Poor	Poor	Poor	Poor	Poor

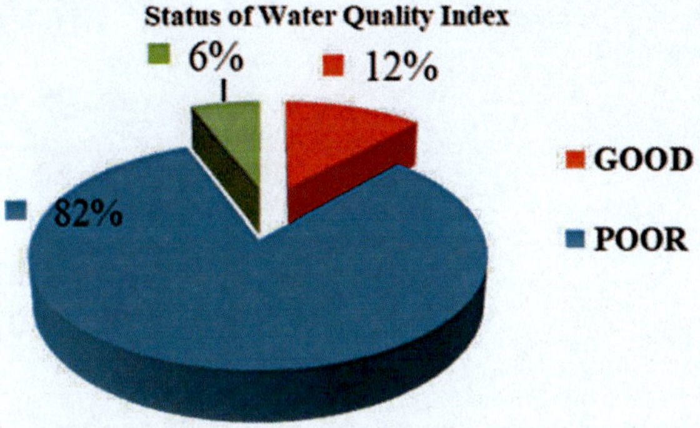

Fig. 1 Status of water quality index for temple pond water

Table 6 Regression equation of observed parameters

Dependent variable (Y)	Independent variable (X)	R^2	Regression equation ($Y = aX + b$)
TDS	SO_4	0.940	$Y = 6.731\ X + 101.6$
	Cl	0.978	$Y = 3.277\ X + 84.40$
Fe	SO_4	0.380	$Y = 0.001\ X + 0.506$
	Cl	0.358	$Y = 0.000\ X + 0.507$
pH	SO_4	0.495	$Y = 0.006\ X + 7.329$
	Cl	0.531	$Y = 0.003\ X + 7.305$

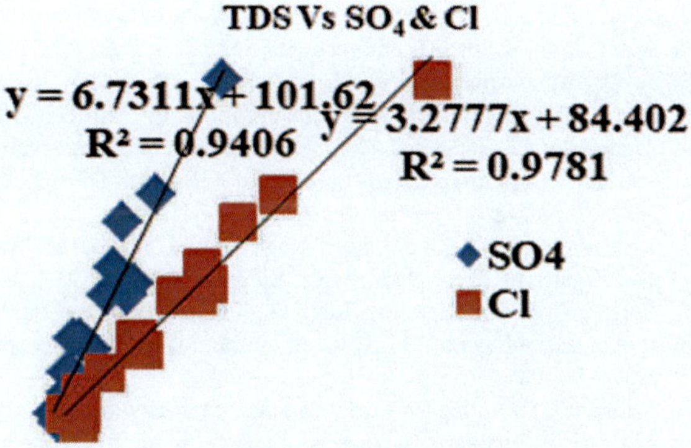

Fig. 2 Linear regression between TDS with SO_4, Cl.

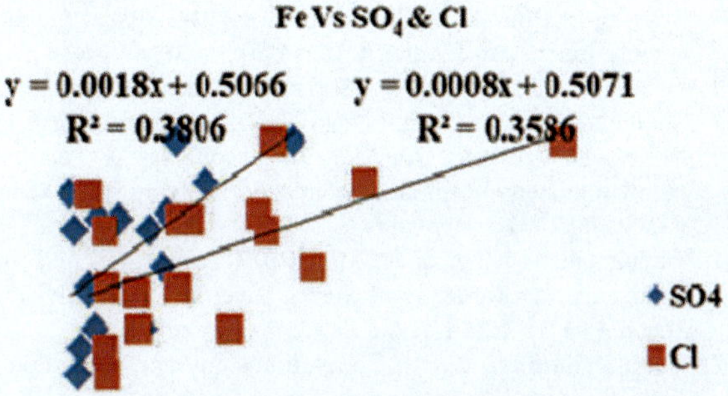

Fig. 3 Linear regression between Fe with SO_4, Cl

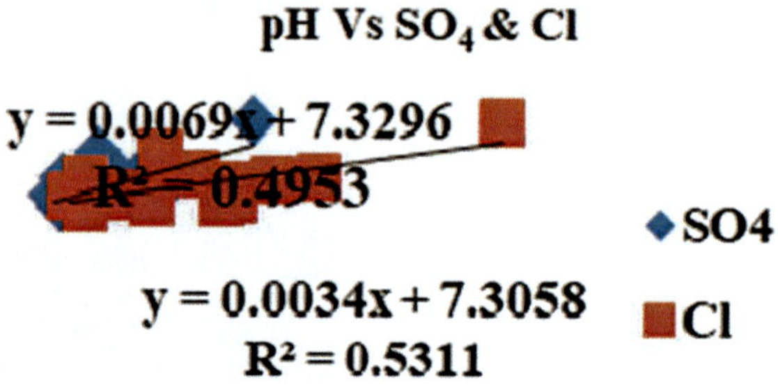

Fig. 4 Linear regression between pH with SO_4, Cl

DO obtained from the results does not have any impact on the surface water bodies. The parameters such as pH and TCC were very high in Dharmalingeswarar temple pond and make it as very poor quality. Hence, this pond water needs full-fledged treatment process for further use and the remaining ponds require normal treatment process for other than domestic needs.

References

1. Bharti N, Katyal D (2011) Water quality indices used for surface water vulnerability assessment. Int J Environ Sci 2(1):154–173
2. Jena V, Dixit S, Gupta S (2013) Assessment of water quality index of industrial area surface water samples. Int J Chem Tech Res 5(1): 278–283
3. Jeyaraj M, Nirmaladevi G, Magudeswaran PN (2014) Assessment of water quality index of Sulur pond, Coimbatore–Tamil Nadu, India. Int J Emerg Trends Sci Technol 1(7):1200–1204
4. Manoharan D, Isaiarasu L (2016) Observation of the seasonal changes in the physical, chemical and nutrient parameters of three temple ponds at Thirupparankundram Near Madurai, Tamil Nadu, India. Int J Curr Sci Res 2(8):823–847
5. Ameeth Basha I, Shanthi T, Nagalakshmi R (2016) Water quality analysis of Chinna Kanchipuram, Tamil Nadu. India. Int J Chem Sci 14(2):431–438
6. Smitha Asok V, Jobai NM, Reghunath R (2015) Water quality index determination of ponds in Anad Panchayath, Thiruvananthapuram, Kerala—a geospatial approach. Int J Sci Res Publ 5(12): 94–98
7. Sridhar SGD, Kanagaraj G, Sharma KK, Mahalingam S, Priyalakshmi R (2013) Hydrochemical analysis and evaluation of quality of ground water in Kancheepuram District, Tamil Nadu, India. Asian J Sci Technol 4(9):19–27
8. Tyagi S, Sharma B, Singh P, Dobhal R (2013) Water Quality Assessment in Terms of Water Quality Index. Am J Water Resour 1(3):34–38
9. Deb PSM, Chakraborty H (2015) Assessment of groundwater quality index in and around Sholinganallur Area, Tamil Nadu. Int J Sci Technol 8(36): 1–7
10. Harichandan A, Patra HS, Sethy KM, Mohanty SK, Panda R (2016) Water quality index and correlation study of religious ponds in The Temple City, Bhubaneswar, Odisha. Am J Sci Ind Res 7(2): 22–31
11. Tajmunnaher T, Chowdhury MAI (2017) Assessment and correlation analysis of water quality parameters: a case study of Surma River at Sylhet Division, Bangladesh. Int J Eng Trends Technol 53(3): 126–136
12. Dutta B, Sarma B (2018) Correlation study and regression analysis of ground water quality assessment of Nagaon Town of Assam, India. Int J Eng Res Technol 7(6): 320–331

13. Shrestha AK, Basnet N (2018) The correlation and regression analysis of physicochemical parameters of River water for the evaluation of percentage contribution to electrical conductivity. J Chem 8369613: 1–9
14. Sharma S, Chhipa RC (2013) Interpretation of ground water quality parameter for selected area of Jaipur using regression and correlation analysis. J Sci Ind Res 72:781–783
15. Thilagavathi N, Subramani T (2015) Physico-chemical and quality assessment of ground water in Chalk Hills, Mine area, Salem, Tamil Nadu. Ind J Geo Marine Sci 44(9):1428–1435
16. Nandigam J, Subba Rangaish G, Geddada MNR (2016) A study on seasonal changes of microalgae in relation to the physico-chemical parameters of Satyavaram Pond, Srikakulam Dist, India. Ind J Geo Marine Sci 45(12): 1660–1668

Comparative Corrosion Behaviour of Ferrous and Non-ferrous Metals in Bio-lubricant and Bio-diesel Environment

R. Ellappan, S. Arumugam, R. Sundararajan, and K. Venkatesh

Abstract Major cause for the deterioration of metals in automotive engine components is due to the corrosive effect of bio-fuels as well as bio-lubricant. This study compares the corrosive behaviour of rapeseed oil bio-lubricant and petroleum lubricant on ferrous and non-ferrous metals. Rapeseed oil bio-lubricant and petroleum lubricant with 10% by volume of rapeseed bio-fuel blend are used for conducting static immersion tests for the period of 2880 h at room temperature. The changes in properties of oils like density and kinematic viscosity were examined for the different metals exposed under bio-lubricant and petroleum lubricant. The corrosion rates of metallic coupons are evaluated by weight loss basis. Degradation of metallic surfaces was characterized by digital camera (visual testing) and SEM. Static immersion test studies show that the non-ferrous metals are more prone to corrosion comparatively ferrous metals exposed in bio-based fuels and lubricants.

Keywords Corrosion · Bio-lubricant · Corrosion rate · Degradation

1 Introduction

Bio-fuels and bio-lubricants have better degradation tendency compared with conventional fuels and lubricants. Savita kaul et al. [1] observed that the usage of bio-diesel shows notable corrosion on metallic parts of diesel engine compared with conventional diesel fuel. Numerous literatures reported that the usage of bio-fuel promotes corrosion on engine component materials and fuel supply systems [2–10]. Haseeb et al. [2] concluded that the copper and leaded bronze show higher corrosion rate in bio-diesel as compared with neat diesel, and the evidence of degradation is noticed by increased TAN and oxidation rate and hygroscopic nature of bio-fuel. Fazal et al. [3] observed that the weight loss and corrosion rate of copper increased due the presence of oxygen moieties and fatty acids and higher moisture absorption

R. Ellappan (✉) · S. Arumugam · R. Sundararajan · K. Venkatesh
Department of Mechanical Engineering, Sri Chandrasekharendra Saraswathi Viswa Mahavidyalaya University, Enathur, Kanchipuram, Tamil Nadu 631561, India
e-mail: ellsraaj@gmail.com

T. Rajmohan et al. (eds.), *Advances in Materials and Manufacturing Engineering*, Springer Proceedings in Materials 7, https://doi.org/10.1007/978-981-15-6267-9_49

nature of bio-fuels. Increase in temperature leads to promote corrosion in both neat diesel and bio-diesel [4].

Haseeb et al. [5] explored that the metallic component of automobile can undergo tribo-corrosion attacks and also elastomers seems to degrade to a greater extent in bio-diesel. Fazal et al. [6] studied that the bio-diesel feasibility in automobiles with materials compatibility and concluded that long-term test leads to loss of lubricity due to its corrosive and oxidative nature. Enzhu hu et al. [7] compared that the exposure of ferrous and non-ferrous meats over rapeseed oil bio-fuel. The corrosion rate of copper is slightly higher than the rest of the metals with immersion time and temperature. Singh et al. [9] reviewed that the stainless steel is having significant corrosion resistant in bio-diesel compared with other metals.

Palm bio-diesel promotes corrosion due to the presence of functional groups and greater amount of water content [11]. Chew et al. [12] noticed higher corrosion rate of magnesium compared to that of aluminium exposed in bio-diesel, increase in TAN limit proved the degradation of metals in bio-diesel. Static immersion test reveals that the B100 fuel blend shows increased TAN and B0 fuel blend measured lowest TAN value [13]. Arumugam et al. [14] conducted static immersion test under high temperature environment for ferrous and non-ferrous metals exposed under rapeseed-based bio-lubricant and petroleum base lubricant, on ferrous metals found more corrosive than ferrous metals in terms of corrosion rate.

2 Materials and Methods

2.1 Material Selection and Sample Preparation

Selection of materials for corrosion studies under lubricant atmosphere is mainly based on the major constituent of engine component materials, therefore, the ferrous materials in diesel engine cast iron as liner and medium carbon steel as timing gears and non-ferrous metals like copper, aluminium and brass were used in big end bearings of crank shaft were taken for testing. All the samples were cut from round bars and polished by using 80–1600 grade silicon carbide emery sheets. Before the start immersion, all the metal coupons were cleaned for removing the oiliness and other contaminants by using acetone and ultrasonic cleaning with absolute ethanol as per ASTM G1 std [11], and rapeseed-based bio-lubricant and synthetic lubricant were added with 10% B20 bio-diesel to correlate the probability of fuel dilution in the normal diesel engines for corrosion studies (Fig. 1).

Fig. 1 Test coupons before testing

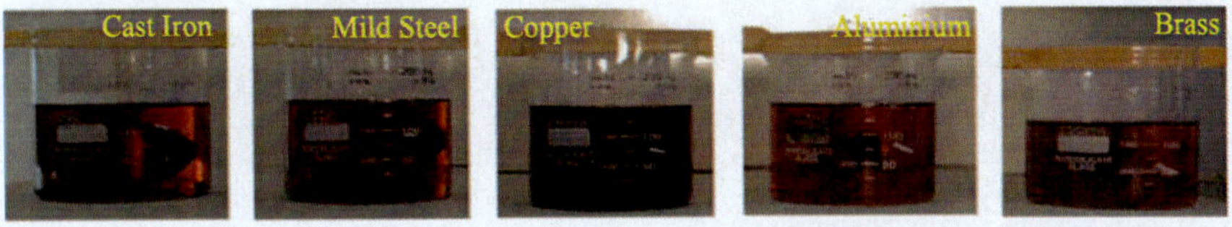

Fig. 2 Materials immersed in petroleum lubricant diluted with bio-diesel

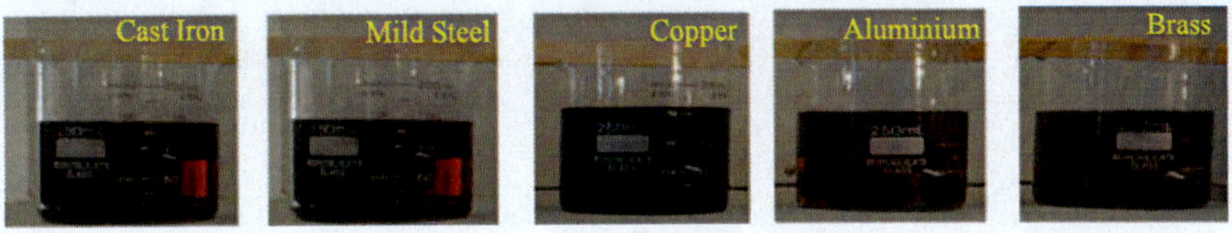

Fig. 3 Materials immersed in bio-lubricant diluted with bio-diesel

2.2 Experimental Methods

Each metal coupon for test was immersed in the cylindrical glass beakers contains bio-lubricants and petroleum lubricants separately for corrosion studies under static condition at room temperature for the duration of 2880 h [11]. The metal coupons immersed in both the lubricant were cleaned to remove the corrosion products on the metal surfaces by exposing the metal coupons about twice for 5–10 s in 10–15% HCL with Clarke solution, consequently, the metal coupons were cleaned using acetone and dried and instantly weighed of weigh balance with 5 decimal accuracy to account the weight loss for evaluating corrosion rate (Figs. 2 and 3).

3 Result and Discussions

3.1 Visual Examinations

Figures 4 and 5 show that the change of colour on the tested ferrous and non-ferrous metal coupons in both bio-lubricant and petroleum lubricant. Among all the change of

Fig. 4 Test coupons exposed in petroleum lubricant

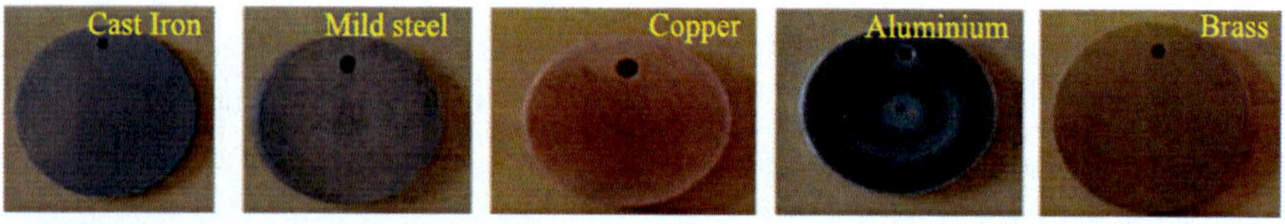

Fig. 5 Test coupons exposed in bio-lubricant

colour in the copper coupon shows comparatively darker than the synthetic lubricant environment.

3.2 3.2 Corrosion Rate Evaluation

Corrosion rate is the amounts of metals which deteriorate in specific environment condition. Metal degradation was evaluated as corrosion rate and it is expressed in MPY (mils per year). The corrosion rate can be calculated using the equation given below,

$$\text{Corrosion Rate (MPY)} = \frac{w \times 534}{D \times T \times A}$$

where w—weight loss in the metal in mg, D—density of the metal in g/cm^3, A—exposed surface area of the metal in in^2, T—time of exposure of the metal in lubricant in $h = 2880$ h. The density of each metal at 20 °C is taken from the heat transfer data book and the weight loss is calculated from the weight of the specimens measured before and after testing, which is converted into corrosion rate by substituting the values in the above equation, and the results are tabulated in Table 1. From the table, it can be found that for the coupons tested in bio-lubricant shows slightly increase corrosion rate compared with petroleum lubricant in the order given below.

3.3 Surface Morphology Analysis

Scanning electron microscope (SEM) is used to analyse the qualitative differences in the surfaces before and after the static immersion test in bio and petroleum lubricant

Table. 1 Corrosion rate of bio-lubricant and petroleum lubricant

			Bio-lubricant			Petroleum lubricant		
S. No.	Metal	Density at 20 °C (g/cm^3)	Area (in^2)	Weight loss W (mg)	Corrosion rate (MPY)	Area (in^2)	Weight loss W (mg)	Corrosion rate (MPY)
1	Cast iron	7.20	3.5145	0.005	0.000036	3.4988	0.004	0.000029
2	Mild steel	7.85	1.1501	0.003	0.000061	1.1550	0.002	0.000040
3	Copper	8.96	0.8752	0.013	0.00030	0.8747	0.003	0.000070
4	Aluminium	2.78	1.4517	0.002	0.00009	1.4611	0.001	0.000045
5	Brass	8.73	1.7985	0.006	0.000070	1.7972	0.003	0.000035

through high-resolution imaging technique. The SEM images of ferrous and non-ferrous metals before and after testing in bio and petroleum lubricants are shown and discussed below.

3.3.1 Cast Iron

Cast iron surface exposed in bio-lubricant shows some non-metallic deposits on cast iron surface. There is no significance change on the surface exposed in petroleum lubricant. Mild steel surface exposed in petroleum lubricant shows severe colour change, and the surface exposed in bio-lubricant shows that the deposits of non-metallic compounds through the surface. In case of brass surface exposed in bio-lubricant as well as petroleum lubricant found more number of small pits can be found throughout the surface which shows that pitting corrosion has taken place. Copper surface exposed to petroleum lubricant shows more number of pits throughout the surface similar to brass. Whereas the coupon exposed to bio-lubricant shows drastic change in base metal of copper indicates that copper surface is more prone to corrosion. Aluminium surface exposed in petroleum lubricant shows changes in colour and slight distortion on the surfaces, but the coupon exposed in bio-lubricant shows severe colour change on the surface and mild distortion on base metal (Fig. 6).

3.3.2 Mild Steel

See Fig. 7.

3.3.3 Brass

See Fig. 8.

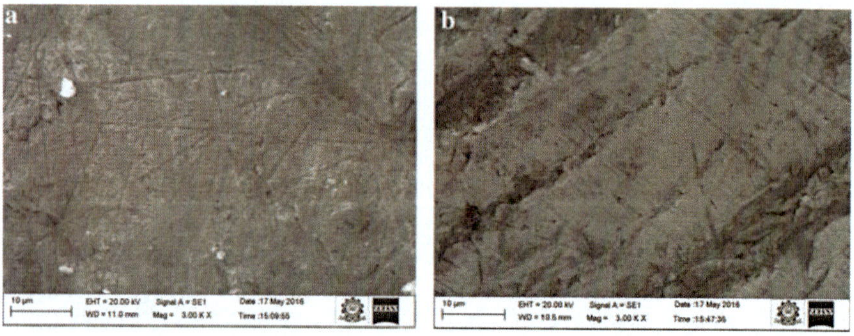

Fig. 6 SEM images of cast iron, **a** surface in petroleum lubricant, **b** surface in bio-lubricant

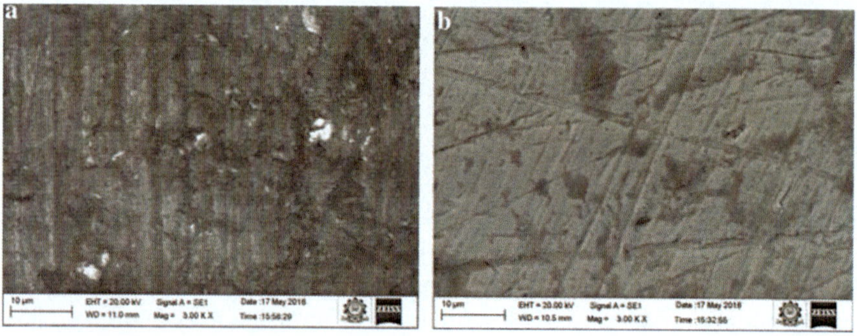

Fig. 7 SEM images of mild steel. **a** Surface in petroleum lubricant, **b** surface in bio-lubricant

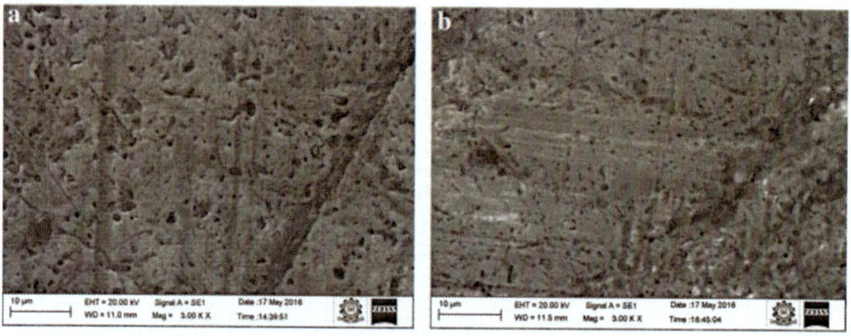

Fig. 8 SEM images of brass. **a** Surface in petroleum lubricant, **b** surface in bio-lubricant

3.3.4 Copper

See Fig. 9.

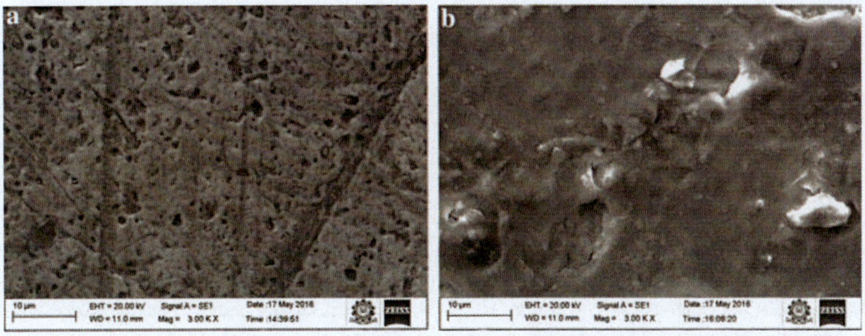

Fig. 9 SEM images of copper. **a** Surface in petroleum lubricant, **b** surface in bio-lubricant

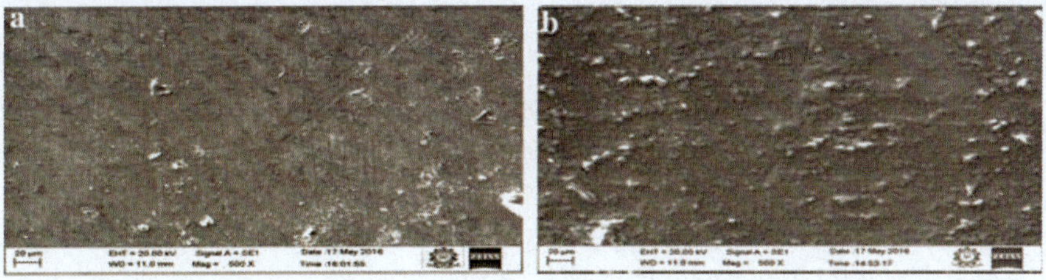

Fig. 10 SEM images of aluminium. **a** Surface in petroleum lubricant, **b** surface in bio-lubricant

3.3.5 Aluminium

See Fig. 10.

4 Conclusions

The effect of lubricant on different metals which comprises in the engine components associated with lubrication oil was analysed under static immersion test. From the above study, corrosion rate, surface morphology and lubricant properties were examined, comparing the results before and after testing shows that the corrosion in bio-lubricant is slightly higher than the petroleum lubricant. Upon the exposure to different metals tested copper is more prone to corrosion in bio-lubricant. Other metals such as cast iron and mild steel, brass and aluminium does not show significant corrosion and also found that more corrosion resistant in both bio and petroleum lubricant environment. To correlate with the real-time application of lubricant further study is needed for bio-lubricant in high temperature environment will provide the effect of temperature that promotes the oxide films leading to corrosion on the metallic surfaces due to degradation of lubricant.

References

1. Savita Kaul, Saxena, R.C., Ajay Kumar, Negi, M.S., Bhatnagar, A.K., Goyal, H.B., Gupta, A.K.: Corrosion behaviour of biodiesel from seed oils of Indian origin on diesel engine parts. Fuel Process. Technol. 88, 303–307 (2007).
2. Haseeb ASMA, Masjuki HH, Ann LJ, Fazal MA (2010) Corrosion characteristics of copper and leaded bronze in palm biodiesel. Fuel Process. Technol. 91:329–334
3. Fazal MA, Haseeb ASMA, Masjuki HH (2010) Comparative corrosion characteristics of petroleum diesel and palm biodiesel for automotive materials. Fuel Process. Technol. 91:1308–1315
4. Fazal MA, Haseeb ASMA, Masjuki HH (2011a) Effect of temperature on the corrosion behaviour of mild steel upon exposure to palm biodiesel. Energy. 36:3328–3334
5. Haseeb ASMA, Fazal MA, Jabirus MI, Masjuki HH (2011) Compatibility of automotive materials in biodiesel: A review. Fuel 90:922–931
6. Fazal MA, Haseeb ASMA, Masjuki HH (2011b) Biodiesel feasibility study: An evaluation of material compatibility;performance;emission and engine durability. Renew Sustain Energy Rev 15:1314–1324
7. Enzhu Hu., Yufu., Xianguo Hu., Lijun Pan., Shaotong Jiang.: Corrosion behaviours of metals in biodiesel from rapeseed oil and methanol. Renew Energ, 37, 371–378, (2012).
8. Fazal MA, Haseeb ASMA, Masjuki HH (2012) Degradation of automotive materials in palm biodiesel. Energy 40:76–83
9. Singh, B., John Korstad., Sharma, Y.C.: A critical review on corrosion of compression ignition (CI) engine parts by biodiesel and biodiesel blends and its inhibition. Renewable and Sustainable Energy Reviews, 16, 3401–3408, (2012)
10. ASTM G: 1.Standard Practice for Preparing, Cleaning, and Evaluation Corrosion Test Specimens.
11. Fazal MA, Haseeb ASMA, Masjuki HH (2013) Corrosion mechanism of copper in palm biodiesel. Corros. Sci. 67:50–59
12. Chew KV, Haseeb ASMA, Masjuki HH, Fazal MA, Gupta M (2013) Corrosion of magnesium and aluminium in palm biodiesel: A comparative evaluation. Energ 57:478–483
13. Akhabue CE, Aisien FA, Ojo CO (2014) The effect of jatrropha oil biodiesel on the corrosion rate of aluminium and mild carbon steel. Bio fuels 5(5):545–550
14. Arumugam S, Ellappan R, Sangavi S, Sriram G, Ramakrishna P (2018) Feasibility Analysis of Biodegradable Automotive Lubricant: An Evaluation of Material-Lubricant Compatibility in a Corrosion Perspective. Arab J Sci Eng 43:1345–1368

Multi-response Optimization and Mechanical Behaviour of Al–Cu/Al–Mg–Si Alloys by Dissimilar Friction Stir Welding

R. Dinesh Kumar, S. Pradeep, and S. Muthukumaran

Abstract AA2024 (Al–Cu alloy) and AA6061 (Al–Mg–Si alloy) are joined by friction stir welding (FSW) technique. The process parameters are selected using Taguchi L9 orthogonal array and optimization was carried out in two modes, namely single response by ANOVA and multi-response by grey relational analysis (GRA). ANOVA results revealed that rotational speed has the maximum influence on ultimate tensile strength (UTS) and yield strength (YS), whereas traverse speed dominates in percentage elongation and impact strength. The weld trial with parameter 60 mm/min, 1000 rpm and threaded shoulder is found to be optimal through GRA with the responses such as 221 MPa ultimate tensile strength, 184 MPa yield strength, 15.1% elongation and 7.86 J impact strength. The validation of predicted and experimental response shows very minimal error of 2%. The microstructural study reveals finer grains in the weld nugget and formation of an onion ring. The tensile studies indicated the failure of welded joints at the heat affected zone of AA6061 alloy which is further confirmed by hardness studies.

Keywords Dissimilar material · Multi-response optimization · Onion ring · Hardness

1 Introduction

Aluminium alloys are widely used in the structural components such as construction of frames, pipelines and storage tanks in the aerospace and marine industries [1]. The fusion welding of aluminium alloys leads to defects like hot cracking, gas porosity and thermal distortion [2]. Al–Cu and Al–Mg–Si aluminium alloys are generally termed as nonweldable/heat treatable aluminium alloys that could lose its stability and strength gained by heat treatment in fusion welding [3, 4]. Friction stir welding (FSW) is a revolutionary solid-state welding process invented by Wayne Thomas of The Welding Institute (TWI), UK and patented in 1991 [5]. The shoulder profile

R. Dinesh Kumar · S. Pradeep · S. Muthukumaran (✉)
Department of Metallurgical and Materials Engineering, National Institute of Technology, Tiruchirappalli, Trichy, India
e-mail: smuthu@nitt.edu

T. Rajmohan et al. (eds.), *Advances in Materials and Manufacturing Engineering*, Springer Proceedings in Materials 7, https://doi.org/10.1007/978-981-15-6267-9_50

and tool pin profile that helps to improve the strength by forging action on the first mode and regulating the metal flow pattern in first and second mode of metal transfer [6]. AA2024 on advancing side leads to the formation of elongated particles such as Al2Cu and Mg-Si which increases the strength subsequently leaving it on retreating side generates Al–Si and Mg–Si which offers better corrosion resistance in the weld nugget [7]. The fatigue behaviour of the joint is found improved on fixing harder material on the advancing side [8]. Most of the research concentrate on advancing and retreating side, apart from that very few researchers have worked in presenting both individual and combined effect of a process parameter such of rotation speed, traverse speed and tool profile, and this research mainly focus optimizing various process parameter to improve the weld quality and microstructural correlation with strength and hardness.

2 Experimental Procedure

The AA2024 and AA6061 were cut into rectangular samples of dimensions 110 × 60 × 6 mm and welding was done using FSW machine. The chemical composition results for the base material were shown in Table 1.

The welding tool used in this study was D3 tool steel, which was heat treated and tempered before welding. The various tool profiles used in this work include concave shoulder taper threaded pin, flat shoulder taper threaded pin and threaded shoulder taper threaded pin as shown in Fig. 1. The process parameters are used in

Table 1 Chemical composition of the Al alloys [1]

Alloy (wt%)	Mg	Mn	Cu	Fe	Si	Cr	Zn	Ti	Al
AA2024	1.5	0.6	4.35	0.5	0.5	0.10	0.25	0.15	Remaining
AA6061	1	0.15	0.27	0.7	0.6	0.19	0.6	0.15	Remaining

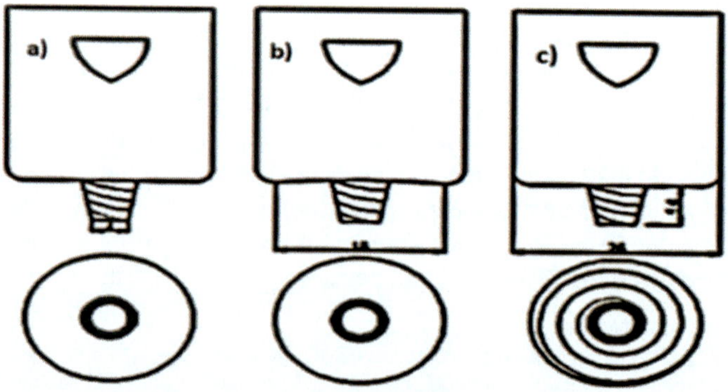

Fig. 1 Tool profile **a** flat shoulder, **b** concave shoulder and **c** threaded shoulder

Table 2 Mechanical properties of AA2024 and AA6061 base metals [3]

Alloy	UTS (MPa)	YS (MPa)	Elongation (%)	Hardness (Hv)
AA2024	464	344	16	137
AA6061	317	286	10	75

welding were shown in Table 3. AA2024 was placed on the advancing side (AS) because of its greater mechanical strength. Then, the welded samples were cut using wire electric discharge machining for tensile testing as per ASTM E8 standard and for characterization studies. The tensile studies were carried out with the strain rate of 1 mm/min. Vickers microhardness test was carried out with a load of 500 mg with a dwell time of 15 s across the weld sample. V-Notch specimens were cut as per ASTM E370 standard and the tests were conducted to estimate the impact strength of the welded sample. The macrostructure and microstructure analysis of the weldments were done by stereo and optical microscopes which are properly etched with Poulton's reagent (Table 2).

3 Results and Discussion

Tensile and impact results are calculated after testing are shown in Table 3. MINITAB software is used to predict the optimized parameter and to analyse the influence of different process parameters on the weld quality. There are two methods adopted in this process to calculate the best setting ANOVA technique to estimate the single response, i.e. effect of every input parameter over the weld strength and grey relational analysis for multi-response which gives the optimal parameter to get better weld quality.

Table 3 Tensile testing results after testing

Exp. No.	Traverse speed (mm/min)	Tool shoulder profile	Rotational speed (rpm)	UTS (MPa)	YS (MPa)	Elongation (%)	Impact strength (Joules)
1	30	Flat	800	186	163	12.8	7.86
2		Threaded	1000	175	134	10.5	3.85
3		Concave	1200	180	141	11.9	5.82
4	45	Flat	1000	188	124	15.3	7.01
5		Threaded	1200	215	178	14.7	3.03
6		Concave	800	218	174	15.2	6.12
7	60	Flat	1200	207	186	14.4	6.85
8		Threaded	800	221	184	15.1	7.86
9		Concave	1000	170	160	8.81	8.51

3.1 ANOVA

ANOVA is a major tool which investigates the impact of every parameter used in the process from sequential responses gained. The MINITAB 15 software is used to examine the importance of welding parameters like rotational speed, traverse speed and tool shoulder profile individually. The responses like UTS, YS, % elongation, impact strength and its % contribution on the parameters are studied and the combined effect on responses are studied by considering the response factor larger the better criterion is given in Eq. (1).

$$\frac{S}{N} = -10\log_{10}\left(\frac{1}{n}\sum_{k=1}^{n} 1\frac{1}{Z_{ij^2}}\right) \tag{1}$$

where Z_{ij} is the different parameters and n is the number of levels.

From Table 4, the rotational speed has maximum influence on UTS, and tool profile has the least impact on UTS. Figure 2 shows that when the traverse speed increases, UTS also increase to a certain limit and drops down; when the rotational speed increase, UTS falls over certain rpm and again raises and flat shoulder provides better UTS than others.

From Table 5, rotational speed has the maximum influence on YS and traverse speed has the least impact. Figure 3 shows a relative trend that when traverse speed increases, the YS also increases and for rotational speed it drops down over specific interval and again increases, and flat pin profile provides better yield property than others. Yield strength plays a vital role in an aerospace domain where a change from elastic to plastic limit which will not lose its property (Fig. 4).

From Table 6, the traverse speed has maximum influence on elongation and welding rotational speed has the least impact on elongation. Figure 4 shows that when traverse speed increases, the elongation also increases till 45 mm/min and continues to drop till 60 mm/min, whereas, for rotational speed, it drops down till 1000 rpm and again increases, and concave shoulder profile gave better elongation than others. Elongation has maximum control in the formability of the welded product

Table 4 ANOVA of ultimate tensile strength

Factors	DOF	Sum of square	Mean sum	*F* value	Probability of significance	% of contribution
Traverse speed (mm/min)	2	1130.89	565.44	7.91	0.112	36.17
Tool shoulder profile	2	324.22	162.11	2.27	0.306	10.37
Rotational speed (rpm)	2	1528.22	764.11	10.70	0.086	48.83
Error	2	142.89	71.44			4.57
Total	8	3126.22				

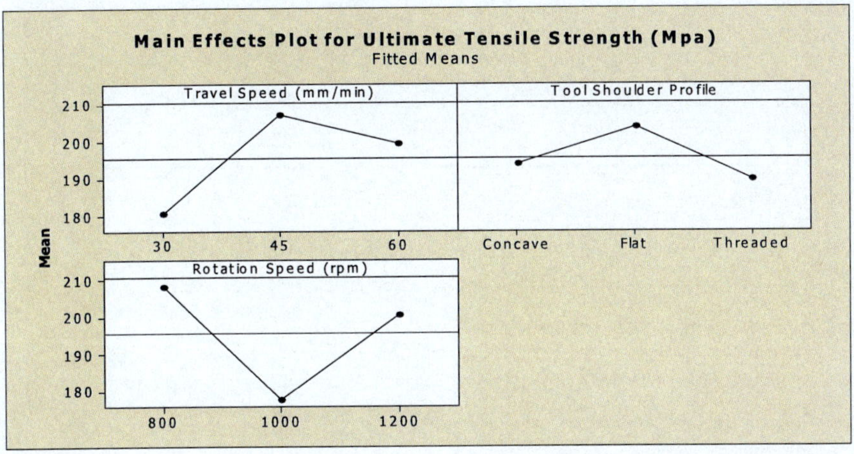

Fig. 2 Main effects plot for UTS

Table 5 ANOVA of yield strength

Factors	DOF	Sum of square	Mean sum	*F* value	Probability of significance	% of contribution
Traverse speed (mm/min)	2	1424.9	712.4	2.68	0.271	34.65
Tool shoulder profile	2	108.2	54.1	0.20	0.831	2.63
Rotational speed (rpm)	2	2048.2	1024.1	3.86	0.206	49.80
Error	2	530.9	265.4			12.91
Total	8	4112.2				

in the field of aerospace where a lot of complicated structures involved which can only be formed into a required shape.

From Table 7, both traverse speed and tool shoulder profile have the maximum and almost equal influence on impact strength and rotational speed has the least on impact strength. Figure 5 reveals that when traverse speed increases, the yield strength decreases up to certain value and get sudden hike at 60 mm/min and when rotation speed increases, impact strength decreases, and concave shoulder provides better impact strength than others. It is always important to have the maximum impact strength for an aerospace component because when the spacecraft moves with maximum velocity the sudden strike of a foreign body will lead to collapse of the system.

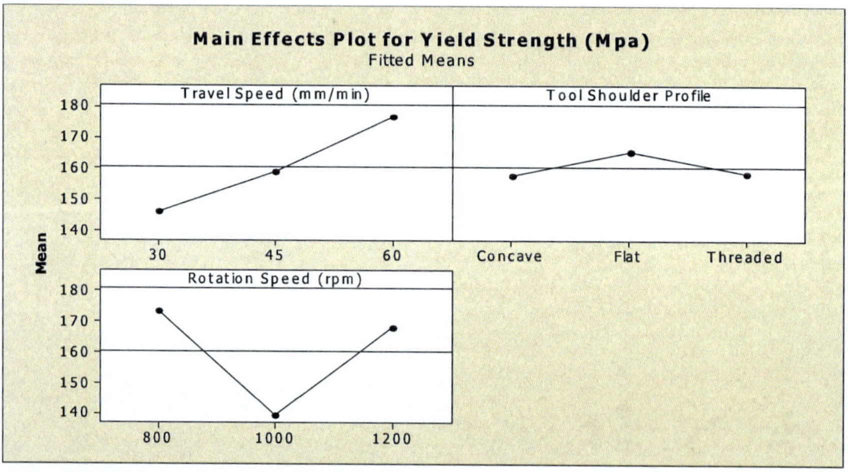

Fig. 3 Main effects plot for yield strength

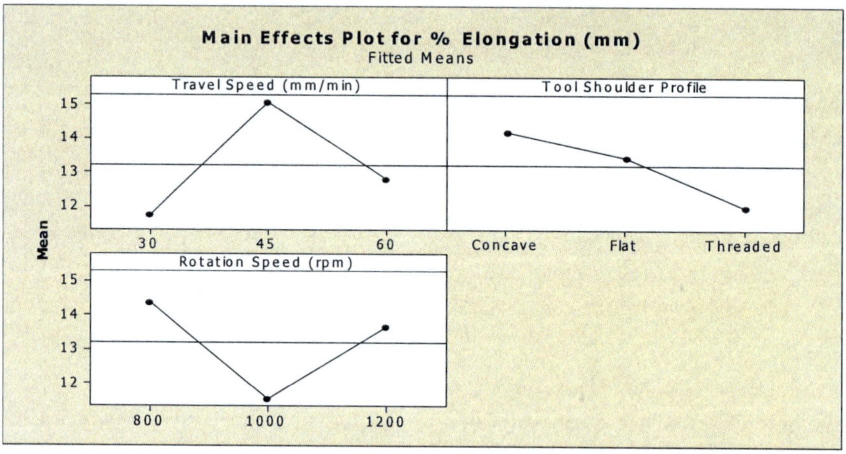

Fig. 4 Main effects plot for % elongation

3.2 *Grey Relational Analysis (GRA)*

Grey relational analysis (GRA) which uses the normalization of data to calculate grey relational coefficients (GRC) and grey relational grades (GRG). The normalizing process by calculated using Eq. (2),

$$z_i^*(k) = \frac{z_i(k) - \min z_i(k)}{\max z_i(k) - \min z_i(k)} \tag{2}$$

Table 6 ANOVA of elongation

Factors	DOF	Sum of square	Mean sum	*F* value	Probability of significance	% of contribution
Traverse speed (mm/min)	2	17.46	8.730	2.85	0.260	39.57
Tool shoulder profile	2	7.504	3.752	1.23	0.449	17.0
Rotational speed (rpm)	2	13.036	6.518	2.13	0.320	29.54
Error	2	6.12	3.060			13.87
Total	8	44.121				

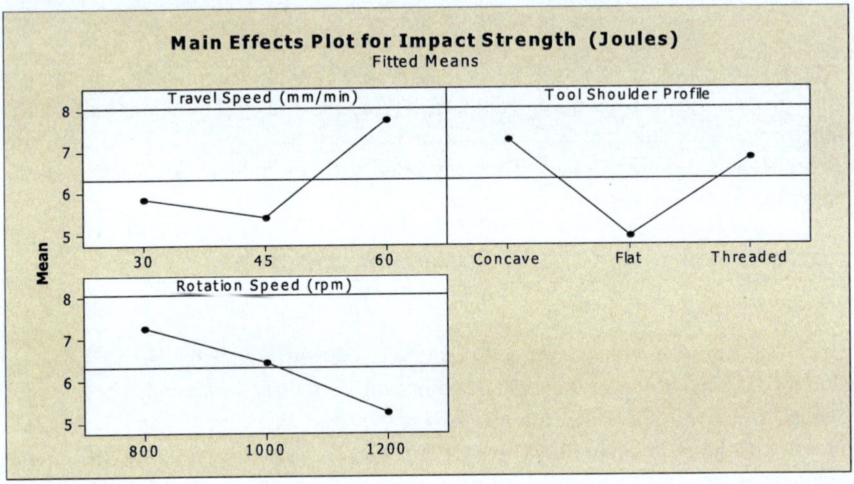

Fig. 5 Main effects plot for impact strength

Table 7 ANOVA of impact strength

Factors	DOF	Sum of square	Mean sum	*F* value	Probability of significance	% of contribution
Traverse speed (mm/min)	2	9.344	4.672	3.61	0.217	33.96
Tool shoulder profile	2	9.215	4.608	3.56	0.219	33.95
Rotational speed (rpm)	2	6.363	3.182	2.46	0.289	23.12
Error	2	2.589	1.294			9.41
Total	8	27.511				

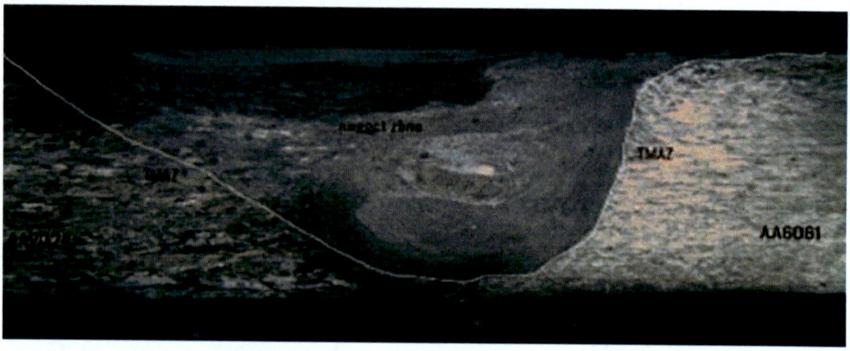

Fig. 6 Macrostructure of the weldment

where $I = 1, 2, 3 \ldots n$; $k = 1, 2, 3 \ldots m$; $n =$ no. of data generated; $m =$ no. of factors; $z_i(k)$ is the corresponding value; max $z_i(k)$ and min $z_i(k)$ are maximum and minimum value recorded; $z_i^*(k)$ is the normalization factor.

Grey Relational Coefficient. The grey relational coefficient (GRC) was calculated by the following Eq. (3)

$$\varepsilon_i(k) = \frac{\Phi \min + \omega \Phi \max}{\Phi_{\mathrm{oi}}(k) + \omega \Phi \max} \tag{3}$$

where $\varepsilon_i(k)$ is the grey relational coefficient; Φ_{oi} is the difference between $z_i^*(k)$ and $z_o^*(k)$; $z_o^*(k)$ is reference value and $z_i^*(k)$ normalized value; Φmax is highest value of Φ the $_{\mathrm{oi}}(k)$, Φ min is least value of $\Phi_{\mathrm{oi}}(k)$; the ω is maintained as 0.5; $n =$ total number of response in GRC and $i(k)$ represent a level. The corresponding values are listed in Table 8.

Table 8 Grey relation coefficient (GRC) and (GRG)

Exp. No.	Grey relational coefficient				Grey relational grade	Rank
	UTS	YS	Elongation	Toughness		
1	0.4215	0.5741	0.5648	0.8083	0.5922	6
2	0.3566	0.3735	0.4034	0.3703	0.3759	9
3	0.3835	0.4079	0.4883	0.5046	0.4461	8
4	0.4359	0.3333	1.0000	0.6462	0.6039	5
5	0.8095	0.7949	0.8440	0.3333	0.6954	4
6	0.8947	0.7209	0.9701	0.5341	0.7800	2
7	0.6456	1.0000	0.7829	0.6227	0.7628	3
8	1.0000	0.9394	0.9419	0.8083	0.9224	1
9	0.3333	0.5439	0.3333	1.0000	0.5526	7

Table 9 Optimal process parameters

Parameter and symbol	Level 1	Level 2	Level 3	(max-min)	Rank	Optimal level
Traverse speed (A)	0.4714	0.6931	**0.7459**	0.2745	2	3
Tool profile (B)	0.6529	**0.6646**	0.5929	0.0717	3	2
Rotational speed (C)	**0.7648**	0.5108	0.5647	0.2540	1	1

Table 10 Validation of optimal process parameter

Optimum parameter	Mode of analysis	UTS (MPa)	YS (MPa)	Elongation (%)	Impact strength (Joules)
$A_3B_2C_1$	Predicted	220.22	196.22	15.12	7.99
$A_3B_2C_1$	Experimental	221	184	15.1	7.86

Grey Relational Grade. The grey relational grade (GRG) was calculated by the following Eq. (4)

$$\varepsilon_i = \frac{1}{n}\sum_{k=1}^{n} i(k) \tag{4}$$

The larger GRG obtained is considered to have an optimal parameter for the output generated as shown in Table 8. Sample 8 with maximum GRG value is found to be the optimal parameter with traverse speed of 60 mm/min, rotational speed of 1000 rpm and threaded shoulder which is further confirmed by analysing the maximum grade value in Table 9.

Validation weld is carried out for the optimal parameter and tested for results. From Table 10, it is found that there is no much deviation between predicted and experimental values with an error rate of less than 2%.

3.3 *Metallographic Analysis*

From the macrographic studies, different regions of weldments are identified and it represents the active stir of both the base material in the nugget zone as shown in Fig. 6. The stir zone is dominated by AA2024 since it was placed on the advancing side. The microstructure of the different regions of the optimized trail was shown in Fig. 7a–e. The highly refined grains in the stir zone in Fig. 7c could enhance the strength of the weldment. There are no significant changes in the microstructure of the base metals even after considerable amount of thermal cycle evident from Fig. 7d, e. On the other side, the thermal cycle influences the HAZ which is evident from the microstructure (Fig. 7a, b).

In the TMAZ, there is significant growth in the grain boundaries which could be due to the plastic deformation and the less heat input. It is apparent from the

Fig. 7 Microstructure of the weldment

microstructure that a distinct grain boundary separates the deformed zones of the TMAZ from the recrystallized zone (weld nugget) where fine equiaxed grains were formed.

3.4 Hardness of Welded Sample

Vickers microhardness test is conducted on optimized parameter trial with 25 indentations in each horizontal row with 1, 3 and 5 mm from the top surface of the weld nugget is shown in Fig. 8. The hardness of the stir zone was comparatively lower than that of AA2024 and higher than AA6061 base material and TMAZ. It is also found that hardness is higher for the reading measured at 3 mm from the top surface of the weld nugget which confirms the onion ring zone.

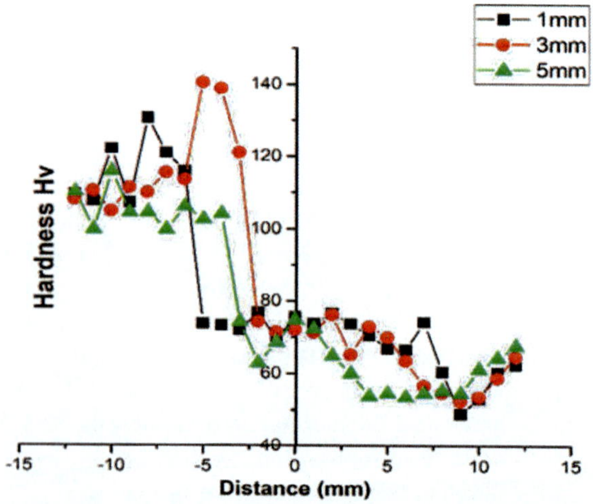

Fig. 8 Hardness distribution in different regions of optimized parameter trial

4 Conclusion

In this analysis, Taguchi L9 array has been used for selection of process parameters. Single response optimization is done by ANOVA technique and multi-response optimization is carried out by grey relational analysis.

1. ANOVA optimization reveals that the rotational speed has the maximum influence on yield strength with 49.80% contribution and 48.83% contribution for ultimate tensile strength and traverse speed has maximum influence on percentage elongation with 39.57% contribution and on impact strength 33.96% contribution.
2. Multi-response optimization by GRA results with maximum GRG value for the 8th sample with parameters 60 mm/min, 1000 rpm and threaded shoulder with taper threaded pin profile marking the responses such as 221 MPa ultimate tensile strength, 184 MPa yield strength, 15.1% elongation and 7.86 J impact strength.
3. The validation trial is made and found to have less than 2% error between experimental and predicted value.
4. Further, macrostructure, microstructure and hardness studies are made on the optimal parameter and found to have better mixing of metal in the weld nugget and increase in hardness due to the presence of onion ring present in the weld region which is confirmed by microhardness test.

References

1. Boşneag A, Constantin MA, Niţu E, Iordache M (2017) Friction stir welding of three dissimilar aluminium alloy used in aeronautics industry. IOP Conf Ser Mater Sci Eng 252(1)
2. Li Y, Murr LE, McClure JC (1999) Flow visualization and residual microstructures associated with the friction-stir welding of 2024 aluminum to 6061 aluminum. Mater Sci Eng A 271(1–2):213–223
3. Dergisi B, Abdulwadood N, Sahin B, Yildirim N (2014) Effect of welding parameters on the mechanical properties of dissimilar aluminum alloys 2024-T3 To 6061-T6 joints produced by friction stir welding. In: KaynaParametrelerinin Sürtünme KariştirmaAluminyum AlaşimlarininMekaniÖzellikleri, pp 25–36
4. Dinesh Kumar R, Ilhar Ul Hassan MS, Muthukumaran S, Venkateswaran T, Sivakumar D (2019) Single and multi-response optimization and validation of mechanical properties in dissimilar friction stir welded AA2219-T87 and AA7075-T73 Alloys Using T-GRA. Exper Tech 43(3): 245–259
5. Vilaça P, Thomas W (2011) Friction stir welding technology. In: Structural connections for lightweight metallic structures. Springer-Verlag Berlin Heidelberg, pp 85–124
6. Muthukumaran S, Mukherjee SK (2006) Two modes of metal flow phenomenon in friction stir welding process. Sci Technol Weld Joining 11(3):337–340
7. Babu N, Susenthirar R, Karunakaran N (2016) Microstructural and mechanical properties of solid state welded dissimilar aluminum alloy joints. Int J Chem Sci 14(4):2958–2966
8. Cavaliere P, De Santis A, Panella F, Squillace A (2009) Effect of welding parameters on mechanical and microstructural properties of dissimilar AA6082-AA2024 joints produced by friction stir welding.Mater Des 30(3): 609–616

Tribological Behaviour of Al 7075 Alloy Reinforced with Nano-Alumina and Silicon Carbide Particulates

T. S. A. Suryakumari, S. Ranganathan, B. Mahendra Reddy, G. Nethaji, and Challam Naveen Kumar

Abstract Aluminium metal matrix composites are being used worldwide in the fields of aeronautical and automotive industries. In this study, Aluminium 7075 metal matrix nano-composites are developed by low-cost stir casting method. Al 7075 alloy is reinforced with 1% nano-alumina and 1% nano-silicon carbide, 2% alumina, 2% silicon carbide nano-particles to fabricate metal matrix nano-composites. Wear tests were conducted on pin on disc apparatus by applying different loads 10, 15N at constant sliding velocity of 1.5 m/s with sliding distance 750 m. Wear rate increased significantly in addition of applied load. The coefficient of friction decreases with increase of applied load for the metal matrix and hybrid nano-composites. The results attributed that wear rate improved, uniform distribution of nano-particles obtained and lesser porosity for the metal matrix composite reinforced with 2% of Al_2O_3 nano-particles.

Keywords Al 7075 · Nano-alumina · Nano-silicon carbide · Wear behaviour

1 Introduction

Aluminium alloys are the combination of different elements as they improve the mechanical properties such as tensile strength, impact strength, wear loss and compressive strength to the base metal. Materials with attractive properties of high strength to weight, stiffness to weight ratio and wear resistance have great demand in manufacturing industries. Aluminium alloys reinforced with SiC, Al_2O_3, B_4C, TiC, TiB_2, etc. particulate to form metal matrix nano-composites and enhance mechanical and tribological properties. In the fields of automobile and aerospace industries, replacement of material is generally taken place because of possessing low wear

T. S. A. Suryakumari (✉) · B. M. Reddy · G. Nethaji · C. N. Kumar
Department of Mechanical Engineering, Saveetha School of Engineering, SIMATS, Chennai, India
e-mail: suria.ramana@gmail.com

S. Ranganathan
Department of Mechanical Engineering, GMR Institute of Technology, (An Autonomous Institution, Affiliated To JNTU, Kakinada), GMR Nagar, Rajam, Andhra Pradesh 532127, India

T. Rajmohan et al. (eds.), *Advances in Materials and Manufacturing Engineering*, Springer Proceedings in Materials 7, https://doi.org/10.1007/978-981-15-6267-9_51

and low coefficient of friction. It leads to loss of economy to the industry. To rectify these problems, tribological properties place a crucial role in development of material sciences [1]. Aluminium alloys reinforced with solid lubricants, hard ceramic particles, whiskers and short fibres form as advanced metal matrix composites to balance mechanical, physical and tribological characteristics precisely [2]. Huge research has been carried out on metal matrix composites with nano-particulate ceramics for manufacturing strengthened composites with proper dispersion of particles [3]. Stir casting and compo casting are the low-cost liquid state technics for the fabrication of aluminium metal matrix composites [4]. Recent studies strongly confirmed that clustering of reinforcements, wettability of reinforcement particles in molten base metal are the important parameters in the production of composites by stir casting process [5]. MMCs reinforced with ceramic particles have been studied in the last two decades and concluded that the wear performance was superior than the unreinforced metal matrix composites. The effect of Al_2O_3 particles on wear rate has been studied using Taguchi method and ANOVA and analyzed the parameters. Wear parameters particle size and sliding distance were having significant effect on wear resistance [6]. Krishna Kant Sahu et al. studied the micro-structure and wear behaviour of Al/Al_2O_3 nano as cast and heat-treated composites. Aluminium nano-composites with various weight percentages of Al_2O_3 were successfully fabricated and investigated for tribological properties [7]. Singla et al. made a study on Al–SiC composite. The maximum hardness observed is 45.5 BHN at Wt. % of 25. Siddique et al. made a study in determining the hardness of Al–SiC composite. Hardness increased with the increase in addition of silicon carbide particulates [8]. In the last two decades, wear performance of MMCs reinforced with ceramic particles has been reported to be superior to that of their respective unreinforced matrix materials

2 Materials and Experimental Procedure

2.1 Raw Materials

The raw materials considered for the fabrication and testing properties like wear properties and coefficient of friction are Al7075 alloy and the reinforced particles are Al_2O_3 and SiC of size 50 nm. The elements that are observed in the base alloy are as shown below. The compositions were made by using non-destructive testing method and contains, Si-04, Fe 0.5, Cu 1.2-2, Mg 2.1-2.9, Mn 0.3, Cr 0.18-0.28, Zn 5.1-6.1, $Ti0_2$, balance aluminium metal. The reinforcements that preferred in this study are Al_2O_3, SiC. The sizes of these reinforcements are 5 0 nm.

Fig. 1 Stir casting setup

2.2 *Experimental Procedure*

The fabrication of the three different namely Al7075+1%Al_2O_3+1%SiC, Al7075+2%Al_2O_3 and Al7075+2%SiC compositions was carried out by using low-cost stir casting method. During stir casting, the base alloy and reinforcements are heated in the furnace to a temperature of 850 °C. After reaching the optimum temperature of the base alloy in the furnace, the reinforced particles were added to the molten metal alloy that was located in the crucible. In order to reduce the emission of harmful gases and to remove the impurities inside the crucible, coverall and degasser were added to molten metal. The composite molten material is reheated to a temperature of 850 °C [4, 5]. The composite material is then poured in their respective cylindrical and rectangular dies. During this process, the die is maintained at a temperature of about 450 °C. For attaining good mechanical and tribological properties of the MMC, the material was undergone heat treatment process. The resultant metal matrix composite is soaked in water for a time of around 7 h. Heat-treated MMC always improves the particle distribution in the base alloy [6] (Fig. 1).

3 Tribological Properties of Metal Matrix Composite

3.1 *Wear Test*

The wear pins were designed as per ASTM standards as shown in Fig. 2. The wear tests were carried out for three composites by varying sliding distance, sliding

Fig. 2 Wear pins after testing

velocity and applied load. Three different composites compositions are as follows: Al7075 + 1%Al_2O_3 + 1%SiC, Al7075 + 2%Al_2O_3, Al7075 + 2%SiC.

4 Results and Discussions

4.1 Density

The theoretical densities and porosities of the three different composites were calculated and has shown in Table 1.

From the above table, it is observed that the density of the third composite was more compared with the base alloy, i.e. the density of base alloy. Sample 2 porosity was the lowest when compared with the other samples. The mechanical properties like micro-hardness, tensile strength were improved to an extent for sample 2. The wear loss for three samples increased with increased applied load at constant sliding velocity and sliding distance.

Figure 3 indicates the variation of the specific wear rate with respect to an applied load. From the graph, it is observed that the specific wear rate increased with increase in the applied load in sample 1 and 3, whereas in sample 2, the specific wear rate has decreased may be due oxidation alumina nano-particles. The specific wear rate in sample 3 for an applied load of 15 N is more when compared with remaining samples because of micro-cracks in the casting.

Table 1 Measured densities with porosities

S. No.	Composition	Theoretical density (gm/cc)	Experimental density (gm/cc)	Porosity (%)
1	Al7075 + 1%Al_2O_3 + 1%SiC	2.822	2.775	3.79
2	Al7075 + 2%Al_2O_3	2.826	2.782	1.556
3	Al7075 + 2%SiC	2.818	2.769	1.73

Fig. 3 Effect of applied load on wear rate

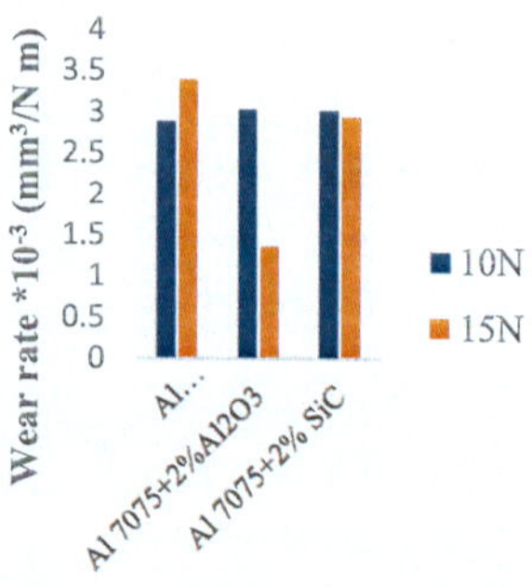

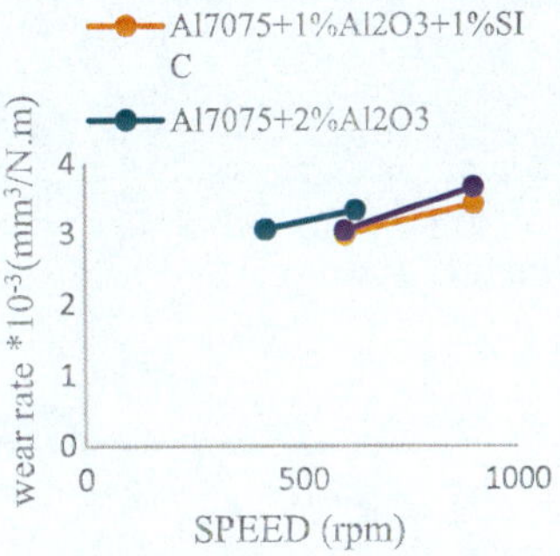

Fig. 4 Effect of speed on wear rate

From Fig. 4, it was understood that the specific wear rate of sample 1 increases as the speed of the disc increases from 597 to 896 rpm. Similarly, the specific wear of sample 3 increases as the speed of the disc increases from 597 to 896 rpm. The specific wear of sample 2 decreases as the speed of disc increases. Due to high hardness, base alloy that is reinforced with nano-alumina is possessing less specific wear rate when compared with other sample 1 and 3. Coefficient of friction is defined as the maximum force that a material can with stand against the applied load.

The variation of the coefficient of friction for an applied load of 10 N was shown above. The maximum coefficient of friction observed in Graph 1 is 0.561, Graph 2 is 0.61 and Graph 3 is 0.485 that is considered for an applied load of 10 N, with a sliding velocity of 1 m/s and a sliding distance of 500 m.

The coefficient of friction decreases for all the three samples as the load increases from 10 to 15 N. It is concluded that coefficient of friction is inversely proportional to the applied load on the disc has shown in Fig. 5.

The microstructures of the above samples are shown in Fig. 6 which indicate the distribution of nano-particles in the matrix phase. The distribution of the particles in sample 1 is not uniform because of improper distribution of silicon carbide particles where as in sample 3. During the experiment, when SiC particles were added to the molten alloy, they tend to float on the surface of the melt. Even though SiC has slightly larger specific density than that of Al 7075 molten alloy, it floats because of high surface tension of the melt and poor wettability between the particles and the melt. SEM analysis for three composites has shown in Fig. 7.

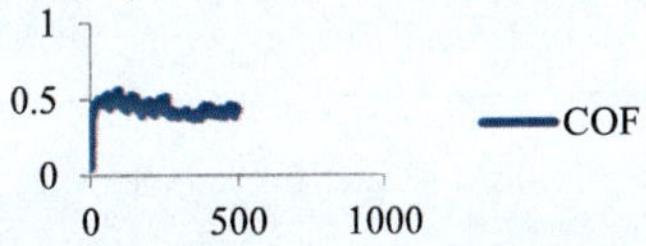

Graph 1 Variation of COF for Specimen 1

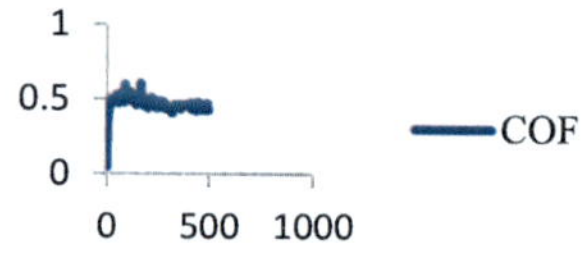

Graph 2 Variation of COF for Specimen 2

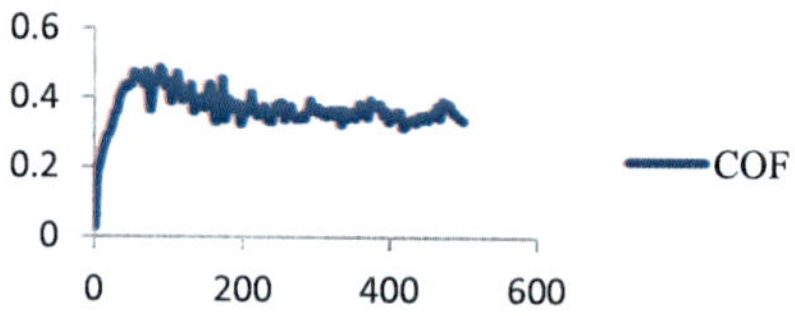

Graph 3 Variation of COF for specimen 3

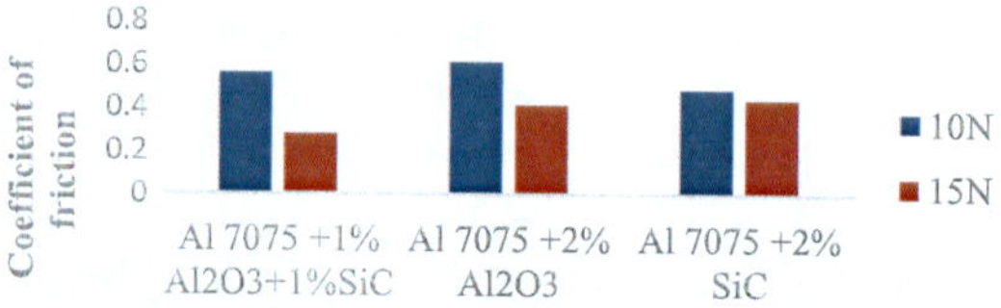

Fig. 5 Effect of applied load on COF

Fig. 6 Microstructure of samples 1, 2, 3

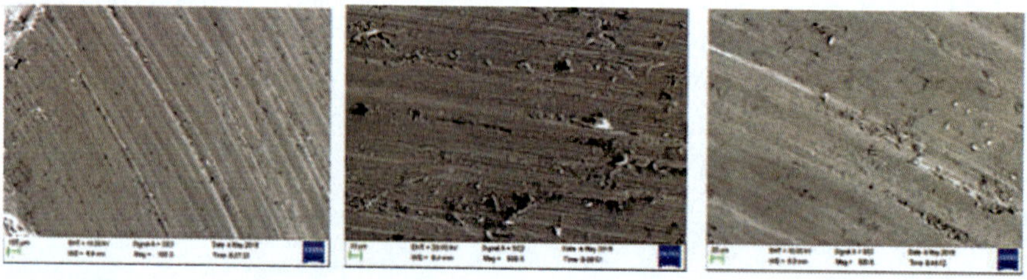

Fig. 7 SEM images for samples 1, 2, 3

5 Conclusions

The conclusions made based on the tribological study of nano-composites fabricated by low-cost stir casting method have given below.

- The three different composite materials were fabricated by stir casting approach. During stir casting, there is a formation of clusters and agglomerations in sample 1 and sample 2.
- The densities of the three composites were calculated and compared with the unreinforced alloy and specimen 2 possesses high density and low porosity.
- An increased wear loss of the composites was observed because of increase in applied load maintaining sliding velocity and sliding distance as constant.
- The specific wear rate has increased when speed of disc increased. Hence, it is concluded that the specific wear rate is directly proportional to the speed of the disc.
- The coefficient of friction decreased when there is an increase in applied load.

References

1. Vencl A, Bobic I, Arostein S, Bobic B (2010) Structural mechanical and tribological properties of A 356 Al alloy reinforced with Al_2O_3, SiC + Graphite Particles. J Alloys Compound 506:631–639
2. Prasad SV, Asthana R (2004) Aluminum metal-matrix composites for automotive applications: tribological considerations. Tribol. Lett. 17:445–453
3. Mortensen A, Conie JA, Flemings MC (2008) J Metals 12B:12–19
4. Mazahery A, Abdizadeh H, Baharvandi HR (2009) Development of high-performance A356/nano-Al_2O_3 composites. Mater Sci Eng A 518:61–64
5. Sijo MT et al (2016) Anlysation on stir casting of aluminum silicon carbide metal matrix composites. Procediatechnology 24:379–385
6. Kok M, Kanca E (2010) Abrasive wear model for Al_2O_3 particle reinforced MMCs using genetic expression programming. CMC 18(3):213–235
7. Krishnkanth S et al (2015) Wear behavior and micro structural study of Al/Al_2o_3 Nano composites before and after heat treatment. Mater Today Proc 2:1892–1900
8. Singla M, Tiwari DD, Singh L, Chawla V (2009) Development of aluminium based silicon carbide particulate metal matrix composite. J Miner Miner Charact Eng 8:455–467

Optimization of Process Parameters on Sliding Wear Behavior of Aluminum 7075 Hybrid Nanocomposite Analyzed Using Desirability Approach-Part II

T. S. A. SuryaKumari and S. Ranganathan

Abstract Aluminum 7075 metal matrix composites are most attracted materials in the production of marine applications, automobile parts and aircraft parts due to their excellent tribological and mechanical properties. Ultra-sonic cavitation method is the most effective method for the production of aluminum nanocomposites which was adopted for the composite fabrication. Pin-on-disk tribometer was utilized to conduct the dry sliding wear tests and three process parameters have been considered. This paper presents the optimization of wear process parameters on cast Al7075/SiC (*p*)/Gr hybrid metal matrix nanocomposites using desirability approach. The specific wear rate and wear loss have been optimized using desirability method. The specific wear rate increases with increased sliding distance and sliding velocity and decreases with increase in applied load at low sliding distance. The applied load and sliding distance contribute more on perdition of specific wear rate and wear loss. The effect of parameters and significance of their interactions on wear loss, specific wear rate has been investigated by ANOVA and the wear pattern was analyzed with 3D surface profilometer.

Keywords Al nanocomposites · Specific wear rate · Wear loss · Graphite particulates · Desirability

1 Introduction

Aluminum nanometal matrix composites find lot of application in engineering industries. Present scenarios, the aerospace, automobile, defense and electrical industries, are utilizing the aluminum-based nanometal matrix composites to enhance the mechanical and tribological properties. Aluminum alloy has low density, superior

T. S. A. SuryaKumari (✉)
Department of Mechanical Engineering, Saveetha School of Engineering, Saveetha Institute of Medical and Technical Sciences, Chennai 602105, India
e-mail: suria.ramana@gmail.com

S. Ranganathan
Department of Mechanical Engineering, GMR Institute of Technology, (An Autonomous Institution, Affiliated To JNTU, Kakinada), GMR Nagar, Rajam, Andhra Pradesh 532127, India

T. Rajmohan et al. (eds.), *Advances in Materials and Manufacturing Engineering*, Springer Proceedings in Materials 7, https://doi.org/10.1007/978-981-15-6267-9_52

damping capacity, and high corrosion resistance, good thermal and electrical conductivity. Mohammad Sharifi and Kaimzadeh presented the dry sliding wear behavior of aluminum metal matrix composites fabricated through powder metallurgy and reported that the high amount of oxide compounds content on the worn surface of nanocomposites have lower wear rate [1]. Gheorghe lacob et al. studied the wear rate and micro-hardness of hybrid composites and the lower wear rate obtained due the presence of Al_2O_3 and graphite which acts as a self-lubricant [2]. Veeresh Kumar et al. studied the mechanical and dry sliding wear of Al 6061-SiC composites and presence of SiC were superior to that of the matrix material [3]. Yulin et al. studied the tribological and wear mechanisms of WC–Co and WC–Fe3 Al hard material under dry sliding conditions [4]. Miranda et al. presented the mechanical and wear behavior of hybrid metallic and ceramic composites [5]. Garcia Rodriguez et al. discussed the dry sliding wear behavior of globular AZ91 magnesium alloy and AZ91/SiCp composites and the presence of SiC particles determines the wear behavior of AZ91 magnesium alloy [6]. Rahaman and Liangchi Zhang investigated the friction and wear mechanisms of aluminum high silicon alloy under dry contact sliding and found that the subsurface deformation and cracks in silicon particles play a major role on friction and wear [7]. Umanath et al. analyzed the dry sliding wear behavior of Al6061 hybrid metal matrix composites and established the statistical regression analysis to estimate the wear [8]. Yunfeng su et al. presented the tribologocal and lubrication mechanisms of self-lubricating ceramic/metal composites and formulated some principles to design self-lubricating ceramic/metal composites [9]. The purpose of this research work was to fabricate the Al7075/SiC/Gr hybrid metal matrix composites to establish the empirical model and regression equations of wear rate and friction force during the dry sliding condition. The desirability approach multiple optimization was carried out to predict the wear loss and specific wear rate of hybrid metal matrix composites.

2 Experimental Procedure

The Al 7075 used as a base metal and addition of 1.2 wt % of nano-SiC particle with 0.5 wt % of graphite was used to fabricate the hybrid metal nanocomposites using ultra-sonic probe assisted stir casting method. The dry sliding wear tests were carried out on pin-on-disk apparatus using EN 31 grade steel disks as per the ASTM G99 standards. The wear tests were carried with an applied of load of 10–30 N, sliding velocity of 1–3 m/s and with sliding distance of 1000–3000 m in three levels to find the wear loss and specific wear rate of hybrid metal matrix composites. Experiments were planned and designed based on central composite design (CCD) using Design Expert 11.0. The desirability-based multiperformance optimization of wear loss and specific wear rate was established.

3 Results and Discussions

Experiments were planned according to CCD and test results obtained by conducting wear test. The relation between wear process responses and parameters was developed as mathematical expressions by using the test results. Regression equations have established as per the interactions of process parameters for wear loss and specific wear rate using design expert software (Table 1).

Statistical interpretations of data and significance of interactions between variables were analyzed by analysis of variance. Wear loss and specific wear rate have been considered as response variables. From Table 2, the regression analysis of wear loss observed that the applied load and sliding velocity are the most influenced factors. P value indicates more than 0.05 for the sliding distance, so it is not so significant factor. Table 3 depicts the parameter applied load is the most influencing factor on specific wear rate whereas sliding velocity moderately significant. Interaction of sliding velocity and sliding distance has less significant. Lack of fit for the responses specific wear rate and wear loss was given as not significant so that the models fit into the data. Regression equation for specific wear rate was given below.

Specific wear rate = $9.43–2.82A-1.41B-0.8375C + 0.5931AB + 0.8896AC + 0.2965BC-0.340A^2-1.60B^2-0.8388\ C^2 + 0.2964\ ABC + 2.00\ A^2B-0.0521\ A^2C-0.1442\ AB^2$.

In the above equation, value 9.43 indicates the overall average model value for all testing conditions. Coefficients having positive sign indicate that the specific wear

Table 1 ANOVA for wear loss

Source	Sum of squares	*df*	Mean square	*F*-value	*p*-value	
Model	0.0001	13	8.17610^{-06}	7.16	0.0119	Significant
A-Applied load	8.00010^{-06}	1	8.00010^{-06}	7.00	0.0382	
B-sliding velocity	8.00010^{-06}	1	8.00010^{-06}	7.00	0.0382	
C-sliding distance	6.12510^{-06}	1	6.12510^{-06}	5.36	0.0598	
AB	6.12510^{-06}	1	6.12510^{-06}	5.36	0.0598	
AC	1.12510^{-06}	1	1.12510^{-06}	0.9848	0.3593	
BC	3.12510^{-06}	1	3.12510^{-06}	2.74	0.1492	
ABC	3.12510^{-06}	1	3.12510^{-06}	2.74	0.1492	
A^2C	9.08210^{-08}	1	9.08210^{-08}	0.0795	0.7874	
AB^2	3.75310^{-06}	1	3.75310^{-06}	3.29	0.1199	
Residual	6.85410^{-06}	6	1.14210^{-06}			
Lack of fit	6.81910^{-06}	1	6.81910^{-06}	974.16	< 0.0001	Not significant
Pure error	3.50010^{-08}	5	7.00010^{-09}			
Cor. total	0.0001	19				

Table 2 ANOVA for specific wear rate

Source	Sum of Squares	df	Mean Square	F-value	p-value	
Model	194.06	13	14.93	8.49	0.0076	Significant
A-applied load	45.02	1	45.02	25.61	0.0023	
B-Sliding velocity	11.25	1	11.25	6.40	0.0447	
C-Sliding distance	3.97	1	3.97	2.26	0.1837	
AB	2.81	1	2.81	1.60	0.2527	
AC	6.33	1	6.33	3.60	0.1065	
BC	0.7035	1	0.7035	0.4002	0.5503	
A^2	1.67	1	1.67	0.9523	0.3668	
B^2	36.84	1	36.84	20.96	0.0038	
C^2	10.14	1	10.14	5.77	0.0532	
ABC	0.7030	1	0.7030	0.3999	0.5505	
A^2B	13.30	1	13.30	7.57	0.0333	
A^2C	0.0090	1	0.0090	0.0051	0.9453	
AB^2	0.0689	1	0.0689	0.0392	0.8496	
Residual	10.55	6	1.76			
Lack of fit	10.45	1	10.45	558.11	< 0.0001	Not significant
Pure error	0.0937	5	0.0187			
Cor. total	204.61	19				

rate increases with increase in the factors, whereas negative sign indicates that the specific wear rate decreases with increases in the factors. Factor applied is −2.82 which indicates that the specific wear rate decreases with increase in applied load and VIF value more than one which indicates severe correlation between the factors.

3.1 *Optimization of Responses*

Desirability approach is the multiple parameter optimization technique which determines the optimum sliding wear parameters for the multiple responses by maximizing the desirability of composite material. Each response will be converted as an individual desirability function and validated for its target. Desirability function ranges zero to one.

If the response is near the target, the function value is one and otherwise it is zero. This zero indicates the response values are not in an acceptable region. The desirability function of composite is the geometric mean weightage of individual responses. The current study maximizes the composite desirability. Figure 1 shows

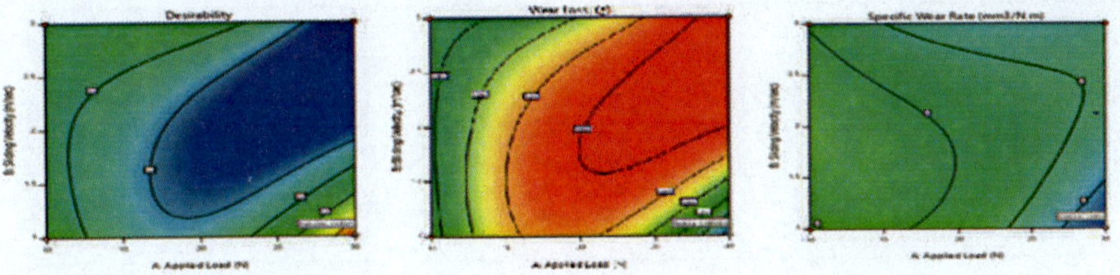

Fig. 1 Effect of applied load and sliding velocity on responses

the effect of applied load and sliding velocity on wear loss and specific wear rate (Fig. 2).

Counter maps describe the desirability results in detail. Desirability nearing 1 at higher applied load irrespective of sliding distance. At higher applied loads pin oxidized and specific wear rate has decreased as shown in Fig. 3. Out of 69 experimental runs, optimum parameters obtained as applied load 30 N, sliding velocity 1 m/s and sliding distance 2000 m as shown in the ramp graph. Figure shows the desirability results for wear loss and specific wear rate individually and combined for both. The overall solution of desirability function is 1 and 0.9224 desirability value describes the closeness of responses toward the target. Wear loss value is 0.002583 g, whereas specific wear rate 1.90835×10^{-5} mm^3/N-m at optimum parameters (Figs. 4 and 5).

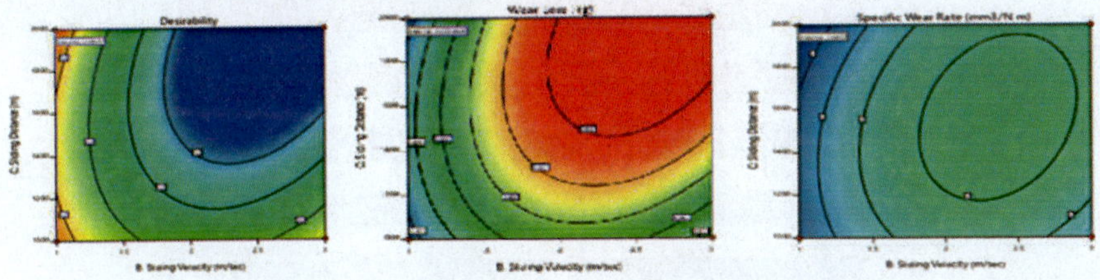

Fig. 2 Effect of sliding velocity and sliding distance on responses

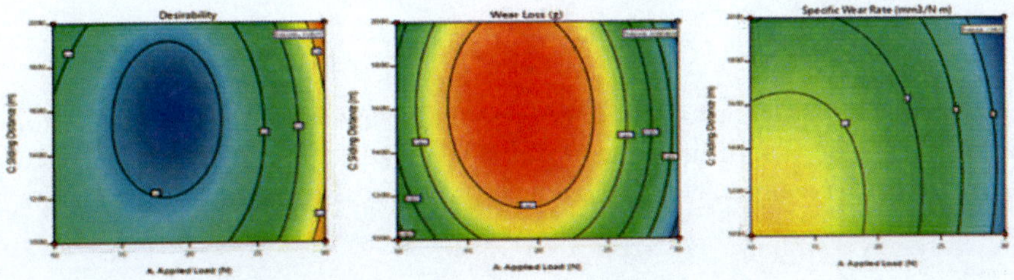

Fig. 3 Effect of applied load and sliding distance on responses

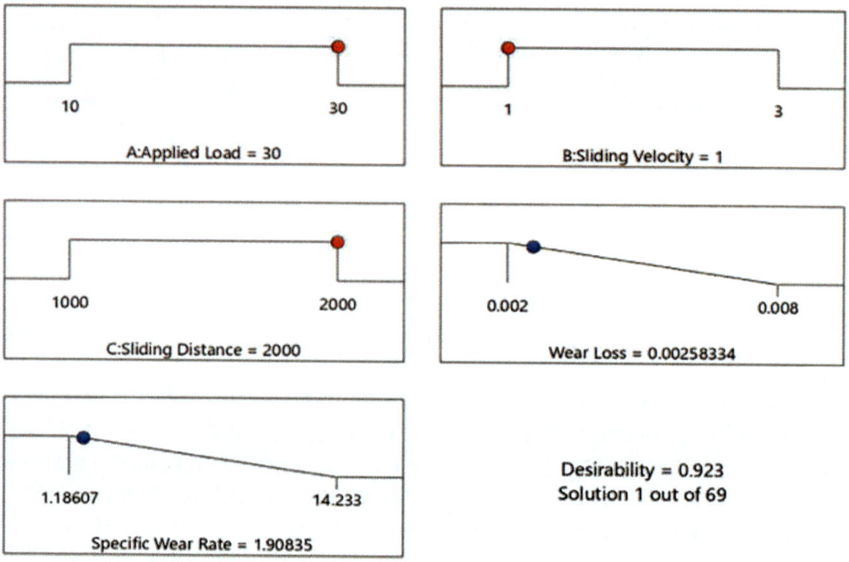

Fig. 4 Ramp graph

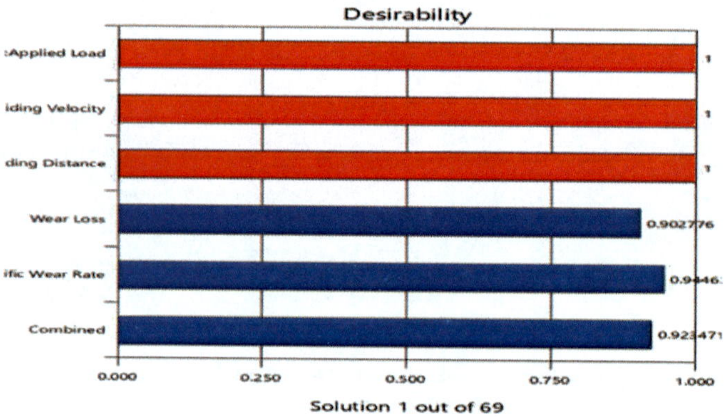

Fig. 5 Bar graph

Confirmation experiments were conducted for the optimum wear parameters for the verification of performance of the factors. The results are very close to predicted values, so that error percentage is minimum for wear loss and specific wear rate.

4 Conclusion

The effect of sliding wear parameters on wear loss and specific wear rate was investigated. Multiple response optimization was applied using desirability approach. Optimum parameters determined by the simulation using design experts software. Adequacy of the model was analyzed by ANOVA. Applied load and sliding velocity are the more influencing factors and sliding distance was less influencing. Desirability value for combined responses obtained as 1, which indicates responses are within the acceptable region. Optimum parameters were determined as applied load, 30 N, sliding velocity 1 m/s and sliding distance, 2000 m. Validation experiments were conducted at optimum conditions and error percentage 3.021 for wear loss and 2.325 for specific wear rate. This indicates the agreement toward predicted values. 3D profile for wear specimen was analyzed.

References

1. Mohammed Shriffi E, Karimzadeh F (2011) Wear behaviour of aluminium matrix hybrid nanocomposites fabricated by powder metallurgy. Wear Wear behaviour of aluminium matrix hybrid nanocomposites fabricated by powder metallurgy. 271:1072–1079.
2. Gheorghe L, Valcriu GG et al (2015) Studies on wear rate and micro hardness of the $Al/Al_2O_3/Gr$ hybrid composites produced via powder metallurgy. Comp Part B 69:609–611s
3. Veeresh Kumar GB, Swamy C, Rao P et al (2012) Studies on mechanical and dry sliding wear of Al6061-SiC composites. Comp Part B 43
4. Liu Y, Cheng J, Yin B et al (2017) Study of the tribological behaviours and wear mechanisms of WC-Co and Wc-Fe3Al hard materials under dry sliding condition. Tribol Int 109:19–25
5. Miranda G, Buciumeanu M et al (2015) Hybrid composites–Metallic and ceramic reinforcements influence on mechanical and wear behaviour. Comp Part B 74:153–165
6. Garcia-Rodrguez S, Torres B et al (2017) Dry sliding wear behaviour of globular $AZ91/SiC_p$ composites. Wear 390–393, 1–10
7. Rahaman ML, Zhang L (2017) An Investigation into the friction and Wear mechanisms of aluminium high silicon alloy under contact sliding. Wear 376–377, 940946
8. Umakanth K, Palanikumar K, Selvamani ST, Analysis of dry sliding wear behavior of $Al6061/SiC/Al_2O_3$ hybrid metal matrix composites. Compos Part B 53:159–168
9. Su Y, Zhang Y et al (2017) Tribological behavior and lubrication mechanism of self–lubricating ceramic/metal composites: the effect of matrix type on the friction and wear properties. Wear 372–373,130–138
10. Mordyuk BN, Prokopenko GI et al (2014) Wear assessment of composite surface layers in Al-6Mg alloy reinforced with AlCuFe quasi crystalline particles: effects of particle size, micro structure hardness. Wear 319:84–95

Investigation of Process Parameters on Sliding Wear Behaviour of Aluminium 7075 Hybrid Nanocomposite Analyzed Using Response Surface Methodology-Part I

T. S. A. SuryaKumari and S. Ranganathan

Abstract This study presents the abrasive wear behaviour of Aluminium 7075 metal matrix nanocomposite (MMNC) reinforced with hard ceramic SiC and solid lubricant graphite. The nanocomposite was prepared using ultrasonic cavitation method to improve the wettability and dispersion of reinforcements. Wear behaviour of composite was examined by conducting unlubricated pin-on-disc tests as per the experimental design, by varying the parameters such as applied load, sliding distance and sliding velocity. Significant control factors and their interactions on output features were identified with analysis of variance (ANOVA). Further, tribology characteristics have been estimated by mathematical formulation using response surface method.

Keywords Al7075 nanocomposites · Wear loss · ANOVA · RSM

1 Introduction

Aluminium metal matrix composites have more attention in engineering materials and have wide application in aerospace, automotive, defense, sports goods and ship hull production industries due to their distinct mechanical properties such as Young's modulus, tensile strength with lightweight, wear resistance and coefficient of thermal expansion. Generally, metal matrix composites attain very good mechanical properties with microparticulate reinforcements, but ductility increases with increase in percentage of reinforcements. Nano particulates with less weight fraction, significantly improves the matrix strength and maintains ductility than microparticulate composites [1–3]. The performance of mechanical properties improved significantly by the addition of ceramic reinforcements such as SiC, Al_2O_3,B_4C, TiO_2 and wear

T. S. A. SuryaKumari (✉)
Department of Mechanical Engineering, Saveetha School of Engineering, Saveetha Institute of Medical and Technical Sciences, Chennai 602105, India
e-mail: suria.ramana@gmail.com

S. Ranganathan
Department of Mechanical Engineering, GMR Institute of Technology, (An Autonomous Institution, Affiliated To JNTU, Kakinada), GMR Nagar, Rajam, Andhra Pradesh 532127, India

T. Rajmohan et al. (eds.), *Advances in Materials and Manufacturing Engineering*, Springer Proceedings in Materials 7, https://doi.org/10.1007/978-981-15-6267-9_53

resistance of metal matrix by incorporating a hard ceramic phase to a relatively soft matrix alloy. Ultimate tensile strength and yield strengths were improved by 23% with the addition of nano-TiB_2 with SiC particles when compared with alloy reinforced with mono SiC [4]. Technical difficulties involved in distribution of reinforcements uniformly for Al/SiC alloy were discussed and good wettability was achieved by stir casting method [5]. Mechanical milling is one of the methods to obtain proper dispersion of nanoparticles in the matrix and better mechanical properties than tempered T6 aluminium alloys. Reinforcing the alloy with more than 2 weight percentage of nanoparticles does not have much effect on microhardness [6]. Kannan et al. investigated the tribological characteristics under static and dynamic loading conditions for single and hybrid aluminium nanocomposites reinforced with 2wt% of Al_2O_3, 4wt%of SiC. Wear performance enhanced by the usage of nanocomposites and hence life of product will be increased. The influence of SiC particles on wear resistance of Al6061/SiC composite at dry sliding wear atmosphere and the effects of applied load and temperature were studied. The results show that the matrix containing SiC particles exhibit more wear resistance [7]. By varying weight percentages of alumina and MoS_2 of Al6061/Al_2O_3 hybrid composites, the BHN and tensile tests were conducted. Mechanical properties were decreased with increased weight percentage of molybdenum and optimum weight fractions obtained as 12wt% alumina, 2wt% molybdenum. Wear rate and coefficient of friction were minimized at 3.25 m/s sliding velocity, 15 N load and 6wt% of MoS_2 [8]. Selvakumar et.al. conducted experiments by varying weight fractions of TiC and molybdenum dioxide to identify the effects of reinforcements on tribological properties using L27 Taguchi experiential design. Wear loss and coefficient of friction were decreased up to the weight fraction 7.5% of MoS_2 and then increased for magnesium hybrid composites. Brittle fracture occurred due to decrease in fracture toughness and increase in molybdenum dioxide [9]. A356 alloy powder and alumina particulates mixed using ball milling process; further, MMCs were fabricated by vortex method. Compressive and tensile stresses were improved significantly at 2.5, 1.5 volume percentages of alumina and porosity levels increased with increase in volume fractions. Refined microstructure was obtained by this process [10]. Addition of nanoparticles to the alloy improved the wear resistance significantly and delamination wear was controlled for magnesium metal matrix nanocomposites [11]. Wear loss reduced with increased weight percentage of nano-SiC particulate reinforcement and exhibits improved mechanical properties for Al7075/SiC/graphene composites [12].

Abrasive wear behaviour aluminium MMCs have been studied and reported. The current research presented on dry sliding was behaviour of Al7075/SiC/Gr hybrid nanocomposite fabricated with ultrasonic probe assisted vacuum stir casting.

2 Experimental Procedures

2.1 Composite Preparation

The composites were fabricated by ultrasonic probe assisted stir casting method. Aluminium 7075 used as base alloy and 1.2 wt% of nano-SiC particles with 0.5 wt% of graphite were used as reinforcements. Nano-SiC particles were imported from the US research nanomaterials with an average size of 50 nm and graphite less than 20 μm were used for hybrid MMNC. Motorized mechanical stirrer was used for proper distribution of particles and reinforcements were preheated to improve the wettability, dispersion in molten metal. Particulates reinforcements were fed through steel tube while stirring. A bottom poured stir casting equipment with vacuum graphite crucible used for the fabrication of composites.

2.2 Wear Test

Wear behaviour of MMNC samples on hardened steel discs was investigated using pin-on-disc apparatus as per the ASTM G99 standards. Specimen was prepared by using conventional machining processes and surfaces of pins; discs were polished with emery paper. The pin dimensions were considered as 8 mm diameter and 32 mm height. Sliding track diameter of discs was 26 mm. Dry sliding wear parameters: applied load, sliding velocity, sliding distance were varied from 10–30 N, 1 to 3 m/s and 1000–2000 m, respectively. The traces on cylindrical pin and circular disc were cleaned after each test with acetone solvent. Weights of pins were weighed before and after wear tests using an electronic weighing machine with an accuracy of 0.1 mg and wear loss was obtained for each pin.

2.3 Design of Experiments and Results

ExperimessssetsmechOnts were planned and designed based on box Behnken quadratic model. This is the most popular assessment response surface methodology for identifying the mathematical relationship between process parameters and responses. This is a two-level (high, +1, low, −1) full factorial design with eight factorial points and six axial face centered points. Corresponds to anα value of 1. This is designed for 20 experimental observations at three input variables. The design details of experiments with response as given in Table 1.

Table 1 Experimental results

Ex. No	Load in N	Sliding velocity in m/s	Sliding distance in m	Wear loss in gm
6	10	1	1000	0.003
5	30	1	1000	0.002
2	10	3	1000	0.003
19	30	3	1000	0.003
8	10	1	2000	0.004
15	30	1	2000	0.002
10	10	3	2000	0.004
20	30	3	2000	0.008
1	10	2	1500	0.003
14	30	2	1500	0.007
16	20	1	1500	0.007
12	20	3	1500	0.003
9	20	2	1000	0.004
3	20	2	2000	0.0075
11	20	2	1500	0.008
18	20	2	1500	0.005
13	20	2	1500	0.007
4	20	2	1500	0.0055
7	20	1.5	1500	0.008
17	20	1.5	1500	0.005

3 Results and Discussion

The influence of independent variables and their interactions with an output responses mathematically modelled using analysis of variance (ANOVA). Proper fit of the quadratic polynomial mathematical model obtained in this wear behaviour of hybrid nanocomposite as shown in ANOVA Table 2. Significance of individual parameters, significance of regression model and significance of lack of fit tests can be summarized using ANOVA. F value can be determined by finding the ratio of regression mean square and error of the mean square. The significance of model individual factors and the variance and the error can be estimated at the desired level of significance [13].

3.1 Influence of Testing Parameters on Wear Loss

The above table shows that the interactions between (i) applied load and velocity, (ii) Load and distance, (iii) velocity and distance. The model F value obtained as 7.16

Table 2 ANOVA for cubic model for wear loss

Source	Sum of Squares	df	Mean Square	F-value	p-value	
Model	0.0001	13	8.17610^{-06}	7.16	0.0119	Significant
A-Applied Load	8.00010^{-06}	1	8.00010^{-06}	7.00	0.0382	
B-Sliding Velocity	8.00010^{-06}	1	8.00010^{-06}	7.00	0.0382	
C-Sliding Distance	6.12510^{-06}	1	6.12510^{-06}	5.36	0.0598	
AB	6.12510^{-06}	1	6.12510^{-06}	5.36	0.0598	
AC	1.12510^{-06}	1	1.12510^{-06}	0.9848	0.3593	
BC	3.12510^{-06}	1	3.12510^{-06}	2.74	0.1492	
A^2	0.0000	1	0.0000	23.04	0.0030	
B^2	0.0000	1	0.0000	23.04	0.0030	
C^2	0.0000	1	0.0000	14.88	0.0084	
ABC	3.12510^{-06}	1	3.12510^{-06}	2.74	0.1492	
A^2B	0.0000	1	0.0000	12.36	0.0126	
A^2C	9.08210^{-06}	1	9.08210^{-06}	0.0795	0.7874	
AB^2	3.75310^{-06}	1	3.75310^{-06}	3.29	0.1199	
Residual	6.85410^{-06}	6	1.14210^{-06}			
Lack of Fit	6.81910^{-06}	1	6.81910^{-06}	974.16	< 0.0001	Not significant
Pure Error	3.50010^{-06}	5	7.00010^{-06}			
Cor Total	0.0001	19				

implies that the designed model is significant. If the multiple regression coefficient, R^2 value nearing the unity shows that the less difference exists between predicted data and actual data. Adequate precision (AP) value is higher than 4 shows the model discrimination [14].

Higher values obtained for R^2 and adequate precision in this model. The obtained values are $R^2 = 0.9394$ and AP = 7.1663. The value of P less than 0.05 shows 95% confidence level and the model is statistically significant. The response equation of wear loss for the model given below is derived from regression fit to confirm their validity conformation test must be conducted. From Table 2, calculated error value is observed as small and error between the experimental and predicted values is less.

Wear loss = 0.008 + 0.0012A -0.0012B + 0.0010C + 0.0009AB + 0.0004AC + 0.0006BC-0.0014 A^2-0.0014 B^2-0.0011 C^2 + 0.0006ABC + 0.0021A^2B-0.0002 A^2C-0.0011 AB^2.

The effect of individual factors can be discussed by considering the regression equation. The constant value 0.008 indicates the overall mean value of the dry sliding wear of composite under all the testing conditions. Further, the coefficient 0.0012 associated with applied load is positive, which shows that the wear loss increased with increase in load due to contact pressure increased on the sliding surface which resulted as high wear loss at high load [14]. Further, coefficients describe that the wear loss decreases with increased sliding velocity and increases with increased sliding distance. Figure 1 shows that the influence of applied and sliding velocity on wear loss. Wear loss decreased with increased sliding velocity due to the formation of mechanical mixed layer (MML) in hybrid nanocomposite [15].

Interaction of sliding velocity and sliding distance on were loss is shown in Fig. 2. This response shows the increase in wear loss with sliding distance due to weak tribolayer. Figure 3 shows the response of applied load with sliding distance on wear loss. Reduction in wear loss observed at higher applied load may be effect of the formation of mixed layer. When load increases, temperature may be increased and by that oxidation of metallic surfaces takes place which leads to MML.

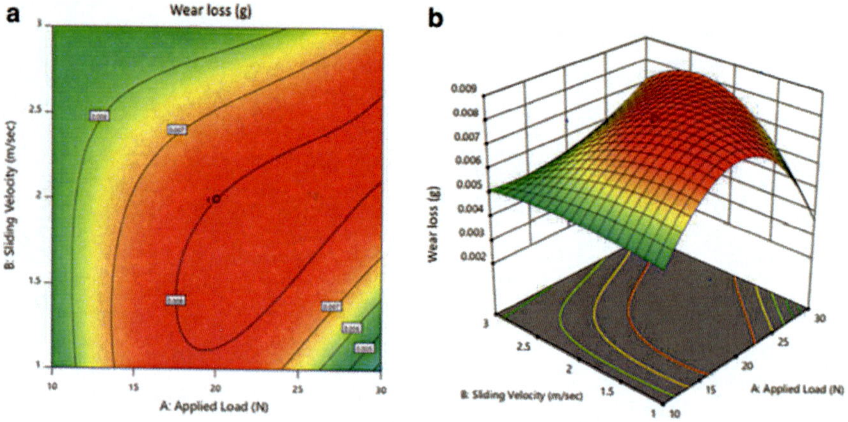

Fig. 1 **a**, **b** Effect of sliding velocity and applied load interaction on wear loss

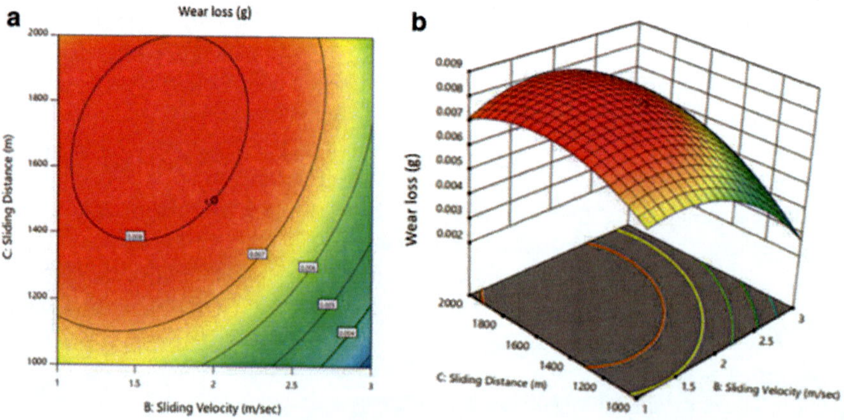

Fig. 2 **a**, **b** Effect of Sliding velocity and sliding distance interaction on wear loss

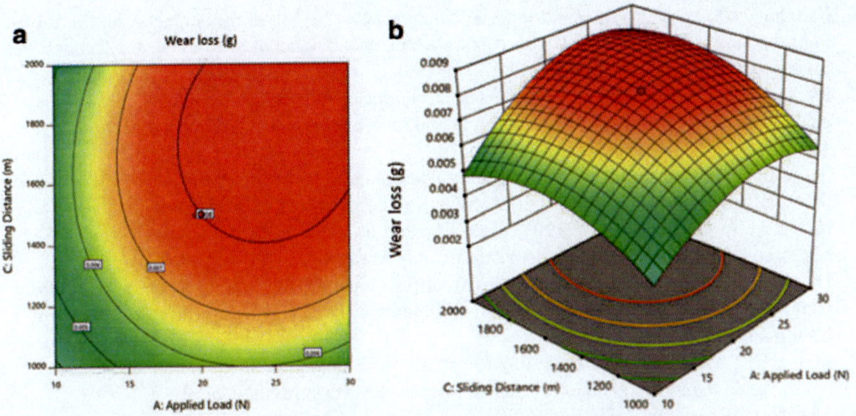

Fig. 3 **a**, **b** Effect of applied load and sliding distance interaction on wear loss

4 Conclusion

The Al7075/SiC/Gr hybrid nanocomposite was fabricated by ultrasonic probe vacuum stir casting method. The wear resistance of hybrid composite has improved. Experimental design was planned according to the RSM and mathematical modelling was done by analysis of variance to confirm the confirmation test significance. From ANOVA, it is observed that the applied load and sliding velocity are more influencing factors and sliding distance is the least influencing factor on wear loss of hybrid composite. Wear loss was reduced with increase in applied load with the effect of mechanical mixed layer and increase in sliding velocity decreases the wear by the support of MML. Confirmation tests resulted that the error between experimental and predicted values of wear was very less.

References

1. Sivasankaran S, Sivaprasad S, Naranasamy K, KumarIyer R (2010) An investigation on fflowability and compressibility of AA6061 100-x-xwt%TiO_2 micro and nanocomposites powder prepared by blending and mechanical alloying, Powder Technol 201:70–82.
2. Suresh S, Moorthi NSV, Vettivel SC, Selvakumar N, Jinu GR (2014) Mater Sci Eng A 612:16–27
3. Choi SM, Awaji M (2005) Nanocomposite–anew material design concept. Sci Technol Adv Mater 6:2–10
4. Tan A, Teng J, Zeing C, Fu D, Zhang H (2017) Fabrication of aluminium matrix hybrid composites reinforced with SiC microparticles and nano TiB_2 nano particles by powder metallurgy. Powder Metall 1–7
5. Hashim J, Looney L, Hasmi MSJ (1999) Metal matrix composites: production by the stir casting method. J Mater Process Technol 92:1–7
6. Flores-Campos R, Mendoza-Ruiz DC, Amezaga-Madrida P, Estrada-Guel I, Miki-Yoshidaa M (2010) Microstructural and mechanical characterization in 7075 aluminium alloy reinforced by silver nanoparticles dispersion. J Alloy Compd 495:394–398

7. Kannan C, Ramanujam R, Venkatesan K, Dheeraj NV (2018) An investigation on the tribological characteristics of Al 7075 based single and hybrid nanocomposites. Mater Today Proc 5:12837–12847
8. Pitchayyapillai G, Seenikannan P, Raja K, Chandrasekaran K (2016) Al6061 hybrid metal matrix composites reinforced with alumina and molybdenum disulphide. Adv Mater Sci Eng 1–9
9. Selvakumar N, Naranasamy P (2016) Optimization and effect of weight fraction of MoS2 on the tribological behavior of Mg-TiC-MoS2 hybrid composites. Tribol Trans 1–46
10. Mazahery A, Ostadshabani M (2011) Investigation on mechanical properties of nano–Al_2O_3 reinforced aluminium matrix composites. J Comp Mater 1–8
11. Du XM, Zheng KF, Zhao T, Liu FG (2018) fabrication and characterization of al 7075 hybrid composite reinforced with graphene anf SiC nanoparticles by powder metallurgy. Digest J Nanomater Biostruct 13(4):11–33
12. Myers RH, Montgomery DC (1995) Response surface methodology: process and product optimization using designed experiments, Response Surface Method 91601–8
13. Sharma D, Khanduja S, Sharma S (2016) dry sliding wear investigation of Al6082/Gr metal matrix composites by response surface methodology. JMR&T 5(1):29–36
14. Manivannan I, Ranganathan S, Gpalakannan S, Suresh S (2018) Dry sliding wear behaviour of cast Al/Al_2O_3/Gr hybrid nano composite using response surface methodology. Mater Sci Eng 390

Effect of Process Parameters on Drilling of Carbon Fiber Reinforced Polymers

D. Vijayan, A. Tamilarasan, and B. Vignesh Aravind

Abstract This paper presents the experimental investigation on the implementation of Taguchi technique to optimize the drilling process parameters of nano-filled carbon fiber reinforced polymers (CFRP) to minimize the thrust force. The experiments were performed by using three drill bit types by varying point angles at three levels of drilling speeds, feed rates, percentage of CNT and point angles. The responses, namely thrust force, were measured by using Kistler multicomponent dynamometer. The drilling process parameters which directly influenced the performance characteristics were optimized using Taguchi technique. The analysis of variance (ANOVA) results clearly indicated that the feed rate was the significant factor which affected the responses. Thereafter, scanning electron microscopy (SEM) analysis was used to discuss the effect of drilling parameters on the microstructure of the investigated hybrid CFRP.

Keywords Taguchi method · Process parameters · ANOVA · Carbon fiber reinforced polymer (CFRP) · Drilling

1 Introduction

Carbon fiber reinforced polymer (CFRP) composites are extensively used in automobile, aerospace, marine, sports equipment applications because of high strength to weight ratio. CFRP is an excellent replacement of conventional materials when compared to well-known materials such as aluminum alloys, stainless steel and high carbon steels. Moreover, the CFRP composites are said to have good corrosion resistant when mixing with proper resins. Thus, CFRP composite materials are widely used in aircraft, automobiles, marine and many more areas where lightweight and high strength are major factors. The CFRP composites are often fabricated by using drilling operation; however, it is extremely difficult to drill due to bonding strength between fiber and resin. By selecting an appropriate optimal process parameter, the

D. Vijayan (✉) · A. Tamilarasan · B. Vignesh Aravind
Sri Chandrasekharendra Saraswathi Viswa Mahavidyalaya, Enathur, Kanchipuram, Tamil Nadu 631561, India
e-mail: Vijaiand2012@gmail.com

T. Rajmohan et al. (eds.), *Advances in Materials and Manufacturing Engineering*, Springer Proceedings in Materials 7, https://doi.org/10.1007/978-981-15-6267-9_54

CFRP composites can be drilled without defects. Since, delamination is a major defect associated in drilling of CFRP composite which mainly cause for poor structural rigidity and performance. In the past decades, Usually, trial and error method to identify the optimal process parameters of the process, however it is a time-consuming process some time require numerous resources. Alternatively, optimization techniques such as Taguchi method, gray relational analysis, response surface method, etc. can be used to obtain optimal process parameters. In the past decades, numerous investigations have been proposed by the researchers based on the applications of optimization techniques. For instance, Ozcelik and Ozbay [1] investigated the effects of process factors on the mechanical properties of polymers by using Taguchi method. Shunmugesh and Panneerselvam [2] optimized the drilling parameters for CFRP composites using Taguchi-based gray relational analysis. Prakash et al. [3] used the Taguchi-based gray relational analysis and ANOVA to investigate the effect of feed rate, spindle speed and drill diameter on medium density fiber drilling. Siddiquee et al. [4] used Taguchi technique to optimize deep drilling parameters to minimize AISI 321 steel surface roughness and concluded that the surface roughness is minimized by the effectiveness of the approach. Sahoo and Pradhan [5] used Taguchi technique to optimize process parameters in metal matrix composites machining and reported statistically significant mathematical models for performance characteristics. Gowda et al. [6] conducted drilling experiments on Al–Si_3N_4 metal matrix composite and investigated the effect of process parameters on circularity, cylindricity and surface roughness. They concluded that by wobbling the drill bit, circularity error increases at high speed and cylindricity increases. In drilling carbon fiber-reinforced plastic (CFRP), Tsao [7] proposed a Taguchi method for optimizing machining parameters with multiple performance characteristics. From the confirmation results, it can be understood that by finding the optimal process parameters in the drilling of CFRP laminates, the delamination factor can be effectively reduced. In the drilling of Al6061/SiC/Al_2O_3 composite by Adalarasan and Santhanakumar [8], surface finish and burr height were observed. Taguchi L18 orthogonal array is used to design the following parameters: feed rate, spindle speed and tool point angle. Sahu et al. [9] have optimized multi—objective optimization of upsetting parameters for Al—TiC metal matrix composites using gray Taguchi approach. The authors observed that Taguchi technique is a best technique for design the experiment in order to optimize the levels of input parameters. The authors concluded that the Taguchi method is best suited to finding the optimum drilling process parameter setting. It is evident from the above literature review that Taguchi technique is a best optimization technique and it can be applied to find the optimal process parameters of drilling. Therefore, in the present investigation, Taguchi method is used by considering significant process parameters involved in drilling such as drilling speed, feed rate, percentage of CNT and point angle.

2 Experimental Setup

A hybrid composite consisting of reinforcing carbon fibers and carbon nanotubes embedded in an epoxy matrix is used in the present investigation. Multiwall carbon nanotubes (MWCNTs), are about 50–80 nm, were added as a filler in the composite. The drilling experiment was carried out by using computerized vertical machining center as shown in Fig. 1. Kistler tool dynamometer is used to measure the thrust force during drilling of CFRP composite. The experiment was carried out in dry condition without using coolant oil. The drilling parameters and their levels are shown in Table 1. The drilling parameters such as drilling speed, feed rate, point angle are considered as significant process parameter in the present investigation. Three-point angles chosen based on the pilot study conducted for the present investigation such as 110°, 120° and 130° for a standard carbide drills of 6 mm diameter have been used. In order to measure the effects of thrust force on the drilling of CFRP composites, every new tool was allowed to drill only four holes of a composite laminate. The drilling

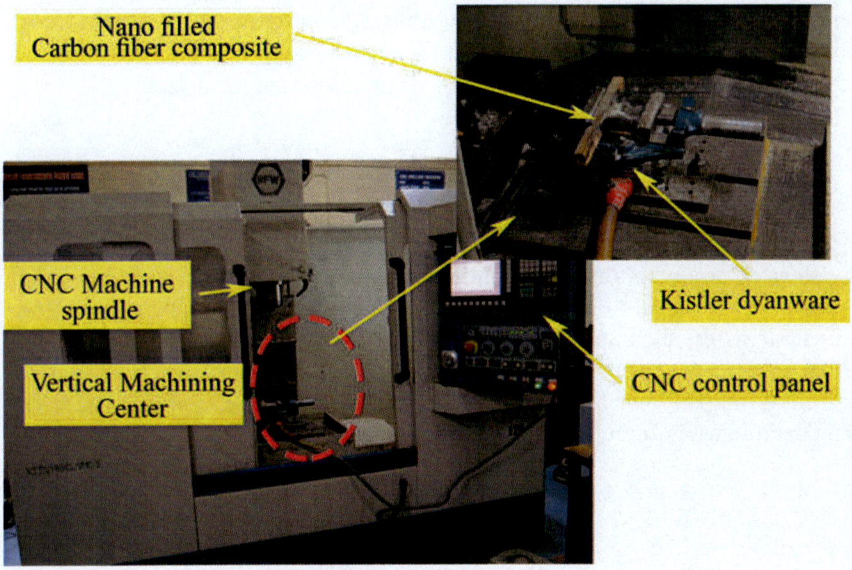

Fig. 1 Composite drilling experimental setup

Table 1 Factors and their levels

Factors	Units	Low actual	High actual
Drilling speed (N)	rpm	500	1100
Feed rate (*f*)	mm/rev	4	8
% of CNT (%)	%	3	5
Point angle (ɵ)	°	110	130

Table 2 Experimental results

S No	Drilling speed, N (rpm)	Feed rate, f (mm/rev)	% of CNT, (%)	Point angle, ɵ (°)	Thrust force, TF (N)
1	500	4	3	110	15.75
2	500	6	4	120	12.36
3	500	8	5	130	16.36
4	800	4	4	130	33.17
5	800	6	5	110	45.85
6	800	8	3	120	58.68
7	1100	4	5	120	65.67
8	1100	6	3	130	85.3
9	1100	8	4	110	95.5

experiments were conducted as per the order given in Taguchi L_9 orthogonal array. The experimental results are presented in Table 2. For each experiment, three trails were conducted in order to ensure the reliable measurements and less error in the experiment. After completing all the drilling experiments, each hole was observed through scanning electron microscope (SEM) to understand the defects.

3 Result and Discussion

The obtained results for the quality attributes and the mean outcomes are displayed in Table 3. The signal-to-noise ratio is a significant quality character utilized by designer to assess the impact on the product quality index of process parameters. follows: Since the objective of the investigation has to minimize the thrust force of the drilling, therefore smaller S/N ratio to be considered. The S/N ratio has estimated using the following formula,

$$\frac{S}{N} = -10 \times \log\left(\frac{1}{n}\sum_{i=1}^{n} y_i^2\right) \tag{1}$$

Table 3 Response table for means

Level	Drilling speed	Feed rate	Point angle	% of CNT
1	52.37	14.82	38.2	47.24
2	37.9	38.23	49.5	47.01
3	46.61	83.82	49.18	42.63
Delta	14.46	69.00	11.31	4.62
Rank	2	1	3	4

If n is the number of quality index data sets equal to 9, and y_i is the value of the quality index for the ith data set.

Table 3 shows the results of the response means. Table 3 is utilized to find the ideal control factor settings for minimizing the thrust force. The optimum input parameters can be obtained by choosing the appropriate levels from each process factor. Therefore, using MINITAB—statistical software, the polynomial coefficients are determined.

The coded and un-coded regression equations are given below,

$$\textbf{Thrust force (TF)} = +47.63 + 33.67 \times \text{A} + 9.33 \times \text{B} - 5.31 \times \text{C} - 3.71 \times \text{D} \tag{2}$$

$$\textbf{Thrust force (TF)} = -4.35278 + 0.11222 \times \text{Drilling speed} + 4.66250 \times \text{Feed rate} - 5.30833 \times \%\text{ of CNT} - 0.37117 \times \text{Point angle} \tag{3}$$

The obtained statistical analysis results show the best combination of input parameter for obtaining minimum thrust force. The obtained levels are as follows, A_1, B_1, C_2 and D_2. Further, to demonstrate the relationship between the feed rate and drilling speed on thrust force, 3D surface plot is introduced in Fig. 2a and c. It very well may be seen that the thrust force of the drill increases with increasing feed rate. Increasing feed rate subsequently increases the heat that raises more thrust force of the drilling. Hence, more delamination was found on the composite surfaces as appeared in Fig. 2a when increasing feed rate. And, it was observed that adding more quantity of CNT particle builds toughness therefore which improve more resistance on the shear plane of drilling. Thus, much thrust force was observed during the drilling while increasing % of CNT in the present examination as appeared in Fig. 2b. The effect of point angle and feed rate on thrust force is introduced in Fig. 2c. It is seen that large point angles of the drill tool produce more thrust force particularly when at high cutting speed. At high cutting speed, the interfacial temperature between the composite material and drill tool has observed more therefore which deform the materials rapidly that result more delamination was observed in the fabricated CFRP composite material as appeared in Fig. 3. Analysis of variance (ANOVA) is used to estimate the contribution of each factor involved in the present investigation. The obtained ANOVA results are introduced in Table 4. The contribution of each process factor is as follows, drilling speed = 7%, feed rate = 90%, percentage of CNT = 2% and point angle = 1%. From the results, it is recognized that the feed rate is assuming an independent factor that decides the quality of drilled hole surface of the fabricated composite material. The graphical results of obtained ANOVA are plotted in Fig. 4.

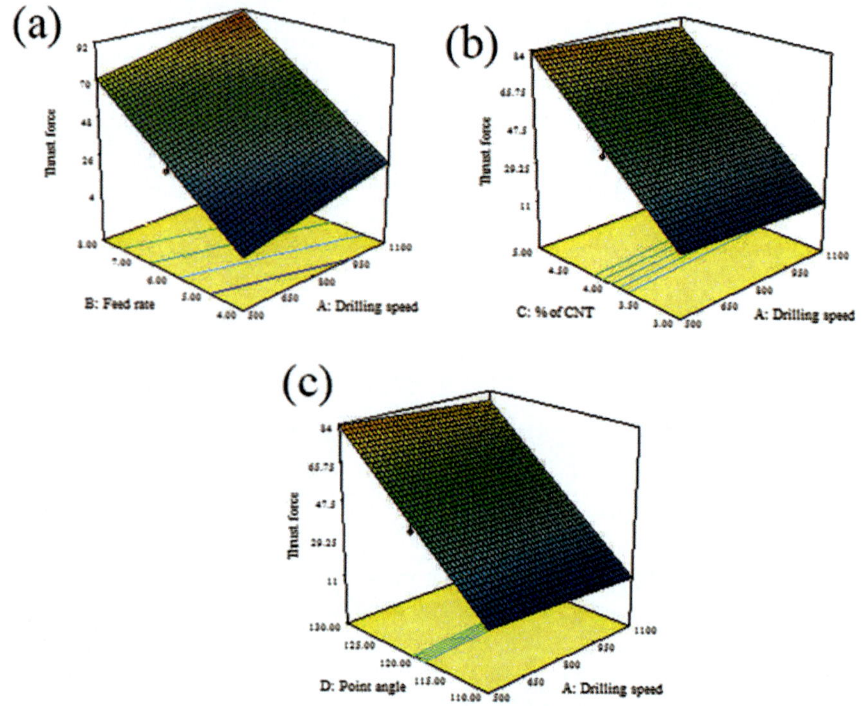

Fig. 2 3D Surface plots for drilling speed, feed rate, % of CNT and point angle on thrust force

Fig. 3 Delamination and Broken fibers in CFRP

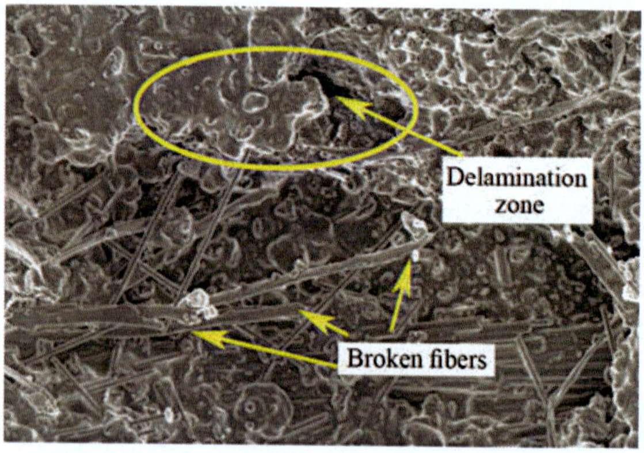

4 Conclusion

The present investigation proposed to minimize the thrust force of the drilling of the CFRP composite material using Taguchi method. Four significant parameters considered, namely drilling speed, feed rate, percentage of CNT and point angle.

Table 4 ANOVA results

Source	Sum of Squares	df	Mean Square	*F*-Value	*p*-value Prob > F	
Model	7574.13	4	1893.53	220.43	< 0.0001	Significant
Drilling speed (N)	521.73	1	521.73	60.74	0.0015	
Feed rate (*f*)	6800.67	1	6800.67	791.69	< 0.0001	
% of CNT (%)	169.07	1	169.07	19.68	0.0114	
Point angle (ө)	82.66	1	82.66	9.62	0.0361	
Residual	34.36	4	8.59			
Cor Total	7608.49	8	R-Sq = 0.9955	Aedq Precision = 36.929	Adj R-Sq = 0.9910	Pred R-Sq = 0.9626

Fig. 4 ANOVA contribution

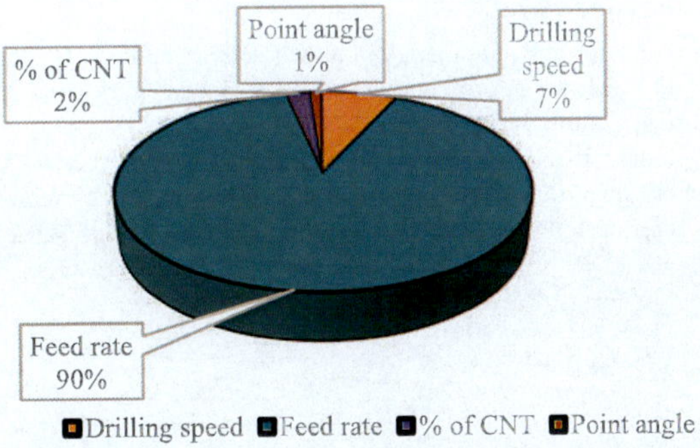

ANOVA is applied to estimate the contribution of the individual process factor. The result shows the feed rate is a dominant factor which decides the final quality of drilled hole and to minimize the thrust force of the drilling process. The mean of mean results are used to identify the optimal input factors. Moreover, 3D surface plots are used to demonstrate the relationship between the input and output parameter also to enhance the stability of the CFRP drilling process. Furthermore, it is observed that by attaining optimal drilling parameters can be improved the quality of drilling hole of the fabricated CFRP composite.

References

1. Ozcelik B, Ozbay A, Demirbas E (2010) Influence of injection parameters and mold materials on mechanical properties of ABS in plastic injection molding. Int Commun Heat Mass Transfer 37:1359–1365
2. Shunmugesh K, Panneerselvam K (2018) Multi-performance optimization of drilling carbon fiber reinforced polymer using Taguchi: membership function. Trans Indian Inst Met 71:1615–1627
3. Prakash S, Mercy JL, Salugu MK, Vineeth KSM (2015) Optimization of drilling characteristics using grey relational analysis (GRA) in medium density fiber board (MDF). Mater Today: Proc 2:1541–1551
4. Siddiquee AN, Khan ZA, Goel P, Kumar M, Agarwal G, Khan NZ (2014) Optimization of deep drilling process parameters of AISI 321 steel using Taguchi method. Procedia Mater Sci 6:1217–1225
5. Sahoo AK, Pradhan S (2013) Modeling and optimization of Al/SiCp MMC machining using Taguchi approach. Measurement 46:3064–3072
6. Gowda BMU, Ravindra HV, Gurupavan HR, Ugrasen G, Prakash GVN (2014) Optimization of process parameters in drilling Al-Si_3N_4 metal matrix composites material using Taguchi technique. Procedia Mater Sci 5:2207–2214
7. Tsao CC (2015) Multi–response optimisation of machining parameters during candlestick drilling composite material using membership function and Taguchi method. Int J Surf Sci Eng 9:189–200
8. Adalarasan R, Santhanakumar M (2015) Application of taguchi based response surface method (TRSM) for optimization of multi responses in drilling Al/SiC/Al_2O_3 hybrid composite, J Inst Eng (India) Series C 96:65–71
9. Sahu MK, Valarmathi A, Baskaran S, Anandakrishnan V, Pandey RK (2014) Multi-objective optimization of upsetting parameters of Al-TiC metal matrix composites: a grey Taguchi approach. Proc Inst Mech Eng Part B J Eng Manuf 228:1501–1507

An Experimental Study on Drilling of Titanium Alloy Using Taguchi-Based Fuzzy Logic Approach

D. Vijayan, T. Rajmohan, and A. Raajesh Kanna

Abstract In this study, a drilling experiment was conducted on titanium alloy material by varying point angle of a carbide drill bits. The variables in the experiment were drilling speed, feed rate and point angle. A hybrid optimization technique, Taguchi-based fuzzy logic approach, is used to organize the drilling experiment. Totally, 27 experiments were conducted as per Taguchi L27 orthogonal array. The adequacy of experiment is further validated using normal probability values and residuals involved in the experiment. Moreover, the analysis of variance is used to identify the significant variable involved in the experiment. A fuzzy logic approach is used to validate the experiment by developing rules logically in a fuzzy inference system. Three membership functions for the inputs and nine membership functions for the outputs were considered in order to improve the accuracy of the predicted results. It is observed that feed rate and point angle are playing a vital role in the conducted experiment. And the overall effectiveness of the drilling experiment has improved by using Taguchi-based fuzzy approach.

Keywords Titanium · Taguchi · Fuzzy · Drilling · Optimization · ANOVA

1 Introduction

Drilling is commonly utilized for assembling of mechanical and electrical parts [1, 2]. Drilling macro or miniaturized holes with a high perspective proportion in titanium combinations is beyond the abilities of the conventional drilling procedure. Since, titanium alloys (Ti–6Al–4V), an alpha–beta combination having chemical combination of 6% aluminum, 4% vanadium and remaining titanium, in such a way that, it is broadly utilized for the majority of the regular uses just as bio-medical applications due to bio-compatibility [3, 4]. Utilization of these materials has rapidly increased because of the fact that they have such qualities as a high strength to weight proportion at high temperatures, superior corrosion resistance, longer service

D. Vijayan (✉) · T. Rajmohan · A. Raajesh Kanna
Department of Mechanical Engineering, Sri Chandrasekharendra Saraswathi Viswa Mahavidyala, Enathur, Kanchipuram, Tamil Nadu 631561, India
e-mail: vijaiand2012@gmail.com

T. Rajmohan et al. (eds.), *Advances in Materials and Manufacturing Engineering*, Springer Proceedings in Materials 7, https://doi.org/10.1007/978-981-15-6267-9_55

life and compatibility with composite structures. Also, it is broadly utilized due predominant attributes in various technologically advanced industries, for example, aviation, marine, substance handling, restorative and vehicle areas [5]. There are numerous components, for example, airframes, fastener parts, vessels, cases, hubs, forgings, bone plates, bars, disposable ribs pens, finger and toe substitutions, spinal combination enclosures and dental inserts, cylinders and cylinder rings. Therefore, machining of titanium materials winds up important for getting net-shaped parts. In the drilling of titanium and its alloys, it becomes even more difficult because of their unfavorable properties, for example, poor thermal conductivity and lower elastic modulus. Another restriction of conventional drilling is the required tool material hardness, which ought to be no less than 30–50% more than that of work material. Drilling of such alloys some time poses difficulty due to the fact that the material is work hardened and the tool rubs against the hardened zone, causing rapid tool wear. Excessive wear and poor surface finish exist when the tool and work piece are in contact, if the tool has not held enough hardness [6]. Thus, researches have been to a great extent focused on mechanical and thermal properties of the titanium and its alloys. The majority of the items required drilled holed for assembly of parts. Nowadays, a great deal of research work was completed for decreasing the experimental work. In the past research, a considerable lot of the specialists utilized the experimentation, based trial and error [7, 8]. It requires a trial time and associated costs. And few scientists utilized the design of experiment ideas for developing experiments through factorial design [9, 10]. These designs were utilized to reduce the cost and time of analysis. However, this work attempts further decrease on the trial time and cost by utilizing fuzzy logic concepts.

2 Research Methodology

2.1 Experimental Setup

The present investigation begins with experimental approach and the experimental data are analyzed with fuzzy logic concepts and Taguchi method. At last, the modeling data are validated with experimental data. The drilling analyses were conducted in a radial drilling machine as shown in Fig. 1. The tools were used 8-mm TiAlN-coated carbide twist drill. Ti–6Al–4 V was utilized for this investigation. The sample work piece size of 120 × 120 mm was used. The experimental design follows three levels of spindle speed (1500, 1750, and 2000 rpm) and feed rate (4, 6, 8 mm/rev) and cone angle (110,115 and 120°). These cutting conditions were chosen dependent on tool manufacturer recommendation and machine tool capacity. The combinations between these two factors were created, and each experiment observed two replicate values. Altogether, 27 experiments were conducted as per Taguchi L_{27} orthogonal array. The experimental design is as presented in Table 1. The work piece was clamped in the special fixture to measure the force generated during drilling. The thrust force

Fig. 1 Drilling setup

mainly depends on feed rate, drill bit and the work piece thickness. The thrust force generated was measured by using a Kistler® digital dynamometer recorder for all 27 holes. The digital drill tool dynamometer is used to measure the thrust force (Fig. 1).

2.2 *Fuzzy Logic and Regression Modeling for Thrust Force Prediction*

Anand Alagumurthi [11] integrated the fuzzy rule-based reasoning into Taguchi loss function. The most effective combination of a parameter found from fuzzy–Taguchi integrated approach for water jet technology was planned. Similarly, this work assumes to combine the fuzzy logic and Taguchi to reduce the experimental run and cost for drilling operation. The experimental outputs were used to estimate the structure of input and output membership functions [12]. This modeling technique was carried out to calculate the various inputs and output combinations. The fuzzy logic model is built based on experimental results. The output data generated from fuzzy logic model are analyzed with Taguchi. Finally, the results of fuzzy model and fuzzy–Taguchi model are compared. Generally, the fuzzy logic procedure involves evaluation of fuzzy rule, universe partitioning, making relationship between input and output membership function and rules combination [6, 13].

Fuzzy logic provides foundations for approximate reasoning with imprecise fuzzy propositions using fuzzy set theory as a main rule [14]. Fuzzy logic could be used for modeling these potential threats more effectively. Fuzzy logic rules are developed in linguistic terms that address the relationship between the inputs and the outputs from a system. The data collected from experiments are ordered in a logic control system. In this experiment, the input parameters considered such as drilling speed, feed rate, cone angle and the output response is thrust force. The developed fuzzy model is as presented in Fig. 1. The input machining parameter is divided into three membership function called low, medium and high. The output is divided into nine sets as presented in Table 2. The modeling of the thrust force followed the triangular type from the available membership functions like Gaussian, trapezoidal, etc. The developed input and output membership function is presented in Figs. 2 and 3. The

Table 1 Experimental design

Std	Spindle speed (N)	Feed rate (f)	Cone angle (ϴ)	Thrust force (*TF*)	
				Actual	Predicted from fuzzy
1	1500	4	110	15.87	15.91
2	1500	4	110	14.28	13.98
3	1500	4	110	12.33	12.45
4	1500	6	115	11.90	11.92
5	1500	6	115	10.49	11.14
6	1500	6	115	9.67	9.21
7	1500	8	120	8.66	8.76
8	1500	8	120	7.57	7.98
9	1500	8	120	7.47	7.82
10	1750	4	115	11.90	12.47
11	1750	4	115	10.74	9.47
12	1750	4	115	9.27	9.66
13	1750	6	120	10.98	11.64
14	1750	6	120	9.27	8.41
15	1750	6	120	8.05	8.26
16	1750	8	110	10.74	9.47
17	1750	8	110	9.39	9.21
18	1750	8	110	8.90	7.47
19	2000	4	120	10.13	9.54
20	2000	4	120	8.15	8.45
21	2000	4	120	7.58	7.92
22	2000	6	110	9.89	10.45
23	2000	6	110	8.05	8.66
24	2000	6	110	7.78	7.91
25	2000	8	115	8.34	9.54
26	2000	8	115	7.32	9.54
27	2000	8	115	6.34	8.34

fuzzy rules note down in the form of if–then rules and connect the input membership function to output membership function. The rules are as follows (Figs. 4 and 5),

Rule 1 If spindle speed is low, then thrust force is high.

Rule 2 If spindle speed is medium, then thrust force is medium.

Rule 3 If spindle speed is high, then thrust force is low.

The input membership function is partitioned regarding minimum and maximum values that allowed controlling the system. Fuzzy relation builds the relation between

Table 2 Output membership function

Condition	Range	Membership function
Ultra small	−0.375–0.375	Triangular membership function
Very small	0–0.75	
Small	0.375–1.125	
Low medium	0.75 – 1.5	
Medium	1.125–1.185	
High medium	1.5–2.25	
Low	1.185–2.625	
Very low	2.25–3	
Ultra-low	2.63–3.38	

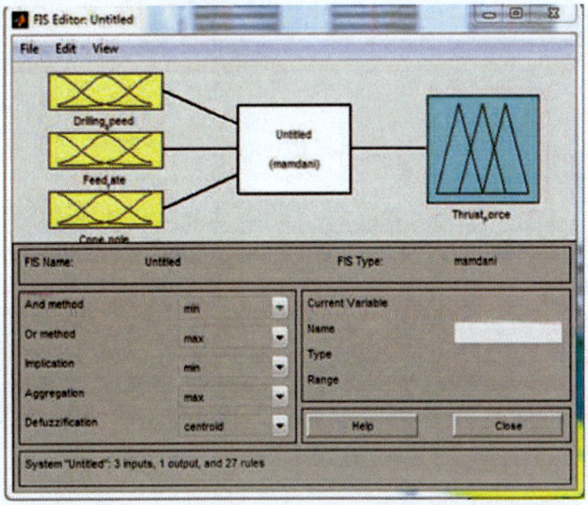

Fig. 2 Fuzzy inference

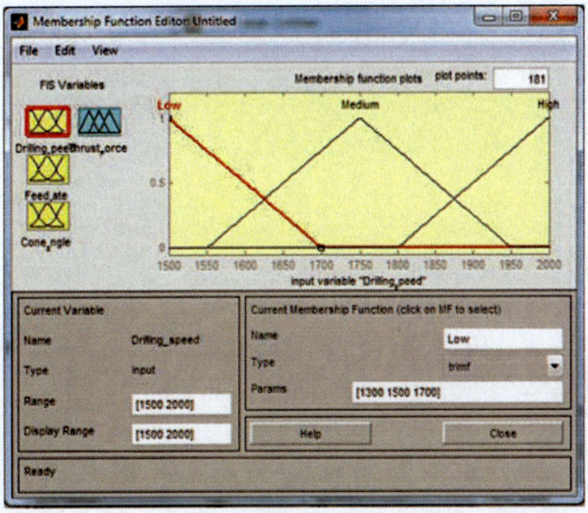

Fig. 3 Input membership function

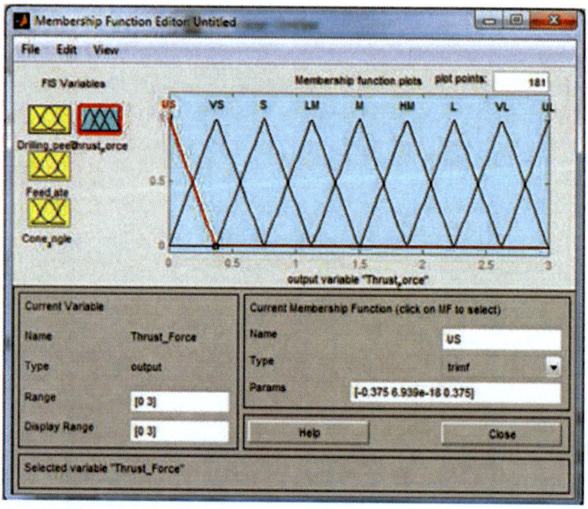

Fig. 4 output membership function

input membership function to output membership function. Then the fuzzy algorithm as follows,

If spindle speed = high THEN thrust force = low OR.

If spindle speed = medium THEN thrust force = medium OR.

If spindle speed = low THEN thrust force = high.

This can be represented in the relation with thrust force (*TF*) which has a membership function as shown in below,

$$N,\ \mathrm{TF} = \min\ N_1\,\mathrm{TF},\ N_2\,\mathrm{TF},\ N_3\,\mathrm{TF}$$

where.

N = Mean drilling speed.

TF = Mean Thrust force.

N_1, N_2 and N_3 = Drilling speed at desired levels.

TF_1, TF_2 and TF_3—Thrust force based on desired levels of drilling speed. Therefore totally 27 fuzzy.

rules were formed based on IF: Then rules in FIS system as shown in Fig. 5.

3 Results and Discussion

Minitab v16 is used to perform the Taguchi statistical analysis and to experimental data is validated using ANOVA. The residual plot obtained from Taguchi analysis is presented in Fig. 6. From the normal probability plot, it can be observed that all the experimental data are nearer to 45° linear line which indicate adequate model is obtained from the statistical analysis [15]. Further, the residuals are scattered in the residual plot which the residual line throughout the experiment. And the main effects

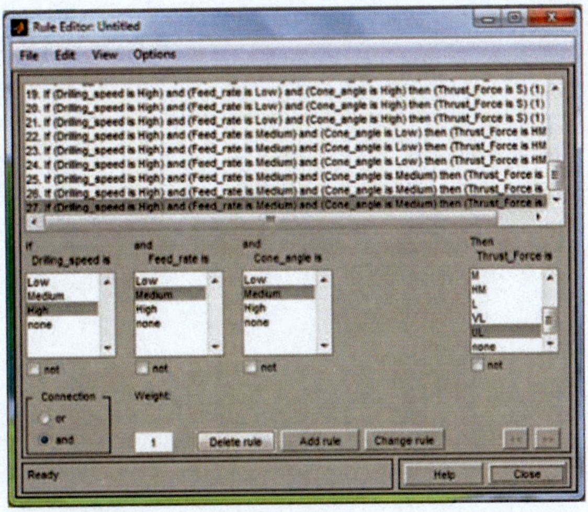

Fig. 5 Rule editor

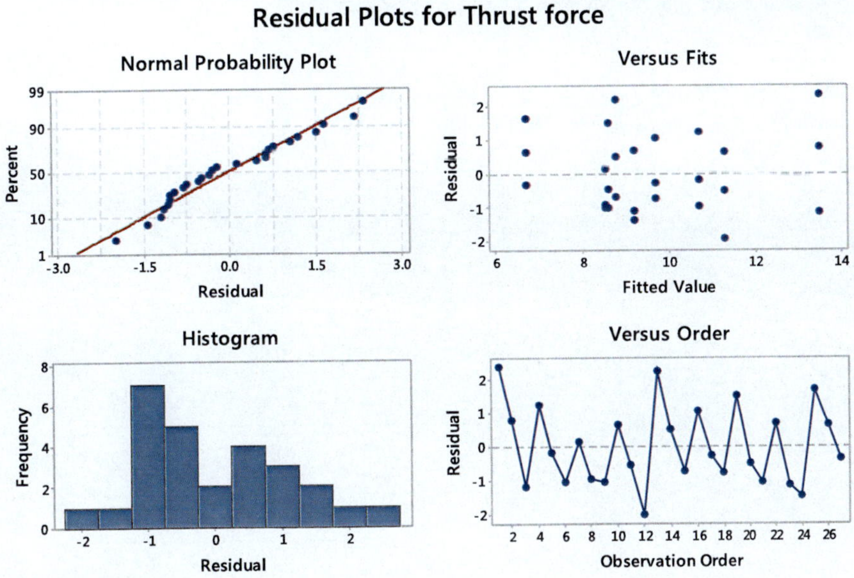

Fig. 6 Residual plot for thrust force

plot from SN ratio is plotted in Fig. 7. The steep slope of SN ratios graph indicates the more influence of drilling parameters in the performance characteristics. In the present investigation, the effect of feed rate has greater influence as can be seen from their steep slope. Moreover, the experiment is validated using ANOVA. The obtained ANOVA values are presented in Table 3. It is observed that the model is significant and all the selected process parameters are significant and playing an important role

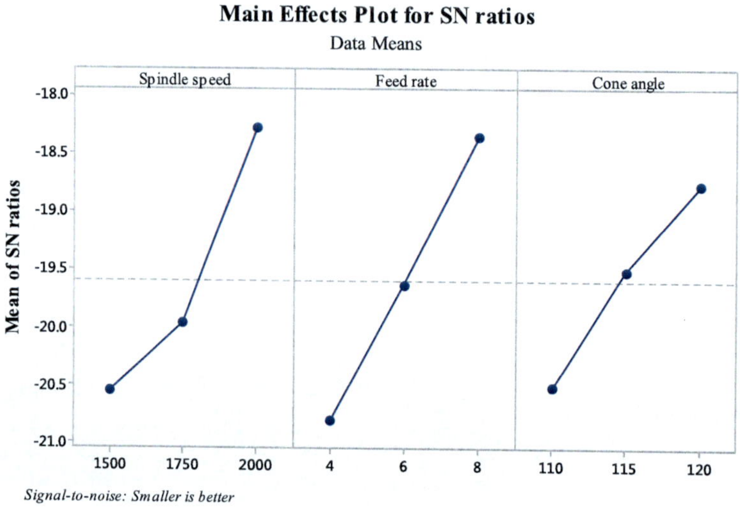

Fig. 7 Main effects plots for SN ratios

Table 3 ANOVA results

Source	Sum of squares	df	Mean square	*F* value	*p*-value Prob > *F*	
Model	99.012	3	33.004	23.008	< 0.0001	Significant
A-Spindle speed	28.506	1	28.506	19.872	0.0002	
B-Feed rate	51.731	1	51.731	36.063	< 0.0001	
C-Cone angle	18.774	1	18.774	13.088	0.0014	
Residual	32.993	23	1.434			
Lack of fit	13.697	5	2.739	2.555	0.0647	Not significant
Pure error	19.296	18	1.072			
Cor total	132.005	26				

in minimizing the thrust force of the drilling process. MATLAB tool is used for estimating the Taguchi-fuzzy output. Triangular membership function is applied for the output response (thrust force) with nine membership functions. A set of rules were written for activating the fuzzy inference system (FIS) and the FIS is evaluated to predict the Taguchi fuzzy reasoning grades for all 27 experiments. Based on the predicted data from Taguchi method and fuzzy is presented in Table 1. It is observed from the results the uncertainty of data is reduced. Figure 8 shows the comparison of predicted results of Taguchi methods and the respective fuzzy for all the experiments.

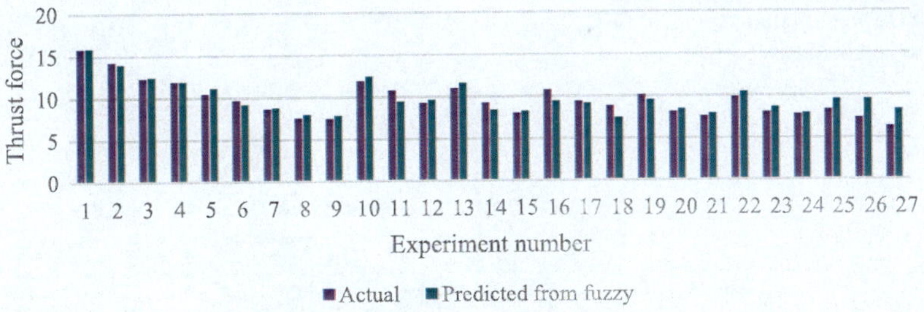

Fig. 8 Comparison of predicted Taguchi and fuzzy results

Fig. 9 Effect of feed rate, speed on thrust force

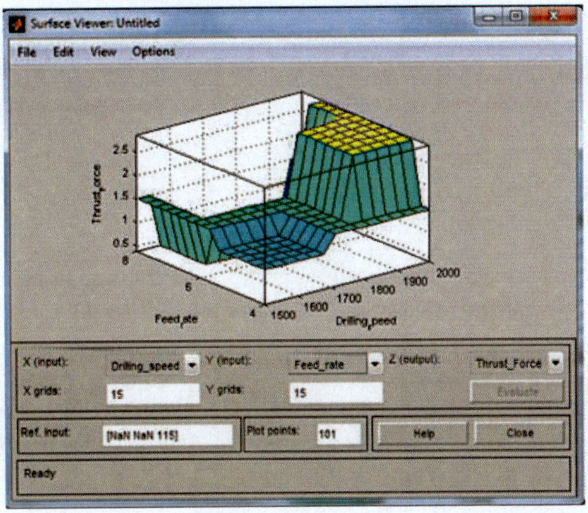

It is obvious that there is an improvement in the fuzzy predicted data as compared to that of Taguchi results.

The effect of feed rate and point angle on thrust force is presented in Figs. 9 and 10. It seems that the thrust force is increases with increasing feed rate. Increase in feed rate increases the heat generation during drilling therefore which increases thrust force during drilling process. This increase in thrust force creates shear bands on the drilling surfaces to a larger extent resulting in larger burr formation. In addition to that the feed rate is directly proportional with thrust force and inversely proportional with drilling speed. Hence, maximum thrust force was observed with increasing feed rate during drilling. On the other hand, varying point angle from 110 to 12 °C induced more strain deformation so thus produced a primary adiabatic shear zone on the drilling area results increase in thrust force during drilling.

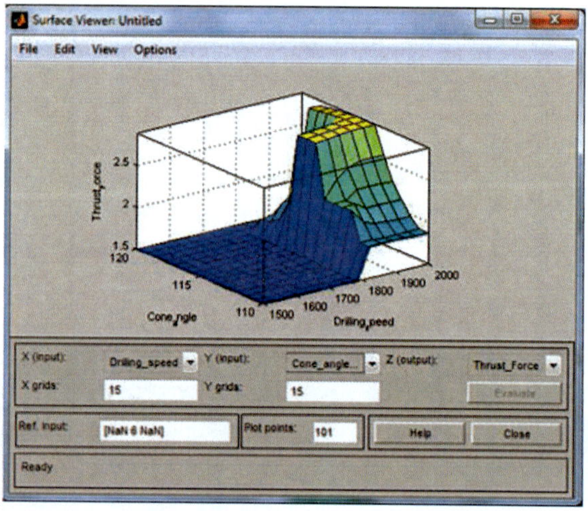

Fig. 10 Effect of point angle, speed on thrust force

4 Conclusion

In the present investigation, a hybrid optimization technique, Taguchi-based fuzzy logic approach, was used to analyze the drilling parameters of titanium alloys. Totally, 27 experiments were conducted using Taguchi *L*27 orthogonal array. The important drilling parameters such as drilling speed, feed rate and point angle were used to organize the experiment and thrust force was considered as a response. ANOVA is used to validate the experiment. Based on the investigation conducted, the following conclusions can be drawn,

(a) Feed rate is the drilling parameters which is playing a vital role in the experiment. Thrust force is observed maximum when feed rate is high in the drilling process. There is no much variation in thrust force when drilling speed is even at maximum or minimum.
(b) Increasing point angle increase thrust force during drilling which develops cracks and adiabatic shear zone on the drilling areas.
(c) Taguchi-based fuzzy is an effective approach on improving effectiveness of the drilling performance and it can be further improved by using a greater number of variables and experiments. Moreover, the application of this model to the optimization of slightly different industrial problems is envisaged.

References

1. Liu Z, Liu Y, Han X, Zheng W (2018) Study on super-long deep-hole drilling of titanium alloy. J Appl Biomater Funct Mater 16:150–156

2. Vijayan D, Abhishek P, Manoj Kumar YG, Balaji P, Siva kumar Reddy P (2018) Optimization of drilling parameters of carbon fiber composites using RSM based desirability function. In: IOP conference series: materials science and engineering, 390
3. Beranoagirre A, Urbikain G, Marticorena R, Bustillo A, López de Lacalle L (2019) Sensitivity analysis of tool wear in drilling of titanium aluminides. Metals 9
4. Beranoagirre A, Urbikain G, Calleja A, Lopez de Lacalle LN (2018) Drilling process in gamma-TiAl intermetallic alloys. Materials (Basel) 11
5. Samuel Raj D, Karunamoorthy L (2019) Performance of cryogenically treated WC drill using tool wear measurements on the cutting edge and hole surface topography when drilling CFRP. Int J Refract Metals Hard Mater 78:32–44
6. Kumar BS, Baskar N (2012) Integration of fuzzy logic with response surface methodology for thrust force and surface roughness modelling of drilling on titanium alloy. Int J Adv Manuf Technol 65:1501–1514
7. Vijayan D, Seshagiri Rao V (2016) Parametric optimization of friction stir welding process of age hardenable aluminum alloys—ANFIS modeling. J Central South Univ 23:1847–1857
8. Goyal R, Dubey AK (2016) Modeling and optimization of geometrical characteristics in laser trepan drilling of titanium alloy. J Mech Sci Technol 30:1281–1293
9. Vijayan D, Rao VS (2014) Friction stir welding of age-hardenable aluminum alloys: a parametric approach using RSM based GRA coupled with PCA. J Inst Eng (India) Series C 95:127–141
10. Vijayan D, Rao VS (2015) Parametric optimization of age hardenable aluminum alloys using TGRA coupled with PCA. Appl Mech Mater 813–814:613–619
11. Anand G, Alagumurthi N, Elansezhian R, Palanikumar K, Venkateshwaran N (2018) Investigation of drilling parameters on hybrid polymer composites using grey relational analysis, regression, fuzzy logic, and ANN models. J Brazilian Soc Mech Sci Eng 40
12. Abhishek K, Datta S, Mahapatra SS (2016) Multi-objective optimization in drilling of CFRP (polyester) composites: Application of a fuzzy embedded harmony search (HS) algorithm. Measurement 77:222–239
13. Rajmohan T, Palanikumar K, Prakash S (2013) Grey-fuzzy algorithm to optimise machining parameters in drilling of hybrid metal matrix composites. Compos B Eng 50:297–308
14. Vijayan D, Rao VS (2017) Optimization of friction stir welding process parameters using RSM based Grey—fuzzy approach. Saudi J Eng Technol 2:12–25
15. Vijayan D, Rao VS (2014) A Multi Response Optimization of Tool Pin Profile on the Tensile Behavior of Age-hardenable Aluminum Alloys during Friction Stir Welding. Res J Appl Sci Eng Technol 7:4503–4518

Determination of Optimum Tensile Strength of Friction Stir Welded AA2219 Aluminum Alloys Using Taguchi's Method

D. Vijayan, V. Seshagiri Rao, and V. S. Anirudh

Abstract The heat-treated aluminum alloys of AA2219 have superior strength-to-weight ratio; therefore, the industries such as aerospace, automobile, and marine industries are widely fabricating their structures using conventional welding techniques. Joining AA2219 aluminum alloys by conventional welding techniques is extremely difficult due to formation of hot cracks and residual stresses on the weld zone. To overcome these difficulties, AA2219 can be welded friction stir welding (FSW). Therefore, in the present investigation, FS welding technique is optimized to join AA2219 aluminum alloys. Significant FS process parameters such as tool rotational speed, welding speed, axial force, and tool pin profile were used to conduct the experiment. Four factors and three levels of Taguchi L_9 orthogonal array are used to design the experiment. Totally, nine experiments were conducted to validate the weld experiments. The fabricated weld samples yield the maximum tensile strength of 316 MPa. The optimal conditions were identified such as tool rotational speed of 1700 rpm, welding speed of 90 mm/min, axial force of 3 kN, and square tool pin profile is best suited for obtaining a maximum tensile strength. ANOVA results indicate that the tool rotational speed and square tool pin profile is playing a vital role in the conducted experiment. Scanning electron microscope is used to reveal the effect on stirred zone of the fabricated weld joint. It was observed that few defects, such as pinholes, and voids have mainly reduced the tensile strength of the fabricated AA2219 weld joints.

Keywords Friction stir welding · Aluminum · Alloys Taguchi · Tensile strength · Optimization

---

D. Vijayan (✉) · V. S. Anirudh
Sri Chandrasekharendra Saraswathi Viswa Mahavidyalaya University, Enathur, Kanchipuram, Tamil Nadu 631561, India
e-mail: vijaiand2012@gmail.com

V. Seshagiri Rao
St.Joseph's College of Engineering, Jeppiaar Nagar, Chennai, Tamil Nadu 600119, India
e-mail: raosvaddi@gmail.com

T. Rajmohan et al. (eds.), *Advances in Materials and Manufacturing Engineering*, Springer Proceedings in Materials 7, https://doi.org/10.1007/978-981-15-6267-9_56

1 Introduction

Friction stir welding (FSW) is a novel solid-state welding technique invented by Thomas [1]. At The Welding Institute (TWI), UK, FSW is mainly used for aluminum and its alloys and is gradually replacing traditional aluminum arc welding process nowadays. FSW has not deliver any radiation and harmful gases as similar in traditional welding techniques. A rotating tool having a unique design parts called "shoulder" and "pin" directly plunged into an abutting surface to be welded. Due to the frictional contact between the shoulder and surface, the heat is generated and allows metal flow usually from retreating side to advancing side by tool travel along the weld line thus producing a weld joint. The heat-treated aluminum AA2219 has superior strength-to-weight ratio and best suited to aerospace structural and cryogenic fuel tank applications. The tensile property of the fabricated weld joints is depending upon the following process parameters such as tool rotational speed, welding speed, axial load, tool tilt angle, shoulder-to-diameter ratio, shoulder design, or pin design. However, these parameters are to be optimized for maximizing the tensile strength of weld joint. Moreover, positioning of weld plate is playing significant role in determining the strength of the weld joint. Numerous researches have been carried out to investigate and improve the process performance.

Manurung [2] developed a model for improving the mechanical properties of friction stir welded AA6061—T651 aluminum alloys using the Taguchi method against the response weld quality, hardness, and tensile strength. They found that tool pin profile and shoulder diameter are the factors that influence more on maximizing the tensile strength. Salem [3] developed a mathematical model for surface roughness of friction stir welded AA2017 aluminum alloys. It was reported that increase in rotational speed improves surface roughness whereas decreases while increasing traverse speed. Hussein [4]. used Taguchi's technique to optimize the friction stir welding of AA2024 and AA5754 aluminum alloys and found that tool pin profile influencing more on improving the tensile strength than the tool rotational speed and plunging time. Parida [5] studied the influence of input parameters of FS welded commercial aluminum alloys using Taguchi-based full fractional experiment. It was reported that the rotational speed is a significant factor contributed much on improving ultimate tensile strength of the fabricated FS welded specimens.

The dissimilar FSW was performed by Sundareswaran [6] using AA6061-to-AA7050 aluminum alloys and found that the tool shoulder diameter and pin diameter are the dominant factors of the process. From the above investigations, it can be found that the FSW process parameters are quite complex welding process, and therefore, it is difficult to find the optimal parameters by conducting the numerous trials. Sometimes conducting numerous trails delays the production and consumes more production cost. Alternatively, optimal parameters can be evaluated using statistical techniques like Taguchi, grey relational analysis, response surface methodology, etc. Usually, statistical techniques require less resources and time to estimate the optimal parameters even though if the process is too complex. Hence, in the present investigation, a statistical technique called Taguchi's method is used. In Taguchi's method,

L_9 orthogonal array was used to obtain the optimal parameters of FSW. Taguchi's technique developed by Genichi Taguchi has been used widely in engineering problems to optimize the performance characteristics and material properties within the combination of design parameters. Taguchi's technique is a potent tool in finding the optimal parameters especially if the problem is complex.

2 Materials and Methods

Aluminum alloys of AA2219 plate having 150 mm × 75 × 6.25 mm each used to fabricate the FS welded samples. AA2219 material is basically the combination Al-Cu-Mn based alloy which is very difficult to join by conventional welding techniques. The chemical composition and mechanical properties of AA2219 aluminum alloys are presented in Tables 1 and 2. A computerized friction stir welding is used to fabricate the FS welded joints. Later, the finishing operation was done on the cut pieces with emery paper up to 2000 grit. Each sample of AA2219 aluminum alloys was fixed simultaneously, and then the welding operation was initiated ensuring that the pin depth in the weld joint was about 4 mm. The FSW machine setup is as presented in Fig. 1. Friction during rotation created heat flow in the workpieces and turned the proposed area for welding into plastic state. There was no melting during this process; therefore, no protection gas was required during welding. The significant FSW parameters are involved in the present investigation, and their corresponding levels are presented in Table 3. There are three pin profiles that are considered in the present investigation, namely cylinder (1), square, (2) and taper (3). For the statistical analysis, each pin is indicated by the numeral numbers as stated in Table 3. As per the Taguchi L_9 orthogonal array, totally nine samples were fabricated. Each sample is visually inspected and machined as per ASTM E-8M standard using WEDM process. Later, all the fabricated welded samples were allowed for tensile test to evaluate ultimate tensile strength (UTS).

A 100-tonnage automatic universal testing machine (UTM) is used for finding UTS value in each sample. Each sample was manually clamped in mechanical clamps which are provided in the UTM machine. Since to confirm good quality

Table 1 Chemical composition of AA2219-T81 aluminum alloy

Component	Cu	Mn	Ti	V	Zr	Al
wt%	5.8–6.8	0.2–0.4	0.02–0.1	0.05–0.15	0.1–0.25	Remaining

Table 2 Mechanical properties of AA2219–T81 aluminum alloy

Material	Ultimate tensile strength	Tensile yield strength	Modular of elasticity
AA2219–T81	476 MPa	393 MPa	73.1 GPa

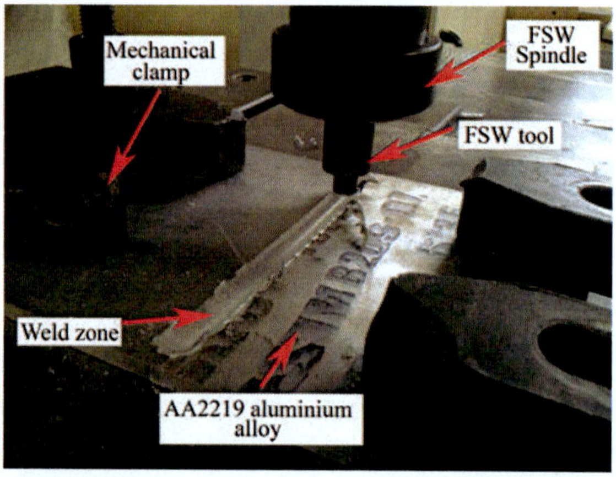

Fig. 1 FSW machine setup

Table 3 FSW parameters range and their levels

FSW parameter	Units	Range
Tool rotational speed	rpm	1300–1700
Welding speed	mm/min	30–70
Axial load	kN	3–9
Tool pin profile	Cat	1–3

weld, each sample was thoroughly observed during the tensile operation till the fracture occurred. Finally, all the samples were visually inspected to analyze the macrodefects in the fractured weld zones. The fabricated weld samples of AA2219 are presented in Fig. 2, and the obtained UTS values from each weld sample are presented in Table 4.

The scanning electron microscopy (SEM) is used to perform the microanalysis of each weld sample. It was observed that each fabricated weld sample has few voids, fine dimples, and pinholes along the weld zone. The identified defects through SEM are presented in Fig. 3. The fine dimples along the weld region indicate all the fabricated samples fractured under ductile mode of fracture that confirms high tensile strength on the weld joints.

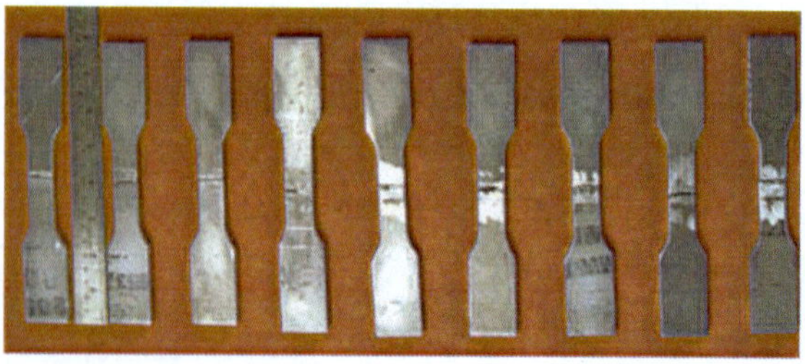

Fig. 2 Fabricated FSW samples of AA2219 aluminum alloys

Table 4 Taguchi L_9 orthogonal array and experimental results

Std	Tool rotational speed, N (rpm)	Welding speed, *f* (mm/min)	Axial force, *F* (kN)	Tool pin profile, PP	Ultimate tensile strength (N/mm^2)
1	1500	30	3	1	178
2	1500	50	6	2	316
3	1500	70	9	3	268
4	1700	30	6	3	295
5	1700	50	9	1	165
6	1700	70	3	2	132
7	1900	30	9	2	224
8	1900	50	3	3	189
9	1900	70	6	1	134

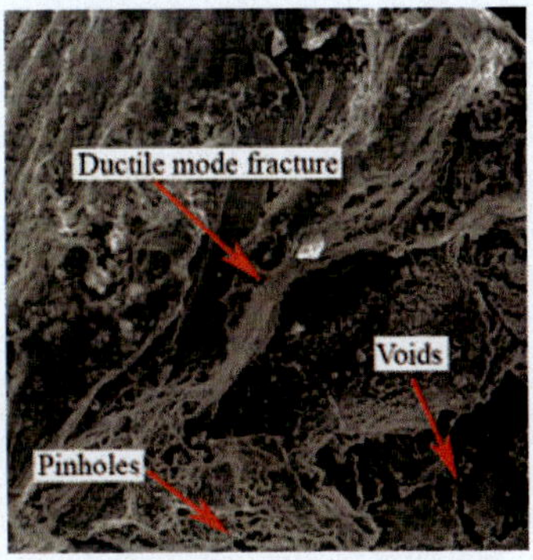

Fig. 3 SEM indicates fractured zone of AA2219 aluminum alloy

3 Results and Discussion

When the speed of rotation of the tool was below 1500 rpm, the tunnel defect was observed in the middle of the retreating side of the welding area, which may be due to insufficient heat generation and insufficient metal transport as shown in Fig. 4a. If the speed of the tool rotation was higher than 1900 rpm, the pinhole defect at the center of the retreating side may be due to excess turbulence caused by a higher speed of the tool rotation. When the welding speed was lower than 30 mm/min, the tunnel defect on the retreating side was observed due to the excess heat input per unit length of the weld. When the welding speed was higher than 70 mm/min, the tunnel at the retreating side and center of the welding area was observed due to inadequate material flow. And, when the axial force was lower than 3 kN, the pinhole defect was observed as shown in Fig. 4b. On the other hand, when axial force increased beyond 3 kN, tunnel defects on both sides of the retreat and progression and excessive thinning were caused by higher heat input on the fabricated AA2219 aluminum weld joints.

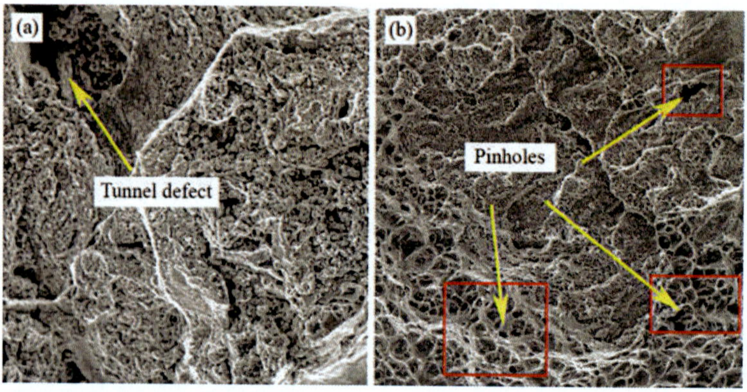

Fig. 4 **a** Tunnel defect. **b** Pinholes

In order to analyze the influence of each factors involved in the process, mean of means and signal-to-noise ratios (S/N) for each process factor was calculated.

Signals are usually indicating the effect on mean responses, and noises are measures of deviations from experiment output. Suitable S/N ratio must be selected based on prerequisite knowledge and experience in the process. For the present investigation, S/N ratio was selected according to criterion, larger-the-better, in order to maximize the tensile strength. S/N ratio (η_j) in jth experiment can be expressed as,

$$\eta_j = -10\log\left(\frac{1}{n}\sum\left(\frac{1}{Y_{ijk}^2}\right)\right)$$

where n is number of tests and Y_{ijk} is experimental value of ith quality characteristics in jth experiment at kth test. In the present investigation, UTS values are to be maximized, and therefore, the UTS values are transformed into SN ratio and means. The obtained means and SN ratios are presented in Table 5. Usually, a larger SN ratio indicates better quality attributes. Consequently, optimum parameter is the highest level in the SN ratio. Means and SN ratios are calculated using MINITAB

Table 5 SN ratio and means of each factor

Level	Tool rotational speed		Welding speed		Axial force		Tool pin profile	
	SN ratio	Means	SN ratio	Means	SN ratio	Means	SN ratio	Means
1	47.85	254.0	47.14	232.3	44.32	166.3	43.97	159.0
2	45.39	197.3	46.62	223.3	47.31	248.3	46.47	224.0
3	45.03	182.3	44.51	178.0	46.64	219.0	47.83	250.7
Delta	2.83	71.7	2.63	54.3	2.99	82.0	3.86	91.7
Rank	3		4		2		1	

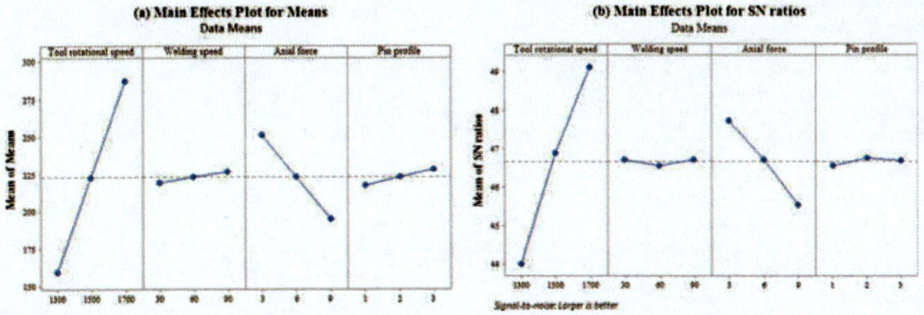

Fig. 5 Graphical results of **a** means and **b** SN ratios

Fig. 6 Percentage contribution of each process factor

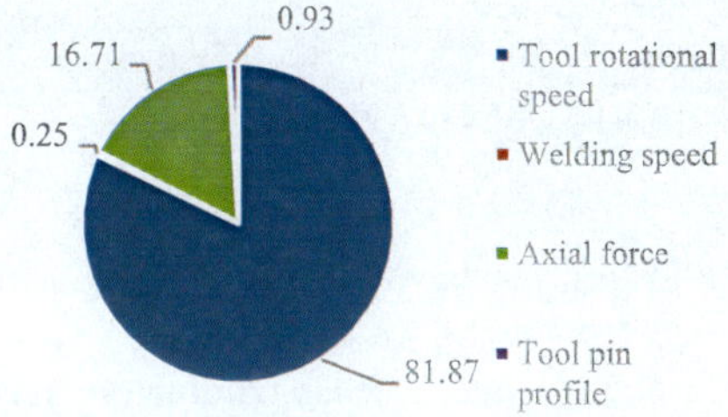

V16—statistical software. The obtained optimum values of means and SN ratios are as follows: tool rotational speed = 1900 rpm welding speed = 90 mm/min, axial force = 9 kN, and tool pin profile = 3 (Square). The graphical results of obtained means and SN ratios are presented in Fig. 5a, b. Analysis of variance (ANOVA) is performed to estimate the contribution of individual factor involved in the investigation. ANOVA results indicate that the selected FSW factors are highly significant factors affecting UTS of FS Weld joints and the effects of interaction between process factors are not significant. The graphical representation of percentage of contribution of each process factor is presented in Fig. 6. It is found that, the rotational speed plays a vital role in the conducted FS welding process. The tool rotational speed contributed 81.87% in the whole process which significantly decides ultimate tensile strength of the fabricated AA2219 aluminum alloys followed by 16.71% of axial force on the weld joint performance (Table 6).

4 Conclusion

The following conclusions can be made based on the present investigation on FSW of AA2219 aluminum alloys.

1. AA2219 aluminum alloys were successfully FS welded with different parameters using Taguchi's optimization techniques.

Table 6 ANOVA results

Source	Sum of squares	df	Mean square	F value	$p$value Prob. > F	% of Cont.	
Model	31,396.8	4	7849.21	734.91	<0.0001		Significant
A-tool rotational speed	25,741.50	1	25,741.50	2410.13	<0.0001	81.87	
B-welding speed	80.67	1	80.67	7.55	0.0515	0.25	
C-axial force	5280.67	1	5280.67	494.42	<0.0001	16.71	
D-pin profile	294.00	1	294.00	27.53	0.0063	0.93	
Residual	42.72	4	10.68				
Cor total	31,439.56	8					

2. Aluminum having advancing side of the FSW resulted in better welds with fewer defects in the stir zone.
3. The optimal process parameters such as tool rotational speed = 1900 RPM, welding speed = 90 mm/min, axial force = 9 kN, and tool pin profile = 3 (Square) have been found.
4. Ductile fracture was observed on the fabricated weld specimens under tensile loading during tensile test. The maximum tensile strength of joints welded was 316 MPa, which was higher than the other fabricated in the present investigation using AA2219 aluminum alloys.

References

1. Thomas W, Nicholas E, Needham JC, Murch M, Templesmith P, Dawes C (1991) Friction stir welding. In: International patent application no. PCT/GB92102203 and Great Britain patent application
2. Mohamed MA, Manurung YHP, Berhan MN (2015) Model development for mechanical properties and weld quality class of friction stir welding using multi-objective Taguchi method and response surface methodology. J Mech Sci Technol 29:2323–2331
3. Boulahem K, Ben Salem S, Bessrour J (2015) Surface roughness model and parametric welding optimization in friction stir welded AA2017 Using Taguchi method and response surface methodology. In: Design and modeling of mechanical systems-II, lecture notes in mechanical engineering
4. Abbass MK, Hussein SK, Khudhair AA (2016) Optimization of mechanical properties of friction stir spot welded joints for dissimilar Aluminum Alloys (AA2024-T3 and AA 5754–H114). Arabian J Sci Eng 41:4563–4572
5. Alkayem NF, Parida B, Pal S (2016) Optimization of friction stir welding process parameters using soft computing techniques. Soft Comput 21:7083–7098

6. Koilraj M, Sundareswaran V, Vijayan S, Koteswara Rao SR (2012) Friction stir welding of dissimilar aluminum alloys AA2219 to AA5083—optimization of process parameters using Taguchi technique. Mater Des 42: 1–7

Analysis of Thrust Force in Drilling of Titanium Alloy Using Taguchi's Method

D. Vijayan, T. Rajmohan, and V. Nithin

Abstract Achieving good quality of hole by drilling is extremely difficult in hard-to-difficult cut material such as titanium alloys and nickel alloys. Hence, numerous investigations have been organized using these alloys by the researchers in the past decades. Thrust force is the main factor associated in drilling that affects the final quality of the hole. Minimizing thrust force in drilling is still questionable and extremely difficult. Therefore, considering as an objective function of the present investigation, thrust force is significant factor to be minimized. Based on the thorough literature survey, it is identified that spindle speed, feed rate, and point angle are the significant factors inducing thrust force during drilling. Identifying appropriate levels of each input factors can minimize the thrust force. Therefore, Taguchi's method is applied to identify the optimal input process factors to minimize the thrust force during drilling of titanium alloys. The results show increasing feed rate subsequently increases the thrust force of the drilling. However, from the analysis results, it is observed that, the point angle of the drill tool is factor advances thrust force during drilling when comparing feed rate.

Keywords Titanium · Taguchi · Fuzzy · Drilling · Optimization · ANOVA

1 Introduction

Drilling is the dominant machining technique in metals and alloys assemblies due to its clean operation, excessive efficiency, and low price. Titanium and its alloys have high melting points and high heat resistance, making them suitable for applications involving high operating temperatures. Titanium and its compounds are highly resistant to chemical corrosion using conventional inorganic acids and chlorides. In addition, pure titanium and its alloys have the advantage of being biocompatible with human tissues and bones [1, 2]. Due to these advantages, titanium and its alloys are widely used in aviation, construction, shipbuilding, petrochemical, energy, and

D. Vijayan (✉) · T. Rajmohan · V. Nithin
Department of Mechanical Engineering, Sri Chandrasekharendra Saraswathi Viswa Mahavidyala, Enathur, Kanchipuram, Tamil Nadu 631561, India
e-mail: vijaiand2012@gmail.com

T. Rajmohan et al. (eds.), *Advances in Materials and Manufacturing Engineering*, Springer Proceedings in Materials 7, https://doi.org/10.1007/978-981-15-6267-9_57

biomedicine [3, 4]. Especially in aviation sector, titanium alloys are used for the manufacturing of fan, compressor, landing gears, and fuselage applications due to its high strength-to-weight ratio. Besides that, due to non-corrosive properties in nature, titanium alloys are mainly used for on-board kitchen and toilets in construction fields. And the galvanic capability of titanium makes it to combine with carbon fibers during the composite fabrication. Titanium alloys have low coefficient thermal expansion; therefore, it can be combined with polymer matrix composites [5, 6]. And, titanium is also used in the manufacturing of sea-cooled heat exchangers and offshore oil rig frames due to their corrosion resistance and fire resistance properties [7]. In addition, geothermal power plants for the handling of harsh equipment and highly corrosive contaminated fluids. Thus, titanium alloys are gaining much popularity in energy sectors to handle high temperatures and hot chloride applications [8–10]. Though many applications reveal the application of titanium alloys in various industrial sectors such as automobile, aviation, energy, marine, and space research, the cost of titanium alloys is quite expensive still [11, 12]. Therefore, this limits the extension of titanium alloys application in other industrial sectors. In addition to that, specifically hole making within very closer tolerances is critical step while fabricating titanium machined parts during post-processing is very difficult and sometimes takes more time to complete the fabrication. Drilling is usually a difficult undertaking for the manufacturing industries particularly while the use of titanium alloys because of its low thermal conductivity, which leads to high heat accumulation at the machining area [9, 10, 13]. Thus, extensive investigations were centered at the traditional drilling methods. Since, there may be an increasing demand for fabrication of miniaturized gadgets and additives of titanium alloys by way of drilling, wherein plated smaller than 10 mm are fabricated via drilling technique. Titanium alloy is receiving increasingly more interest in chemical, aerospace, and biomedical fields due to its low density, good mechanical properties, and excellent corrosion resistance. However, drilling of titanium alloy has to be taken carefully. Chiang [14] revealed that titanium alloy reacted actively with nitrogen and oxygen at excessive temperature. As a traditional drilling approach, drilling should isolate atmospheric gases successfully during high feed rate implementation. Mechanical properties of the titanium after drilling need to be considered for a comprehensive estimation. Optimization method is a consequence vital for a better composite hole quality. Taguchi's method becomes advanced applying signal-to-noise ratio and orthogonal array for optimization of single response. Analysis of variance (ANOVA) and response table are used to optimize the technique.

2 Experimental Setup

Commercially pure titanium grade II materials having dimensions of 120 × 120 x 10 mm are used for the experiment. Titanium alloy grade II possesses good weldability, strength, ductility, and formability for which it is quite difficult to machine. They were sheared into the dimensions of 120 × 120 mm for the drilling experiments, as shown in Fig. 1. Detailed chemical composition of titanium alloy is listed in Table 1. Mechanical and chemical methods had been both beneficial in cleaning impurities earlier than drilling. Mechanical cleansing is initially followed by way of a hard brush for a rough elimination of coarse oxides and contaminants. Chemical cleansing was then used to obtain a stable and good quality of surface plate. Titanium alloys were etched with a blended solution of nitric acid (45%), hydrofluoric acid (20%), and water (35%) for 2–3 min. After wiped clean by way of running water, specimens were positioned in a ventilated environment. Taguchi L_{27} orthogonal array is the most popular experimental design was adopted in this study [15, 16]. The process factors and their levels are depicted in Table 2. Experimental design matrix and their corresponding results as per the Taguchi L_{27} orthogonal array are given in Table 3. The experimental design follows three levels of spindle speed (1500, 1750, and 2000 rpm) and feed rate (4, 6, 8 mm/rev) and cone angle (110, 115 and 120°). These cutting conditions have been selected primarily based on tool manufacturer advice and machine device capability.

In total, 27 experiments were conducted. The workpiece became clamped within the unique fixture to measure the force generated in the course of drilling. The thrust force specially relies upon on spindle speed, feed rate, drill bit, and the workpiece thickness. The thrust force generated was measured by the use of a digital dynamometer recorder for all 27 holes. The digital drill device dynamometer is used to measure the thrust force. After the data for every experiment are summarized, the subsequent step in data analysis has been (1) to develop the mathematical model, (2) to estimate the effect of every control component at various levels on thrust force under study, and (3) to perform ANOVA.

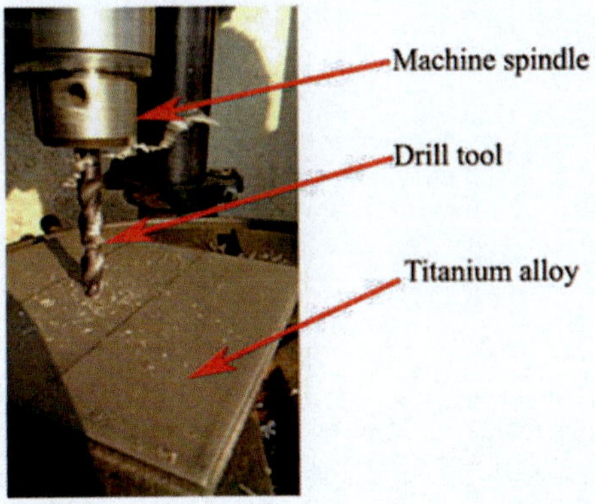

Fig. 1 Experimental setup of titanium alloy drilling

Table 1 Chemical composition of titanium alloy

Material	Iron	Oxygen	Carbon	Nitrogen	Hydrogen	Titanium
CP titanium alloy	<0.3	<0.25	<0.1	<0.03	<0.015	Balance

Table 2 Process factors and their levels

Factor	Factor name	Units	Low	Medium	High
A	Spindle speed	rpm	1500	1750	2000
B	Feed rate	mm/rev	4	6	8
C	Cone angle	deg	110	115	120

Table 3 Design matrix and their results

Std.	Spindle speed (N)	Feed rate (f)	Cone angle (θ)	Thrust force (T_z)		Error in %
				Actual	Predicted	
1	1500	4	110	15.87	14.866	0.063
2	1500	4	110	14.28	13.776	0.035
3	1500	4	110	12.33	11.326	0.081
4	1500	6	115	11.90	11.396	0.042
5	1500	6	115	10.49	9.486	0.096
6	1500	6	115	9.67	9.166	0.052
7	1500	8	120	8.66	7.656	0.116
8	1500	8	120	7.57	7.066	0.067
9	1500	8	120	7.47	6.466	0.134
10	1750	4	115	11.90	11.396	0.042
11	1750	4	115	10.74	9.736	0.094
12	1750	4	115	9.27	8.766	0.054
13	1750	6	120	10.98	9.976	0.091
14	1750	6	120	9.27	8.766	0.054
15	1750	6	120	8.05	7.046	0.125
16	1750	8	110	10.74	10.236	0.047
17	1750	8	110	9.39	8.386	0.107
18	1750	8	110	8.90	8.396	0.057
19	2000	4	120	10.13	9.126	0.099
20	2000	4	120	8.15	7.646	0.062
21	2000	4	120	7.58	6.576	0.132
22	2000	6	110	9.89	9.386	0.051
23	2000	6	110	8.05	7.046	0.125
24	2000	6	110	7.78	7.276	0.065
25	2000	8	115	8.34	7.336	0.120
26	2000	8	115	7.32	6.816	0.069
27	2000	8	115	6.34	5.336	0.158

3 Results and Discussion

3.1 Developing Mathematical Model

The response function represents the thrust of the titanium alloy during the drilling process. The drilling process is a function of the drilling speed (N), feed rate (f), and acute angle (θ) and can be expressed as:

$$\text{Thrust force (TF)} = f(N, f, \theta) \tag{1}$$

The coded and uncoded empirical equations are developed using Design Expert®. The coded and uncoded equations are as follows,

$$\text{TF}_{\text{coded}} = +9.82 - 1.26 * A - 1.70 * B - 1.02 * C \tag{2}$$

$$\text{TF}_{\text{un-coded}} = +47.201 - 5.03378E-003 \times \text{Spindle speed} - 0.84764 \times \text{Feedrate} - 0.20426 \times \text{Cone angle} \tag{3}$$

Table 3 shows the thrust force predicted from the model and the corresponding deviation. The validity of the developed model is tested by ANOVA. Table 4 shows a summary of the ANOVA, the variance ratio (F-value) of the thrust force in the drilling process. The variance ratio, indicated by F in the ANOVA table, is the ratio of the mean square for the factor and error mean square. A large value of F means that the influence of the factor is large and estimates the variance of the error. In addition, the higher the value of F, the more important factor which influencing more on the thrust of the system. When the "R^2" value is 1.0, the predicted model values correspond exactly to the experimental results, and the lower the standard error (SE) of the model indicate the obtained model is more appropriate and it can be used predict the thrust force. Further, the model is validated with scatter diagram as presented in Fig. 2. It

Table 4 ANOVA results

Source	Sum of squares	df	Mean square	F value	p-value Prob. > F	
Model	99.012	3	33.004	23.008	<0.0001	Significant
A-spindle speed	28.506	1	28.506	19.872	0.0002	
B-feed rate	51.731	1	51.731	36.063	<0.0001	
C-cone angle	18.774	1	18.774	13.088	0.0014	
Residual	32.993	23	1.434			
Lack of fit	13.697	5	2.739	2.555	0.0647	Not significant
Pure error	19.296	18	1.072			
Cor. Total	132.005	26				

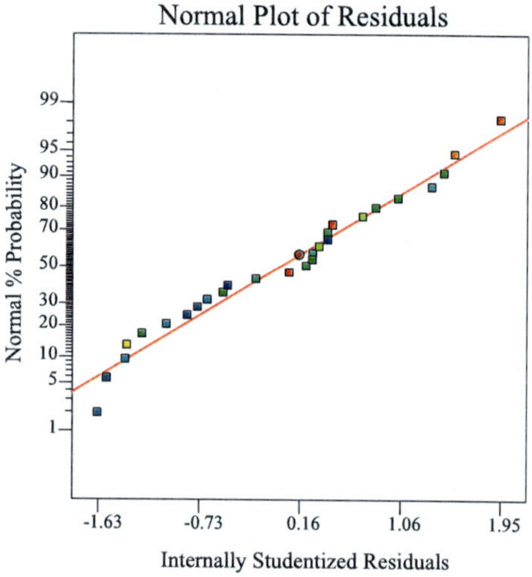

Fig. 2 Scatter diagram for the developed model of thrust force

can be observed that all the experimental values are scattered uniform on both side and close to 45° line which confirms the perfect fitness of the model.

3.2 *Effect of Thrust Force on Process Parameters.*

The effect of thrust force on process parameters of drilling speed, feed rate, and cone angle are presented in Fig. 3a, b. It can be observed from Fig. 3a that thrust force increases with increasing feed rate of drilling. Increasing feed rate increases heat flow on the workpiece that decreases plastic flow of the material around the tool. Hence, high thrust force was observed during drilling. Higher thrust force that caused cracks

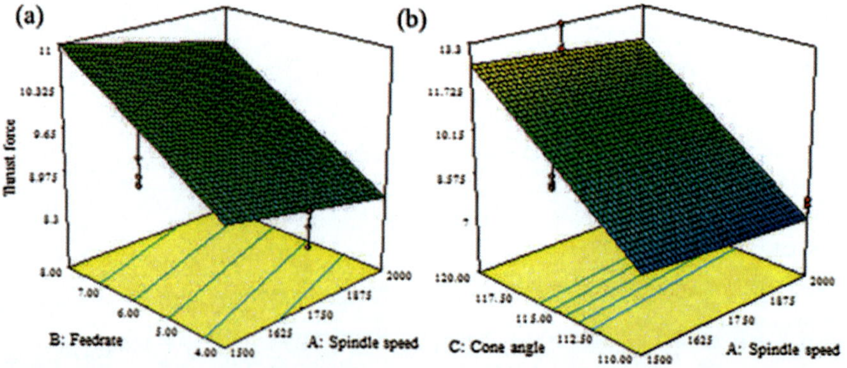

Fig. 3 3D response surface plot for thrust force on **a** spindle speed versus feed rate, **b** spindle speed versus point angle

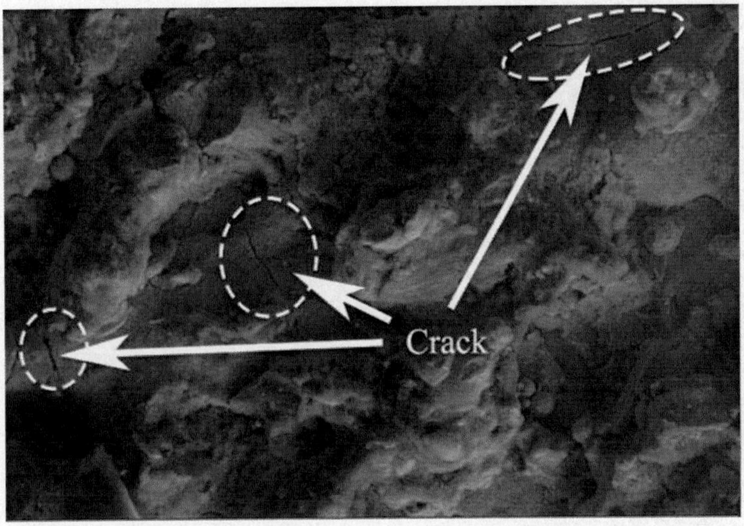

Fig. 4 Formation of cracks after drilling

seem to propagate along the adiabatic shear band due to extremely shear deformation by high strain rates as seen in Fig. 4. On the other hand, varying point angle from 110 to 120° induced more strain deformation, thus producing a primary adiabatic shear zone on the drilling area that results an increase in thrust force during drilling as shown in Fig. 5. Higher point angle (120°) enhances poor material flow around the drilling when at low drilling speed, therefore, more thrust force was observed. Whereas, when at low point angle (110°) with high drilling speed, generates more heat on the cutting zone results low thrust force on the drilling.

4 Conclusion

In the present investigation, the drilling of titanium alloy was carried out using the Taguchi statistical optimization technique. L_{27} orthogonal array was used to complete the experiment effectively. Based on the objective of the experiment, the thrust force is during drilling of commercially pure titanium alloy. The mathematical model was developed for the thrust force based on factors involved in the process such as drilling speed, feed rate, and cone angle of drill bit. Further, the results were further analyzed with ANOVA. The following findings were observed from the Taguchi optimization and the characterization results of drilling made with feasible solutions.

1. At maximum drilling speed, lower thrust force values were obtained due to increasing plastic flow of material.
2. The thrust force of drilled holes increased with increase of feed rate. Increasing point angle increases the thrust force of drilling operation. Therefore, more thrust

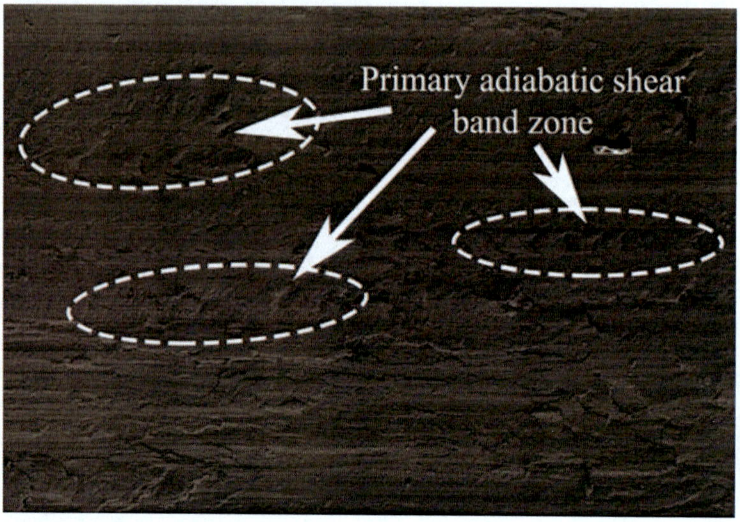

Fig. 5 Zone of adiabatic shear band

force values of holes drilled with high cone angle were measured than that drilled with standard drill tools (120).

3. Depending on the Taguchi results, the optimal drilling parameter combination for minimum surface roughness was determined as drill point angle of 120°, feed of 8 mm/rev, and cutting speed of 2000 mm/rev.
4. According to the Taguchi and ANOVA results, the most effective parameter was found as drill point angle and the minimum effective parameter was found as feed.

References

1. Chatterjee S, Mahapatra SS, Mondal A, Abhishek K (2018) An experimental study on drilling of titanium alloy using CO_2 laser. Sādhanā 43
2. Beranoagirre A, Urbikain G, Calleja A, Lopez de Lacalle LN (2018) Drilling process in gamma-TiAl intermetallic alloys. Materials (Basel) 11
3. Pervaiz S, Anwar S, Qureshi I, Ahmed N (2019) Recent advances in the machining of titanium alloys using minimum quantity lubrication (MQL) based techniques. Int J Precis Eng Manuf Green Technol 6:133–145
4. Boyer RR (1996) An overview on the use of titanium in the aerospace industry. Mater Sci Eng A 213:103–114
5. Beranoagirre A, Urbikain G, Marticorena R, Bustillo A, López de Lacalle L (2019) Sensitivity analysis of tool wear in drilling of titanium aluminides. Metals 9
6. Inagaki I, Takechi T, Shirai Y, Ariyasu N (2014) Application and features of titanium for the aerospace industry. Nippon Steel Sumitomo Metal Tech Rep
7. Zhu Z, Sui S, Sun J, Li J, Li Y (2017) Investigation on performance characteristics in drilling of Ti6Al4V alloy. Int J Adv Manuf Technol 93:651–660

8. Uddin M, Basak A, Pramanik A, Singh S, Krolczyk GM, Prakash C (2018) Evaluating hole quality in drilling of Al 6061 alloys. Materials (Basel) 11
9. Niketh S, Samuel GL (2018) Drilling performance of micro textured tools under dry, wet and MQL condition. J Manuf Process 32: 254–268
10. Liu Z, Liu Y, Han X, Zheng W (2018) Study on super-long deep-hole drilling of titanium alloy. J Appl Biomater Funct Mater 16:150–156
11. Sun D, Lemoine P, Keys D, Doyle P, Malinov S, Zhao Q, Qin X, Jin Y (2016) Hole-making processes and their impacts on the microstructure and fatigue response of aircraft alloys. Int J Adv Manuf Technol 94:1719–1726
12. Ahmed LS, Govindaraju N, Pradeep Kumar M (2015) experimental investigations on cryogenic cooling in the drilling of titanium alloy. Mater Manuf Process 31: 603–607
13. Vijayan D, Abhishek P, Manoj Kumar YG, Balaji P, Siva Kumar Reddy P (2018) Optimization of drilling parameters of carbon fiber composites using RSM based desirability function. IOP Conf Ser Mater Sci Eng 390
14. Liu N-M, Chiang K-T, Hung C-M (2012) Modeling and analyzing the effects of air-cooled turning on the machinability of Ti–6Al–4V titanium alloy using the cold air gun coolant system. Int J Adv Manuf Technol 67:1053–1066
15. Vijayan D, Rao VS (2015) Parametric optimization of age hardenable aluminum alloys using TGRA coupled with PCA. Appl Mech Mater
16. Vijayan D, Rao VS (2014) Friction stir welding of age-hardenable aluminum alloys: a parametric approach using RSM based GRA coupled with PCA. J Inst Eng India Ser C 95:127–141

RSM-Based Optimization of Process Parameters in Synthesis of Pentaerythritol Ester of Rapeseed Oil

P. Vithya, G. Sriram, and S. Arumugam

Abstract This study was intended to optimize the reaction parameters of formulation of the rapeseed oil-based pentaerythritol ester (PE) biolubricant using response surface methodology (RSM) based on the central composite design (CCD).The parameters chosen for the reaction process to synthesize pentaerythritol ester were catalyst concentration (0.5–1.5 wt%), temperature (140–160 °C), and rate of reaction time (1–5 h). The outcome results of process variables revealed that an optimum condition for the biolubricant synthesis was at the temperature of 145.09 °C, and 1.45 wt% of catalyst capacity and 1.89 h of reaction time were acquired with the 81.6% yield of pentaerythritol ester. The catalyst capacity and reaction process time were found to be the highly considerable interaction variables. The yield predicted after the optimization process has produced adequate results with the experimental value with a coefficient of determination (R^2) of 0.9667.

Keywords Response surface methodology · Optimization · Rapeseed oil · Pentaerythritol ester · Biolubricant

1 Introduction

The rapid expansion and utilization of innovative resource of lubricants for various industrial sectors are increasing every year [1]. The petroleum oil-based synthetic lubricants may have a contamination hazard with respect to ecosystems, agricultural land, and groundwater reserves owing to the toxicity and non-biodegradable nature [2, 3]. Vegetable oils are the primary outrider by its uses, and because of their sustainable energy resources, they offer energy independence [4]. Fatty acid esters derived from vegetable oil have gained importance as an alternative lubricant for various applications. Plant oil transesterification with polyol produce biolubricants with high thermo-oxidative stability and pour point for the reason of glycerol replacements by the polyol [5, 6].

P. Vithya (✉) · G. Sriram · S. Arumugam
Department of Mechanical Engineering, Sri Chandrasekharendra Saraswathi Viswa Mahavidyalaya University, Enathur, Kanchipuram, Tamil Nadu 631561, India
e-mail: vithyagautham@gmail.com

T. Rajmohan et al. (eds.), *Advances in Materials and Manufacturing Engineering*, Springer Proceedings in Materials 7, https://doi.org/10.1007/978-981-15-6267-9_58

Extensive research works on formulation of biolubricants had been conceded out previously on polyols especially pentaerythritol ester [7]. Hamizah Ammarah et al. [8] analyzed the synthesis optimization of trimethylolpropane (TMP) ester and pentaerythritol (PE) ester through the esterification of oleic acid with polyhydric alcohols and stated that optimum condition for the TMP and PE synthesis was the temperature at 150 and 180 °C, catalyst content of 1.5 wt%, molar ratio 3.9:1 and 4.9:1, and reaction time of 5 h and 6 h, respectively, obtained with the yield of 91.2% ester TMP and 92.7% ester PE. Musa et al. [9] investigated the optimization process of transesterification of jatropha methyl ester with TMP using response surface methodology. The model was satisfactorily predicting the yield of jatropha biolubricant with a coefficient of determination (R^2) of 0.9068. Venu and Goud [10] analyzed the optimization of RSM-based castor oil synthesis and inferred that the optimal condition was the temperature of 52.81 °C, catalyst capacity of 15.14 wt%, 1.65:1 molar ratio, and reaction time of 2.81 h. The present work aimed to synthesize the rapeseed oil-based pentaerythritol ester by successive transesterification process. The reaction parameters of transesterification process were optimized, and different interaction parameters were also examined by using response surface methodology (RSM) and further CCD was used to determine the effect of process variables using analysis of variance (ANOVA).

2 Materials and Methods

2.1 *Transesterification Reaction of Pentaerythritol Ester of Rapeseed Oil*

Methanol, sodium hydroxide (NaOH), pentaerythritol (99%), para-toluene sulfonic acid (p-TSA), and xylene were purchased from M/s. Sigma Aldrich, USA, was used in the synthesis of pentaerythritol ester. Rapeseed oil extracted from rapeseed was transesterified to prepare the rapeseed oil methyl ester. A mixture of two-liter of rapeseed oil with 880 ml of methanol and 18.46 g of NaOH catalyst was poured into a round bottom flask equipped with a mechanical stirrer. The reaction mixture was heated up to 55 °C in stirring condition of 300 rpm for 1 h. The mixture was later cooled and drained in a funnel separation of methyl ester. To remove the excess methanol and excess catalyst, the obtained methyl ester was heated to 70 °C and washed with distilled water. The rapeseed oil methyl ester of one liter, 85 g of pentaerythritol, and 55 g of para-toluene sulfonic acid was mixed with xylene. This reaction product was then heated in the nitrogen atmosphere up to 160 °C for 5 h. Finally, the product obtained from rapeseed oil was designated as pentaerythritol ester [11].

Table 1 Coded and actual levels of variables for design

Factor	Level		
	Low	Center	High
Temperature (°C)	140	150	160
Catalyst (wt%)	0.5	1.0	1.5
Duration (h)	1	3	5

2.2 *Experimental Design*

The CCD design was used in the design expert software to find out the three factors which were the capacity of catalyst (sodium hydroxide, weight %), temperature, and the reaction process duration. Table 1 depicts the coded and actual levels of variables for design determined based on earlier literature findings. The temperature was assorted between 140 and 160 °C, catalyst concentration between 0.5 and 1.5% w/w, and rate of reaction time between 1 and 3 h. Three factors in 23 full factorial CCD with three levels resulted in 20 runs of experiments ($2k + 2k + 6$), and k stands for number of independent variables/factors chosen. Center point experiments of six runs have analyzed the pure error augmented with 8 factorial and 6 axial experimental runs. The response evaluated from the experiments was the percentage of pentaerythritol ester composition obtained in terms.

2.3 *Statistical Analysis*

The regression coefficients of the second-order polynomial models were utilized to effectuate the pentaerythritol ester biolubricant yield from the CCD optimization data.

$$Y_{\text{yeild}} = b_0 + \sum_{i=1}^{k} b_0 X_i + \sum_{i=1}^{k} b_{ij} X_i^2 + \sum_{ij>1}^{k} \sum_{j}^{k} b_{ij} X_i X_j + e \tag{1}$$

where Y_{yield} was the response variable (composition of pentaerythritol ester), i and j indicate linear and quadratic coefficients, regression coefficients were expressed by b_0, b_i, b_{ij}, and b_{ij}, k was the number of factors studied and optimized in the experiments, while e denoted the random error [12]. The Design–Expert software was used to produce the regression analysis of the obtained data, and analysis of variance (ANOVA) was utilized to find out the model adequacy.

3 Results and Discussion

3.1 Optimization of Process Parameters using RSM

The CCD based on RSM was used to optimize the reaction parameters for the formulation of rapeseed oil-based pentaerythritol ester. The temperature, time, and catalyst concentration were selected as independent variables. Their interaction between the parameters was examined for finding out the optimum yield of biolubricant. The design layout and experimental results are shown in Table 2. The composition of pentaerythritol ester was determined using an estimated response model, and final equation with respect to actual factors was given in Eq. (2).

$$\begin{aligned} Y = &-330.013 + 4.844X_1 + 74.017X_2 + 5.699X_3 \\ &- 0.01493X_1^2 - 0.7441X_2^2 - 0.625X_3^2 - 0.403X_1X_2 \\ &+ 0.0004X_1X_3 - 2.313X_2X_3 \end{aligned} \quad (2)$$

Table 2 Design layout and experimental results for response surface analysis

Run	Temperature (°C)	Catalyst concentration (wt%)	Reaction time (hour)	% yield	Actual	Predicted	Residual
1	140	0.5	1	68.19	68.19	68.002	0.187
15	160	0.5	1	71.24	71.24	71.249	−0.009
5	140	1.5	1	81.51	81.51	81.702	−0.192
13	160	1.5	1	76.9	76.9	76.884	0.015
7	140	0.5	5	71.32	71.32	71.419	−0.099
2	160	0.5	5	74.81	74.81	74.701	0.108
16	140	1.5	5	75.79	75.79	75.864	−0.074
3	160	1.5	5	70.81	70.81	71.081	−0.271
19	140	1	3	77.12	77.12	76.940	0.179
12	160	1	3	76.33	76.33	76.172	0.157
10	150	0.5	3	75.15	75.15	75.336	−0.186
9	150	1.5	3	80.9	80.9	80.376	0.523
20	150	1	3	76.89	76.89	78.050	−1.160
14	150	1	5	75.29	75.29	74.953	0.336
6	150	1	3	78.38	78.38	78.050	0.329
8	150	1	3	78.38	78.38	78.050	0.329
18	150	1	3	78.38	78.38	78.050	0.329
11	150	1	3	78.38	78.38	78.050	0.329
4	150	1	3	78.38	78.38	78.050	0.329
17	150	1	3	76.89	76.89	78.050	−1.160

where Y was pentaerythritol ester yield composition, X_1, X_2 and X_3 were the temperature, catalyst capacity, and reaction duration, respectively. The highest percentage of pentaerythritol ester (PE) yield of 81.51% was obtained at 150 °C, catalyst content of 1.89%, and reaction time of 1.89 h [13].

Figure 1 shows the actual versus predicted values plot. The uniformly scattered points in the plot showed that the points were very closer to the straight lines which represent the better agreement of the experimental values with the predicted values of the response. Figure 2 depicts the normal probability of residual plot. The residuals distribution was established by the normal probability of residual plot. The data normality is confirmed by the straight lines [14].

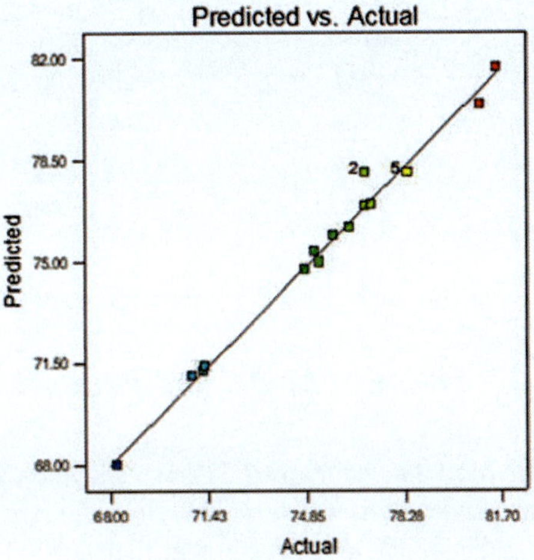

Fig. 1 Actual verses predicted values plot

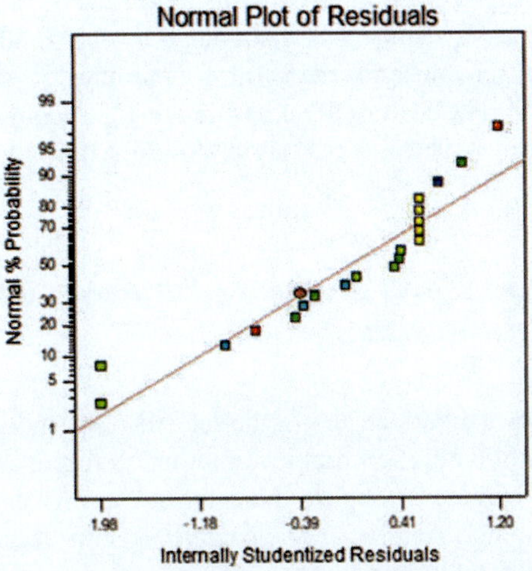

Fig. 2 Normal probability plot of residuals

Table 3 ANOVA for pentaerythritol ester synthesis

Source	Sum of squares	Degrees of freedom	Mean square	*F* value	*p*-value Prob. > *F*	R^2	Adj-R^2	Adequate precision (AP)
Model	217.686	9	24.1870	62.208	<0.0001	0.9824	0.9666	31.07
A-temp	1.474	1	1.4740	3.792	0.0801			
B-catalyst	63.504	1	63.504	163.329	<0.0001			
C-duration	2.980	1	2.9800	7.665	0.0198			
AB	32.522	1	32.522	83.645	<0.0001			
AC	0.00061	1	0.00061	0.0015	0.9691			
BC	42.827	1	42.8270	110.150	<0.0001			
A^2	5.260	1	5.2600	13.529	0.0043			
B^2	0.088	1	0.0883	0.227	0.6439			
C^2	11.111	1	11.1110	28.577	0.0003			
Residual	3.888	10	0.3880					
Lack of fit	0.716	4	0.1790	0.338	0.842		Not significant	
Pure error	3.171	6	0.5280					
Cor. total	221.57	19						

3.2 *Fitting of Model and Variance Analysis (ANOVA)*

The Model *F*-value of 62.21 has validated the importance of the model. In ANOVA analysis-catalyst, duration, *AB*, *BC*, A^2, C^2 have considerable model variables shown in Table 3. The *P*-values higher than 0.1000 disclose the model terms are not noteworthy. The lack of fit *F*-value of 0.34 complimented to the pure error. It may be 84.28% possibility with this large lack of fit *F*-value could arise as a result of noise. The fitted model is guaranteed by the good non-significant lack of fit. There is a reasonable agreement with the predicted R^2 of 0.9480 and adjusted R^2 of 0.9667 [13, 15]. The desirability is assured by the adequate precision (signal-to-noise) ratio more than 4. The acquired signal-to-noise ratio of 31.072 point outs an adequate signal.

3.3 *Effect of Process Parameter on Transesterification Reaction*

The 3D surface plot of the second-order model was used for the reaction parameters to find out the variables' interaction response on the biolubricant yield.

Figure 3a depicts the catalyst capacity and temperature interactive effect on the biolubricant yield. The catalyst quantity considerably increases the pentaerythritol ester biolubricant yield. This is due to the reason for the inclusion of more catalyst that accelerated the higher reaction process in the lesser duration of time. The increasing

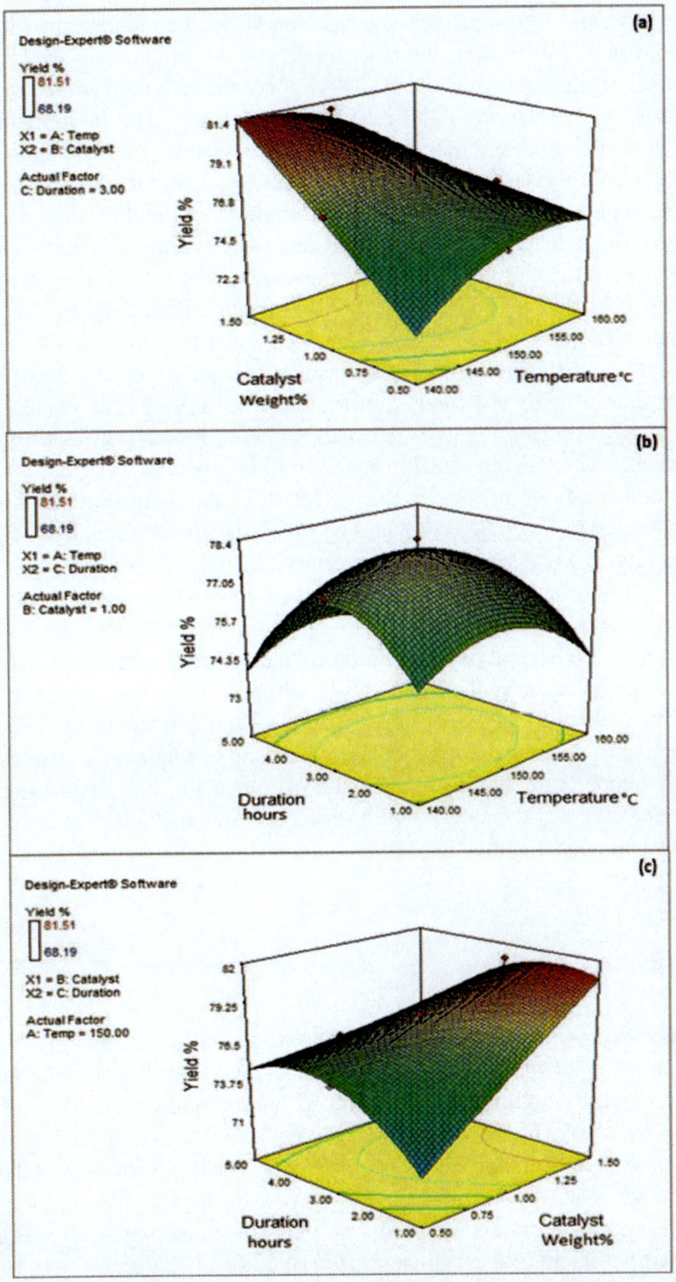

Fig. 3 3D contour plot on PE yield between **a** catalyst and temperature, **b** duration and temperature, **c** catalyst and duration

catalysts' effect was observed at the reaction condition; the temperature of 140 °C, catalyst intensity of 1.45% w/w, and reaction time of 3 h produce the 81.51% yield, and the pentaerythritol ester yield of 75.19% was recorded at a temperature of 150 °C, catalyst capacity of 0.5% w/w, and reaction time of 3 h. The biolubricant yield decreases with temperatures increase owing to the speeding up of saponification reaction rate which leads to the decrease of transesterification reaction yield [9, 12].

Figure 3b depicts the interactive effect on biolubricant yield of temperature and reaction time. The pentaerythritol ester yield decreases as temperature increases and is attributable to reduced rate of reaction speed resulting in lower pentaerythritol ester conversion [12]. The reduced effect of temperature demonstrates that 77.95% yield was perceived at a temperature of 160 °C, the reaction time of 5 h, and 1.0% catalyst quantity, whereas 72.67% yield was recorded at a temperature of 140 °C, the reaction time of 3 h, and catalyst intensity of 1.0% w/w. The results showed that the pentaerythritol ester yield increases with the increase in duration of the reaction process. This raising effect of reaction time produces the yield of 74.12% for the reaction condition of temperature at 160 °C, catalyst quantity of 1.0%, and 5 h reaction time. However, 77.95% yield was recorded at a temperature of 140 °C, catalyst quantity of 1.0%, and reaction time.

Figure 3c shows the interaction response of catalyst concentration and reaction process duration on biolubricant yield. The results show that the higher significant effect of the increase in catalyst concentration on the biolubricant yield. This is due to the more catalyst capacity that promotes the higher conversion of pentaerythritol ester [16]. The yield of 81.51% was substantiated at a temperature of 145 °C, catalyst intensity of 1.45%, and reaction process period of 1.89 h, whereas 61.19% yield was observed at a temperature of 135 °C, the reaction time of 1 h, and catalyst strength of 0.5% [17]. The effect of the catalyst with respect to time had a higher considerable impact contributing to the fitted model.

4 Conclusion

The outcome of this investigation is detailed below.

The formulation of the rapeseed oil-based pentaerythritol ester (PE) biolubricant was successfully optimized. The obtained optimum reaction's conditions are temperature at 145.09 °C, the catalyst capacity of 1.45 wt% and process reaction time of1.89 h with the pentaerythritol ester yield of 81.6%. The catalyst effect with time had a higher considerable impact contributing to the fitted model. The yield foreseen after the optimization process produces good agreement with the experimental values with a coefficient of determination (R^2) of 0.9667.

References

1. Kodali DR (2002) High performance ester lubricants from natural oils. Ind Lubr Tribol 54(4):165–170
2. Erhan SZ, Asadauskas S (2000) Lubricant basestocks from vegetable oils. Ind Crop Prod 11(2–3):277–282
3. Wagner H, Luther R, Mang T (2001) Lubricant base fluids based on renewable raw materials: their catalytic manufacture and modification. Appl Catal A-Gen 221(1–2):429–442
4. Arbain NH, Salimon J (2011) Synthesis and characterization of ester trimethylolpropane based Jatropha Curcas oil as biolubricant base stocks. J Sci Tech 2(2):47–58
5. Hwang HS, Erhan SZ (2006) Synthetic lubricant basestocks from epoxidized soybean oil and Guerbet alcohols. Ind Crop Prod 23(3):311–317
6. Kamil RNM, Yusup S (2010) Modeling of reaction kinetics for transesterification of palm-based methyl esters with trimethylolpropane. Bioresour Technol 101(15):5877–5884
7. Kyzimova NS (2008) Thermostable lubricating composition based on pentaerythritol esters. Chem Technol Fuels Oils 44:169–171
8. Mahmud HA, Salimon J (2014) Optimization of Esterification of Oleic acid and Trimethylolpropane (TMP) and Pentaerythritol (PE). AIP Conf Proc 1614:230–236
9. Umaru M, Aris MI, Munnir SM, Aliyu AM, Aberuagba F, Isaac AJ (2016) Statistical optimization of biolubricant production from jatropha curcas oil using trimethylolpropane as a polyol. ProcWorld Congr Eng Comp Sci 2:1–6
10. Borugadda VB, Goud VV (2015) Response surface methodology for optimization of biolubricant basestock synthesis from high free fatty acids castor oil. Energy Sci Eng 3(4):371–383
11. Vithya P, Sriram G, Arumugam S (2018) Tribological analysis of biodegradable refrigeration oil for vapour compression refrigeration system using four ball tribometer. IOP Conf Ser Mater Sci Eng 390:012070. https://doi.org/10.1088/1757-899X/390/1/012070
12. Aziza NAM, Yunusa R, Rashida U, Syama AM (2014) Application of response surface methodology (RSM) for optimizing the palm-based pentaerythritol ester synthesis. Ind Crop Prod 62:305–312
13. Shrivastsvs A, Sandagar P, Baja I, Singhal R (2008) Media optimization for the production of U-linolenic acid by cunninghamella Echinulata varielegans MTCC 522 using response surface methodology. Int J Food Eng 4(2):1–32
14. Ocholi O, Menkiti M, Auta M, Ezemagu I (2018) Optimization of the operating parameters for the extractive synthesis of biolubricant from sesame seed oil via response surface methodology. Egyp J Petro 27:265–275
15. Russell VL (2009) Response surface methods in R, using RSM. J Stat Softw
16. Yunus R, Fakhrul-Razi A, Ooi TL, Iyuke SE, Idris A (2003a) Development of optimum synthesis method for transesterification of palm oil methyl esters and trimethylolpropane to environmentally acceptable palm oil-based lubricant. J Oil Palm Res 15:35–41
17. Yunus R, Fakhrul-Razi A, Ooi TL, Iyuke SE, Idris A (2003b) Preparation and characterization of trimethylolpropane esters from palm kernel oil methyl esters. J Oil Palm Res 15:42–49

Physical Characteristics of Keyhole in 316L Stainless Steel Joint During an Autogenous Pulsed Laser Beam Welding

A. Jayanthi, K. Venkataramanan, and K. Suresh Kumar

Abstract An attempt made to investigate the physical characteristics of a keyhole during pulsed laser welding in AISI 316L stainless steel joints for better understanding of keyhole and integrity of the joints. The experimentally prepared joints are examined using a metallurgical microscope across the heat-affected zone (HAZ) and melt regions. Then, the observations are compared with results predicted by a volumetric heat source model proposed by Kuang et al., and numerical model for heat transfer is during pulsed laser welding of 316L stainless steel joints. Hence, it was found that predictions and experimental observations are close with the solid-melt and melt-vapor boundaries of keyhole in weld joints. However, in case of complete penetration, that cone-shaped bottom of the keyhole has an expansion known as downward-expanding vapor region, which differs from the proposed volumetric heat source model. The effects of pulse irradiation by laser beam on the centerline weld (melt region) and at the melt-solid boundaries are discussed; hence, heat flow in the melt region and keyhole is examined.

Keywords Pulsed laser welding · Stainless steel · Keyhole · Melt pool · Vapor region

A. Jayanthi (✉)
Sri Chandrasekharendra Saraswathi Viswa Mahavidyalaya University, Enathur, Kanchipuram 631561, Tamil Nadu, India
e-mail: lectsuresh25@gmail.com

Department of Physics, Jeppiaar Institute of Technology, Chennai, India

K. Venkataramanan
Department of Physics, Sri Chandrasekharendra Saraswathi Viswa Mahavidyalaya University, Enathur, Kanchipuram 631561, Tamil Nadu, India

K. Suresh Kumar
P.T. Lee Chengalvaraya Naicker College Engineering and Technology, Kanchipuram, India

T. Rajmohan et al. (eds.), *Advances in Materials and Manufacturing Engineering*, Springer Proceedings in Materials 7, https://doi.org/10.1007/978-981-15-6267-9_59

1 Introduction

Laser welding has received much attention as a promising joining technology because of its less heat input and low distortion. 316L stainless steel is used extensively for weldment where its immunity to carbide precipitation during welding assures optimum corrosion resistance [1] that can promote their wide range of applications in many industries. Laser welding on 316L stainless steel provides deep penetration welding with very small heat-affected zones at high travel speeds. Hence, it is necessary to have a clear understanding of the physical processes involved in laser–material interaction and keyhole formation during pulsed laser welding.

The theoretical solution for a moving heat source was developed for laser welding and material processing technologies based on the heat conduction equations [2]. Then, the first 3D double ellipsoidal moving heat source was developed using finite element model [3], a three-dimensional volumetric heat source model consisting of a rotary Gaussian volumetric heat source and a double ellipsoid heat source model using control volume method [4],the finite volume method used to solve the hydrodynamic equations governing the compressible vapor flow model [5], weld pool transport phenomena during the transition from conduction mode to keyhole mode laser spot welding [6], a three-dimensional transient model for simulating continuous laser keyhole welding process [7], the continuum formulation model solid [8], the volume-of-fluid (VOF) method [9] etc.... were investigated and reported about the heat and mass flow in heat affected, melt and vapor zones of keyhole mode welding with laser as a heat source. The effects of different beam intensity distributions for deep penetration welding are investigated based on a pressure and energy equilibrium of the initial keyhole radii and different depths of keyhole [10]. It has found that the ratio between weld and keyhole stays almost the same independent of the beam diameter with both bead on plate and butt joint, and however, it is possible to control the penetration and weld shape with the focal point position [11]. Further, many more authors have used numerical techniques to evaluate different problems in heat transfer and their reviews on laser welding and related process studies up to the year 2002 [12]. It was briefed about the keyhole mode laser welding that how useful to follow the previous work to appreciate the complexity of the simultaneous physical processes in recent decades [13].

The article is to investigate the keyhole characteristics during pulsed laser welding process using an analytical model proposed by Jao-Hwa Kuang and a numerical model developed using COMSOL Multiphysics code by validating with the keyhole of 316L stainless steel butt joint welding using pulsed Nd:YAG laser. The metallographic examinations across the weld joints evolve the features of keyhole, weld pool, and HAZ of AISI stainless steel. The trajectories of laser pulses appeared at the centerline weld and at edge of the melt pool of the joints are examined and discussed.

Fig. 1 Micrograph of a keyhole formed in weld joint of AISI stainless steel

2 Materials and Methods

The sample considered for all the experiment, simulation, and analytical methods is AISI 316L stainless steel plates having dimensions of 150 mm (length) × 50 mm (width) × 2 mm (thickness). AISI 316L stainless steel plates are taken for welding, which was processed on fully automated workbench using pulsed Nd: YAG laser as a heat source. The chemical composition of the materials, operational parameters, experimental specifications, and environmental conditions is detailed in [14]. The micrographs taken to investigate the evolved features of the weld pool across the butt joint interfaces of AISI 316L stainless steel joint at the magnification of 100× using a PC-controlled metallurgical microscope are shown in Fig. 1, which is to be considered as actual reference for discussions and for comparison with the predicted results from analytical and numerical models.

3 Numerical Model for Laser Welding Process

The COMSOL Multiphysics code is used to estimate the temperature distributions of keyhole mode pulsed laser welding for AISI316L stainless steel joints. The physics of heat conduction equations, materials geometry, thermophysical properties, and boundary conditions for laser welding process and laser spot positions on the butt welding are detailed [15]. The simulated keyhole during laser welding is presented as shown in Fig. 2 to compare with an analytical model and experimentally obtained keyhole.

Fig. 2 Simulated keyhole shape with heat flow directions in 316L stainless steel joint

4 Analytical Model for Keyhole Mode Welding Process

The volumetric heat source model introduced by Kuang et al. [16] adopted for prediction and validation of experimentally obtained keyhole, and hence, the region of the keyhole assumed to have a cylindrical shape, which comprises an annular liquid zone (melt region) and a cone-like vaporization zone (vapor/plasma region).

'r_l' the radius of the annular liquid region of the cylindrical heat source is given by

$$r_l^2 = -\left(\frac{\frac{1}{3}\pi r_v^2 \rho h L_v r_0^2}{1.582(\Delta E) + (\pi^2 h\rho(\Delta T_s C_s + L_m + \Delta T_l C_l + L_V) r_0^2}\right) \quad (1)$$

'r_v' the radius of the conical vapor region of the cylindrical heat source is given by.

$$r_v^2 = -\left(\frac{4.746(\Delta E)}{\pi h\rho(\Delta T_s C_s + L_m + \Delta T_l C_l + L_V)}\right) r_0^2 \quad (2)$$

where $\Delta T_s = T_m - \text{T}_o$ and $\Delta T_1 = T_v - T_m$, in which T_o—ambient temperature, T_m—melting point and T_v—vaporization temperature, ρ—density, h—thickness (height of the volumetric heat source), L_m and L_v are the latent heats of fusion melting and vaporization, C_s and C_l are the specific heat at the solid and liquid phase of AISI 316L stainless steel respectively. 'r_o'—radius of the beam waist and 'ΔE'—input power (in Peak Power) of laser source used in practice.

The volumetric heat source model introduced by Jao-Hwa Kuang was adopted for prediction of keyhole that comprises a melt region and a vaporization zone. Thermophysical properties of the materials were considered and predicted as per IAEA, 2008 materials database with the local coordinates *x*, *y*, and *z* as input using Microsoft Excel code and subsequently plotted using Orgin® code. The predicted

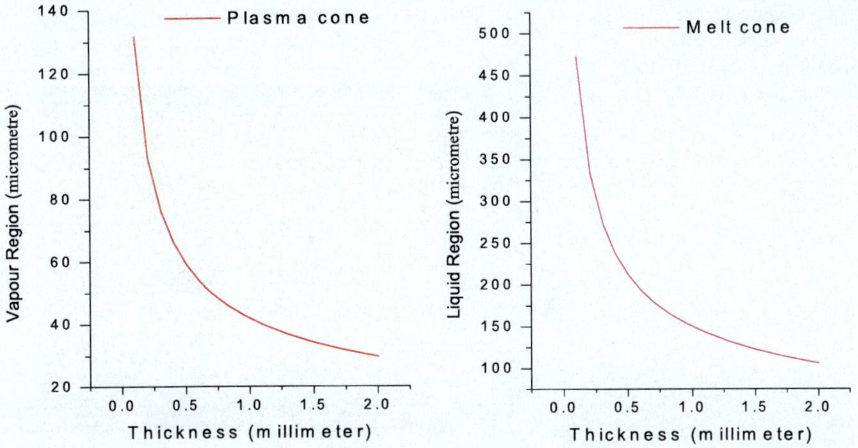

Fig. 3 Predicted vapor and melt region for a 316L stainless steel at average peak power 2100 W based on volumetric heat source model [16]

vapor and melt regions of keyhole of a 316L stainless steel based on volumetric heat source are as shown in Fig. 3.

5 Result and Discussion

5.1 The Process of Formation of Keyhole

The physics of converting laser into heat energy is the basic principle of laser welding. Repetition of laser beam is irradiated with very high amount of localized heat input as average peak power (APP) = 2100 watts, i.e., about 1.31 MW/cm^2 in terms of power density on the butt joint that leads to increase in temperature far above the melting point 1703 K quickly. Therefore, the stretched molecular bonding linkages no longer capable of exhibiting mechanical strength, and thus, melting occurs. The heat energy is quickly absorbed in the region in and around the beam waist, resulting in the liquidation of a small area of the material called melt pool. On further heating, the melt pool receives more energy, which enables deep penetration in the joint, known as keyhole as shown in Fig. 1. The predicted melt region of a keyhole compared with an actual keyhole of 316L stainless steel found very close association with experimental observations as shown in Fig. 4.

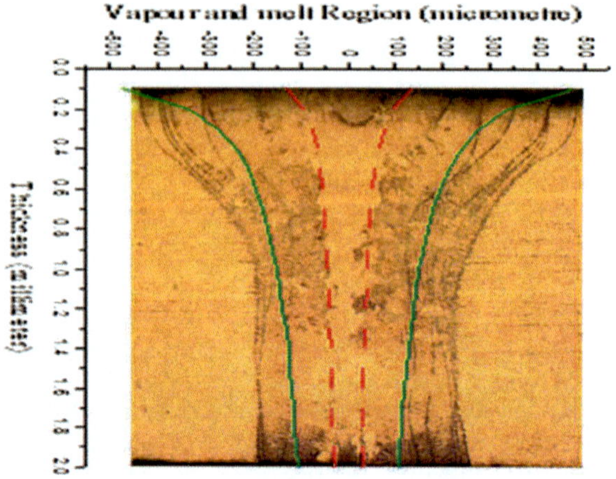

Fig. 4 Comparison of estimated vapor and melt region with keyhole of 316L stainless steel joint

5.2 *The Process of Formation of Vapor/Plasma*

The keyhole behaves like an optical black body when the high-power laser pulse is subject to multiple reflections, hence absorptivity reaches 100%. Due to the Kerr effect and Pockel's effect, suddenly the temperature of region around the keyhole increases. This raise in temperature suddenly reaches to the melting point of 316L stainless steel by the principle of Clapeyron–Clausius on latent heat, which leads to the strong convectional movements of melt in the weldpool called 'Marangoni effect.' This thermocapillary convection in the keyhole ejects the melt to the upper portion of the keyhole and hence the depth increases accordingly. An increase in keyhole depth of the metal joint leads to even higher absorptivity due to multiple reflections, evaporation, and re-radiations that take place and further increase heat energy above 7450 kJ/kg that crosses vaporization temperature 5000K and breaks the metallic bonds of molten 316L stainless steel; vaporization takes place. The set of the evaporated atoms with a mixture of ablated electrons and ions (electron–ion-atom) is formed over the surface through the process of inverse bremsstrahlung phenomenon, known as plasma. In such intensity of evaporation, a local negative pressure cavity is generated, and hence, the wall thickness of the keyhole become thinner. Then, recoil pressure is formed that drags molten liquid toward low-pressure cavity and ejects the melt centripetally around the keyhole. When laser pulse irradiation stopped, the hot plasma starts to radiate its heat energy to the keyhole wall. Since the plasma is unstable, the temperature drops very quickly. At the same time, the heat conduction from the keyhole wall to the metal region is very strong due to the high aspect ratio and high temperature gradient. Hence, the magnitude of temperature of the keyhole wall drops very quickly; then, its size became very small at the bottom end as shown in Fig. 4. Thus, a cycle of keyhole formation took place during pulsed laser welding.

Arrival of new pulses at the impacted surface can easily get into melt pool that leads to the local ablation and generates a recoil pressure. When the local recoil pressure increases beyond the surface tension, the keyhole wall propagates inside

Fig. 5 Comparison of numerically simulated keyhole with 316L stainless steel joint

the melt pool because of Marangoni effect. A similar process occurs when this pressure exerted on the rear side. In that case, the melt pool collapses, which would become like a keyhole. By this way, each translating pulse irradiated onto the melt pool and makes propagation inside the preserved keyhole front repetitively [17]. Figure 4 shows the comparison of the predicted vapor region with an actual keyhole profile of 316L stainless steel, and Fig. 5 compares the numerical simulation of the keyhole structure with an actual keyhole. It has found that theoretical prediction of the heat source model proposed by Jao-Hwa Kuang et al. has very close association with computed keyhole and as well as with an actual keyhole.

5.3 *Identification of Downward Expansion of Vapor Region*

In the actual keyhole, the sequences of irradiated pulses almost appeared in parallel to the depth direction and shows strong oscillations, especially at the keyhole front. These waves lead to a periodical change of keyhole diameter and moved to the lower part of the keyhole as shown in Fig. 6b, c. Since there is no direct laser irradiation at the bottom of the keyhole that leads to the low pressure, the top of the keyhole resulted in a downward-flowing vapor. This downward flowing vapor region reversing upwards to the top and met the downward-flowing vapor oscillations due to the pulse irradiations from the top creates a whirlpool that drove the suspended droplets randomly again to bottom that forms a downward-expanding vapor region [18] as shown in Fig. 6b, c.

It may be due to the temperature distribution at the lower part that become two-dimensional, i.e., no further temperature distribution in depth direction while it reaches full penetration, but strongly in radial directions. However, it is interesting to point out that no signs of either vapor region or downward-expanding vapor region identified in Fig. 6a even it has reached close to full penetration. Therefore, input power of laser has an important role to play in formation of such downward expanded

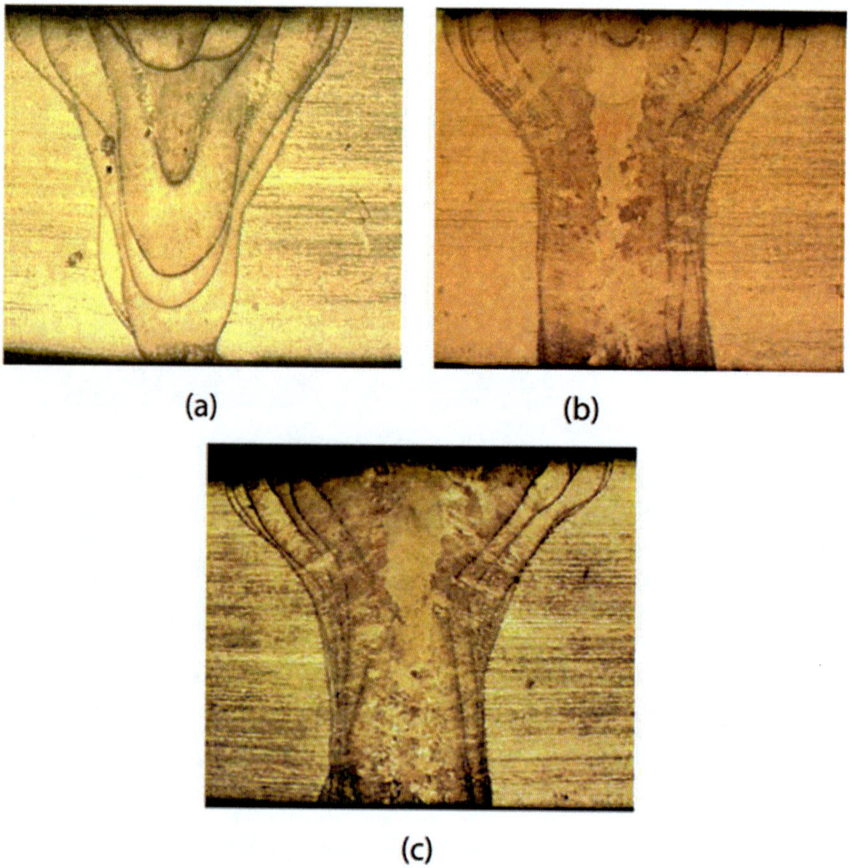

Fig. 6 Keyhole of 316L stainless steel at various average peak power **a** 2000 W, **b** 2100 W and **c** 2150 W

vapor region. This result is deviated from the cone-like vapor zone as proposed in the volumetric heat source model [19].

5.4 *Meltflow in Molten Region*

From the novelty of literatures, the schematic of complex dynamics of the melt during the translational motion of laser beam along the weld has been presented and imposed with the micrograph of melt flow in the fusion zone around keyhole of 316L stainless steel joint as shown in Fig. 7. According to the law of conservation of mass and energy, the shape of the weld pool must remain unchanged, with or without convection. Consequently, the weld pool becomes wider and deeper. The convection

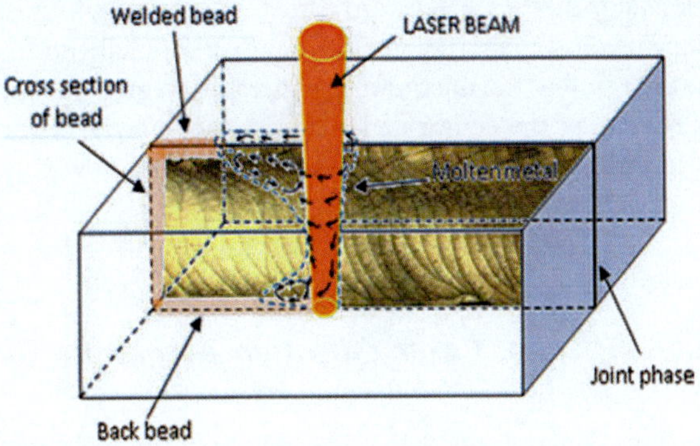

Fig. 7 Schematic of melt flow directions around keyhole compared with 316L stainless steel joint

in weld pool fluid is interrelated to cause phase delay between the net heat input and the temperature gradient.

Figure 8 compares the numerically simulated pulse trajectories with the pulse trials of 316L stainless steel joint. During the pulsation, the top of the weld pool drove and extended in the direction that is opposite to welding direction due to Marangoni effect. Hence, the keyhole appeared widest at the top; a necking is observed in the middle and become narrow in the end. This is revealed by the deviations in the path of every pulse trails in the simulation result and in an actual weld.

The small volume of melt region behind the lower part of the keyhole squeezing out melt to the upward due to the recoil pressure and sideway flow occurs in the broad region behind the keyhole at the top as shown in Fig. 8. The simulated melt flow directions of molten pool are discussed in [20], and this actual melt flow around the keyhole in 316L stainless steel joint reveals the similar melt flow directions at various regions around the keyhole.

In the absence of pulse, the large recoil pressure increased due to fast cooling of vapor cavity by evaporation and hence the surface tension increased that drives

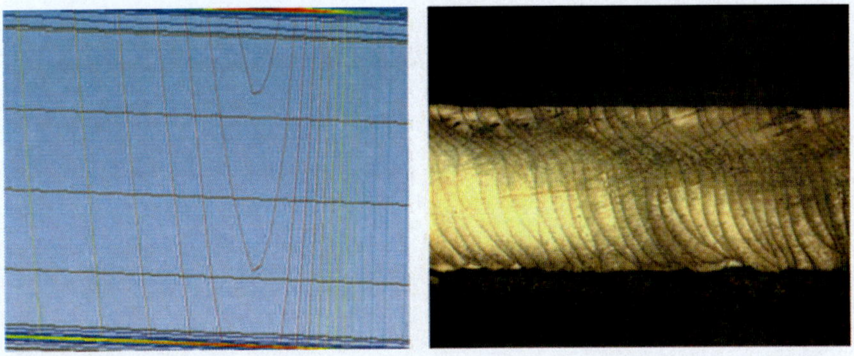

Fig. 8 Comparison of pulse trajectories in 316L stainless steel joint by numerical simulation and experiment

the vapor flow in downward direction. The recoil pressure and the surface tension became closer, and the center flow pattern changed to upward direction. This flow pattern shifted the maximum temperature region from the center of the pool to a ring surrounding the center and a ring-shaped vortex observed as shown in Fig. 1, and it has remained visible and the up/downward flow pattern transition confined to its interior as discussed in [6].

5.5 *Effect of Laser Pulsation Across the Weld Joint*

As Figs. 7 and 8 give better insight about the melt flow directions of the weld pool around the keyhole during pulsed laser welding, Fig. 9 shows the chances of solidification and the weld pool that appears in dark and white region along with many lines and arcs. These solidification lines and arcs having periodical gap and larger deviation at the upper part of the weld pool are due to the time delay between the pulses and the different dimensions of solidification that occur in molten 316L stainless steel. The periodical gap between the solidification pulses trails appeared on 316L stainless steel was adequate to initiate the phase delay in the solidification processes. During the initial phase of solidification of pulse, trails are formed around the keyhole due to the formation of primary dendritic. This phenomenon is similar to the standard thin film growth technique, called as Vollmar-Weber nucleation mechanism [21]. Hence, very thin layers of solidifications lines are noticed in the weld pool for each pulse. These lines appeared almost as parallel to the depth direction in the conduction zone and it has become roughly a semi-circular in the transition zone. Finally, it becomes parabolic when the keyhole reaches full penetration at the end [22].

The trajectories of pulse irradiation appeared at edge of weld pool—heat-affected zone boundary are shown in Fig. 10a. As discussed above, these trails are found as evidence for different modes of energy transfer during pulsed laser irradiations at the

Fig. 9 Trajectories of pulses and solidification lines in weld pool of 316L stainless steel joint

Fig. 10 Comparison of longitudinal cross section of pulse trails at melt pool—solid interface (**a**) and at centerline weld (**b**)

weld pool—heat-affected zone boundary region. But, no such variations noticed on the centerline weld as shown in Fig. 10b. These different stages of the solidification lines occurred due to the translational motion of the laser pulses during welding process. Thus, pulsation/pulse repetition rate plays a vital role in the characteristics of keyhole and good quality of weld bead that is adequate for producing efficient welding joint [2].

In the cases of laser beam welding, the formation of porosity highly depends on the depth-to-width aspect ratio of the keyhole [12]. In higher aspect ratio, porosity can be formed easily. In the present work, the aspect ratios of the keyhole were measured as 0.8396, 0.7838, and 0.7791 for the given laser pulse energy 24.6, 25.2, and 25.8 J, respectively, as shown in Table 3. These smaller aspect ratios increase the back filling rate of the melt around the keyhole faster that prevents the formation of cracks; hence, no cracks were absorbed anywhere across the weld region.

Table 3 Depth-to-width ratio of the keyholes for different input power

Welding trials	Average peak power (Watt)	Bead width (mm)	Penetration depth (mm)	Aspect ratio	Comment
I	2050	2.3819	~2.000	0.8396	Incomplete penetration
II	2100	2.5516	=2.000	0.7838	Full penetration
III	2150	2.5668	>2.000	0.7791	Full penetration

6 Conclusion

The pulsed Nd:YAG laser beam with its optimized parameters such as peak power density, pulse duration, and pulse repetition rate was used to weld 2 mm thick AISI 316L stainless steel plate joints successfully. Hence, investigations were made as follows,

1. The melt and vapor regions were predicted for 316L stainless steel joint to validate a heat source model proposed by Jao-Hwa Kuang et al. These predicted melt and vapor region compared with an actual keyhole structure of 316L stainless steel were found very close matching with each other.
2. The formation of melt pool and keyhole in the 316L stainless steel joints is discussed. The computed keyhole profile was compared with an actual keyhole structure and found that the computed keyhole structure has good agreement with an actual.
3. However, in case of full penetration, the cone-shaped vapor and melt regions near the bottom are slightly deviated from validated model due to a downward-expanding vapor region. Also, it was found that vapor region undergone different shapes for different heat inputs.
4. The melt flow directions in and around the keyhole were investigated with computational results and observations from fusion zone of 316L stainless steel joint. The results confirm larger flow at the wider top and lesser at the narrower bottom in the fusion zone.
5. These different stages of the solidification lines occurred at the centerline weld and at the weldpool—HAZ interface due to the translational motion of the laser pulses during welding process were investigated.
6. No signs of defect was found anywhere across the weld joints due to the lower aspect ratio of the keyhole in 316L stainless steel joints.

Acknowledgements The authors wish to thank the authorities of 'Indira Gandhi Centre for Atomic Research,' Kalpakkam, Tamil Nadu, India, for permitting to use its laser welding and characterization facilities.

References

1. Kirillov PL (2006) Thermo physical properties of materials for nuclear engineering. Institute for Heat and Mass Transfer in Nuclear Power Plants, Obninsk
2. Tzeng Y-F (2000) Parametric analysis of the pulsed Nd: YAG laser seam-welding process. J Mater Process Technol 102:40–47
3. Goldak J, Chakravarti A, Bibby M (1985) A double ellipsoid finite element model for welding heat sources, IIW Doc no. 212-603-85, International institute for welding, Strausbourg, France
4. Wang H, Shi Y, Gong S (2006) Numerical simulation of laser keyhole welding processes based on control volume methods. J Appl Phys 39:4722–4730

5. Amara EH, Fabbro R, Bendib A (2003) Modeling of the compressible vapour flow induced in a keyhole during laser welding. J Appl Phys 93(7):4289–4296
6. Cho J-H, Farson DF, Milewski JO, Hollis KJ (2009) Weld pool flows during initial stages of keyhole formation in laser welding. J Phys D Appl Phys 42:502–513
7. Daneshkhah R, Najafi M, Torabian H (2012) Numerical simulation of weld pool shape during laser beam welding. Int Res J Appl Basic Sci 3(8):1624–1630
8. Zhou J, Tsai H-L, Wang P-C (2006) Transport phenomena and keyhole dynamics during pulsed laser welding. J Heat Transf. https://www.asme.org
9. Wang R, Lei Y, Shi Y (2010) Numerical simulation of transient temperature field during laser keyhole welding of 304 stainless steel sheet. Optics and Laser Technology
10. Volpp J, Vollertsen F (2013) Analytical modelling of the keyhole including multiple reflections for analysis of the influence of different laser intensity distributions on keyhole geometry. Phys Procedia 41:460–468
11. Vänskäa M, Abt F, Weber R, Salminen A, Graf T (2013) Effects of welding parameters onto keyhole geometry for partial penetration laser welding. Phys Procedia 41:199–208
12. Zhou J, Tsai H-L (2007) Effects of electromagnetic force on melt flow and porosity prevention in pulsed laser keyhole welding. Int J Heat Mass Transf 50(2007):2217–2235
13. Rai R, Elmer JW, Palmer TA, Deb Roy T (2007) Heat transfer and fluid flow during keyhole mode laser welding of tantalum, Ti–6Al–4V, 304L stainless steel and vanadium. J Applied Phys D 40:5753–5766
14. Suresh Kumar K, Jayanthi A (2015) Modeling of heat and mass transportation in the keyhole of 316L stainless steel and steel joints during pulsed Nd: YAG laser welding. Int J Appl Eng Res 10(5):3979–3982
15. Suresh Kumar K, Vanitha S, Jayanthi A (2015) Modeling and analysis of transient thermal responses during welding of 316L stainless steel and low carbon steel joint using pulsed Nd: YAG laser. Int J Appl Eng Res 10(5):3979–3985
16. Kuang J-H, Hung T-P, Chen C-K (2012) A keyhole volumetric model for weld pool analysis in Nd: YAG pulsed laser welding. Opt Laser Technol 44:521–1528
17. Suresh Kumar K, Jayanthi A (2014) Formation of laser induced downward expanding vapour region in autogenous welding of AISI 316L stainless steel joint using pulsed Nd:YAG laser. National Laser Symposia [NLS–23], S.V University, 3–6 Dec 2014. ISBN: 9788190332156, CP-06-06.
18. Fabbro R (2010) Melt pool and keyhole behaviour analysis for deep penetration laser welding. J Appl Phys D 43:445501–445510
19. Li S, Chen G, Zhang M, Zhou Y, Zhang Y (2014) Dynamic keyhole profile during high-power deep-penetration laser welding. J Mater Process Technol 214:565–570
20. Cho W-I, Na S-J, Thomy C, Vollertsen F (2012) Numerical simulation of molten pool dynamics in high power disk laser welding. J Mater Process Technol 212:262–275
21. Bäuerle D (2011) Laser processing of physics and chemistry, 4th edn. Springer Series
22. Chelladurai AM, Gopal KA, Murugan S, Venugopal S, Jayakumar T (2014) Energy transfer modes in laser beam welding (Accepted manuscript). https://doi.org/10.1080/10426914.2014965829

Influence of Chemical Treatment on Natural Fibers: A Review

G. Venkatakoteswara Rao, R. Vinayagamoorthy, K. Abinesh, M. Sudharsan, S. Ponmeganathan, and L. S. Deepak Kumar

Abstract Natural fiber composites are proven to be biodegradable nowadays and are applicable to several sectors. Many natural fibers are used as reinforcements in polymeric composites with and without pretreatments. Many studies have reported that the chemical pretreatment to natural fibers enhance the surface properties of the fibers and which in turn improves the strength of the composite. Though there are several treatment techniques, sodium hydroxide is majorly used as the chemical for treating the fiber. Apart from this, there are several other treatments namely benzoylation, peroxidation, silane treatment, etc. The present study gives a clear picture of various chemical pretreatments given to the natural fibers and their processing methodologies. These treatments are bond between reinforcement surface and also increases its strengthening mechanism.

Keywords Natural reinforcements · Compound treatments

1 Introduction

Natural fibres like hemp, sisal, flax, kenaf, and jute are extremely hydrophilic because of the presence of hydroxyl group teams (OH) of anhydroglucose repetition unit in polysaccharide structure. However, these fibres are coated with cellulose and waxy materials, thus hindering the hydroxyl group teams from reacting with chemical compound matrices. This will result in the formation of ineffective interfaces between the fibres and matrices, with resultant problems like debonding and voids in ensuing composites [1–5]. Chemical treatments provide a vital and effective means that to get rid of non-cellulosic elements in cellulose fibres and add practical teams to change higher bonding in chemical compound composites. Additionally, treatment will alter the crystalline structure of the polysaccharide as well as fibre tensile properties [6–12]. The main disadvantage of victimization fibres in chemical

G. Venkatakoteswara Rao (✉) · R. Vinayagamoorthy · K. Abinesh · M. Sudharsan · S. Ponmeganathan · L. S. Deepak Kumar
Sri Chandrasekharendra Saraswathi Viswa Mahavidyalaya University, Enathur, Kanchipuram, Tamil Nadu 631561, India
e-mail: venkatakoteswararao02@gmail.com

T. Rajmohan et al. (eds.), *Advances in Materials and Manufacturing Engineering*, Springer Proceedings in Materials 7, https://doi.org/10.1007/978-981-15-6267-9_60

compound composites is that the incompatibility between natural fibre and chemical compound matrix because of the nearness of compound components (e.g., cellulose, lignin, hemicellulose, and wax substances). The natural fibers having various properties like biodegradable are comfortable to wear and not harmful for environment. Advantages of common fibers over counterfeit fibers embody tenuity, minimal effort, recyclability, and biodegradability [6, 13–17]. These benefits build characteristic fibers potential swap for glass fibers in composite materials. Mechanical properties of normal fibers, especially flax, hemp, jute, and sisal, are super band which should contend with glass fiber in specific quality and modulus. Table 1 records the mechanical properties of some common and fake fibers. Fiber reinforced composites are utilized as a replacement for plastics and hence they are used for several applications. They are made by reinforcing fibers inside the matrix element and are known to be fibre-strengthened composites. The pieces of normal fibers embody polyose, hemicelluloses, lignin, gelatin, waxes, and water solvent substances [6, 18–23].

2 Chemical Treatments

2.1 *Alkaline Treatment*

The treatment on regular filaments by sodium hydroxide (NaOH) is as a rule broadly used to alter fiber structure. Characteristic fiber assimilates dampness because of the nearness of hydroxyl bunches in the nebulous district of cellulose, hemicellulose, and lignin constituents. Amid soluble base treatment, alkalized gatherings (NaO–H) respond with these hydroxyl bunches (–OH) of the fiber and produce water particles (H–OH) which are therefore expelled from the fiber structure. At that point, the remaining alkalized gatherings ($Na–O^-$) respond with the fiber cell divider and produce fiber-cell-O–Na gatherings. Therefore, cellulose microfibrils are presented to the fiber surface. Thusly, treatment changes the introduction of the exceptionally stuffed crystalline cellulose request, shaping a formless region. Alkaline treatment additionally isolates the basic strands from their fiber packages by evacuating the covering materials, hence expanding the powerful surface zone of fiber for framework bond and improving the fiber scattering inside the composite. Treated fiber surfaces become rougher, which can additionally improve fiber-grid attachment by giving extra fiber locales to mechanical interlocking [1–6].

$$\text{Fiber} - \text{OH} + \text{NaOH} \rightarrow \text{fiber}-\text{O} - \text{Na} + \text{H}_2\text{O} \tag{1}$$

2.2 *Acetylation Treatment*

Strands are acetylated with and without a corrosive impetus to join the acetyl bunches onto the cellulose surface. By and large, acidic corrosive does not respond adequately with the filaments. Subsequently, it is important to go through an impetus to speed the acetylation procedure. Acidic anhydrides, pyridine, sulphuric corrosive, potassium and sodium acetic acid derivation, and so forth are generally utilized impetuses for acetylation process. Be that as it may, solid corrosive impetuses cause hydrolysis of cellulose which results in the harming of the fiber structure. For this, choice of impetus is an imperative factor for the acetyl treatment. The acetylation process is additionally affected by the response time. Longer response time permits acidic anhydride to get to fiber constituents. The reagent at that point responds with hemicellulose and lignin constituents and expels them from the fiber, bringing about the opening of cellulose surface to permit response with the lattice atoms [1–6].

$$\begin{aligned} &\text{Fiber} - \text{OH} + \text{CH}_3{-}\text{C}(=\text{O}){-}\text{O}{-}\text{C}(=\text{O}) \\ &- \text{CH}_{3x} \rightarrow \text{Fiber} - \text{OCOCH}_3 + \text{CH}_3\text{COOH} \end{aligned} \tag{2}$$

2.3 *Benzoylation Treatment*

Benzoyl treatment utilizes benzoyl chloride to diminish the hydrophilic idea of the fiber and improves interfacial grip, in this manner expanding the quality of the composite. Treatment likewise improves the warm security of the fiber. Amid this treatment, extractable materials, for example, lignin, waxes, and oil covering materials, are expelled from the fiber and more hydroxyl bunches (−OH) connected with cellulose are uncovered on the fiber surface. At that point, the filaments are treated with benzoyl chloride. Gracious gatherings of the fiber are additionally supplanted by the benzoyl gathering, and it connects to the cellulose spine. These outcomes in a progressively hydrophobic nature of the fiber and improves bond with the framework [1–6].

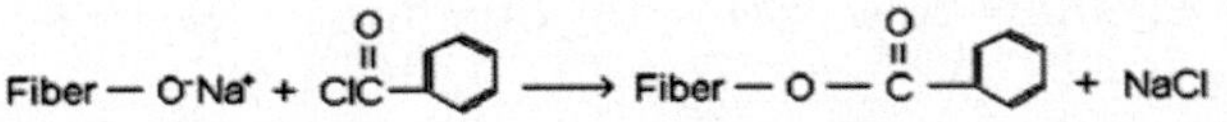

(3)

2.4 *Peroxide Treatment*

Interface properties of fiber and lattice can be improved by peroxide treatment. The peroxide-incited uniting of polyethylene holds fast to the fiber surface. Moreover, peroxide starts free radicals that respond with the cellulose surface of the fiber too similarly as with the framework. Therefore, great fiber network attachment along the interface happens. This treatment additionally lessens the dampness ingestion limit of the fiber and improves warm soundness [1–6].

$$\mathrm{RO{-}OR} \rightarrow \mathrm{2RO} \tag{4}$$

$$\mathrm{RO} + \mathrm{PE} - \mathrm{H} \rightarrow \mathrm{ROH} + \mathrm{PE} \tag{5}$$

$$\mathrm{RO} + \mathrm{Cellulose{-}H} \rightarrow \mathrm{ROH} + \mathrm{Cellulose} \tag{6}$$

$$\mathrm{PE} + \mathrm{Cellulose} \rightarrow \mathrm{PE{-}Cellulose}. \tag{7}$$

2.5 *Silane Treatment*

Silane is a multifunctional particle which is utilized as a coupling specialist to alter fiber surfaces. Silane particles structure a substance interface between the fiber surface and the framework through a siloxane connect. Silane experiences a few phases of hydrolysis, buildup, and bond development amid the treatment procedure of the fiber. Silane frames silanols within the sight of fiber dampness. This coreactivity gives subatomic progression over the interface of the composite. It likewise gives the hydrocarbon chain that limits the fiber swelling into the network. Amid the silane treatment, hydroxyl bunches on the fiber surface are secured by silane atoms. Because of this, hydroxyl bunches that presents in hemicellulose and lignin constituents cannot assimilate the air dampness. Accordingly, dampness retention limit of the treated strands is decreased [1–6].

$$\mathrm{CH_2CHSi\ (OC_2H_5)_3} \xrightarrow{\mathrm{H_2O}} \mathrm{CH_2CHSi(OH)_3} + \mathrm{3C_2H_5OH} \tag{8}$$

$$\mathrm{CH_2CHSi(OH)_3} + \mathrm{Fiber{-}OH} \rightarrow \mathrm{CH_2CHSi(OH)_2O} - \mathrm{Fiber} + \mathrm{H_2O}. \tag{9}$$

2.6 *Permanganate Treatment*

Permanganate treatment on regular filaments is led by potassium permanganate ($KMnO_4$) in $CH3)_2CO$ arrangement. This treatment shapes profoundly responsive permanganate particles (Mn_3+) which respond with the cellulose (hydroxyl gatherings) and structure cellulose manganate and starts to join copolymerization. This treatment upgrades compound interlocking at the interface and furnishes better fiber attachment with the grid. Treatment likewise responds with the OH bunches in lignin and expels them from the fiber, and therefore, the hydrophilic idea of the fiber is diminished. Higher convergences of $KMnO_4$ (over 1%) cause overabundance delignification from the fiber structure and corrupt its qualities.[1–6].

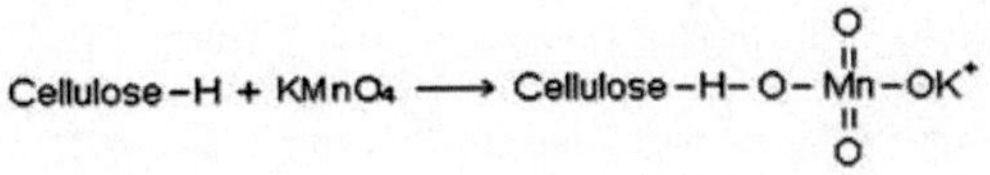
(10)

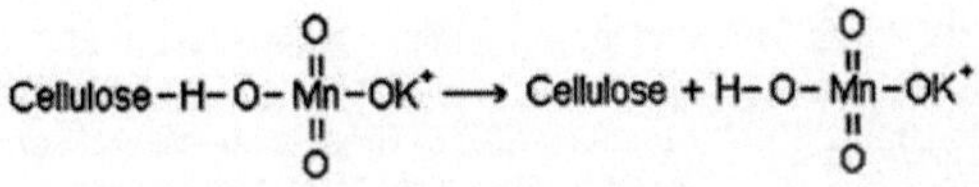
(11)

3 Conclusion

Regular fibers are considered as potential trade for man-made fibers in composite materials. Albeit common fibers have points of interest of being ease and low thickness, they are not absolutely free of issues. A major issue of regular fibers is their solid polar character which makes inconsistency with most polymer lattices. Surface medications, in spite of the fact that negatively affecting financial aspects, are possibly ready to conquered the issue of contrariness. Synthetic medications can expand the interface attachment between the fiber and lattice and decline the water ingestion of fibers. Consequently, concoction medications can be considered in altering the properties of characteristic fibers. A few mixes are known to advance grip by artificially coupling the cement to the material, for example, sodium hydroxide, silane, acidic acid, and permanganate, peroxide, and so on. Fiber modification techniques examined in this paper have distinctive efficacy in causing grip between the lattice and the fiber. However, most compound medications have made different dimensions of progress in improving fiber quality, fiber fitness, and fiber–lattice bond in characteristic fiber-fortified composites.

References

1. Li X, Tabil LG, Panigrahi S (2007) Chemical treatments of natural fiber for use in natural fiber-reinforced composites a review. J Polym Environ 15:25–33
2. Sawpan MA, Fernyhough KL, Fernyhough A (2011) Effect of various chemical treatments on the fibre structure and tensile properties of industrial hemp fibres. Compos Part A Appl Sci Manuf 42(8):888–895

3. Hashim MY, Amin AM, Marwah OMF, Othman MH, Yunus MRM, Huat NC (2017) The effect of alkali treatment under various conditions on physical properties of kenaf fiber. J Phys Conf Ser 914:012030
4. Paul S, Joseph K, Mathew G, Pothen L, Thomas S (2010) Influence of polarity parameters on the mechanical properties of composites from polypropylene fiber and short banana fiber. Compos A Appl Sci Manuf 41(10):1380–1387
5. Rahman MM, Mallik AK, Khan MA (2007) Influences of various surface pre-treatments on the mechanical and degradable properties of photo grafted oil palm fibers. J Appl Polym Sci 105(5):3077–3086
6. Vinayagamoorthy R, Rajmohan T (2018) Machining and its challenges on bio-fibre reinforced plastics: a critical review. J Reinf Plast Compos 37:1037–1050
7. Vinayagamoorthy R, Venkatakoteswara Rao G (2019) Synthesis and property analysis of green resin based composites. J Thermoplast Compos Mater. https://doi.org/10.1177/0892705719828783
8. John M, Anandjiwala R (2008) Recent developments in chemical modification and characterization of natural fiber reinforced composites. Polym Compos 29:187–207
9. Rajmohan T, Vinayagamoorthy R, Mohan K (2018) Review on effect machining parameters on performance of natural fibre–reinforced composites (NFRCs). J Thermoplast Compos Mater. https://doi.org/10.1177/0892705718796541
10. Vinayagamoorthy R (2019) Influence of fiber surface modifications on the mechanical behavior of Vetiveria zizanioides reinforced polymer composites. J Nat Fib 16:163–174
11. Vinayagamoorthy R (2018) A review on the machining of fiber-reinforced polymeric laminates. J Reinf Plast Compos 37:49–59
12. Kalia S, Kaushik VK, Sharma RK (2011) Effect of benzoylation and graft copolymerization on morphology, thermal stability, and crystallinity of sisal fibers. J Nat Fibers 8:27–38
13. Vinayagamoorthy R (2017a) A review on the polymeric laminates reinforced with natural fibers. J Reinf Plast Compos 36:1577–1589
14. Vinayagamoorthy R, Koteshwar TN, Madhav VV, Sai TK, Konda V (2019) Drilling associated parametric investigations on chemically treated natural fiber composite. Mater Today Proc 16:277–283
15. Mylsamy K, Rajendrann I (2010) Investigation on physio-chemical and mechanical properties of raw and alkali-treated agave Americana fiber. Journal 29(19):2925–2935
16. Vinayagamoorthy R, Konda V, Tonge P, Koteshwar TN, Premkumar M (2019) Surface roughness analysis and optimization during drilling on chemically treated natural fiber composite. Mater Today Proc 16:567–573
17. Vinayagamoorthy R, Manoj IV, Narendra Kumar G, Sai Chand I, Sai Charan Kumar GV, Suneel Kumar K (2018) A central composite design based fuzzy logic for optimization of drilling parameters on natural fiber reinforced composite. J Mech Sci Technol 32: 2011–2020
18. Wang B, Panigrahi S, Tabil L, Crerar W (2007) Pre-treatment of flax fibers for use in rotationally molded bio composites. J Reinf Plast Compos 26:447
19. Vinayagamoorthy R, Sharma A, Iyer V, Navneeth G (2018) Investigation of surface damages in hole making on luffa/jute/glass reinforced plastics. Adv Manuf Process Lect Notes Mech Eng 521–532
20. Vinayagamoorthy R (2020) Influence of fibre pretreatments on characteristics of green fabric materials. Polym Polym Compos. https://doi.org/10.1177/0967391120943461
21. Vinayagamoorthy R, Manoj IV, Narendra Kumar G, Sai Chand I, Sai Charan Kumar GV, Suneel Kumar K (2018) Challenges on the synthesis, characterization and machining of green fiber plastics: a review. IOP Conf Ser Mater Sci Eng 390: 012029
22. de Oliveira PF, Marques MFV (2014) Comparison between coconut and curaua fibers chemically treated for compatibility with PP matrixes. J Reinf Plast Compos 33(5): 430–439
23. Vinayagamoorthy R (2019) Trends and challenges on the development of hybridized natural fiber composites. J Nat Fib. https://doi.org/10.1080/15440478.2019.1598916

Optimization on Tribological Characteristics of Waste Ayurvedic Oil Blends as an Engine Lubricant Additive

R. Balakumar, G. Sriram, S. Arumugam, S. Jagannath, R. Vamsi Krishnaa, and P. Venkatesh

Abstract The fast exhaustion of fossil fuels and problems with environment are activating the researches on alternative fuels. The exploitation of wastes will lead the saving in the original feed stocks used. The waste oil will be the source for extra damages to the environment when it discarded in the ground or in water sources. Out of many waste oils (waste cooking oils, waste transformer oils, waste engine oils), the waste ayurvedic oil (WAO) observes a minimal attention as alternate fuels utilized. In a year, more than one lakh liters of ayurvedic oil after treatment is getting wasted in India, and disposing the oil becomes very crucial process. The WAO has converted as methyl ester using transesterification process as waste ayurvedic oil methyl ester (WAOME). Furthermore, the lubricating engine oil, i.e., SAE20W40, was contaminated with WAO as an identified percentage of 15 and 30%. Further, the utilization of real-time cylinder liner/piston ring tribo pair material is investigated using pin-on-disc tribometer under standard test condition. The tribological experimentations were done as per the standard ASTM G99 with an applied load of 50, 100, and 150 N; sliding speed 500 rpm; sliding distance 1000, 2000, and 3000 m; and for the period of totally five hours. For this present investigation, response surface methodology-based D-optimal design was used to get the optimized results. The test results clear that the WAOME contaminated lubricant gives the reduction in friction and wear

R. Balakumar (✉) · G. Sriram · S. Arumugam · S. Jagannath · R. Vamsi Krishnaa · P. Venkatesh
Department of Mechanical Engineering, Sri Chandrasekharendra Saraswathi Viswa Mahavidyalaya, Enathur, Kanchipuram, Tamil Nadu 631561, India
e-mail: balavhd@gmail.com

G. Sriram
e-mail: drg.sriram@gmail.com

S. Arumugam
e-mail: aru_amace@yahoo.co.in

S. Jagannath
e-mail: jaganathsriram@gmail.com

R. Vamsi Krishnaa
e-mail: vamsikrish1997@gmail.com

P. Venkatesh
e-mail: venkatesh2810@gmail.com

T. Rajmohan et al. (eds.), *Advances in Materials and Manufacturing Engineering*, Springer Proceedings in Materials 7, https://doi.org/10.1007/978-981-15-6267-9_61

rate when compared with synthetic oil. After the investigation, the wear surfaces of pin were examined by scanning electron microscopy (SEM). The complete results of this investigation expose that the addition of 30% waste ayurvedic oil with the synthetic-based lubricant results a healthier performance and good anti-wear characteristics. The waste ayurvedic oil blend can be utilized as an alternate fuel which is ecofriendly in the nature and partial replacement of petroleum-based fossil fuels.

Keywords Waste ayurvedic oil · RSM · D-optimal · Pin on disc · Friction coefficient · Specific wear rate

1 Introduction

In ever growing industrialization and the present world condition, it largely depends on petroleum-based fossil fuels in a wide areas of power generation, fuels for vehicles, and ever burning area of agriculture sectors. The hike in the fuel price, partial standby of petroleum oils, and constantly fluctuating emission regulations enforced the investigators which made keen to travel on alternative fuels [1]. The cumulative attention in the research on alternate fuels given that it not meant only for the escape from the petroleum-based fossil fuel crisis and similarly reduce the gases being discharged in the process of combustion of these fuels [2]. The conventional oil from petroleum oil producing fields presently establishes roughly 85% of the global liquid fuel mix and is predictable to weakening at a rate of 4.07% per year after 2010 [3]. Obviously, researchers have found different biomass-based liquid fuels such as alcohol, biodiesel, and other liquid fuels synthesized and characterized from the various crops, plants, and different waste oil sources as alternate fuel [4]. On the other hand, the use of edible and non-edible vegetable oil makes the method of production of biofuels that are more expensive. An investigation on production of biodiesel from several resources reveals that the input source itself costs nearly 75% of the entire cost of production of biodiesel [5]. The utilization of low-cost feedstock such as waste frying oils, waste vegetable oils, waste transformer oils, waste engine oils, and non-edible oils in biodiesel process should help as inexpensive in price with petroleum-based diesel [6–8]. Among many researches based on the many oil and waste oil studies, the waste ayurvedic oil (WAO) has gained a miniature attention to the researchers. Balakumar et al. [9] investigated the most common way to produce biodiesel via transesterification process, which refers to a catalyzed chemical reaction connecting to WAO and an alcohol to fatty acid methyl esters and glycerol. In IC engines, the various engine components produce friction between the metal–metal contact, which in-turn reduces the engine life, and hence, lubricity behavior of fuel is one of the most predominant factors in extending engine reliability [10].

Masarof et al. [11] investigated the wear and frictional characteristics of palm and calophyllum inophyllum biodiesel blends in addition to performance and emission behavior. They confirmed that 20% of palm oil biodiesel blend displayed the favourable engine performance, lower emissions together with better lubricating performance. Dae et al. [12] investigated the utmost typical factor of friction and wear loss in engine cylinder liner and piston ring combination pair with the synthetic oil lubricant SAE5W40. Holmberg et al. [13] approximately 70% of the mechanical power loss in the engines is especially due to friction. The more frictional losses of about 30% of engine is engine bearings and cylinder liner-piston ring pair. There are many parameters that are influencing the friction and wear characteristics in pin on disc to optimize the process parameters, and the design expert software was used to optimize the optimum values. Arumugam and Sriram [14] investigated the rapeseed oil-based biolubricant, which decreases the coefficient of friction and wear as compared to biodiesel and diesel-contaminated lubricating oil and synthetic lubricant (SAE20W40). By viewing the surface, the rapeseed oil-based biolubricant produces evener surface than biodiesel-contaminated lubricating oil. Myers et al. [15] and Rajmohan et al. [16] investigated about the response surface methodology (RSM) is an important division of experimental design. An optimal experimental plan can be calculated by the use of the methodology of optimal experimental design for parameter approximation. From the above studies, it is clear that the RSM-based mathematical model is used for modelling and analyzing the tribological characteristics of cylinder liner and piston ring tribo pair lubricated by synthetic oil contaminated by WAOME.

2 Materials and methods

The input waste source, i.e., WAO, was received from Sri Jayendra Saraswathi Ayurveda College and Hospital, Chennai, India. The WAO was filtered to eliminate unwanted impurities. Initially, to convert the WAO, the methoxide of 400 ml and WAO 1400 ml was mixed entirely and heated in a triple neck flask at 55 °C for one hour using transesterification test setup, and the heated mixture was transferred to a separating funnel to settle for half a day to separate the glycerol. Then, the detached methyl ester was collected and heated for 60 °C and above to eliminate the additional alcohol. The bubble wash was done with distilled water to remove the excess alcohol, catalyst. Finally, the WAOME was contaminated with SAE20W40 blends that were prepared. In this investigation, WAO was added with a synthetic lubricant (SAE20W40) in the ratios of SAE20W40; WAO15 (SAE20W40+WAOME15%); WAO30 (SAE20W40+WAOME30%). Table 1 shows the properties of WAO biodiesel blends and diesel fuel.

Table 1 Properties of WAO biodiesel blends and diesel fuel

Properties	SAE20W40	WAO15	WAO30
Specific gravity at 20 °C	0.843	0.856	0.897
Viscosity @40 °C, cSt	164	171	118
Flash point, °C	218	232	258
Total acid number	0.5672	0.5604	0.6812

Table 2 Numerical factors, categorical factor, and levels considered during the design of experiments

Factors	Level 1	Level 2	Level 3
Load (N)	50	100	150
Sliding distance (m)	1000	2000	3000
Lubricating oil	SAE20W40	WAO15	WAO30

The typical pair of cylinder liner/piston ring pair material is studied using pin-on-disc tribometer. The disc was made of AISI-316 material and the pin was made of AISI-304 were machined as specified by the tribometer machine supplier, and the heat treatment was carried out for the pins to achieve the essential hardness. Specifications of the pin-on-disc tribometer are: maximum load = 50–150 N; maximum sliding distance = 1000–3000 m; maximum disc rotating speed = 500 rpm;

The purposes of using RSM is to improve the quality, including decrease in variability and improvement in the process and product performances. Further, the D-optimal design is used as another method to formulate the optimal design when some categorical factors are considered in the testing. An RSM-based D-optimal design was used for designing the tribological experiments to be investigated using pin-on-disc tribometer to optimize the output responses of coefficient of friction (COF) and specific wear rate (SWR). An experimental plan of three factors like two numerical and one categorical factor in D-optimal design based on RSM was used to do the experiments using pin-on-disc tribometer. The load and sliding distance were nominated as the numerical factors, and the lubricating oil was categorical factor. Both the factors were considered at three levels as given in Table 2, and as per the D-optimal design, 22 experimentations were carried out and listed in Table 3.

3 Results and discussion

In order to discuss about the tested materials, it needs to analyze the coefficient of friction, specific wear rate, and analyze its surface roughness and SEM morphology. The values of statistical correlation coefficients, 'R^2', value 0.831264972 and specific wear rate, 'R^2' value 0.785877281 for the optimized model clearly gives the clue about the correctness of the recommended model. In this investigation, the concluding quadratic models given in the real terms about coefficient of friction and specific wear rate are represented in Eqs. 1–2. Each categorical factor has a separate equation that estimates the output responses based on the lubricant samples and the numerical factors. The positive and negative values indicate that the increase in the COF and SWR in association with the variable input parameters.

$$\begin{aligned}(\text{WAO})\text{COF} = {} & 0.28573917 + 2.5075E^{-05} \\ & \times \text{Load} - 5.93472E^{-05} \times \text{Slidingdistance} \\ & + 1.98487E^{-07} \times \text{Load} \times \text{Slidingdistance} \end{aligned} \quad (1)$$

Table 3 D-optimal design matrix

Run	Load (N)	Sliding distance (m)	Lubrication oil	COF	SWR (mm^3/Nm)
1	150	1000	SAE20W40	0.209	0.00029
2	150	1000	WAO15	0.197	0.00031
3	91	3000	WAO30	0.154	0.00013
4	50	3000	SAE20W40	0.251	0.00023
5	150	3000	WAO15	0.236	0.00035
6	50	1838	WAO30	0.174	0.0002
7	150	3000	SAE20W40	0.298	0.00026
8	50	1000	WAO15	0.211	0.00011
9	150	2164	WAO30	0.239	0.00014
10	50	1000	SAE20W40	0.173	0.00033
11	100	2000	WAO15	0.192	0.00012
12	108	1000	WAO30	0.265	0.00028
13	100	1535	SAE20W40	0.223	0.00025
14	71	2938	WAO15	0.211	0.00018
15	100	2043	WAO30	0.202	0.00021
16	100	2516	SAE20W40	0.238	0.00027
17	111	2813	WAO15	0.216	0.00036
18	108	1000	WAO30	0.225	0.00032
19	91	3000	WAO30	0.164	0.00031
20	50	1000	WAO15	0.202	0.00019
21	150	1000	SAE20W40	0.178	0.0003
22	50	1838	WAO30	0.232	0.00024

$$\begin{aligned}(\text{WAO})\text{SWR} = 0.28573917 + 2.5075E^{-05} \\ \times \text{Load} - 5.93472E^{-05} \times \text{Slidingdistance} \\ + 1.98487E^{-07} \times \text{Load} \times \text{Slidingdistance}\end{aligned} \quad (2)$$

The discrepancy in COF and SWR with applied load and sliding distance for the different contaminated lubricant with WAOME and synthetic lubricant mixtures is represented as a 3D graph in Fig. 1a, b. From Fig. 1a, it can be seen that the COF rises with increase in sliding distance and increasing load and reaches the extreme level. For maximum sliding distance with maximum load, the COF is high due to more pressure acting between the disc and pin materials. The COF value fluctuates between 0.154 and 0.298. The synthetic lubricant gives more COF than the WAOME due to more ester content present in the mixture. The results were consistent with Singh et al. [17].

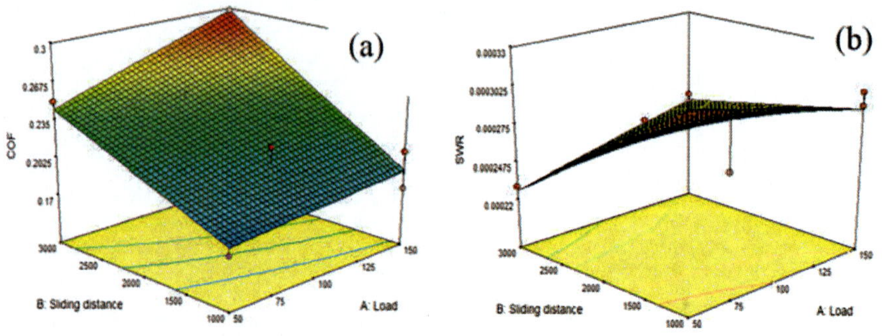

Fig. 1 3D response surface plot for **a** COF, **b** SWR

3.1 Specific Wear Rate

Figure 1b represents the 3D response surface plot of specific wear rate regarding sliding distance and load configuration. From the graph when the applied load and sliding distance increases, the SWR reaches the maximum level. However, for higher load, the SWR is high, and it is minimally reduced with increase in the sliding distance. The SWR value fluctuates between 0.00011 and 0.00036 mm^3/Nm. The reduction in the SWR is due to WAOME acting as a better anti-wear additive between the mating parts [17].

Figure 2 shows the ramp function that signifies each input and output parameter responses. The dot on each ramp specifies the desired input parameters corresponding to the minimal COF and specific wear rate. From the ramp function plots, it is clear that a load of 750 N, sliding distance of 3000 m, lubricating oil of SAE20W40+30% WAOME were essential to reduce the COF as 0.153995 and specific wear rate of 7.4283×10^{-5} mm^3/Nm and the corresponding desirability value is 0.917.

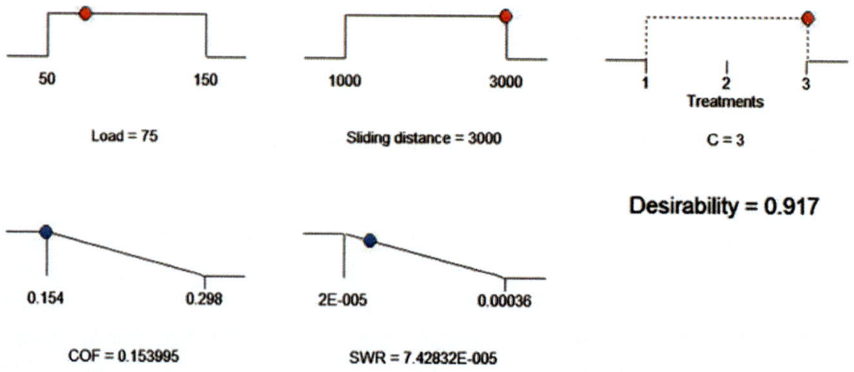

Fig. 2 Ramp function

Table 4 Surface roughness values

	WAO30	WAO15	SAE20W40
Ra (μm)	3.164	3.491	3.536
Sa (nm)	107.41	196.7	427.1
Sq (nm)	163.88	260.17	569.2
Sp (nm)	1081.7	1262.4	1779.2

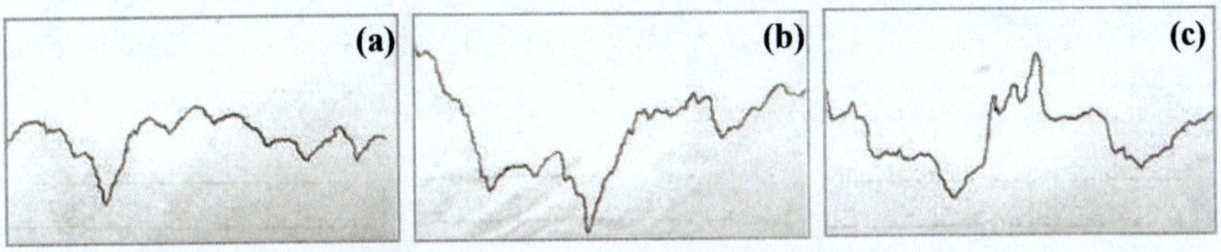

Fig. 3 Surface profile **a** WAO30, **b** WAO15, **c** SAE20W40

In accumulation to the SEM analysis, the surface profile for the investigated pin was studied to estimate the surface roughness such as synthetic lubricating oil and contaminated oil. The surface profilometer evaluates the texture of the surface profiles and produces many parameter such as average roughness (Ra), root mean square height (Sq), arithmetic mean height (Sa), and maximum peak height (Sp) as given in Table 4.

The surface profiles of tested pins are shown in Fig. 3a–c. From the surface profilometry, it is understood that average roughness values are minimal for WAO30 lubricant about 3.164 μm compared to WAO15 and SAE20W40. Similarly, for other lubricants, viz WAO15 and SAE20W40, the Ra value is about 3.491 μm and 3.536 μm, respectively.

3.2 *SEM analysis*

After the optimization process, the experiments with various parameters were investigated, and the optimal pin samples were examined using scanning electron microscope (SEM) which is shown in Fig. 4a–c. There were more pits, and deep grooves and cracks were observed in the synthetic lubricant. However, the WAO contaminated lubricant gives better scuffing between the mating parts which leads less cavities and shallow grooves [11]. From the SEM images, it is clear that when the ester concentration increases which means the higher viscosity that results to maintain a good fence between the mating parts like disc and pin materials.

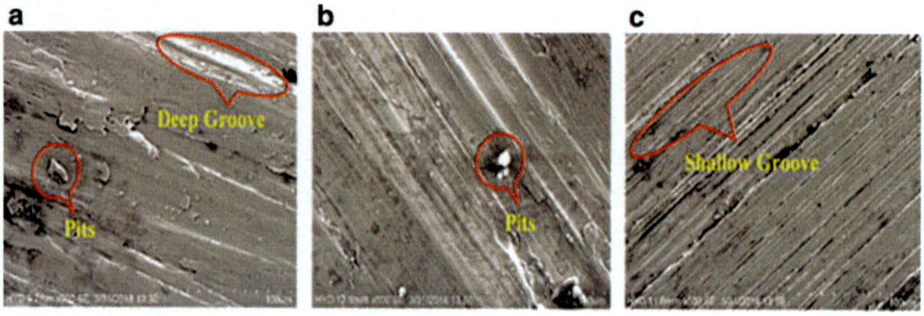

Fig. 4 SEM images of pin **a** SAE20W40, **b** WAO15, **c** WAO30

4 Conclusions

The response surface methodology-based D-optimal method was applied to optimize the coefficient of friction and specific wear rate using pin-on-disc tribometer. The results were concluded as follows:

- The RSM-based D-optimal model was used to analyze the WAO contaminated lubricant which gives the desirability value of 0.917.
- The optimized values of load, sliding distance, and lubricants were 75 N, 3000 m, and WAO30, and the values of COF and SWR were noted as 0.153995 and 7.42832E-005.
- As per the surface profile, the WAO30 produces the minimal average roughness value as 3.164 μm than the other blends.
- The morphology of the wear surface of tested pins were investigated using SEM. The worn surface of the pin material lubricated with WAO30 is shows as smoother surface as compared to other blended lubricant and synthetic lubricant.

References

1. Ong HC, Silitonga AS, Masjuki HH (2013) Production and comparative fuel properties of biodiesel from non-edible oils: jatropha curcas, sterculia foetida and ceiba pentandra. Energ Convers Manage 73:245–255
2. Brennan L, Owende P (2010) Biofuels from microalgae—a review of technologies for production, processing, and extractions of biofuels and co-products. Renew Sustain Energ Rev 14:557–577
3. Owen NA, Inderwildi OR (2010) The status of conventional world oil reserves hype or cause for concern. Energy Policy 38(8):4743–4749
4. Devan P, Mahalakshmi NA (2009) study of the performance, emission and combustion characteristics of a compression ignition engine using methyl ester of paradise oil–eucalyptus oil blends. Appl Energy 86:675–680
5. Yadav SPR, Saravanan CG, Vallinayagam R, Vedharaj S, Roberts WL (2015) Fuel and engine characterization study of catalytically cracked waste transformer oil. Energ Conv Manage 96:490–498

6. Zheng S, Kates M, Dube MA, McLean DD (2006) Acid-catalyzed production of biodiesel from waste frying oil. Biomass Bioenerg 30:2670–2672
7. Balakumar R, Sriram G, Arumugam S (2018) Waste ayurvedic oil as a engine fuel—a review. Adv Sci Eng Med 10(3):249–254
8. Zhang Y, Dube MA, McLean DD, Kates M (2003) Biodiesel production from waste cooking oil: 1. Process design and technological assessment. Bio resour Technol 89:1–16
9. Balakumar R, Sriram G, Arumugam S, Shree Navin A (2018) Analysis on tribological characteristics of waste ayurvedic oil biodiesel blends using four ball tribometer. IOP Conf Ser Mater Sci Eng 390: 012080
10. Fazal M, Haseeb A, Masjuki H (2011) Biodiesel feasibility study: an evaluation of material compatibility; performance; emission and engine durability. Renew Sustain Energy Rev 15:1314–1324
11. Mosarof MH, Kalam MA, Masjuki HH, Alabdulkarem A, Ashraful AM, Arslan A, Rashedul HK, Monirul IM (2016) Optimization of performance, emission, friction and wear characteristics of palm and Calophyllum inophyllum biodiesel blends. Energy Convers Manag 118:119–134
12. Dae HC, Sung AL, Young ZL (2010) The effects of surface roughness and coatings on the tribological behavior of the surfaces of a piston skirt. Tribol Trans 53:137–144
13. Holmberg K, Andersson P, Erdemir A (2011) Global energy consumption due to friction in passenger cars. Tribol Int 47:221–234
14. Arumugam S, Sriram G (2012) Effect of bio-lubricant and biodiesel-contaminated lubricant on tribological behavior of cylinder liner–piston ring combination. Tribo Trans 55:438–445
15. Myers RH, Montgomery DC (2002) Response surface methodology. Process and Products Optimization Using Designed Experiments. Wiley, New York
16. Rajmohan T, Palanikumar K (2013) Modeling and analysis of performances in drilling hybrid metal matrix composites using D-optimal design. Int J Adv Manuf Technol 64:1249–1261
17. Singh Y, Garg R, Kumar S (2016) Effect of load on friction and wear characteristics of Jatropha oil bio-lubricants. Biofuels 8:125–133

Effect of MWCNT on Mechanical Properties of Glass-Jute Fiber Reinforced Nano Composites

K. Mohan, T. Rajmohan, and R. Prasath

Abstract In the present scenario, the need for ecofriendly and biodegradable materials are growing significantly. Natural fiber composite materials are having the priority choice of various applications. Polymer matrix composite's containing with nano additives are being developed and projected for diverse applications in various fields of engineering like automotive and aeronautical sectors. The present investigation deals with the effect of multi-wall carbon nanotube (MWCNT) on the mechanical properties of glass–jute fiber reinforced epoxy composites. Composites were prepared using compression molding technique with dispersing different wt% of MWCNT in to the epoxy resin. In this work, the modified resin was prepared by dispersing the MWCNT into the epoxy resin using ultrasonic probe sonicator. As per ASTM standards, mechanical properties like tensile, compression, and flexural test as has been carried out. The composites which are prepared by using modified resin exhibiting good mechanical properties. Morphological characteristics of fractured mechanical testing specimens were performed using scanning electron microscope (SEM) with energy dispersive X-rays (EDS) to understand the bonding between fiber/matrix. XRD analysis were performed to ensure the presence of natural fiber, glass fiber, and MWCNT in the nano composites through the confirmation of cellulose, silicon, and carbon contents.

Keywords Natural fiber reinforced composites: multi-wall carbon nanotubes (MWCNT) · Scanning electron microscopy (SEM) · XRD · Mechanical properties

K. Mohan (✉) · T. Rajmohan
Centre for Composite Science and Tribology, Sri Chandrasekharendra Saraswathi Viswa Mahavidyalaya, Enathur, Kanchipuram, Tamil Nadu 631561, India
e-mail: mohank84@gmail.com

T. Rajmohan
e-mail: rajmohanscsvmv@yahoo.com

R. Prasath
Department of Mechanical Engineering, Sri Chandrasekharendra Saraswathi Viswa Mahavidyalaya, Enathur, Kanchipuram, Tamil Nadu 631561, India

T. Rajmohan et al. (eds.), *Advances in Materials and Manufacturing Engineering*, Springer Proceedings in Materials 7, https://doi.org/10.1007/978-981-15-6267-9_62

1 Introduction

The natural fibers like jute, sisal, banana, pineapple, bamboo, etc., are focused for an development of natural fiber reinforced composites which is replacing the conventional materials like metals, woods, etc. Recently, the polymer-based composites materials applications are increased due to their lightweight and good mechanical properties [1]. Depa ray et al. investigated the flexural strength of jute fiber vinyl estar composites. The jute fibers are treated with NAOH (5%) solution at 30 °C for 0, 2, 4, 6, 8 h, respectively. Flexural strength and modulus were improved by 20 and 23% for jute fiber reinforced composite at 35 wt% for 4 h treating with NAOH solution [2]. Cicala et al. investigated the performance of hybrid glass/natural fiber composite which is used in applications of curved pipes. Natural fibers are generally light in weight and cheaper compared to synthetic fibers. But, they have lower mechanical strength than glass fibers. This issue is sorted by doing hybridization. More number of studies have been conducted using natural fibers concerned with single reinforcement. For making hybrid composite, the natural fibers are combined with glass fibers which is comparatively cheaper and easy for use [3]. M. Ramesh et al. evaluated the mechanical properties of sisal/glass fiber and jute/glass fiber reinforced composites. They have concluded that better tensile strength is exhibited by sisal/GFRP than the jute/GFRP, and the flexural property of jute/GFRP is higher than the sisal/GFRP composites. It is holding the maximum flexural load of 1.03 kN compared to sisal/GFRP composites [4]. Dash B. N. et al. evaluated the mechanical properties of bleached jute fiber composite and control jute fiber composites. The control jute fiber composite shows highest tensile strength, and bleached jute fiber composites indicate highest flexural strength [5].

Ajith Gopinath et al. investigated jute—epoxy and jute—polyester composite tread with 5% and 10% NAOH, respectively. From the results, they concluded that over all mechanical property is improved in jute epoxy composites treated with 5% NAOH, and the impact energy for jute polyester composite is better than jute epoxy composites [6]. He Tian et al. studied mechanical behavior bagasse fiber reinforced composites containing high volume of fly ash and in conjunction with steel wires. While increasing fly ash, the compressive strength decreases and ductility of the composite increases. Due to the high porosity in the matrix, the mechanical performance of the composite declined continuously while increasing the bagasse fiber content [7]. Boopalan et al. investigated and concluded that the hybrid jute and banana epoxy composite has 44% more tensile strength in 50/50 weight ratio, and also this combination of fiber weight ratio possesses the better thermal properties and less water absorption capacity compared to 25/75, 75/25 fiber weight ratio. By addition of banana, fibers shows that there is an improvement in tensile, flexural, and impact strength [8]. M. Ramesh et al. fabricated the samples glass/jute, glass/sisal, and glass/jute/sisal, respectively. They have concluded that maximum tensile strength is obtained up to 229.54 MPa for glass/jute composites. Glass/sisal composite can have the maximum impact strength up to 18.67 J. Jute sisal and glass fiber composites are having the maximum flexural strength with 14.5 mm displacement and 3 kN load [9]. R. Vinayagamoorthy et al. reported an equal proportions of jute, loofah, and glass fibers that are exhibiting the good tensile, flexural, and impact strength. Compressive strength is not compromised when 10% of glass fibers substituted by loofah fibers [10]. Sabeel et al. reported the mechanical properties of woven jute-coir fiber reinforced composite by changing their stacking sequence. They have concluded that keeping the glass fibers at the extreme ends improves the mechanical properties [11]. Mohan et al. investigated the mechanical properties of natural fiber reinforced composites filled with an MWCNT. They have noticed that 1%MWCNT composites with jute at the extreme ends enhance the mechanical strength rather than the flax at the extreme ends reinforced composites [12]. Islam et al. have prepared nano composites using coir and wood fiber with MMT nano clay. They observed that the tensile strength is reduced due to the similar cellulose content present in the fibers. By adding the MMT, the composites show the better tensile strength and highest rate of water absorption [13]. The duration of sonication time on dispersion of MWCNT into the epoxy resin affects the mechanical properties [14, 15]. Boopalan et al. fabricated the jute and sisal fiber reinforced epoxy composites with and without treatment of sodium hydroxide. They have concluded that treated fiber with 20% sodium hydroxide exhibits the better mechanical properties rather than a raw jute and sisal reinforced composites [16]. Mohan et al. investigated mechanical properties of sisal fiber reinforced nano composites. While increasing the wt% of MWCNT, tensile strength is increased and flexural strength of the composites is decreased [17].

From the literature, the works that have been carried out using nano composites are very limited. In this present work by addition of multi-walled carbon nanotube (MWCNT) into the epoxy resin, a newly modified resin is prepared and used as a matrix reinforced with jute fiber and glass fiber. The developed composite used with modified resin is subjected to various mechanical testing and morphological characteristics that are analyzed using SEM.

2 Experimental

2.1 Materials Used

In this work for fabricating the samples, jute and sisal are procured from Eco craft enterprises, Osur. Glass fiber, epoxy resin araldite LY556, and hardener aradur HY951 are obtained from M/S. Sun Tech fibers, Chennai. Multi-wall carbon nanotube (MWCNT) manufactured by M/S US Research Nano materials Inc, USA, was used in this composites as an additive in epoxy resin.

Table. 1 Composition of various composites

S. no.	Particulars	Glass/Jute 1	Glass/Jute 2	Glass/Jute 3
1	Epoxy resin	60	59.5	59
2	Glass fiber	25	25	25
3	Jute fiber	15	15	15
4	MWCNT	0 wt%	0.5 wt%	1 wt%

2.2 *Resin Preparation*

In this present investigation, modified resin is prepared by adding the MWCNT at 0%, 0.5%, and 1% by wt%, respectively, into the epoxy resin. Ultrasonic probe sonicator is used to uniformly mix the MWCNT into the epoxy resin. Initially, the desired wt% of MWCNT and epoxy resin are taken into the beaker. Then, the beaker is kept into the ultrasonic probe sonicator and allowed to vibrate 3 h. For better dispersion, using rotary shaker, the same blend is allowed to stir well for 2 h to ensure homogeneous mixing of MWCNT into the epoxy resin without agglomeration. The resin and hardener were mixed at 10:1 ratio [18, 19].

2.3 *Composite Preparation*

For fabricating, the composite specimen's compression molding technique is used. In this investigation, the jute and glass fibers were taken in 30 cm length. Those fibers are arranged sequentially, whereas the glass fibers are used to keep at the top and bottom of the composite. Here, resin and fiber ratio 60:40 is considered [20]. According to that the required amount of resin as taken and respective wt% of MWCNT is incorporated with the resin. Then the modified resin is used to fabricate the specimen using compression molding machine. For curing, the specimen is kept in the compression machine for 24 h at the pressure range of 350 Psi. The various compositions of samples are as given in Table 1.

3 Results and Discussion

3.1 *Tensile Properties*

According to the ASTM D638, tensile test specimens were prepared. The specimens has cut from the laminate using band-saw and finished with the corners using file and emery sheet. There are three different variation specimens that are prepared. The specimens are having the content of MWCNT in the variation of 0 wt%, 0.5

wt%, and 1 wt%, respectively. The test has been carried out in the universal testing machine. The testing specimen is placed in the machine and subjected to tension until its fracture. Then, the tensile strength is recorded. The different composite specimens are subjected to tensile load in the universal testing machine. Figure 1 shows that the tensile strength variation for 0, 0.5, and 1% MWCNT incorporated glass-jute reinforced composites. The ultimate tensile strengths are plotted in the graph for different specimen. The result indicated that glass–jute composite incorporated with 1% MWCNT shows better tensile strength than the other two specimens. It has the tensile strength of 444.44 MPa followed by 401.36 and 373.57 MPa for glass–jute 0.5 and 0% incorporated MWCNT composites. Due to the higher tensile strength of MWCNT and also by keeping the glass fibers at the extreme layers of the reinforced composites enhance the tensile strength of the composites. Similar results are reported by Rajmohan et al. [18]. The tested tensile specimens are shown in Fig. 2, and the specimen which is having 1% MWCNT shows the more elongation than the others. From the SEM analysis, it is conformed that the fractured fibers of the specimen is elongated more during the tensile test. The uniform dispersion of MWCNT in to the epoxy resin also enhances the strength of the composites.

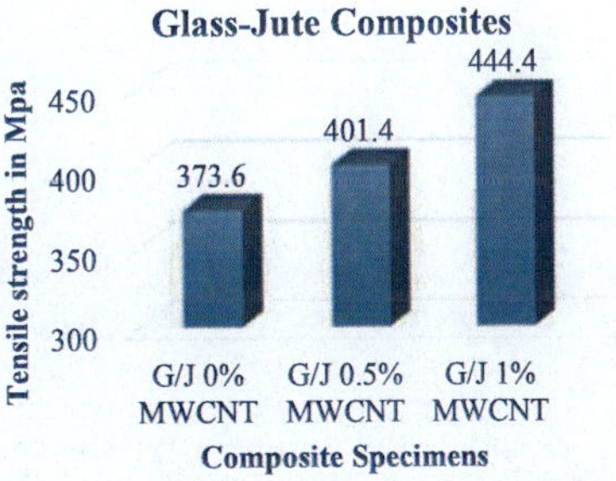

Fig. 1 Effect of wt% of MWCNT on tensile

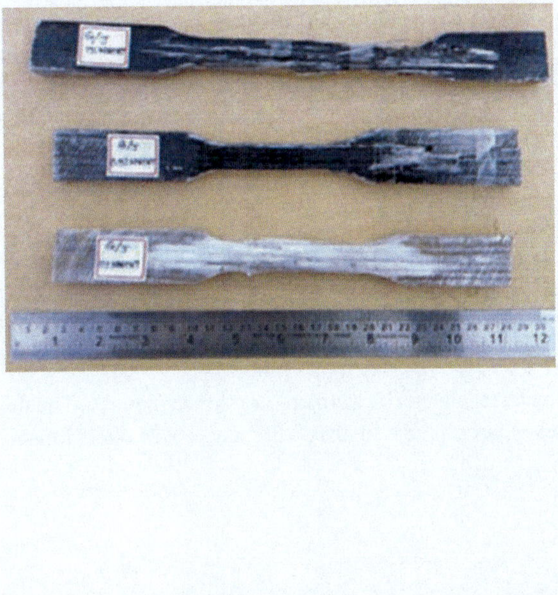

Fig. 2 Tested tensile specimen

Fig. 3 Effect of wt% of MWCNT on compressive

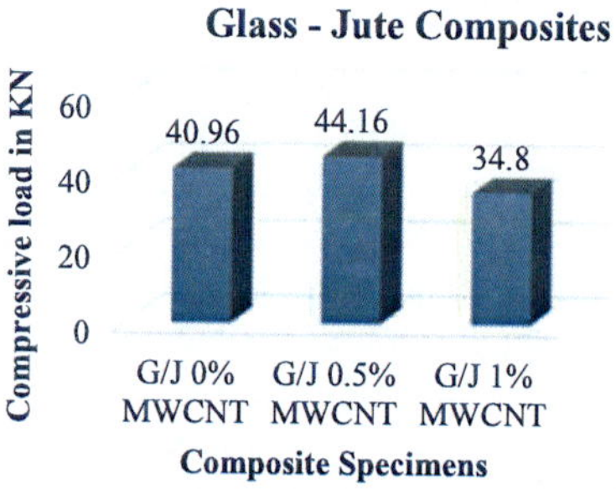

Fig. 4 Tested compressive specimen

3.2 Compressive Properties

Fig. 4 shows the compressive tests specimens. The ASTM standard D695 is used to carry out the test. From Fig. 3, it is clearly seen that the specimen which is incorporating the MWCNT with 0%, 0.5%, and 1% has the compressive strength of 40.96, 44.16, and 34.83 kN, respectively.

The compressive strength has increased 7.24% initially for 0.5% incorporated MWCNT. Then, the strength is decreased 22.12% for 1% incorporating the MWCNT. This may be due to the air voids presence in the composites, and the cluster formation of MWCNT also may reduce the strength of the composites. The cluster formation is also identified from the SEM images. Proper dispersion of MWCNT into the resin can increase the compressive strength of the composites.

3.3 Flexural Properties

The tested flexural specimens are shown in Fig. 6. The flexural strength of the composites incorporated with 0%, 0.5%, and 1% MWCNT has 7.55KN followed by 10.05 kN and 6.23 kN, respectively. It as shown in Fig. 5. From that, it is concluded that the flexural strength of the composite increases while incorporating with 0.5% MWCNT, and further by increasing the wt% of MWCNT, the strength of the composite is decreased. The reduction in flexural strength of composites may be due to the improper blending of MWCNT in to the epoxy resin, this also can lead to the cluster formation of MWCNT in the composites. The flexural strength not only depends upon addition of nanoparticles but also an arrangement and sequence of fiber layers, bonding between fibers and matrix. The stress transfer between fiber and matrix is increased when the adhesion between fiber and matrix increased [19]. Since jute is high modulus material, for the same deformation, higher fiber concentration demands the higher stress. The flexural strength also depends upon the interfacial bonding between fiber and matrix, other than the strength of the top and bottom layer of the reinforcement composites. Poor flexural properties are exhibited when the interfacial bonding between fibers and matrix is weak.

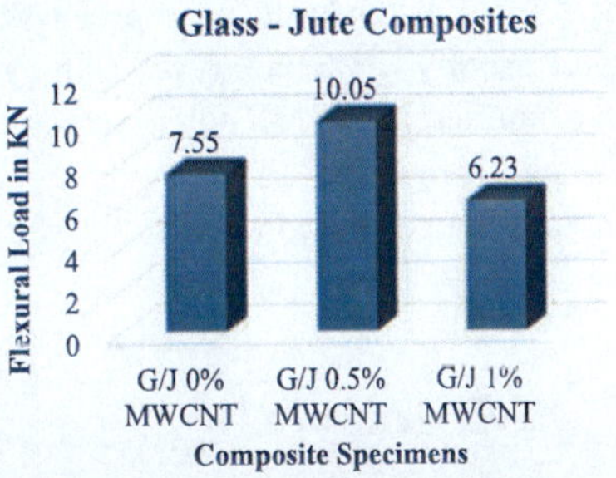

Fig. 5 Effect of wt% of MWCNT on flexural

Fig. 6 Tested flexural specimen

4 Microstructure Analysis

The effect of MWCNT on microstructure, pulled out fibers during tensile test, bonding between fibers/matrix, and dispersion of nanoparticles into the resin is analyzed using scanning electron microscope (SEM). Figs. 7a and 8a show the fractured pulled out tensile specimens of the composite incorporated with 0.5 and 1% MWCNT. The composite which is having 0.5% MWCNT shows more breakage of fibers compared to 1% MWCNT composite. Fig. 9a confirms the more elongation of fibers during tensile test with increasing the wt% of MWCNT which enhances the tensile strength of the composites. Figs. 7b and 8b show the dispersion of MWCNT into the resin. The 0.5% incorporated MWCNT is dispersed uniformly into the resin rather than 1% MWCNT. The formation of clusters in the composites while increasing the wt% of MWCNT into the resin may be due to the improper blending. The fibers are well bonded with the matrix in the reinforcement which are identified in Figs. 7c and 8c and also confirms the uniform distribution of the matrix on the fibers during fabrication processes. Figs. 7d and 8d show the energy-dispersive spectroscopy (EDS) analysis of the glass–jute fiber reinforced composites. The 1% incorporated MWCNT composites show the maximum carbon peaks than the other. It is conformed that the composites are having the presence of MWCNT. However, the presence of silicon conforms of glass fibers in the reinforced composites.

5 XRD Analysis

Diffraction pattern was conducted in Kalsalingam University using Carl ZEISS instruments (9 KW) at approximately 2 min intervals, all XRD analyses were matched with the standard XRD patterns, and all the phases present in the composite were determined. From the XRD pattern, cellulose is found in the composites as shown in Fig. 9, and the diffraction peaks at $2\theta = 13.375°$ (range between 12 and 15) confirms the presence of cellulose which confirms presence of natural fibers.

6 Conclusion

On the basis of the investigation carried out in the present work, it is concluded that:

- Hybrid polymer matrix reinforced with glass–jute fibers incorporated MWCNT was successfully synthesized using compression molding process.
- Various mechanical testings were conducted, and from that it is concluded that by increasing the wt% of MWCNT, the tensile strength of the composite is increased.
- The flexural and compressive strength of the composites increases initially while increasing from 0 to 0.5 wt% of MWCNT. For further increasing with 1 wt% of

Fig.7 SEM images of 0.5% MWCNT composite

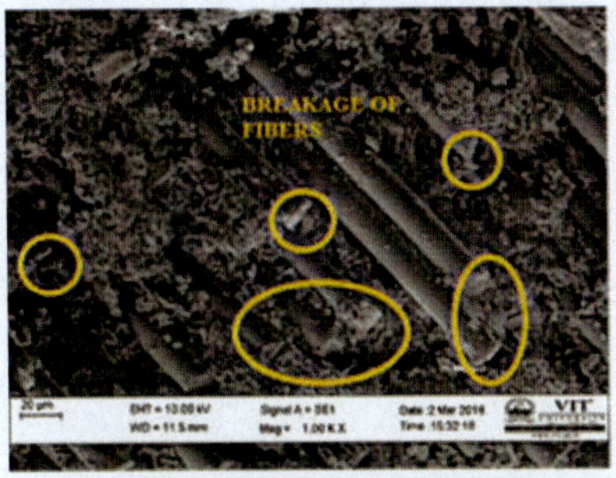

(a) Pulled out fiber during tensile

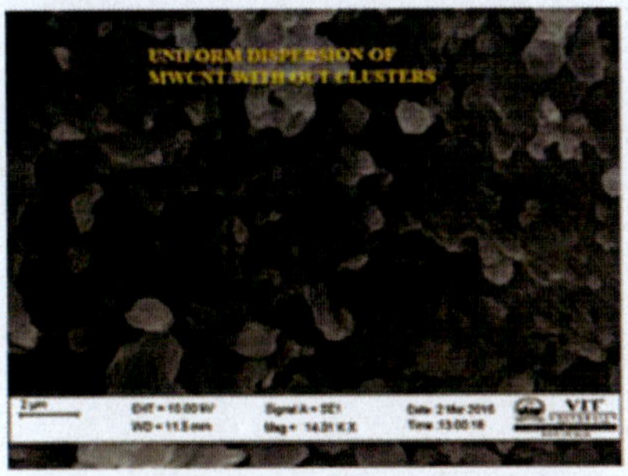

(b) Dispersion of MWCNT into Matrix

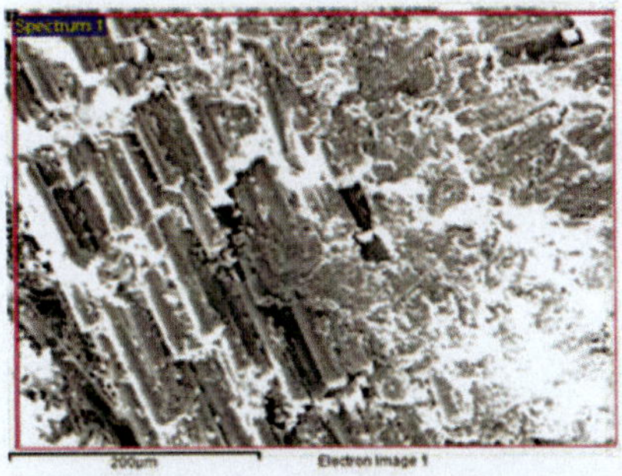

(c) Bonding between fiber and matrix

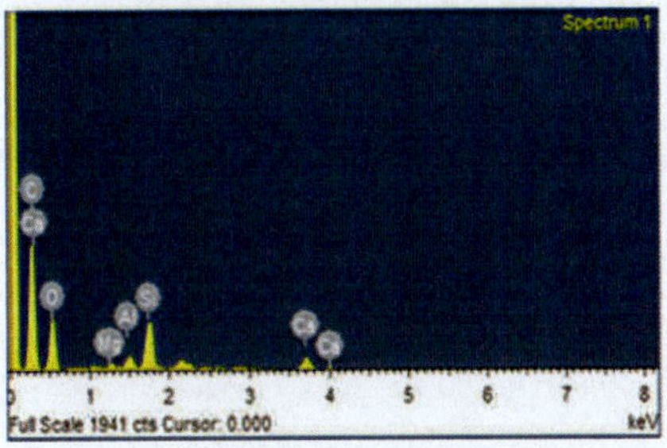

(d) Image of EDAX analysis

Fig. 8 SEM images of 1% MWCNT composite

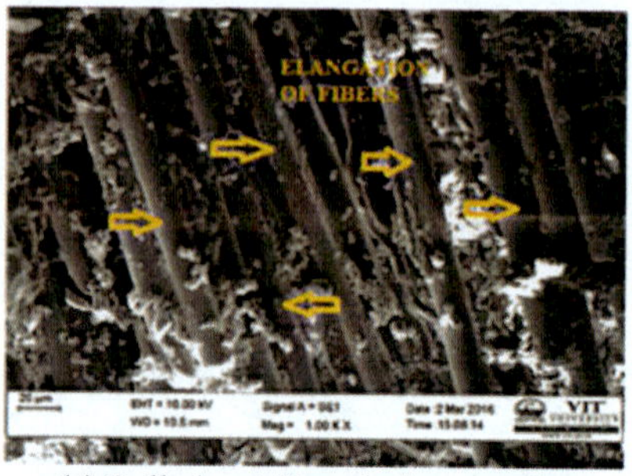

(a) Pulled out fiber during tensile

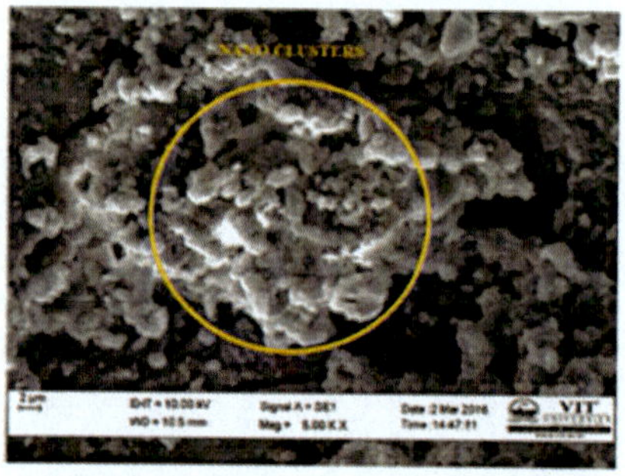

(b) Dispersion of MWCNT into Matrix

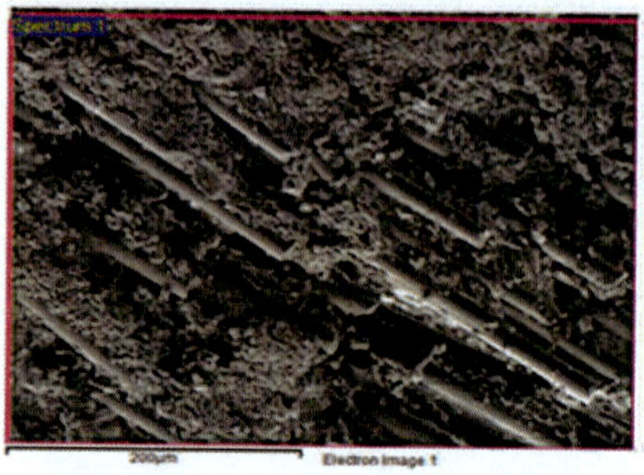

(c) Bonding between fiber and matrix

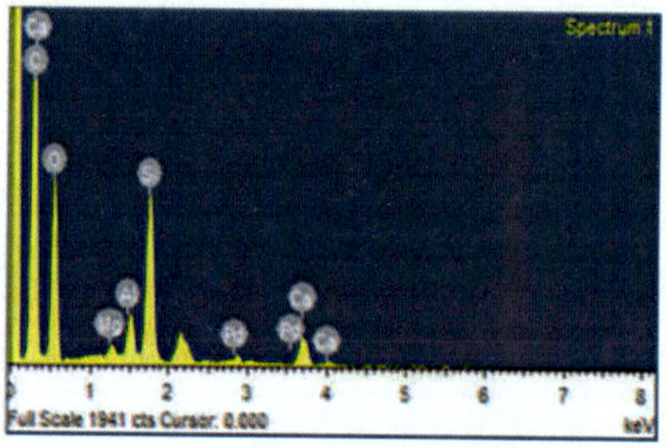

(d) Image of EDAX analysis

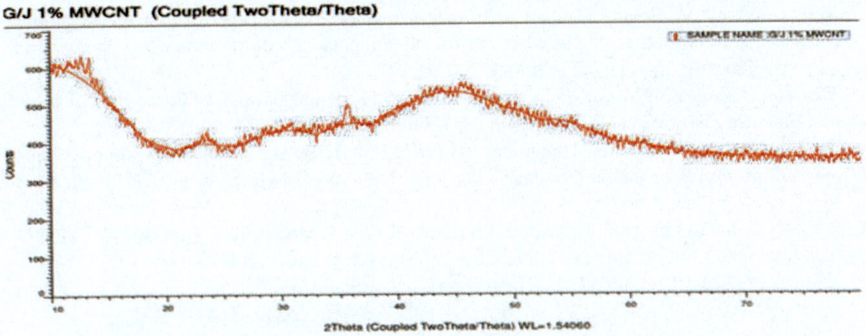

Index	Name	Scan	Angle	d Value	Net Intensity	Gross Intensity	Rel. Intensity
0	Peak #1	G-J 1% MWCNT.raw#1	13.375	6.61461	77.8	618	100.0 %
1	Peak #2	G-J 1% MWCNT.raw#1	28.340	3.14663	61.1	453	78.6 %
2	Peak #3	G-J 1% MWCNT.raw#1	32.786	2.72941	58.4	465	75.1 %

Fig. 9 XRD image of glass-jute composites with 1 wt% MWCNT

MWCNT, the composites are exhibiting the reduction in flexural and compressive strength. This may due to nano clusters and air voids present in the composites.

- While increasing the wt% of MWCNT into the resin, it should ensure the proper sonification and mechanical stirring to avoid the nano clusters formation. During fabrication of the composites, it should be concentrated to avoid the air gap formation.
- SEM images confirmed more elongation of the fibers during tensile test, and it shows the good bonding between the matrix and fibers.
- The EDS analysis confirms the presence of MWCNT and differentiates the composites which is having 0.5 and 1% MWCNT by their carbon peak values. The XRD pattern confirms the presence of jute, glass fibers, and MWCNT in the reinforced nano composites.

References

1. Mishra AK et al (1999) Novel low-cost jute-polyester composites part 1: processing, mechanical properties and SEM analysis. Polym Compos 20(1):62–71
2. Ray D et al (2001) Effect of alkali treated jute fibres on composite properties. Bull Mater Sci 24:129–135
3. Cicala G et al (2009) Properties and performances of various hybrid glass/natural fibre composites for curved pipes. Mater Des 30:2538–2542
4. Ramesh M et al (2012) An investigation of mechanical and sliding wear behaviour of glass fiber reinforced polymer composite with or without addition of silica (SiO2). In: International conference on PFAM XXI, IIT Guwahati
5. Dash BN, Reddy R (2013) Comparative evaluation on properties of hybrid glass fiber-sisal/jute reinforced epoxy composites. Procedia Eng 51:745–750

6. Ajith Gopinath M, Senthil Kumar A, Elayaperumal A (2014) Experimental investigations on mechanical properties of jute fiber reinforced composites with polyester and epoxy resin matrices. Procedia Eng 97:2052–2063
7. Tian He, Zhang YX, Ye L, Yang C (2015) Mechanical behaviours of green hybrid fibre-reinforced cementitious composites. Constr Build Mater 95:152–163
8. Boopalan M, Niranjanaa M, Umapathy MJ (2013) Study on the mechanical properties and thermal properties of jute and banana fiber reinforced epoxy hybrid composites. Compos B 51:54–57
9. Ramesh M, Palanikumar K, Hemachandra Reddy K (2013) Mechanical property evaluation of sisal–jute–glass fiber reinforced polyester composites. Compos Part B 48:1–9
10. Vinayagamoorthy R et al (2015) Characteristic investigations on loofah, jute and glass fiber reinforced sandwich polymeric composites. Appl Mech Mater 813–814:14–18
11. Sabeel Ahmed K, Vijayarangan S (2008) Tensile, flexural and inter laminar shear properties of woven jute and jute-glass fabric reinforced polyester composites. J Mater Process Technol 207:330–335
12. Mohan K, Rajmohan T (2017) Fabrication and characterization of MWCNT filled hybrid natural fiber composites. J Nat Fibers. https://doi.org/10.1080/15440478.2017.1300115
13. Islam S et al (2015) Natural fiber reinforced hybrid polymer nano composites: effect of fiber mixing and nano clay on physical, mechanical and biodegradable properties. Bio Resour 10(1):1394–1407
14. Montazeri A, Chitsazzadeh M (2014) Effect of sonication parameters on the mechanical properties of Multi Walled carbon Nano Tube/epoxy composites. J Mater Des 56:500–508
15. Mohan K, Rajmohan T (2018a) Effects of MWCNT on Mechanical Properties of glass-flax fiber reinforced nano composites. Mater Today Proc 5(5):11628–11635
16. Boopalan M, Umapathy MJ, Jenyfer P (2012) A comparative study on the mechanical properties of jute and sisal fiber reinforced polymer composites. J Silicon 4:145–149
17. Mohan K, Rajmohan T (2018b) Mechanical behaviour of sisal—glass fiber reinforced hybrid Nano composites. IOP Conf Ser Mater Sci Eng 390:012090
18. Rajmohan T, Mohan K, Palanikumar K (2015) Synthesis and characterization of multi wall carbon nano tube (MWCNT) filled hybrid banana-glass fiber reinforced composites. Appl Mech Mater 766:193–199
19. Mohan K, Rajmohan T, Sankar Raman J, Ramanujam V (2018) Dynamical analysis of nano filled—sisal fiber hybrid reinforced composites. IOP Conf Ser Mater Sci Eng 390: 012059. https://doi.org/10.1088/1757-899X/390/1/012059
20. Vinayagamoorthy R (2017) A review on the polymeric laminates reinforced with natural fibers. J Reinf Plast Compos 36(21):1577–1589

Effect of Stacking Sequence on Mechanical Properties of MWCNT Filled Natural Fiber Reinforced Composites

K. Mohan and T. Rajmohan

Abstract Hybrid composites are generally prepared by uniting two or more fibers by using a common matrix. The hybrid composites may be natural to natural or natural to artificial. The stacking sequence and multi-wall carbon nanotubes (MWCNTs) filled natural fibre reinforced composites have charmed significance of the polymer composite researchers since of its mechanical properties. The investigation presents the mechanical performance of the natural fibre reinforced hybrid composites. The natural fibres such as banana, jute, and flax were hybridized and fabricated by changing their stacking sequence with incorporating various wt% of MWCNT (0, 0.5, 1) using compression molding technique. The MWCNT particles were dispersed in epoxy resin using ultrasonic bath sonicator used as matrix face for natural fibre reinforced composites. The various mechanical properties like tensile, compressive, flexural, and impact were tested as per ASTM standard. The microstructure of the tested composites like fiber fracture, pull out fibers, bonding between fiber, and matrix were studied by scanning electron microscopy (SEM).

Keywords Natural fiber reinforced composites · Multi-wall carbon nano tubes (MWCNT) · Mechanical properties · SEM

1 Introduction

Natural fiber polymer composites are more eco-friendly compared to polymer composites with synthetic fibres reinforced. The main advantages of natural fibres composites over synthetic fiber composites are low-density renewability, biodegradability, and non-toxicity. Hybrid composite materials are broadly used in the field

K. Mohan (✉)
Centre for Composite Science and Tribology, Sri Chandrasekharendra Saraswathi Viswa Mahavidyalaya, Enathur, Kanchipuram, Tamil Nadu 631561, India
e-mail: mohank84@gmail.com

T. Rajmohan
Department of Mechanical Engineering, Sri Chandrasekharendra Saraswathi Viswa Mahavidyalaya, Enathur, Kanchipuram, Tamil Nadu 631561, India
e-mail: rajmohanscsvmv@yahoo.com

T. Rajmohan et al. (eds.), *Advances in Materials and Manufacturing Engineering*, Springer Proceedings in Materials 7, https://doi.org/10.1007/978-981-15-6267-9_63

of engineering applications due to their strength-to-weight ratio, ease of manufacturing, and low cost. Hybrid composites are preferred to achieve a blend of properties such as ductility, stiffness, and strength which cannot be achieved by single fiber reinforced composites examined by Gururaja et al. [1] The mechanical properties of natural hybrid abaca, jute with glass fiber reinforced composite have been investigated and reported, the combination of abaca–glass and jute–glass composites exhibits lesser tensile strength than the combination of abaca–jute–glass fiber composite. The flexural and imapact of abaca–glass fiber composite is higher than the abaca–jute composites [2]. Sathish et al. investigated the mechanical and thermal properties of banana, kenaf, and glass fiber reinforced composites with different volume fraction and fiber orientation. The hybrid composite which is arranged in 45° orientation shows the enhanced mechanical properties [3]. Ramesh et al. fabricated flax and glass fiber reinforced composites with two different fiber orientation as 0° and 90°. From the result, 0° orientation fiber reinforced composites have the better mechanical properties rather than the 90° fiber orientation reinforced composites [4].

The natural fiber properties like cellulose, hemi-cellulose, and lignin play a important role by providing strength and stability of the whole fiber. When the fibers having higher cellulose content, it indicates dramatic improvement in the flexural strength and flexural toughness of the cementitious composites reported by Tonoli [5]. Banana–glass reinforced composites exhibits the ultimate tensile strength followed by flax–glass and flax–banana–glass hybrid reinforced composites. More percentage of elongation is obtained by flax–glass composites rather than other composites. Excellent ability to absorb impact energy is observed in banana–flax-glass hybrid composites than the other composites [6].

Ramakrishna Malkapuram et al. reported that the chemical and physical properties of natural fibers have distinct properties. The presence of cellulose content of these fibers varies from fiber to fiber. If the more presence of moisture content in the fibers and adhesion between matrix and fibers are influencing the mechanical properties of composites. By treating, the fibers can change adhesion properties [7]. The mechanical properties of multi-walled carbon nanotube (MWCNT) filled banana, jute, and flax fiber composites were analyzed by changing their stacking sequence and keeping the jute and flax fiber at the extreme layers of the composites. Jute fiber at the extremities shows the improved tensile, compressive, and hardness properties while increasing the wt% of MWCNT other than the impact properties [8]. Behzad Kord et al. investigated that the dynamic mechanical properties of polypropylene (PP)/reed flour (RF) composites are filled with an MWCNT. The results indicated that the storage modulus is increased and damping properties are decreased while incorporation of MWCNT into the composites. Addition of MWCNT also leads to positive shift in a glass transition temperature [9]. Mohan et al. investigated the mechanical properties of MWCNT filled glass—flax fiber reinforced hybrid composites. The maximum tensile and compressive properties obtained by incorporation are 1 wt% and 0.5 wt% MWCNT, respectively [10]. Various surface chemical-treated vetivera fiber is fabricated and subjected to mechanical tests. Benzoyl chloride-treated composites exhibiting superior mechanical properties among the various surface treatment were reported by Vinayagamoorthy [11]. In the review on effect, machining parameters of natural fiber laminates on its performance have been studied and recognizing the different process parameters that affect the performance of the machined NFRCs [12]. In recent years, the researchers have been focused on machining of natural fiber reinforced composites. Vinayagamoorthy et al. reported that an equal proportions of jute, loofah, and glass fibers are exhibiting the good tensile, flexural, and impact strength. Compressive strength is not compromised when 10% of glass fibers substituted by loofah fibers [13]. From the above literature, a very limited work have been carried out in the combination of various natural fiber reinforcement by varying their stacking sequences and varying the wt% of MWCNTs.

2 Experimental

2.1 *Materials Used*

Epoxy resin araldite LY556 and hardener HY951 are obtained from M/s. Suntech fibers, Chennai, Banana fibers are obtained from Eco Craft, Hosur, and MWCNTs are procured from M/S US Research Nano materials Inc, USA, which are used as constituents for fabricating the composite.

2.2 *Resin Preparation*

Initially, the MWCNT particles taken by wt% of 0, 0.5, and 1%. By using ultrasonic probe sonicator, the MWCNT particles are uniformly dispersed into the resin without agglomeration by the blends which are allowed to vibrate for the period of 3hrs in ultrasonic probe sonicator. To ensure homogeneous mixing of MWCNT particles into resin without agglomeration, the same blends are to be kept in a rotary shaker for additional 2 h [14, 15].

2.3 *Composite Preparation*

Compression molding technique is used to fabricate the composite specimens in the present investigation. Banana, jute, and flax fibers of 30 cm length are used to fabricate the specimen. By using MWCNT filled resin, the composites were fabricated. The various mechanical properties such as tensile, compressive and flexural properties are tested in the composite stacking sequence of JBFBFBFJ in which jute fiber as arranged in extreme layers, banana and flax are arranged as intermediate layers. Similarly in the stacking sequence of FBJBJBJF composites fabricated and tested in which flax fiber as arranged as extreme layers banana and jute fibers were arranged as an intermediate layers. In this work, the composite specimens includes a total of eight layers, in which the sequence of fibers are arranged as BJFJFJFB, the banana fibers as kept in extreme layers. Jute fiber and flax fibers have arranged as intermediate layers. The MWCNT filled epoxy resin is used for fabricating the composites.

3 Results and Discussion

3.1 *Tensile Properties*

Specimens for tensile test has been prudently cut from the laminate using band saw, and it has been finished to accurate size by using emery paper. Tensile test has been carried out using universal testing machine, according to the ASTM standard ASTM D638. The tensile behavior of the composite specimens are shown in Fig. 1. The tensile strength reaches the maximum value of 222.2 MPa for **JBFBFBFJ** specimen with addition of 1 wt% of MWCNT followed by 195.1 MPa and 193 MPa, respectively. The tensile strength is increased up to 31.09% by incorporating 1% of MWCNT than the 0% MWCNT composite for the stacking sequence of **JBFBFBFJ** specimen. For the stacking sequence of **FBJBJBJF** specimen with addition of 1% MWCNT increases the tensile strength only up to 4.44%, and also it is observed that the extreme plies of flax fiber reinforced composites at 0% MWCNT show the higher value than the extreme plies of jute fiber reinforced composite.

The stacking sequence of **BJFJFJFB** composites indicates the lower tensile strength rather than other composites. Even the strength of the composites increased upto 17.8% by incorporating the wt% of MWCNT from 0 to 1 wt%. Due to low hemi-cellulose presence in the banana fibers when it is arranged at extreme plies lead to form quick crack propagation in the composites which leads to lower in

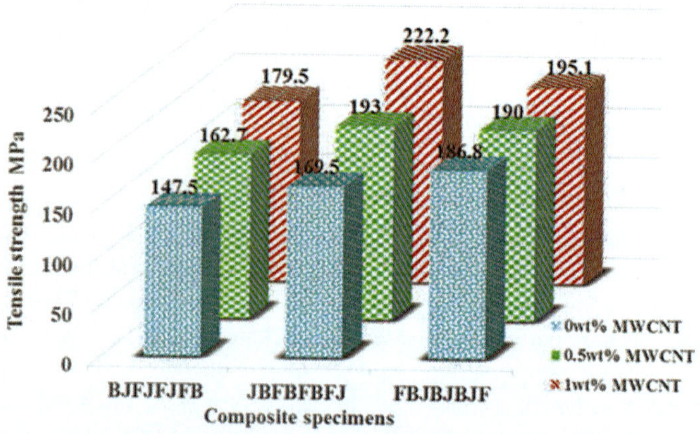

Fig. 1 Effect of stacking sequence and wt% of MWCNT on tensile strength

tensile properties than the others. From the SEM analysis, the fiber delamination was identified. Due to the high toughness and tensile properties of multi-wall carbon nanotube, tensile strength of the composite is increased while increasing the wt% of MWCNT [16] and also the tensile strength of the specimens with jute fibers at the extremities show higher values than the other specimens because of tensile properties of jute and banana fibers which are superior then the flax fibers. Similar results have been obtained in a research on jute/oil palm hybrid composites, with jute layer at extreme portions possess greater tensile strength.

3.2 Compressive Properties

The compressive behavior plot is shown in Fig. 2. It has been carried out according to the ASTM D695 standard. It has been observed that the stacking sequence of **J**BFBFBF**J** specimen is having the maximum compressive load of 20.8 KN, followed by 19.0 kN by incorporating the 1 wt% of MWCNT and 0.5 wt%, respectively, rather than other composites. The stacking sequence of BJFJFJFB shows 18.4 kN and followed by 17.2 kN by incorporating 1 wt% of MWCNT and 0.5 wt%, respectively. An increase in the compressive strength of 18.44 and 15.13% is observed for the stacking sequence of the specimens **J**BFBFBF**J** and **F**BJBJBJ**F** in addition of 1% MWCNT reinforced composites when compared to 0% MWCNT incorporated composites. The jute fiber at the extremities shows that superior compressive strength compared to the flax fiber and banana fiber at the extremities. May the intermediate layer of 10% flax fiber content influencing to enhance the compressive properties in the specimen sequence of **J**BFBFBF**J** compared to the extreme plies of flax fiber and banana fiber reinforced composite. Compressive properties not only depend on stacking sequence and volume fraction of fibers, and apart from this, the extreme

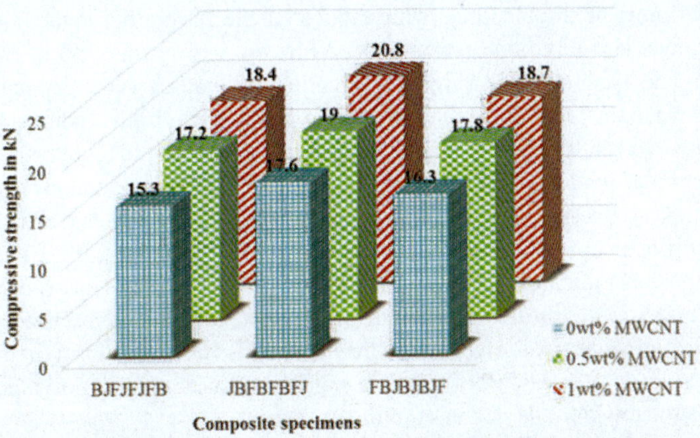

Fig. 2 Effect of stacking sequence and wt% of MWCNT on compressive strength

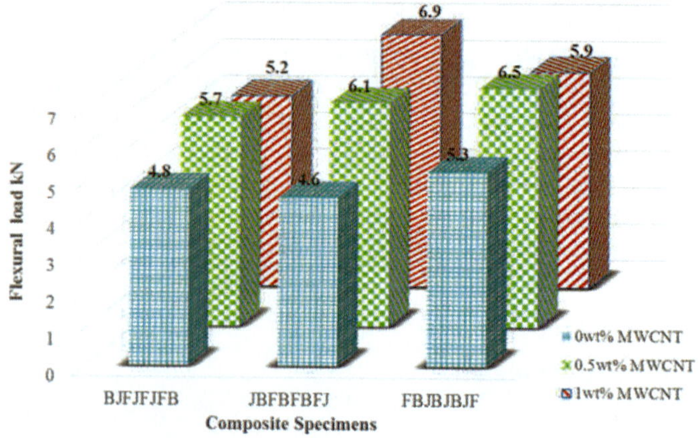

Fig. 3 Effect of stacking sequence and wt% of MWCNT on flexural strength

plies of the composites also affect the compressive properties. Similarly, the 10% flax fiber is present in the specimen sequence of BJFJFJFB composites, but which is having lesser compressive load-carrying capacity than the extremities of jute and flax fibers. Strengthened fibers at the extremities improve the compressive properties.

3.3 *Flexural Properties*

According to the ASTM D790 the flexural specimens were prepared, and the test has been carried out in thee point flexural test machine. The test results are plotted in Fig. 3. The maximum flexural load of the composites is achieved 6.9 kN followed by 6.5 kN and 5.9 kN, respectively. The specimen sequence of JBFBFBFJ composites exhibits maximum flexural load-carrying capacity at the extremities layer of jute fiber than the banana and flax fiber extremities. While increasing the wt% of MWCNT from 0, 0.5, and 1 which leads to increase the flexural load-carrying capacity of the composites when the jute fiber at the extremities. Since jute is high modulus material, for the same deformation, higher fiber concentration demands the higher stress. High matrix-filler interfacial area results because of uniform and homogeneous dispersion of nanoparticles are responsible for changing relaxation behaviour, as well in ensuing the flexural properties. The specimen sequence of BJFJFJFB and FBJBJBJF reinforced composites indicates increasing the flexural properties while incorporating the MWCNT up 0.5 wt%. Further addition of MWCNT leads to decrease in their properties due to the weaker interfacial or adhesion bonds between highly hydrophilic natural fibers and hydrophobic fibers and non-polar organophilic polymer matrix which leads to considerable decrease in the properties.

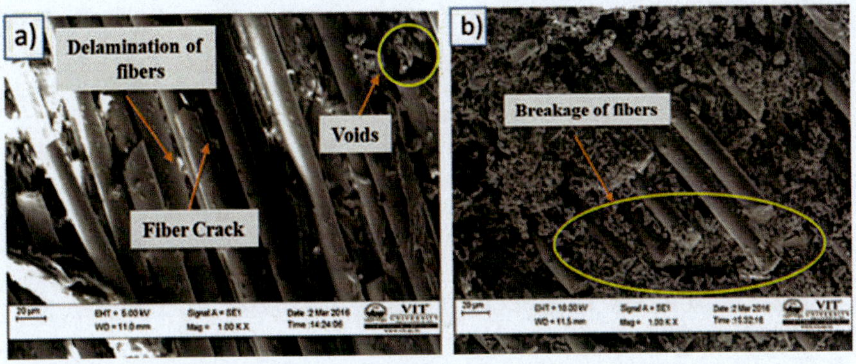

Fig. 4 Microstructure analysis of banana fiber reinforced natural fiber hybrid composite

4 Microstructure Analysis

The microstructure of the nano-filled banana fiber reinforced composites were analyzed using scanning electron microscope (SEM). The banana fibers having low hemi-cellulose when it is arranged at extreme plies lead to form a crack in the fibers which leads to reduce the tensile properties of the composites rather than the jute and flax reinforced composites. From the SEM analysis, Fig. 4a indicates the fiber crack and fiber delamination of the composites. Similarly, the presence of air voids also leads to reduce in their mechanical properties. Fig. 4b shows the fiber breakages during the tensile test. Also, it indicates the well bonding of matrix with the reinforcement.

5 Conclusion

Banana fiber reinforced natural fiber hybrid composites with incorporation of MWCNT are fabricated using compression molding method. Our emphasis has been on effects of varying the stacking sequence and wt% of MWCNT on mechanical properties of the composites like tensile, compressive, impact, and flexural have been evaluated, and also the morphological characteristics are studied using SEM. The following conclusion are drawn.

- The stacking sequence of BJFJFJFB composites shows the increased in tensile properties while increasing the wt% of MWCNT, and the maximum tensile strength of 179.5 MPa is obtained when compared to the other extremities' fiber like **J**BFBFBF**J** and **F**BJBJBJ**F** composites having the higher tensile strength than the **B**JFJFJF**B** composites. The compressive load-carrying capacity of 18.4KN is obtained for **B**JFJFJF**B** composites with incorporating 1 wt% of MWCNT.

The maximum compressive load of 20.8 kN is achieved by the extremities' of **J**BFBFBF**J** composites.

- While increasing the wt% of MWCNT from 0 to 0.5%, the flexural load-carrying capacity of the **B**JFJFJF**B** composites is increased. For further increasing with the MWCNT, there is a reduction in the properties due to the weaker interfacial or adhesion bonds between highly hydrophilic natural fibers and hydrophobic fibers.
- The energy absorbed of the extremities fiber of **B**JFJFJF**B** and **F**BJBJBJ**F** composites is marginally increased while increasing the wt% of MWCNT. The impact energy is reduced while increasing the wt% of MWCNT for stacking sequence of **J**BFBFBF**J** composites due to the reinforcing effect of MWCNT. SEM analysis shows the fiber pull out during tensile and well adhesion of matrix in the reinforcement.

References

1. Gururaja MN, Hari Rao AN (2012) A review on recent applications and future prospectus of hybrid composites. Int J Soft Comput Eng 1(6):2231–2307
2. Vijaya Ramnath B et al (2013) Evaluation of mechanical properties of abaca–jute–glass fibre reinforced epoxy composite. Mater Des 51:357–366
3. Sathish P et al (2015) Effect of fiber orientation and stacking sequence on mechanical and thermal characteristics of Banana-Kenaf hybrid epoxy composite. Silicon. https://doi.org/10.1007/s12633-015-9314-7
4. Ramesh M, Sudharsan P (2017) Experimental investigation of mechanical and morphological properties of flax-glass fiber reinforced hybrid composite using finite element analysis. Silicon. https://doi.org/10.1007/s12633-016-9526-5
5. Tonoli HD et al (2007) Performance and durability of cement based composites reinforced with refined sisal pulp. Mater Manuf Process 22:149–156
6. Srinivasan VS, Rajendra Boopathy S, Sangeetha D, Vijaya Ramnath B (2014) Evaluation of mechanical and thermal properties of banana-flax based natural fibre composite. https://doi.org/10.1016/j.matdes.2014.03.014
7. Malkapuram R, Kumar V, Negi YS (2009) Recent development in natural fiber reinforced polypropylene composites. J Reinf Plast Compos. https://doi.org/10.1177/0731684407087759
8. Mohan K, Rajmohan T (2017) Fabrication and characterization of MWCNT filled hybrid natural fiber composites. J Nat Fibers 14(6):864–874
9. Kord B, Jamshidi M, Hosseinihashemi SK (2016) Effect of multi-walled carbon nanotubes on viscoelastic properties of pp/reed flour composites. J Polym Environ. https://doi.org/10.1007/s10924-016-0909
10. Mohan K, Rajmohan T (2018a) Effects of MWCNT on mechanical properties of glass-flax fiber reinforced nano composites. Mater Today Proc 5:11628–11635
11. Vinayagamoorhy R (2017) Influence of fiber surface modifications on the mechanical behavior of Vetiveria zizanioides reinforced polymer composites. J Nat Fibers. https://doi.org/10.1080/15440478.2017.1410513
12. Rajmohan T, Vinayagamoorthy R, Mohan K (2018) Review on effect machining parameters on performance of natural fiber-reinforced composites (NFRCs). J Thermoplas Compos Mater 2018:1–21. https://doi.org/10.1177/0892705718796541
13. Vinayagamoorthy R et al (2015) Characteristic investigations on loofah, jute and glass fiber reinforced sandwich polymeric composites. Appl Mech Mater 813–814:14–18

14. Mohan K, Rajmohan T (2018b) Mechanical behaviour of sisal—glass fiber reinforced hybrid Nano composites. IOP Conf Ser Mater Sci Eng 390:012090
15. Mohan K, Rajmohan T, Raman S, Ramanujam V (2018) Dynamical analysis of Nano filled - Sisal fiber hybrid reinforced composites. IOP Conf Ser Mater Sci Eng 390: 012059. https://doi.org/10.1088/1757-899X/390/1/012059.
16. Rajmohan T, Mohan K, Palanikumar K (2015) Synthesis and characterization of Multi wall Carbon Nano Tube (MWCNT) filled hybrid banana-glass fiber reinforced composites. Appl Mech Mater 766:193–199

Optimization of Effective Process Parameters During Pentaerythritol Ester Production Using Taguchi Technique

P. Chengareddy, S. Arumugam, G. Sriram, and M. Bhanu Prakash

Abstract The present work is to study the optimization of effective process parameters on pentaerythritol ester (PE) production by the ultrasonicated transesterification process from raw rapeseed oil using L9 orthogonal array Taguchi method. For this work, the parameters were selected as pulse (10, 15, 20 s), amplitude (25, 50, 75%), and catalyst concentration (0.5, 1, 1.5 g). The maximum yield analysis is done on the basis of "larger is better" signal-to-noise ratio (SN Ratio). Among all nine experiments, the optimal conditions are found as pulse (10 s), amplitude (50%), and catalyst (1.5 g). The effect of parameters on PE yield is determined by analysis of variance (ANOVA).

Keywords Taguchi method · ANOVA · Pentaerythritol ester · Transesterification · Ultrasonication

1 Introduction

The major problem in global is polluted by the emission of harmful gases from fossil fuels like petroleum products, natural gas, and coal. Among these, petroleum products play a major role in automotive and other applications. Also, there is a depletion problem which leads to hike in price. On the other side, there is a big problem with petroleum products, i.e., biodegradability [1]. With this connection, researchers are focused on to produce alternative sources for the petroleum products from the vegetable oils. Vegetable oils are non-toxic, biodegradable, lower sulfur, and a smaller amount of emissions [2]. On the other side, vegetable oils have some drawbacks like poor oxidative stability, lower volatility, and low pour point [3]. To overcome these problems, researchers are modifying the vegetable oils by chemically

P. Chengareddy (✉) · S. Arumugam · G. Sriram · M. Bhanu Prakash
Department of Mechanical Engineering, Sri Chandrasekharendra Saraswathi Viswa Mahavidyalaya University, Enathur, Kanchipuram, Tamil Nadu 631561, India
e-mail: chengareddy.p@kanchiuniv.ac.in

S. Arumugam
e-mail: aru_amace@yahoo.co.in

T. Rajmohan et al. (eds.), *Advances in Materials and Manufacturing Engineering*, Springer Proceedings in Materials 7, https://doi.org/10.1007/978-981-15-6267-9_64

via esterification and transesterification process [4, 5] using conventional methods. Also, the modified oils are shown good pour point, improved oxidative stability, better physico-chemical properties like viscosity, viscosity index, etc.

Some downsides with the conventional methods are to produce biodiesels at higher temperatures, stirring, reaction time, etc. To overcome these problems, an ultrasonic power was used as an alternative method. Ultrasonicator consists an acoustic wave which leads the molecules vibrate, and the molecules are getting collided each other which increase the temperature of the reaction medium, thus reducing the time of reaction for the transesterification process from hours to mins [6]. Many researchers are producing biodiesel from various vegetable oils (Rapeseed, sunflower, Jatropha, palm, etc.) using ultrasonication energy [7].

Many works are reported that the process parameters on ultrasonic-assisted transesterification were playing a vital role, which affects the final yield of oil. The selected parameters are methanol ratio, time of reaction, temperature, catalysts, frequency, amplitude, pulse, etc. [8–10]. The studies were also on amplitude, pulse, and intensity of ultrasound-assisted transesterification process. Under the high molar ratio and increasing the amplitude, this shows a positive effect on the reaction yield percentage of biodiesel [11].

Taguchi is a simple technique, systematic, and efficient one to optimize the design of experiments, and also it minimizes the experimental runs [12]. Similarly, another experimental work was conducted by using Taguchi L9 orthogonal array technique with various parameters, namely methanol, a catalyst of KOH, and stirrer speed [13]. Hence, the Taguchi method is found to be the best optimization method to employ the optimal conditions for biodiesel production using the L9 orthogonal array [14].

On the basis of the above literature survey, the present investigation was to optimize the process parameters (catalyst, amplitude, and pulse) on ultrasonic-based transesterification process for the production of pentaerythritol ester of vegetable oil. The optimization process was carried out using L9 orthogonal array Taguchi technique. The formation of PE was confirmed by FTIR and gas chromatography (GC) analysis is used to know the components present in the obtained product.

2 Experimental Work

2.1 *Production of Pentaerythritol Ester via Transesterification Process Using Ultrasonication*

First, the methyl ester was prepared by the conventional method. For this, raw rapeseed oil (2000 ml) and methanol (356.8 ml) were taken in a three-necked round bottomed flask of capacity 3000 ml. The total mixture was stirred well with a speed of 300 rpm under the maintained temperature of 50–60°C to get the complete mixing of oil and methanol. Before adding the catalyst of NaOH (17.84 g), dissolve the catalyst into the methanol for faster reaction. The reaction process was continuing for the time of 60 min. The final quantity of obtaining methyl ester was 1240 ml (after separation and bubble wash) by measuring.

Second, the transesterification process was carried out (for 1 h) in a three-necked glass reactor vessel under nitrogen atmosphere. The whole setup was placed in an ultrasonic chamber and was controlled by the unit processor controller. In an experimental run, the desired quantity of methyl ester, pentaerythritol, and xylene were poured into a reaction vessel. The bath was heated up with a heating mantle at a temperature of 100 °C. Then, the desired amount of catalyst (p-TSA) was added into the vessel. Later, the products were poured into a separating funnel for the purpose of removing glycerol, and the final product was washed with warm water and heated up with a temperature of 70 °C to remove the excess xylene from the oil [15].

2.2 *Experimental Design Using Taguchi Method*

To identify the optimal conditions, a large number of experimental runs are required to investigate the process parameters of the ultrasonicated transesterification process [16]. Taguchi method is a statistical technique, which reduces the number of experimental runs, and also it is a simple and efficient method [17]. The Taguchi technique uses L9 orthogonal array design for experimental work to find the effective process parameters with minimum runs [18]. In this present study, three factors and three levels were selected, which is given in Table 1.

Signal-to-noise ratio (S/N ratio) is one which measures the quality characteristics of the process in various levels, and the quality characteristics are three different categories, namely smaller-the-better, larger-the-better, and nominal-the-best based on the type of the problem. In this work, the maximum yield of pentaerythritol ester is determined by means of larger-the-better S/N ratio. The S/N ratio is determined by using the given Eq. 1 [12].

$$\frac{S}{N}\text{Ratio} = -10\log\left(\frac{1}{n}\sum_{i=1}^{n} 1/y_i^2\right) \tag{1}$$

where y is the PE conversion to the corresponding run, i is the number of replicating, and n is the number of trial runs performed.

Table 1 Selected factors and levels are used in experimental design

Real factors		Level		
		1	2	3
A	Pulse (s)	10	15	20
B	Amplitude (%)	25	50	75
C	Catalyst concentration (g)	0.5	1	1.5

Table 2 L9 orthogonal array Taguchi design with yield (%) as response

Run no.	Real variables			Yield (%)			Mean yield (%)	S/N ratio
	Pulse (s)	Amplitude (%)	Catalyst (g)	Test 1	Test 2	Test 3		
1	10	25	0.5	84.3	86.1	86.4	85.6	38.64
2	10	50	1	93.12	93.89	96.52	94.51	39.50
3	10	75	1.5	83.78	83.2	84.45	83.81	38.46
4	15	25	1	75.4	74.21	68.02	72.54	37.21
5	15	50	1.5	88.72	74.23	72.58	78.51	37.89
6	15	75	0.5	50.23	51.41	52.41	51.35	34.21
7	20	25	1.5	92.37	86.54	87.82	88.91	38.97
8	20	50	0.5	77.82	79.21	78.11	78.38	37.88
9	20	75	1	50.92	50.78	51.21	50.97	34.14

$$\text{PE Yield}(\%) = \frac{\text{Weight of PE Produced}}{\text{Weight of Methyl Ester Used in Reaction}} \tag{2}$$

The nine sets of experiments are conducted by using 100 ml of methyl ester in a three-necked round bottomed vessel, and the corresponding yield percentage is given in Table 2. The yield percentage of pentaerythritol ester is evaluated by the given Eq. 2 [19]. The effect and improvement of each optimal conditions are performed by the study of analysis of variance (ANOVA).

3 Results and Discussions

3.1 Analysis of Taguchi Method to Determine the Optimal Conditions

The effect of variables such as amplitude, pulse, and catalyst concentration on yield was evaluated, and their results are given in Table 2. Among all nine sets of experiments, it is very clear that the highest yield (94.51%) of PE was obtained by the experimental run number 2 with 10 s pulse, 50% amplitude, and 1 gm of catalyst. The lowest yield (50.97%) of PE was obtained by the run number 9, i.e., 20 s pulse, 75% amplitude, and 1 gm of catalyst concentration as given in the table. However, this is not the right method to select the optimal condition parameters with the help of Taguchi technique.

S/N ratio is the one used to measure the quality characteristics from the results of Taguchi design. In this work, the S/N ratio measures characteristics based on the larger-the-better rather than that the other two types of characteristics, i.e., smaller-the-better and nominal-the-best. The results of the S/N ratio for all nine experiments were shown the highest (39.50956) and lowest (34.14629) yield percentage of PE from the same experimental runs, i.e., run number 2 and run number 9 as given in Table 2. By analyzing the results of the above experiments and with the influence of S/N ratio, the optimal conditions for the highest yield percentage are shown in Fig. 1. From Fig. 1 the optimal conditions are identified as amplitude (50%), catalyst concentration (1.5 g), and as the pulse (10 s). Figure 2 shows the residual plots for yield percentage according to the analyzed data from experimental results. Hence, The S/N ratio results of experimental works were meeting the earlier research works, i.e., a moderate percentage of amplitude shows the positive impact on the ultrasonicated transesterification process [20]. Due to the increment of high ultrasonic amplitudes to the mixture, there is a decrement in the final yield of oil because of the formation of cavitation bubbles in high numbers in the mixture [20]. Later, these bubbles are grown together to form large numbers, and they are more stable, which prevents the transfer of acoustic energy through the reaction mixture [21]. From Fig. 1 the ultrasonic amplitude plots show that the highest yield was obtained in 50% of amplitude, 25% amplitude shows lesser than that of 50% amplitude, but 75% amplitude shows very less yield than other two amplitudes. So, the increment of high amplitudes shows the lesser yields.

By increasing the catalyst concentration in the reactants, the yield percentage of the product was increased simultaneously, but further increase of catalyst concentration decreases the conversion of oil [22]. The pentaerythritol ester composition was increased significantly by increasing the molar ratio and catalyst concentration [23]. Reported that the amount of catalyst added to the mixture, then the mixture was

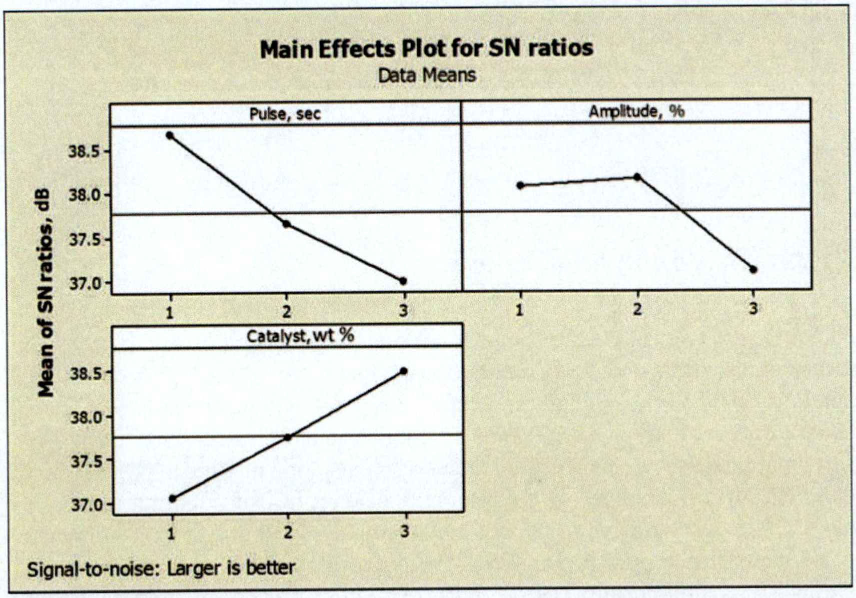

Fig. 1 S/N ratio plots for each variable at different levels

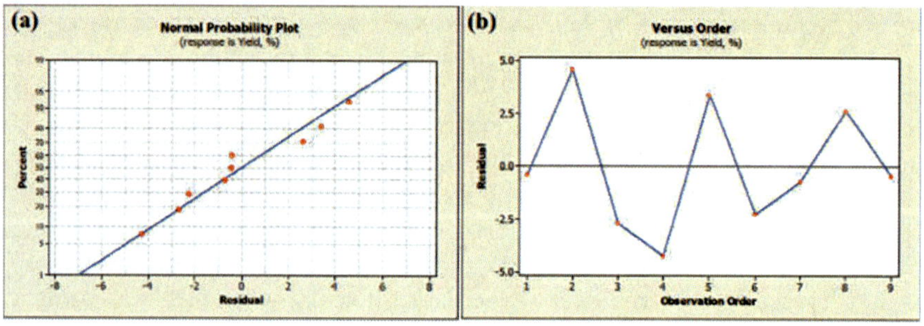

Fig. 2 Residual plots for yield (%)

accelerated, which produced more pentaerythritol ester. From the results high catalyst concentration, i.e., 1.5 g shows higher yields than that of 0.5 and 1.0 g catalyst concentration.

Another work reported that the ultrasonic pulse exceeded about 5 s, and then, the effect on yield was found as more significant [24]. From the pulse plots in Fig. 1, 10 s pulse has been shown higher yields than that of other two pulses (i.e., 15 and 20 s). With this, concluded that the selected variables are proven as real variables and show their contribution in the part of high yield production of PE.

For producing the PE yield by employing the transesterification process, the regression equation was developed by the given Eq. 3. This regression equation gives the relation between the effective process parameters and the response (i.e., yield %). From Eq. 2, the R^2 value is found to be at 95.52%, which is used to measure the effect of transesterification process.

$$\begin{aligned}\text{Yield}(\%) = 107.234 - 1.522\text{pulse} \\ - 0.406133\,\text{amplitude} + 11.9667\,\text{catalyst}\end{aligned} \tag{3}$$

3.2 *ANOVA (Analysis of Variance)*

ANOVA is used to examine the effects of the real variables on PE yield percentage. Based on the analysis, the most contribution parameters are discussed in detail, i.e., amplitude, catalyst, and pulse. Table 3 shows the effect of process parameters on the response, the contribution of each parameter, and its rankings. According to ANOVA results, amplitude shows a significant effect on the reaction mixture to produce high yields of PE. By calculating [14], the contribution percentage of amplitude was more, i.e., 46.2293% as compared to the other two variables, and it is the most effective process parameter to produce PE yield. The order of contribution to produce high yields of PE is amplitude (46.2293%), pulse (35.4143%), and catalyst concentration (13.8794%), by the rank as given in Table 3.

Table 3 ANOVA with contribution % and rank

Source	DF	Seq SS	Adj MS	*F*	*P*	Contribution %	Rank
Pulse	2	680.12	340.06	7.91	0.112	35.4143	2
Amplitude	2	887.82	443.91	10.33	0.088	46.2293	1
Catalyst	2	266.55	133.28	3.1	0.244	13.8794	3
Error	2	85.98	42.99			4.477	
Total	8	1920.47				100	

4 Conclusions

Ultrasonic-based transesterification process is a suitable method to reduce the reaction time for the production of PE. Taguchi method is the statistical technique which is the most efficient one to optimize the process parameters for obtaining high yields of PE with minimal experimental runs for the reduction of time and cost. The highest yield percentage of PE was obtained from experimental run number 2 (i.e., 94.51%). Based on the quality characteristics of larger-the-better from the S/N ratio, the optimal conditions are found as 10 s pulse, 50% amplitude, and 1.5 g of catalyst concentration. Analysis of variance (ANOVA) is used to measure the significance of individual effective process parameters. It is found that the most affecting factor in the conversion of PE is ultrasonic amplitude, and its contribution is more (46.2293%) as compared to the pulse (35.4143%) and catalyst (13.8794%).

References

1. Ajala OE, Aberuagba F, Odetoye TE, Ajala AM (2015) Biodiesel: sustainable energy replacement to petroleum-based diesel fuel–a review. ChemBioEng Rev 2(3):145–156
2. Panchal TM, Patel A, Chauhan DD, Thomas M, Patel JV (2017) A methodological review on bio-lubricants from vegetable oil-based resources. Renew Sustain Energy Rev 70:65–70
3. Cavalcanti ED, Aguieiras ÉC, da Silva PR, Duarte JG, Cipolatti EP, Fernandez-Lafuente R, da Silva JAC, Freire DM (2018) Improved production of biolubricants from soybean oil and different polyols via esterification reaction catalyzed by immobilized lipase from Candida rugosa. Fuel 215:705–713
4. McNutt J (2016) Development of bio lubricants from vegetable oils via chemical modification. J Ind Eng Chem 36:1–12
5. Saboya RMA, Cecilia JA, García-Sancho C, Sales AV, de Luna FMT, Rodríguez-Castellón E, Cavalcante CL Jr (2017) Synthesis of bio lubricants by the esterification of free fatty acids from castor oil with branched alcohols using cationic exchange resins as catalysts. Ind Crops Prod 104:52–61
6. Fan X, Wang X, Chen F (2010) Ultrasonically assisted production of biodiesel from crude cottonseed oil. Int J Green Energy 7(2):117–127
7. Chengareddy P, Arumugam S, Sriram G, Parthasarathy R (2018) Ultrasonication based formulation and characterization of biodegradable compressor oil. Adv Sci Eng Med 10(3):446–450
8. Hoseini SS, Najafi G, Ghobadian B, Mamat R, Ebadi MT, Yusaf T (2018) Ailanthus altissima (tree of heaven) seed oil: characterisation and optimisation of ultrasonication-assisted biodiesel production. Fuel 220:621–630

9. Parida S, Sahu DK, Misra PK (2016) A rapid ultrasound-assisted production of biodiesel from a mixture of Karanj and soybean oil. Energy Sour Part A Rec Utiliz Environ Effects 38(8):1110–1116
10. Kumar G, Kumar D, Johari R (2014) Time reducing process for biofuel production from non-edible oil assisted by ultrasonication. Ultrason Sonochem 21(5):1618–1623
11. Badday AS, Abdullah AZ, Lee KT (2013a) Optimization of biodiesel production process from Jatropha oil using supported heteropolyacid catalyst and assisted by ultrasonic energy. Renew Energy 50:427–432
12. Karabas H (2014) Application of the Taguchi method for the optimization of effective parameters on the safflower seed oil methyl ester production. Int J Green Energy 11(9):1002–1012
13. Sasikumar C, Balamurugan K, Rajendran S, Naveenkumar S (2016) Process parameter optimization in jatropha methyl ester yield using taguchi technique. Mater Manuf Processes 31(6):701–706
14. Kumar RS, Sureshkumar K, Velraj R (2015) Optimization of biodiesel production from *Manilkara zapota* (L.) seed oil using Taguchi method. Fuel 140: 90–96
15. Dhote PS, Ganvir VN, Bhattacharyulu YC (2013) Optimization of Mahua oil methyl ester by using Taguchi experimental design. Int J Adv Eng Technol 6(3):1140
16. Hashem AI, Abou Elmagd WSI, Salem AE, El-Kasaby M, El-Nahas AM (2013) Conversion of some vegetable oils into synthetic lubricants via two successive transesterifications. Energy Sour Part A Rec Utiliz Environ Effects 35(10):909–912
17. Arumugam S, Sriram G, Rajmohan T (2014) Multi-response optimization of epoxidation process parameters of rapeseed oil using response surface methodology (RSM)-based desirability analysis. Arabian J Sci Eng 39(3):2277–2287
18. Mubarak M, Shaija A, Suchithra TV (2016) Ultrasonication: An effective pre-treatment method for extracting lipid from Salvinia molesta for biodiesel production. Resour-Efficient Technol 2(3):126–132
19. Dhawane SH, Kumar T, Halder G (2016) Biodiesel synthesis from Hevea brasiliensis oil employing carbon supported heterogeneous catalyst: optimization by Taguchi method. Renew Energy 89:506–514
20. Badday AS, Abdullah AZ, Lee KT (2013b) Ultrasound-assisted transesterification of crude Jatropha oil using alumina-supported heteropolyacid catalyst. Appl Energy 105:380–388
21. Ho WWS, Ng HK, Gan S, Chan WL (2015) Ultrasound-assisted transesterification of refined and crude palm oils using heterogeneous palm oil mill fly ash supported calcium oxide catalyst. Energy Sci Eng 3(3):257–269
22. Saha R, Goud VV (2015) Ultrasound assisted transesterification of high free fatty acids karanja oil using heterogeneous base catalysts. Biomass Convers Biorefinery 5(2):195–207
23. Aziz NAM, Yunus R, Rashid U, Syam AM (2014) Application of response surface methodology (RSM) for optimizing the palm-based pentaerythritol ester synthesis. Ind Crops Prod 62:305–312
24. Salamatinia B, Mootabadi H, Hashemizadeh I, Abdullah AZ (2013) Intensification of biodiesel production from vegetable oils using ultrasonic-assisted process: optimization and kinetic. Chem Eng Process 73:135–143

A Review on Ultrasonicated Transesterification Process

P. Chengareddy, S. Arumugam, P. H. Pavan Kumar Reddy, and P. Madhan Mohan Reddy

Abstract The downsides of the mineral-based oils/fuels suggest the best choice to synthesize the environmentally friendly lubricants/fuels from plant oils. The present review is to focus on some recent trends to produce the methyl ester for the purpose of reduction in reaction rate and attain high yields. This reduces the power loss and also economic losses. The most effective method is ultrasonic irradiation method to produce high yield percentage of methyl ester at low frequency and minimum reaction time at low cost. The produced products were shown good physico-chemical properties and also improved its performance.

Keywords Mineral-based oils · Biodiesel · Biolubricants · Conventional method · And ultrasonicated transesterification process

1 Introduction

The environmental pressures modeled by the mineral-based fuels are at present a main global concern. These mineral-based fuels are increased environmental pollution by emitting harmful gases and also lead to human diseases. On the other side, consumption of these products leads to depletion problems with the mineral-based oils which rises crude oil prices. By using these, non-renewable fuels show significant effect on environment [1, 2]. By considering this, researchers are focused on to search for alternative bio-based oils/fuels for the substitution of mineral-based products. Vegetable oil is the best alternative one for the mineral oils [3]. Plant oils are eco-friendly, biodegradable, sustainable, and readily available one. But for lubricants, these vegetable oils have some negative sides, i.e., low pour point and poor thermo-oxidative stability [4]. Chemical modification is required to overcome these drawbacks.

The main aim of this paper is to concentration on producing the biolubricants/biodiesel using modern method (i.e., ultrasonic irradiation) for the production

P. Chengareddy (✉) · S. Arumugam · P. H. Pavan Kumar Reddy · P. Madhan Mohan Reddy
Sri Chandrasekharendra Saraswathi Viswa Mahavidyalaya University, Enathur, Kanchipuram, Tamil Nadu 631561, India
e-mail: Chengareddy.p@kanchiuniv.ac.in

T. Rajmohan et al. (eds.), *Advances in Materials and Manufacturing Engineering*, Springer Proceedings in Materials 7, https://doi.org/10.1007/978-981-15-6267-9_65

of high yield percentage at low reaction rate. Also, physico-chemical properties of the produced products are discussed.

2 Petroleum-Based Oils/fuels

Petroleum-based oils & fuels are largely produced by the industries from nineteenth century. As the use of the vehicles are increased globally and the industrial revolution, also consumes the more byproducts during its manufacture process. The raw product taken from the earth is undergoing the distillation fraction and can be formed into the different fuels as the extracted raw petroleum was heterogeneous chemically. Some of the fuel products are liquified petroleum gas, butane, gasoline, jet fuel, kerosene, fuel oil, and diesel fuel [5]. Some of the byproducts are lubricants, wax, paraffin wax, asphalt, and aeromatic petrochemicals.

Petroleum products are obtained naturally, cannot think that it is renewable and environmentally friendly. Usage of the petroleum products have been increased drastically, and during the recent years, there would be a shortage of petroleum-based fuels. Another big reason for encouraging vegetable oils is pollution caused due to them. Some of the environmental effects are ocean acidification which means the decrease of the pH of the ocean water results in the effect of the marine life, global warming causes the increasing of surface temperature by the emission of the greenhouse gases [6]. To avoid these ill effects from the petroleum products, vegetable oils should be replaced.

3 Vegetable Oil

Vegetable oils are generally extracted from the plant seeds or the other form of fruit seeds like watermelon. These oils contain carboxylic acids with long hydrocarbon chains and triglycerides. These are also known as edible oils. They are divided into three types based on saturation. The classifications are saturated, mono-saturated, and poly-saturated [7]. There are so many oils made from different sources like soya bean, cottonseed, peanut, palm oil, sunflower oil, coconut oil, rapeseed oil, rice bran oil, and others. Vegetable oils are mainly used in the food industry as the main ingredient of many food items like pickles, fried dishes, etc., and they are also used in the manufacture of the paint, varnishes, soaps, detergents, etc. But their physical and chemical properties made them useful for the replacement of the petroleum products by the modification of the fatty acids into the fuels and to the lubricants [8, 9]. So, researchers are focused to produce biodiesel from non-edible oils by modifying it. Vegetable oils have some good properties by nature such as high viscosity, viscosity index, high flash point, and fire point, low toxicity, etc. But these have some negatives for the production of biolubricants, i.e., poor thermo-oxidative stability low pour point. High viscosity is a problem for the production of biodiesel [10]. It a biggest challenge to develop biolubricant/biodiesel for the substitute of petroleum products. For the better performance of the petroleum-depended applications, should replace the environmentally lubricants/fuels instead of petroleum products.

4 Modification of the Vegetable Oils

As the straight or the virgin vegetable oil cannot be used for the directly as biodiesel or biolubricant to give satisfactory results. Although straight vegetable oil has the similar properties of diesel, some of the chemical modifications must be made for the best usage of the oils. Generally, the vegetable oils have high viscosity and density, and these are present in the form of the free fatty acids (FFA). For the usage of straight vegetable oil as the biodiesel, it should be converted into the low density and viscosity oil or fatty acid methyl ester (FAME). After many researches carried on the biodiesel production, the conversion of the straight vegetable oils into the biodiesel can be achieved by the four methods pyrolysis/heating, dilution/blending, micro-emulsification, and transesterification [11–13].

Pyrolysis or thermal cracking is the method of chemical change which can be done by the thermal energy for the biodiesel conversion to decrease the viscosity, flash point, pour point & cetane number. The setup of the pyrolysis was the expensive one, and the obtained oils are the process of the distillation were with the sulfur content result in the less eco-friendly [14]. Pyrolyzed process was the first-made process in the biodiesel production, and it is the most complicated because of the different path involve during reaction [15]. Dilution or blending means the mixing of the straight vegetable oil with the ethanol or other synthetic oils. Generally, 4% of diesel is mixed with for increasing the brake power, torque, and thermal efficiency. This also results in the decrease in the viscosity and density of the oil for the better combustion of the oil. Micro-emulsification is the process of decreasing droplet size from 100 to 1000 Am strong. It can be formed by the addition of alcohol, surfactant, and cetane improver for the lowering the viscosity of the biodiesel for better spraying during the burning of the fuel [14, 16]

Out of all the process, transesterification is the most used conventional and the promising process for the good biodiesel yield.

5 Conventional Methods for Biodiesel Production

Transesterification is the process which converts the triglycerides into the biodiesel or fatty acid methyl ester (FAME), it can be done by using the addition of the straight vegetable oil to alcohol, and thus, the biodiesel and the glycerin were formed. Transesterification can be done with or without the catalyst [17, 18]. Earlier, many works were reported about the biodiesel/biolubricant production by modifying chemically via epoxidation, hydroxylation followed by esterification, transesterification processes, etc., using conventional methods. The produced biodiesel/biolubricant shows better results in terms of physico-chemical properties, thermal properties, fuel performance, emissions, tribological properties, biodegradability, etc. But, by using mechanical stirring/conventional methods, the time of reaction rate increase leads to energy consumption and supercritical conditions, consumes more catalyst and alcohol, obtains low yield percentage of final products, and overall it is economically high [19–22].

In catalytic transesterification, three types of the catalyst are used, they are acidic, basic, and enzymic. Acid-catalyzed transesterification reaction is slowest of all the process, and in this type, the mixing of reactants is also very low, and hence, the yield will be low. However, the enzyme catalytic reaction gives the high yield but that was costly due to enzyme preparation. So, the base-catalyzed reaction is more economic and preferred for the low-temperature reactions. The general bases used are KOH, NaOH, etc., but all the conventional reactions are time taking and low reaction rate for the biodiesel production [23, 24].

The conventional transesterification reaction can be speed up by assisting the two ways, one is microwave assisted, another is ultrasound. Microwave-assisted process involves the irradiating the reactants with the microwaves which range from 300 MHz to 30 GHz with the frequency of 2.45 GHz for the laboratory conditions [25, 26]. Though the production of the biodiesel was in short time by this, it cannot be employed for large amount production because of penetration of these waves which is limited to centimeters depending upon the properties of the reactants. It also has the safety drawback for industrial usage [27]. Ultrasound assisted is the suitable alternative for the best transesterification reaction for biodiesel yield at the low temperature and in a short time.

6 Ultrasonicated Transesterification

The term ultrasonicated means the use of the ultrasonic waves or the ultrasound for the extraction of the biodiesel from the straight vegetable oil. The frequency of the ultrasound would be in the range of 20–100 KHz. These sounds are not audible to the human ears, but some animals can identify these frequencies. At the early stages, the waves are used for the medical purposes, and after its wide applications it is used in many industries. The main effect cause due to the use of the ultrasonic waves or ultrasound is the formation of the node and antinode cycles which causes the sonic pressure in the reaction [28], the formation of the bubbles in the reaction or cavitation [29]. The occurrence of this type of cavitation causes the effective blending of reactants in the liquid–liquid reaction, and the mass transfer will also be higher during this cavitation which result in the best yield. The formation and the collapse of the bubbles result in the building up of the pressure and temperature at the small space in the reacting medium. The final effect caused by the ultrasound is acoustic stream blending [30, 31]. Bio-based lubricants also produced by using ultrasonic irradiation with a minimum reaction rate and obtained high yields, and results show better properties of lubricants [32, 33].

7 Optimization on Ultrasonic-Assisted Transesterification

Optimization is one of the best techniques to find the optimal conditions from the set of experimental runs according to the design. There are different design methods which are there to conduct the experiments such as design of expert (DOE), Mini Tab, etc. A central composite design (CCD) was applied to find the optimal conditions for the production of biodiesel with the influence of ultrasonic energy. Reported that 91% of yield is attained in a span of 40 min with the effect of ultrasonic energy at low reaction temperature [34]. Another work was conducted with the influence of ultrasonic irradiation to find the optimal conditions for the production of biodiesel. More than 90% of yield was obtained with the effect of high-frequency ultrasound in 30 min of time [35]. The works were carried out by the effect of various parameters on biodiesel production by optimization using ultrasonic-assisted transesterification process, i.e., catalyst concentration, alcohol, time, temperature, and frequency. They reported that the high yields were obtained with the influence of ultrasonic irradiation at low a reaction rate [36–38].

8 Conclusion

The biggest challenge in the world is to produce biodegradable oils to replace mineral-based oils. Many works are there to produce biodegradable oils/fuels from vegetable oils by modifying chemically. The modified fuels/oils are shown better results in terms of properties and performance. From this paper, the following conclusions were made.

- Ultrasound sound transesterification process is playing a significant role to produce biodiesel with a low reaction rate and high yields.
- No work is carried out for the formulating of biolubricant using ultrasound-assisted transesterification.
- Few works have just started to produce pentaerythritol ester as biolubricant using ultrasound-assisted transesterification.
- Polyol esters (trimethylolpropane ester, pentaerythritol ester, and neopentyl glycol) are used for various industrial and automotive applications in order to minimize the wear and friction. But no work is carried out to produce pentaerythritol ester as biolubricant for air compressor application.

References

1. Sadorsky P (2009) Renewable energy consumption, CO_2 emissions and oil prices in the G7 countries. Energy Econ 31(3):456–462
2. Kampa M, Castanas E (2008) Human health effects of air pollution. Environ Pollut 151(2):362–367
3. Mobarak HM, Mohamad EN, Masjuki HH, Kalam MA, Al Mahmud KAH, Habibullah M, Ashraful AM (2014) The prospects of biolubricants as alternatives in automotive applications. Renew Sustain Energy Revi 33:34–43
4. Borugadda VB, Goud VV (2014) Thermal, oxidative and low temperature properties of methyl esters prepared from oils of different fatty acids composition: A comparative study. Thermochim Acta 577:33–40
5. Speight, J.G.: The chemistry and technology of petroleum. CRC press (2014).
6. Schmutter K, Nash M, Dovey L (2017) Ocean acidification: assessing the vulnerability of socioeconomic systems in small island developing states. Reg Environ Change 17(4):973–987
7. Zambiazi RC, Przybylski R, Zambiazi MW, Mendonça CB (2007) Fatty acid composition of vegetable oils and fats. Boletim do Centro de Pesquisa de Processamento de Alimentos 25(1)
8. Campanella A, Rustoy E, Baldessari A, Baltanás MA (2010) Lubricants from chemically modified vegetable oils. Biores Technol 101(1):245–254
9. Srivastava PK, Verma M (2008) Methyl ester of karanja oil as an alternative renewable source energy. Fuel 87(8–9):1673–1677
10. Canakci M, Sanli H (2008) Biodiesel production from various feedstocks and their effects on the fuel properties. J Ind Microbiol Biotechnol 35(5):431–441
11. Koh MY, Ghazi TIM (2011) A review of biodiesel production from *Jatropha curcas* L. oil. Renew Sustain Energy Rev 15:2240–2251
12. Misra RD, Murthy MS (2010) Straight vegetable oils usage in a compression ignition engine—a review. Renew Sustain Energy Rev 14(9):3005–3013
13. Kannan GR, Anand R (2011) Experimental investigation on diesel engine with diestrol–water micro emulsions. Energy 36(3):1680–1687
14. ARIFIN, S.F.: Production of Biodiesel from Waste Cooking Oil and Rbd Palm Oil Using Batch Transesterification Process. Faculty of Chemical & Natural Resources Engineering Universiti Malaysia Pahang (2009).
15. Parawira W (2010) Biodiesel production from Jatropha curcas: a review. Sci Res Essays 5(14):1796–1808
16. Gashaw A, Getachew T, Teshita A (2015) A review on biodiesel production as alternative fuel. J Prod Ind 4(2):80–85
17. Georgogianni KG, Katsoulidis AP, Pomonis PJ, Kontominas MG (2009) Transesterification of soybean frying oil to biodiesel using heterogeneous catalysts. Fuel Process Technol 90(5):671–676
18. Aranda DA, Santos RT, Tapanes NC, Ramos ALD, Antunes OAC (2008) Acid-catalyzed homogeneous esterification reaction for biodiesel production from palm fatty acids. Catal Lett 122(1–2):20–25
19. Marchetti JM, Errazu AF (2008) Esterification of free fatty acids using sulfuric acid as catalyst in the presence of triglycerides. Biomass Bioenerg 32(9):892–895
20. Georgogianni KG, Kontominas MG, Pomonis PJ, Avlonitis D, Gergis V (2008a) Conventional and in situ transesterification of sunflower seed oil for the production of biodiesel. Fuel Process Technol 89(5):503–509
21. Choedkiatsakul I, Ngaosuwan K, Cravotto G, Assabumrungrat S (2014) Biodiesel production from palm oil using combined mechanical stirred and ultrasonic reactor. Ultrason Sonochem 21(4):1585–1591
22. Georgogianni KG, Kontominas MG, Pomonis PJ, Avlonitis D, Gergis V (2008b) Alkaline conventional and in situ transesterification of cottonseed oil for the production of biodiesel. Energy Fuels 22(3):2110–2115

23. Helwani Z, Othman MR, Aziz N, Fernando WJN, Kim J (2009) Technologies for production of biodiesel focusing on green catalytic techniques: a review. Fuel Process Technol 90(12):1502–1514
24. Noiroj K, Intarapong P, Luengnaruemitchai A, Jai-In S (2009) A comparative study of KOH/Al2O3 and KOH/NaY catalysts for biodiesel production via transesterification from palm oil. Renew Energy 34(4):1145–1150
25. Fini A, Breccia A (1999) Chemistry by microwaves. Pure Appl Chem 71:573–579
26. Taylor M, Atri BS, Minhas S (2005) Developments in microwave chemistry. Evalueserve
27. Yoni G, Aharon G (2008) Continuous flow, circulating microwave system and its application in nanoparticle fabrication and biodiesel synthesis. J Phys Chem C 112:8802–8808
28. Riesz P, Berdahl D, Christman CL (1985) Free radical generation by ultrasound in aqueous and nonaqueous solutions. Environ Health Perspect 64:233–252
29. Kumar D, Kumar G, Singh CP (2010) Ultrasonic-assisted transesterification of Jatropha curcus oil using solid catalyst, Na/SiO_2. Ultrason Sonochem 17:839–844
30. Kumar D, Kumar G, Poonam Singh CP (2010) Fast, easy ethanolysis of coconut oil for biodiesel production assisted by ultrasonication. Ultrasonics Sonochem 17:555–559
31. Stavarache C, Vinatoru M, Nishimura R, Maeda Y (2003) Conversion of vegetable oil to biodiesel using ultrasonic irradiation. Chem Lett 32:716–717
32. Reddy PC, Arumugam S, Babu ML, Teja VK (2018) Synthesis and tribological investigation of compressor's linertribo pair material under the influence of biodegradable compressor oil. IOP Conf Ser Mater Sci Eng 390(1):012041
33. Arumugam S, Chengareddy P, Sriram G (2018) Synthesis, characterisation and tribological investigation of vegetable oil based pentaerythryl ester as biodegradable compressor oil. Ind Crops Prod 123:617–628
34. Badday AS, Abdullah AZ, Lee KT (2013) Optimization of biodiesel production process from Jatropha oil using supported heteropolyacid catalyst and assisted by ultrasonic energy. Renew Energy 50:427–432
35. Mahamuni NN, Adewuyi YG (2009) Optimization of the synthesis of biodiesel via ultrasound-enhanced Base-Catalyzed Transesterification of soybean oil using a multifrequency ultrasonic reactor. Energy Fuels 23(5):2757–2766
36. Tamilarasan S, Sahadevan R (2014) Ultrasonic assisted acid base transesterification of algal oil from marine macroalgae *Caulerpa peltata* Optimization and characterization studies. Fuel 128:347–355
37. Fayyazi E, Ghobadian B, Najafi G, Hosseinzadeh B, Mamat R, Hosseinzadeh J (2015) An ultrasound-assisted system for the optimization of biodiesel production from chicken fat oil using a genetic algorithm and response surface methodology. Ultrason Sonochem 26:312–320
38. Suganya T, Kasirajan R, Renganathan S (2014) Ultrasound-enhanced rapid in situ transesterification of marine macroalgae *Enteromorpha compressa* for biodiesel production. Bioresour Technol 156:283–290

Tribological Investigation of Waste Plastic Oil-Based Methyl Ester Blended Synthetic Lubricant Using Four-Ball Tribometer

S. Baskar, S. Arumugam, G. Sriram, and Venkata Sai Satyanarayana Sastry Sistla

Abstract The present study describes the tribological behavior of waste plastic oil-based methyl ester blended with synthetic lubricant of various concentrations 15 and 30%. The tribological test was conducted by using four-ball tribometer. From the results, synthetic lubricant containing 30% of waste plastic oil methyl ester showed minimum frictional coefficient, lesser mean wear scar diameter, better wear morphology, and high viscosity as compared with the other lubricating oils.

Keywords Four-ball tribometer · Mean wear scar diameter · Bio-lubricant · Waste plastic oil

1 Introduction

Usage of material and energy is becoming a vital global issue. The system of mechanical engineering, friction plays a major role to reduce the energy utilized for mechanical components. Lubricating oil is used to control the friction of mechanical systems. The improving an effective lubrication, base oil selection is very important for a lubricating oil. There has been increasing concern for the use of mineral oil as lubricating oil because of the worldwide interest in environmental issues [1]. The synthetic, mineral, and vegetable oils are used as a lubricating oil. In general, lubricating oil from crude oil is not conceivable with the environment, non-biodegradable, and toxic [2].

S. Baskar
Department of Mechanical Engineering, S.A. Engineering College, Chennai 600077, India
e-mail: baskars@saec.ac.in

S. Arumugam (✉) · G. Sriram · V. S. S. S. Sistla
Sri Chandrasekharendra Saraswathi Viswa Mahavidyalaya, Enathur, Kanchipuram 631561, Tamilnadu, India
e-mail: aru_amace@yahoo.co.in

G. Sriram
e-mail: drg.sriram@gmail.com

V. S. S. S. Sistla
e-mail: mkksai1997@gmail.com

T. Rajmohan et al. (eds.), *Advances in Materials and Manufacturing Engineering*, Springer Proceedings in Materials 7, https://doi.org/10.1007/978-981-15-6267-9_66

The disposal of crude oil-based lubricating oil in the environment creates the issue, and lubricating oil from crude oil exhibits the metal traces and nanoparticles [3, 4]. The lubricant derived from vegetable oil is probable alternate for mineral oil, because they are renewable, biodegradable, and non-toxic [5, 6]. The thin film was formed between metal-to-metal interface with help of the triacylglycerol structure and polar fatty acid composition [7, 8]. The disadvantage of lubricant derived from vegetable oil has lesser thermo-oxidative stability and high flash point [9]. Many researchers were concentrated to modify the drawbacks of lubricant from vegetable oil [10–12]. Furthermore, chemical modification is used to solve the thermo-oxidation stability at lower temperature [13]. The sustainability of lubricant derived from vegetable oil was improved by narrowing their viscosities [14]. Viscosity plays a vital role to control friction and wear between the metal-to-metal interfaces. The eco-friendly viscosity modifiers are used to improve the viscosity of the lubricant. The ethylene-vinyl acetate (EVA) has been established, used as an efficient thickener to the vegetable oil, and converted as a bio-based lubricant [15]. The viscosity of the lubricating oil from vegetable oil was improved by dispersing copolymers and improves the kinematic viscosities at 40 and 100 °C [16]. In present work, tribological behavior of waste plastic oil bio-lubricant was examined by using four-ball tribometer and examined their ability to replace the lubricant from crude oil.

2 Experimental Procedure

2.1 Lubricant Preparations

The methyl ester was derived from the waste plastic oil and blended with the synthetic lubricant as an additive. The various concentrations of (15 and 30%, respectively) methyl ester blended with the synthetic lubricant. The detailed esterification procedure was obtained from Yasvirsingh et al. [17–19]. The properties of lubricating oils are listed in Table 1.

2.2 Experimental Work

A four-ball tribometer was supplied by M/S Ducom Instruments Pvt. Ltd., Bengaluru, used in the present work as shown in Fig. 1. The four-ball tribometer has four balls, one on the top and other three on the bottom side. The bottom side three balls were fixed in ball pot and filled with the test lubricant. The top ball was fixed on the collet and pressed against the ball pot. The dimethyl ketone was used to clean the ball surface. The AISI E-52100 chrome alloy steel was used as a ball of hardness and diameter 64 HRC and 12.7 mm, respectively. The thrust bearing was used to support the ball pot, and normal load was applied by the load cell with plunger arrangement.

Table 1 Properties of various lubricating oils

Properties	Standard	SAE20W40	SAE20W40 with 15% WPOME	SAE20W40 with 30% WPOME
Viscosity @100 °C (cSt)	ASTM D445	15.2	15.4	15.7
Pour point (°C)	ASTM D97	−21	−15	−14
Flash point (°C)	ASTM D92	250	247	246
Viscosity index	ASTM D2270	133	169	173
Specific gravity @ 15 °C	ASTM D287	0.87	0.88	0.89
Wear scar diameter (mm)	ASTM D4172	0.5841	0.3782	0.3571

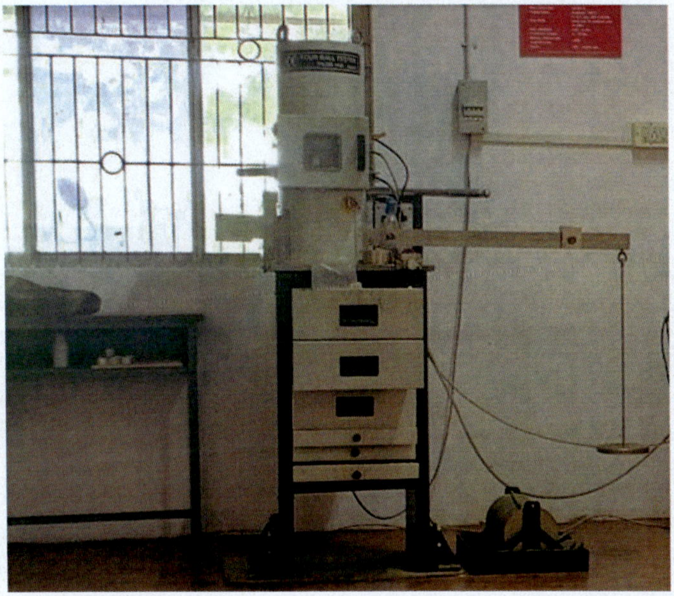

Fig. 1 Four-ball tribometer

The frictional load cell was used to measure the frictional torque of the bottom balls. The temperature sensor was used to measure the temperature of the ball pot. ASTM D 4172 is the standard used to conduct the experiments. The test was conducted at lubricant temperature of 75 °C, speed was 1200 rpm, and normal load was 148 N. The wear scar images of bottom balls were captured by specific CCD microscope. The wear scar diameter of ball was measured by WINDUCOM software. For each test, the average of the wear scar diameter (WSD) of the bottom balls was measured by using WINDUCOM software. Additionally, the viscosity of various lubricating oils was measured using a redwood viscometer and listed in Table 1.

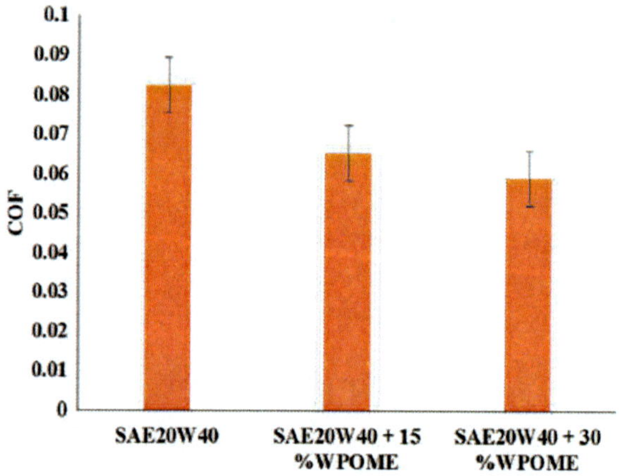

Fig. 2 Friction behavior of various lubricating oils

3 Results

3.1 Friction Behavior

The coefficient of friction of various concentrations of waste plastic oil methyl ester blended synthetic lubricant was examined by using four-ball wear test, and it was compared with synthetic lubricant (SAE20W40). Figure 2 shows the friction behavior of various lubricating oils. From Fig. 2, it is portrayed that an inclusion of various concentrations of methyl ester in synthetic lubricant reduced the frictional coefficient. The frictional coefficient was reduced by about 20.7 and 28.3% for SAE20W40 containing 15 and 30% of waste plastic oil methyl ester, respectively, as compared with SAE20W40. The viscosity of SAE20W40 containing 30% of waste plastic oil methyl ester was higher than the SAE20W40 and SAE20W40 containing 15% of waste plastic oil methyl ester [18, 19] and reduces contact between metal-to-metal interfaces.

3.2 Wear Behavior

The mean wear scar diameter of various lubricants is shown in Fig. 3. From Fig. 3, it is indicated that the mean wear scar diameter of SAE20W40 with 30% of WPOME showed minimum as compared with other lubricating oils. The mean wear scar diameter was reduced by about 35.2 and 38.8% for SAE20W40 containing 15 and 30% of waste plastic oil methyl ester, respectively, as compared with synthetic lubricant. This better wear reduction ability was related to higher viscosity of SAE20W40 containing 30% of WPOME as shown in Table 2. The high viscosity of SAE20W40

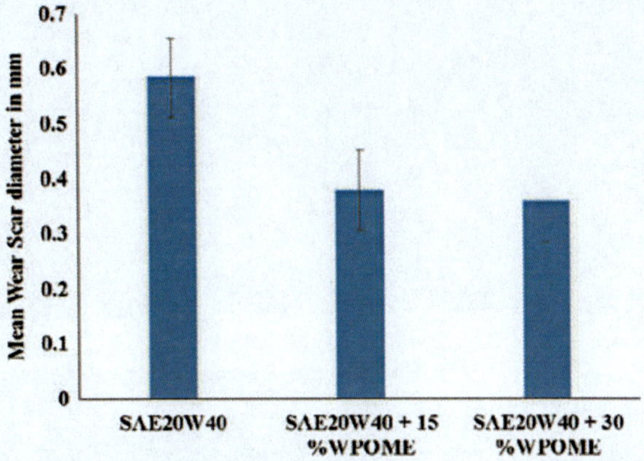

Fig. 3 Mean wear scar diameter of various lubricants

containing 30% of WPOME forms the tribo layer between the metal-to-metal interface, and this tribo layer will serve as a barrier to reduce the frictional coefficient and wear scar value [17].

3.3 *Surface Morphology*

Figure 4a–c shows the worn surface of balls by using scanning electron microscopy. SAE20W40 with 30% of WPOME displays the minimum wear, and SAE20W40 and SAE20W40 containing 15% of WPOME show the maximum wear, respectively. The chemical reaction of an additive plays a significant role to protect the surface from wear. The anti-wear additive is used to protect metal-to-metal interface, resulting to reduce the adhesive wear [18].

4 Conclusions

The tribological characteristics of waste plastic oil methyl ester blended lubricant were analyzed by using four-ball tribometer. The results as follows:

1. From the results, SAE20W40 containing 30% of waste plastic oil methyl ester displayed the minimum frictional coefficient and mean wear scar diameter, respectively.
2. SAE20W40 blended with 30% of waste plastic oil methyl ester displayed the better results among all the other lubricants and also showed an excellent capability to preserve the tribological characteristics.
3. The worn morphology of ball lubricated with SAE20W40 containing 30% of waste plastic oil methyl ester showed the minimum wear scar, lesser voids, and cracks as compared with other lubricating oils.

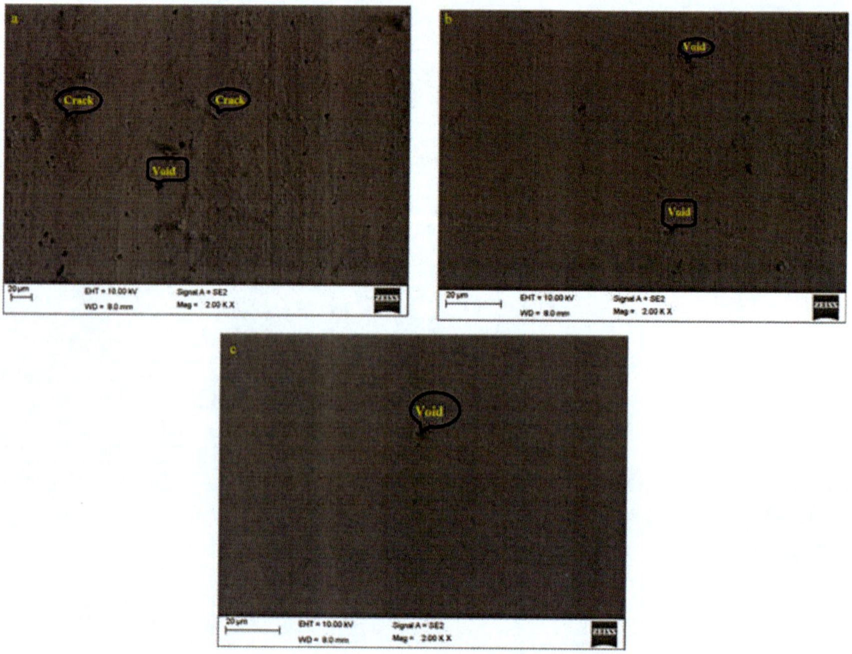

Fig. 4 **a** Worn surface of ball lubricated with SAE20W40, **b** worn surface of ball lubricated with SAE20W40 containing 15% of WPOME, **c** worn surface of ball lubricated with SAE20W40 containing 30% of WPOME

References

1. Bartz WJ (1998) Lubricants and the environment. Tribol Int 31:35–47
2. Adhvaryu A, Liu Z, Erhan S (2015) Synthesis of novel alkoxyllated triacylglycerols and their lubricant base oil properties. Ind Crop Prod 21:113–119
3. Bekal S, Bhat NR (2012) Bio-lubricant as an alternative to mineral oil for a CI engine—an experimental investigation with Jatropha oil as a lubricant. Energ Source Part A 34(11):1016–1026
4. Polmear R, Stark JS, Roberts D, McMinn A (2014) The effects of oil pollution on Antarctic bethic diatom communities over 5 years. Mar Pollut Bull 90(1–2):33–40
5. Hwang HS, Erhan SZ (2002) Lubricant base stocks from modified soybean oil. AOCS Press, Champaign (USA)
6. Saliha N, Salimona J, Yousifb J (2011) The physicochemical and tribological properties of oleic acid based triester biolubricants. Ind Crop Prod 34:1089–1096
7. Havet L, Blouet J, Valloire R, Brasseur E, Slomka D (2001) Tribological characteristics of some environmentally friendly lubricants. Wear 248:140–146
8. Walsh JW (2002) Determination of triglyceride composition of vegetable oils using HPLC and evaporative light scattering detection. In: IFT annual meeting and food expo, Anaheim, California, Session 306
9. Rani S, Joy ML, Nair KP (2015) Evaluation of physiochemical and tribological properties of rice bran oil–biodegradable and potential base stoke for industrial lubricants. Ind Crop Prod 65:328–333

10. Hwang H-S, Erhan S (2001) Modification of epoxidized soybean oil for lubricant formulations with improved oxidative stability and low pour point. J Am Oil Chem Soc 78:1179–1184
11. He Z, Lu J, Zeng X, Shao H, Ren T, Liu W (2004) Study of the tribological behaviors of S, P-containing triazine derivatives as additives in rapeseed oil. Wear 257:389–394
12. Wu Y, Li W, Zhang M, Wang X (2013) Improvement of oxidative stability of trimethylolpropane trioleate lubricant. Thermochim Acta 559:112–118
13. Bokade VV, Yadav GD (2007) Synthesis of bio-diesel and biolubricant by transesterification of vegetable oil with lower and higher alcohols over heteropolyacids supported by clay (K-10). Process Saf Environ Prot 85:372–377
14. Quinchia L, Delgado M, Valencia C, Franco J, Gallegos C (2010) Viscosity modification of different vegetable oils with EVA copolymer for lubricant applications. Ind Crop Prod 32:607–612
15. Martín-Alfonso JE, Valencia C (2015) Tribological, rheological, and microstructural characterization of oleogels based on EVA copolymer and vegetables oils for lubricant applications. Tribol Int 90:426–434
16. Quinchia L, Delgado M, Valencia C, Franco J, Gallegos C (2009) Viscosity modification of high-oleic sunflower oil with polymeric additives for the design of new bio-lubricant formulations. Environ Sci Technol 43:2060–2065
17. Singh Y, Singla A, Singh AK (2017) Tribological characteristics of Mongongo-oil–based biodiesel blended lubricant. Energy Sources Part A Recov Util Environ Effects 39(3):332–338
18. Singh Y, Singla A, Kumar A, Kumar D (2016) Friction and wear characteristics of jatropha oil-based biodiesel blended lubricant at different loads. Energy Sources Part A Recov Util Environ Effects 38(18):2749–2755
19. Baskar S, Sriram G, Arumugam S (2015) Experimental analysis on tribological behavior of nano based bio-lubricants using four ball tribometer. Tribol Ind 37(4):449–454

Mechanical Properties of Biodegradable Calotropis Gigantea–Jute Fibre Hybrid Composite

A. Arun Premnath, Ankur Sharma, A. Arun Kumar, S. Dinesh Kumar, and K. Raghul

Abstract In the current scenario, non-decomposable fibres are the major problem in every field. The present work deals in studying the mechanical properties of Calotropis gigantea and jute fibre reinforced with epoxy resin hybrid natural fibre composites. The composites are fabricated by compression moulding method by varying the Calotropis gigantea fibre by 0, 5, 10% by weight (wt) fraction and keeping the jute with constant 10% by weight fraction. The fabricated composites are tested for its mechanical properties such as hardness, impact strength and compression strength as per ASTM standards. From the results, it is observed that the increase in Calotropis gigantea fibre increases all the mechanical properties of the composites.

Keywords Natural fibre · Jute · Calotropis gigantea · Compression moulding mechanical properties

1 Introduction

The idea of sustainable materials has now become of paramount importance because of the need to guarantee our condition. Natural fibres such as coconut fibre, bamboo, jute, hemp and Calotropis gigantea are currently discovering applications in a large number of companies. Many research studies have been spread over different basic biological characteristic filaments and their compounds, and the present study has aimed to present another regular bio-based basic fibre as one of the natural product and fortification channels in the assembly of a new composite material for lightweight structures. Calotropis gigantea is a medium-sized shrub or small tree, up to 4 m tall, with a large sticky appearance and smooth sap. The stem is shaded, smooth and slag-shaped, extending almost from the base. The leaves are dark green, inverse,

A. Arun Premnath (✉) · S. Dinesh Kumar · K. Raghul
Sri Chandrasekharendra Saraswathi Viswa Maha Vidyalaya University, Centre for Composite Science and Tribology, Kanchipuram, Tamil Nadu, India
e-mail: arun_premnath@kanchiuniv.ac.in

Ankur Sharma · A. Arun Kumar · S. Dinesh Kumar · K. Raghul
Department of Mechanical Engineering, Sri Chandrasekharendra Saraswathi Viswa Maha Vidyalaya University, Kanchipuram, Tamil Nadu, India

T. Rajmohan et al. (eds.), *Advances in Materials and Manufacturing Engineering*, Springer Proceedings in Materials 7, https://doi.org/10.1007/978-981-15-6267-9_67

substitute, waxy, thick and tight. They measure 5–15 cm × 4–10 cm with a short point and a moulded heart-shaped base that holds the stem a little.

Calotropis gigantea is one of the most important accessible plants of our country founded in sandy soils where the rain is on average of: 300–400 mm. It easily grows because the seed of the plant easily accumulates near to road, tides, ponds and local fields. Calotropis gigantea provides a high malleable and rough quality, and these strands have more weight per square metre than cotton fibre. Because Calotropis gigantea is an extremely developed shrub in important parts of India, coconut spatula fibres have an unparalleled elasticity compared to other conventional filaments, such as cotton, coconut and banana filaments. Cellulosic filaments such as coconut, Calotropis gigantea and banana fibre are used as reinforcing specialists for various thermoset and thermoplastic compounds.

Some of the literatures are discussed here. Gopinath et al. [1] concluded that epoxy is the best matrix to use with jute fibre than using polyester as the matrix. They observed that jute–epoxy will be the suitable combination for as to attain superior mechanical properties for the composite material. Singha and Thakur [2] studied the mechanical properties of Hibiscus sabdariffa fibre reinforced with urea-formaldehyde resin green composites. They studied the wear and mechanical properties as a function of fibre loading and found that wear resistance and mechanical properties increase with the increase in the fibre. Rahman et al. [3] observed that synthetic fibres have better mechanical properties than natural fibre composites. But however mechanical properties of natural fibre can be improved by altering some parameters during fabrication such as fibre treatment, length of the fibre and fibre orientation.

From the literature, it is found that limited work is done on Calotropis gigantea fibre. In our study, we have used Calotropis gigantea fibre along with jute in various weight fractions and fabricated three different types of composites.

2 Experimental Details

2.1 Materials

The epoxy resin of 1.15–1.4 g/cc, mixed with a hardener thickness of 0.97–0.99 g/cc, is used to configure the composite plate. The weight ratio of the epoxy and hardener mixture is 10:1. The resin was obtained from a dealer in Chennai. Jute and Calotropis gigantea fibres are used as reinforcement.

2.2 Extraction of Fibre

The stems of Calotropis gigantea were given from unsuccessful lands. The distance through the stems of Calotropis gigantea increased from 8 to 32 mm and decreased to better protect the plant. The thickness of the layer was between 0.8 and 1.4 mm. In crispy stems, there is latex in the middle of the stem and when the stem is cut or leaves are discharged, the latex flows out. The petioles and leaves were expelled, and the stems were left in the field for two days in hot, dry weather. After two days, the latex did not spread from the petiole spot and allowed the rigidity of the filaments to exceed the quality of the bark. The fibre was collected.

2.3 Compression Moulding Technique

The fabrication part was done on compression moulding; the size of the die is 400 × 200 × 10. The compression moulding machine is shown in Fig. 1. First die was cleaned properly. The surface of the die was made free of moisture and cleaned properly. Then, release gel was applied on the surface of the die so as to create a non-sticky surface. Then, the epoxy was applied layer by layer with jute and Calotropis gigantea by considering safety and precautions. Ratio was calculated according to weight fraction where jute fibre kept as constant for all the three specimens. Thc dic was covered with OHP sheet to avoid sticking on top die. Fibres were compressed at a set pressure and temperature of 50 bar and 40 °C, respectively, for 3 h. The fabrication was completed, and specimen has been taken out. The composition of the specimens is shown in Table 1.

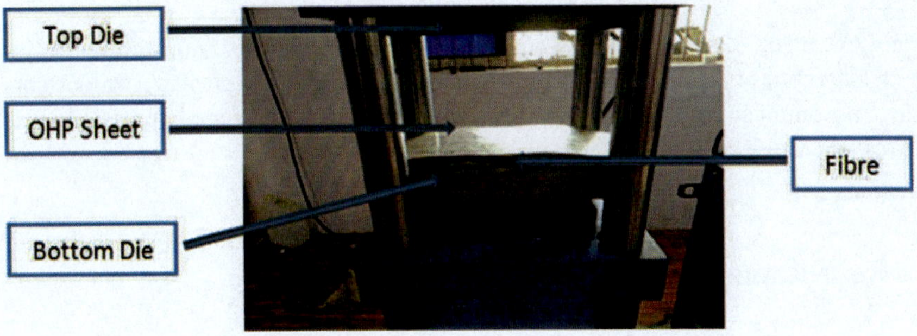

Fig. 1 Compression moulding machine

Table 1 Detailed composition of the specimen

	Matrix epoxy	Jute fibre	Calotropis gigantea fibre
Specimen 1	80	20	0
Specimen 2	75	20	05
Specimen 3	70	20	10

2.4 *Mechanical Testing*

2.4.1 Hardness Test

Samples of 10-mm thickness were cut with a smooth surface for the hardness test. The test was conducted using Rockwell's hardener as per ASTM testing standard using a C-type diamond intender with 150 load.

2.4.2 Impact Test

The impact test specimen was chopped according to the required ASTM standard 55 × 10. The specimen was fixed on the test machine and allowed the pendulum to crack or break the specimen. Using the impact test, the vitality expected to break the material can be effectively estimated and can be used to measure the hardness of the material and the quality of the performance. The effect test dissects the impact of the strain rate on the fracture and the malleability of the material.

2.4.3 Compression Test

There are many applications in which fibre with good compression properties is needed. It is one of the most important mechanical properties which play an important role in the applications. Specimen was chopped in 50 × 50 mm for compression test; the test was done on UTM machine in SCSVMV University.

3 Results and Discussion

3.1 *Hardness Strength*

From the Fig. 2, it is noticed that the Calotropis gigantea content composites have better hardness strength than without Calotropis gigantea composites, by increasing

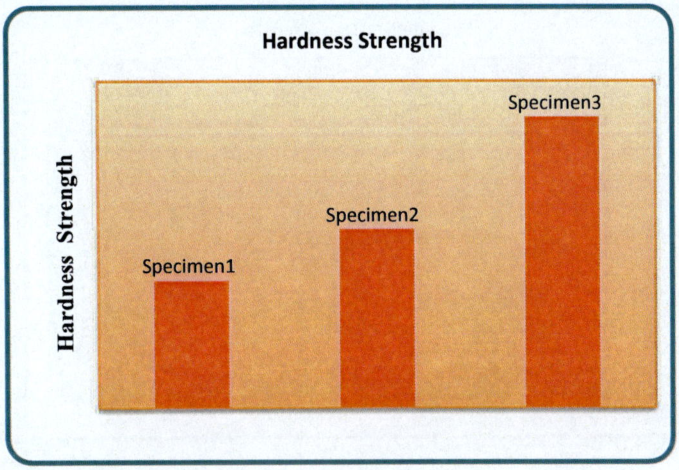

Fig. 2 Hardness strength of specimen

the Calotropis gigantea fibre in composite resulted in an increase in the hardness strength. We observed that after increasing 10% Calotropis gigantea fibre, the hardness strength was increased by 56.56%.

3.2 *Impact Strength*

From the Fig. 3, it is noticed that the Calotropis gigantea content composites have

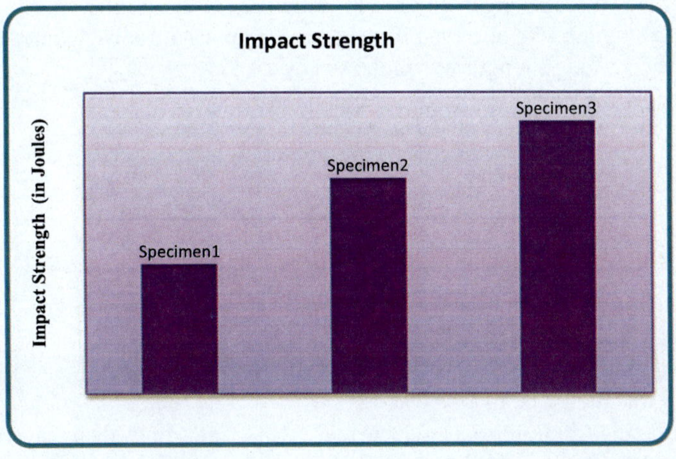

Fig. 3 Impact strength of specimen

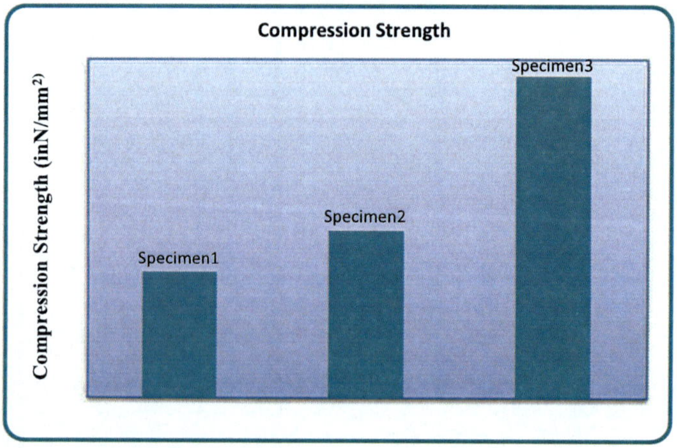

Fig. 4 Compression strength of specimen

better impact strength than without Calotropis gigantea composites, by increasing the Calotropis gigantea fibre in composite resulted in an increase in the impact strength. We observed that after increasing 10% Calotropis gigantea fibre, the impact strength was increased by 52.60%.

3.3 *Compression Strength*

From the Fig. 4, it is noticed that the Calotropis gigantea content composites have better compression strength than without Calotropis gigantea composites, by increasing the Calotropis gigantea fibre in composite resulted in an increase in the compression strength. We observed that increasing the Calotropis gigantea fibre by 10%, the compression strength was increased by 14.62%.

4 Conclusions

From the result obtained, the following conclusions arrived.

1. By comparing 0% Calotropis gigantea with 10% Calotropis gigantea, hardness strength was increased by 56.56%.
2. By comparing 0% Calotropis gigantea with 10% Calotropis gigantea, impact strength was increased by 52.60%.
3. By comparing 0% Calotropis gigantea with 10% Calotropis gigantea, compression strength was increased by 14.62%.

4. Calotropis gigantea composites have better mechanical properties than Calotropis gigantea composites.
5. In future, there is a tremendous investment on natural fibres, to bring a great impact on manufacturing sectors.

References

1. Gopinath A, Kumar MS, Elayaperumal A (2014) Experimental investigations on mechanical properties of jute fiber reinforced composites with polyester and epoxy resin matrices. Proc Eng 97:2052–2063
2. Singha AS, Thakur VK (2008) Mechanical properties of natural fibre reinforced polymer composites. Bull Mater Sci 31(5):791
3. Rahman R, Putra SZFS (2019) Tensile properties of natural and synthetic fiber-reinforced polymer composites. In: Mechanical and physical testing of biocomposites, fibre-reinforced composites and hybrid composites. Woodhead Publishing, pp 81–102

The Concept of Metamaterial Used for the Fabrication of Antenna

K. Sugapriya and S. Omkumar

Abstract This article presents the study of metamaterial characteristics like negative refractive index, process and applications in fabricating the microstrip patch antenna. Metamaterial is a composite material structure that exhibits a special property like negative refractive index or left-handed materials. It gives a polarization in negative direction because of negative μ and negative ε. The proposed metamaterial antenna operates at 2.1 GHZ frequency, can enhance the directivity, gain of the patch antenna and reduce the return loss. The variation in gap size of split ring resonator (SRR) improves the bandwidth of patch antenna. These metamaterial structures miniaturize the antenna, inducing the resonator.

Keywords Metamaterial · SRR · Microstrip patch antenna

1 Introduction

The metamaterial is a material in which the negative characteristics are high, i.e., negative values of permittivity and permeability. The term "Meta" is a Greek word which means something new, advance, altered and new changes in the material, used to design an antenna to get the better directivity, gain enhancement and return loss. This material gives the minimum reflection coefficient and maximum transmission coefficient in antenna design.

Metamaterial design is based on the split ring resonator (SRR) structure, which gives the better radiation as well as it will guide the electromagnetic wave propagation in a proper direction and to produce desired magnetic susceptibility, i.e., magnetic response. The different types of metamaterials are used to design a microstrip patch antenna for different applications.

K. Sugapriya (✉)
SCSVMV University Kanchipuram, Tamilnadu, India
e-mail: dhivyasuga@gmail.com

S. Omkumar
ECE, SCSVMV University Kanchipuram, Tamilnadu, India
e-mail: omkumar1234@gmail.com

T. Rajmohan et al. (eds.), *Advances in Materials and Manufacturing Engineering*, Springer Proceedings in Materials 7, https://doi.org/10.1007/978-981-15-6267-9_68

The SRR structure has a magnetic response without magnetic materials which give the good resonance frequency and gain. The various investigations made using the metamaterial having a negative property which exhibits different parameter characteristics, and analysis was given in Table 1 [Literature survey].

Datta et al. [1] found the enhanced directivity using Yagi-array for wireless applications. Gingrich and Werner et al. [2] determined the low-loss zero index of refraction materials (ZIM) from "Synthesis of Zero Index of Refraction Metamaterials via Frequency-Selective Surfaces Using Genetic Algorithms". This chapter has presented a technique for the synthesis of thin, planar, which are easily optimized for a variety of different design constraints, such as center frequency, desired index of refraction and amount of loss.

Rajab and Mittra et al. [3] analyzed the multilayered left-handed transmission line which increases the wavelength of microstrip patch antenna, and it is suitable for microwave applications. Kim and Varadan et al. [4] determined the effect of capacitive coupling between split ring resonators for X-band applications. The SRR structure makes the inductance and capacitance.

Iñigo et al. [5] found that the highly directive aperture-coupled microstrip patch antenna based on planar metasurface improves the directivity and resonance frequency. Mi et al. [6] determined the reduction of mutual coupling between the patch using waveguide metamaterial used for decoupling in wireless communications. Duong and Venkataraman et al. [7] analyzed ambient energy harvesting using metamaterial, and it is used to focus the harvested energy on a fractal antenna. Lee et al. [8] investigated the CSRR structure which could be useful in design of multiband antennas. Zhou et al. [9] determined the inhomogeneous and an isotropic (IA) zero-index metamaterials (ZIM), meander-line structure which gives the good gain directivity of the antenna. Meng et al. [10] investigated the GRIN metamaterial lens structures for smart user tracking in future mobile WLAN/WPAN communication and localization systems.

2 Antenna Design

The microstrip patch antenna was designed using the metamaterial concept. In modern communication world, very fast and high data rate process is needed because all type of information was conveyed or transmitted through the mobile phone, and for the efficient communication, the antenna was designed and it is fabricated using the different design techniques and materials. Also that gives a good variation of antenna parameters with respect to the materials. In the proposed antenna, the metamaterial structure has been designed with the relative permittivity of $\varepsilon_r = 2.2$. The proposed antenna operates at a resonance frequency of 2.1 GHZ. The 2D and 3D views of the metamaterial antenna are shown in Figs. 1 and 2.

The feed line of metamaterial structure is 50 Ω impedance, which is suitable for the better radiation of antenna. An antenna structure gives a good surface wave suppression for good gain and efficiency. The rectangular patch length L and width

Table 1 Comparative analysis of various metamaterial designs in antenna

S. No.	Design structure	Parameter variation	Applications
1	The metamaterial design based on split ring resonator (SRR) structures	The enhanced directivity was achieved in the Yagi-array design	This type of antenna is used in wireless mode of applications
2	Zero-index materials and negative index material structures through frequency selective surfaces	They got the desired index of refraction, center frequency and good return loss	Potential applications including as a superstrate and substrate for microstrip patch antennas, focusing radomes, phase manipulation and imaging of paraxial beams
3	Multilayered left-handed transmission line increases the wavelength of microstrip patch antenna	MPA size can be reduced by utilizing the loading capacitances and inductances	Microstrip patch can be used for microwave applications
4	SRR structure makes the inductance and capacitance	This paper presents the study about the SRR orientations and inter-particle spacings. They got high resonance frequency for the SRR gap 2 mm	This antenna was designed for X-band applications
5	Metasurface antenna	Improved directivity and resonance frequency with the help of metasurface	It is used for microwave and optical applications and smart antenna systems
6	Wave-guided metamaterial (WG-MTM)structure	Two-element microstrip patch antenna array scheme is proposed for decoupling, and it provides good coupling suppression level for closely spaced microstrip patches	This design was used for decoupling in wireless communications
7	Metamaterial lens that has been designed by the nanoplasmonics	The present work investigates feasibility of enhancing RF energy harvesting using a hyperbolic metamaterial lens	To focus the harvested energy on a fractal antenna

(continued)

Table 1 (continued)

S. No.	Design structure	Parameter variation	Applications
8	The complimentary SRR-loaded microstrip antenna	The experimental results confirm that the CSRR-loaded patch antenna achieves size reduction as well as bandwidth improvement	The proposed antenna could be useful in designing multiband antennas and compact antennas achieving a size reduction without having to sacrifice the antenna bandwidth
9	Inhomogeneous and an isotropic (IA) zero-index metamaterials (ZIM), meander-line structure	The design could improve the gain and directivity in a broad bandwidth	The new Vivaldi IA-ZIM antenna can be applied in practical applications
10	GRIN metamaterial lens	The result gives the enhanced gain and better directivity	Smart user tracking in future mobile WLAN/WPAN communication and localization systems

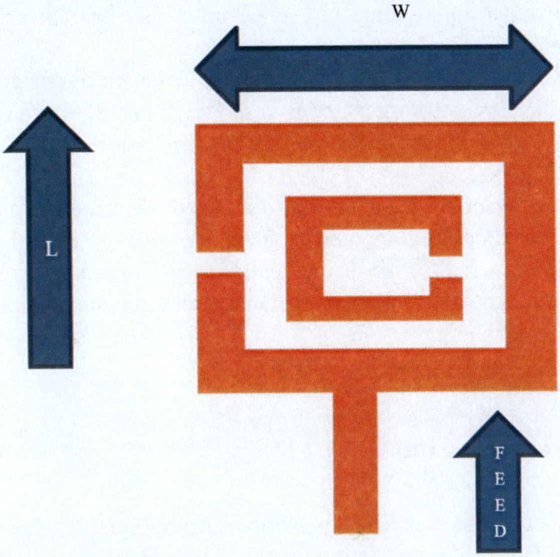

Fig. 1 2D view of microstrip patch

Fig. 2 3D view of patch antenna

W can be chosen according to the excitation of antenna. The gap can be created in the structure based on ground plane.

The designed antenna can be simulated using the advanced design system software, and then, the 3D view of the antenna is shown in Fig. 2, which shows the side view of metamaterial structure. In this design, the simulation gives the S-parameter results.

The proposed antenna gives a good return loss and increased the directivity with the metamaterial structure as compared to the above Table 1. The 50 Ω impedance microstrip feed line is used for the design, and the SRR structure should be created in the rectangular patch, which is like a metamaterial structure, also it provides the good radiation pattern and gain.

3 Result and Discussion

The metamaterial antenna has good performance and smaller size compared to conventional size. Defected ground structure (DGS) is one of the method to reduce the antenna size such as metamaterial and is having the negative permittivity and permeability characteristics which result in good interaction with electromagnetic radiation, and the proposed substrate materials are synthesized by combining magnetic and electric dipole elements. The similar results have been obtained by Li et al. who investigated that the design of a novel rectangular patch antenna with planar metamaterial patterned substrate gives the concept of DGS which arises from the studies of photonic band gap (PBG) structure which deals with light waves. PBG is known as electron band gap (EBG) in electromagnetic application, and the artificial periodic structure gives metamaterial behavior.

The proposed antenna is designed to have the gap of 0.4 mm at the bottom, the result of SRR structure shows that an antenna can operate at the resonance of 2.1 GHZ, its S-parameter value should be calculated at resonance, then the simulated value is −27 dB return loss in the S-parameter plot, and the phase plot can be obtained for the corresponding resonance. Hence, the proposed antenna was suitable for industrial application and wireless applications.

The similar metamaterial antenna can be designed by Devapriya et al. which determined that while dual-band metamaterial antenna using the planar structure for S band and C band applications, they analyzed various thickness return losses and VSWR for resonance 2.7GHZ, −17 dB return loss can be obtained. Hence, the proposed antenna gives a good return loss −27 dB for the 2.1 GHZ in the simulation, and the simulated results are given in Table 2.

Various analyses have been made for the design of antenna using the metamaterial substrate concepts. The proposed antenna gives the linear polarization for the 1.6 mm thickness and microstrip line feed. Hence, it is suitable for wireless applications.

Table 2 Simulated result

Frequency of operation	2.1 GHZ
Return loss (dB)	<−10 dB
Feeding method	Microstrip line
Polarization	Linear

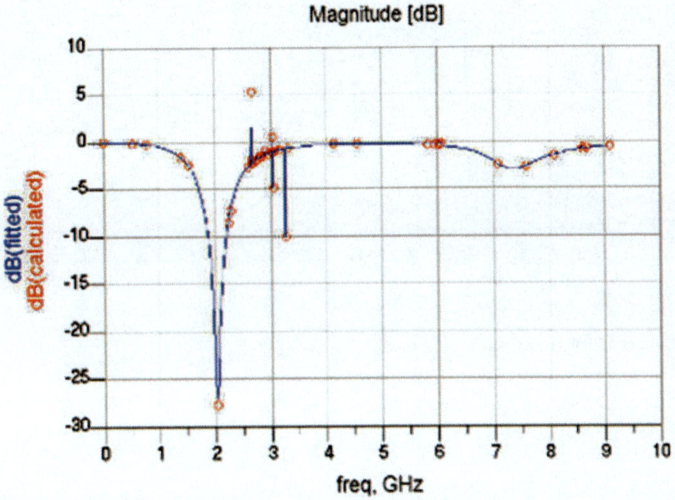

Fig. 3 S-parameter plot-magnitude

4 Conclusions

The microstrip patch antenna should be designed and simulated with the relative permittivity $\varepsilon_r = 2.2$, with the SRR structure. The simulated result gives the return loss of below 10 dB in negative shown in Fig. 3, and it operates in the frequency range of 2.1 GHZ. The array design can be created using metamaterial SRR structure for future development. The simulated result gives the good antenna gain and radiation pattern. Hence, the proposed antenna is suitable for industrial applications like smart antennas and multiband antennas (Fig. 4).

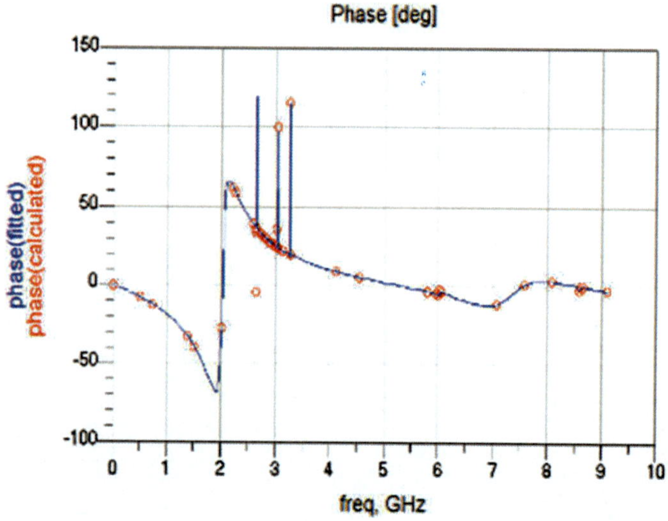

Fig. 4 S-parameter plot-phase

References

1. Datta R, Shaw T, Mitra D (2017) Miniaturization of microstrip yagi array antenna using metamaterial. Progr Electromagn Res C 72:151–158
2. Gingrich MA, Werner DH Synthesis of zero index of refraction metamaterials via frequency-selective surfaces using genetic algorithms. 0-7803-8883-6/05/$20.00 X2005 IEEE
3. Mittra KZRR, Lanagant MT Size reduction of microstrip antennas using metamaterials. 0-7803-8883-6/05/$20.00 @2005 IEEE
4. Kim K, Varadan VV Microwave and optics laboratory for imaging and characterization. Effect of capacitive coupling between split-ring resonator. 978-1-4244-2042-1/08/$25.00 ©2008 IEEE
5. Liberal I, Ederra I, Gonzalo R Public University of Navarre (UPNA). Highly-directive aperture-coupled microstrip patch antenna based on planar meta-surface. 978-1-4244-4968-2/10/$25.00 ©2010 IEEE
6. Yang XM, Liu XG, Zhou XY, Cui TJ (2012) Reduction of mutual coupling between closely packed patch antennas using waveguided metamaterials. IEEE Antennas Wirel Propag Lett 11
7. Duong TN, Venkataraman J, Lu Z Enhancement of ambient energy harvesting using a metamaterial lens. 978-1-4673-5317-5/13/$31.00 ©2013 IEEE
8. Lee Y, Hao Y (2007) Characterization of microstrip patch antennas on metamaterial substrates loaded with complementary split-ring resonators
9. Zhou B, Li H, Zou XY, Cui TJ (2011) Broadband and high-gain planar vivaldi , antennas based on inhomogeneous anisotropic zero-index metamaterials. Progr Electromagn Res 120:235–247
10. Meng F-Y, Liu R-Z, Zhang K, Erni D, Wu Q, Sun L, Li L-W (2013) Automatic design of broad-band gradient index metamaterial lens for gain enhancement of circularly polarized antennas. Progr Electromagn Res 141:17–32

Modeling of Potentially Implementable Configurable Logic Block in Quantum Dot Cellular Automata for Nanoelectronic Device Architecture

R. Jayalakshmi and M. Senthil Kumaran

Abstract As the device density of the existing complementary metal oxide semiconductor keeps on increasing to reach its limitations, there is a need for alternative technologies to the existing technology. One such is the quantum dot cellular automata which performs all the operations performed by the CMOS technology. Moreover, it utilizes very low power, less area and high switching speeds. This paper proposes the design and analysis of potentially implementable configurable logic block in quantum dot cellular automata for nanoelectronic device architecture. The proposed design utilizes coplanar wire crossing with five-input majority voter for decoder along with the memory loop for storing the information. The proposed design is validated using the QCA designer tool.

Keywords Quantum dot cellular automata · Five-input majority gate · Co-planar wire crossing · Look-up table · Configurable logic block

1 Introduction

Quantum dot cellular automata (QCA) is an epitome of the emerging field-coupled nanocomputing techniques [1, 2]. QCA utilizes charge configuration protocol in contrary to the conventional complementary metal oxide semiconductor (CMOS) which utilizes charge dissipation protocol. The position of the electrons in the quantum wells represents the binary values '0' and '1'. Each cell encodes a data state, and when connected together forms different configurations such as wire, inverter and majority voter. The QCA structures are highly pipelined architectures with the homogenous structure which has the potential of nanoscale fabrication. The parameters of QCA technology are the circuit size which is determined by the

R. Jayalakshmi (✉)
Department of ECE, Sri Chandrasekharendra Saraswathi Viswa Maha Vidyalaya, Enathur, Kanchipuram, Tamilnadu 631561, India
e-mail: jayalakshmiecescsvmv@gmail.com

M. Senthil Kumaran
Department of CSE, Sri Chandrasekharendra Saraswathi Viswa Maha Vidyalaya, Enathur, Kanchipuram, Tamilnadu 631561, India

T. Rajmohan et al. (eds.), *Advances in Materials and Manufacturing Engineering*, Springer Proceedings in Materials 7, https://doi.org/10.1007/978-981-15-6267-9_69

number of cells used in the design and the clock frequency which contributes to the overall delay between the fan-in and fan-out between the input and the output cells. The power dissipation is comparatively very less than the conventional CMOS technology which has reached its limits of saturation.

Field-programmable gate array is a main application of QCA technology for implementing universal logic. A programmable switch matrix is presented using programmable interconnects in [3]. Amiri et al. presented a fixed logic using multiplexer tree cells [4]. Using QCA technology, programmable logic arrays are demonstrated by Crocker et al. [5]. Niemier et al. [6] have proposed fixed logic FPGA with programmable interconnects. Pandiammal et al. proposed a 64-bit SRAM with multiplexer-based memory loop [7]. In [8], multiple layers CLB has been implemented. In [9], D-flip-flop with set and reset has been presented with function generator matrix for memory addressing. This chapter proposes a 64-bit look-up table of parallel memory architecture utilizing high input majority voter-based decoders for row and column address decoding. The memory cell is constructed by utilizing LUT-based memory loop implemented using D-flip-flop. The proposed work has reduced latency but with a reduced density because of the increase in hardware count such as cell count and clocking zone.

2 QCA Background

QCA devices are constructed using QCA cells [1]. The four quantum wells are occupied by two electrons in antipodal sites which are positioned in an array of 2 × 2. The electrons occupy their respective positions by columbic law of attraction and repulsion and give two stable states for the quantum cells as binary ‘0’ and binary ‘1’ [10]. When the QCA cells are placed next in a chain of cells, it represents wire which is used as interconnect in larger architectures. Similar to a typical conventional CMOS field-effect transistor, the QCA cells are used for computing by using the majority voter. The majority voter function F is defined as

$$F = AB + AC + BC \tag{1}$$

The other QCA structures are inverters, cross-overs, odd tap and even tap which are used for implementing digital functions. The majority gate can be utilized as a AND gate and OR gate by fixing the polarity of any one of the inputs to −1 or + 1. The gates and wires and the signal flow in QCA circuits are controlled by the clock pulses [11]. The signal amplification is provided by the clock signals which are standard CMOS wires which lie beneath the layer where QCA wires are placed. The clocking is applied in four phases which are switch, hold, release and relax [11]. In hold phase, the QCA cell will retain the current position of the electrons and the clock is in low state. The electrons will tunnel in the rising edge of the clock which is the release state. The next state is the relax state in which the clock reaches its high, and on the falling edge of the clock, the QCA reaches its switch state in which the clock repels the electrons to enter in to new switching state.

3 Decoder and Memory Cell Design

The universal logic is implemented in QCA by utilizing look-up table (LUT), in which either a multiplexer-based or memory-based architecture is used. The LUT gives the inherent potential of flexibility in operation and programmability. The LUT is multi-programmable if the memory contents changed are desirable by the read and write signals [12, 13]. The parallel memory architecture discussed in this paper utilizes a unit memory cell constructed using D-flip-flop. The unit memory cell using D-flip-flop shown in Fig. 1 represents the memory loop which is given as feedback from the present state which is erased every time read/write signal is activated. In Fig. 2, the layout represents the majority gates M1, M2, M3 and M4 which are fixed polarized either as −1 or +1 to function as AND gate and OR gate.

Figure 3 represents the QCA layout of the memory cell which uses coplanar wire crossing with zone-based clocking with enable, read/write signal and the input signal

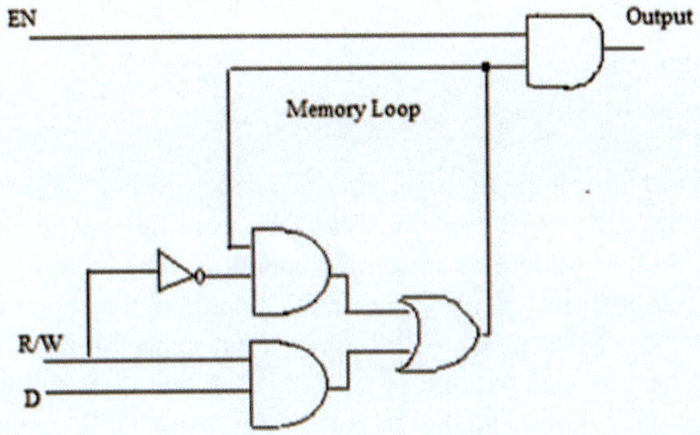

Fig. 1 Memory cell using D latch

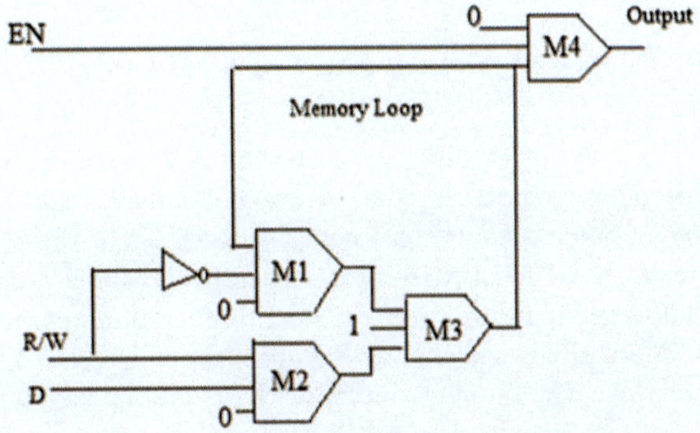

Fig. 2 Memory cell using three-input MV

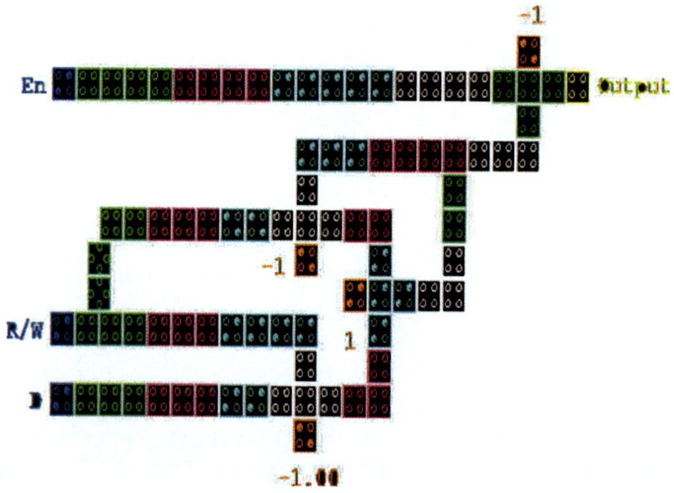

Fig. 3 QCA layout of memory cell and simulated waveform

Table 1 D-latch function

En	R/W	D	Q
1	1	0	0
1	0	1	1
0	X	X	X

to implement the function of D-flip-flop with a cell count of 88 and area of 0.16 μm^2 with a delay of 1.5 clock phases as shown in Table 1. The HIGH logic represents write, and LOW represents read logic operation. The memory addressing is performed by a pre-proposed 3–8 decoder with a enable input implemented by using 8 five-input majority voter along with 8 three-input majority voter. The decoder is constructed using coplanar wire crossing with zone-based clocking with optimized cell count and delay [14]. The design is validated using QCA designer tool [15].

4 Proposed Architecture of 64 Bit LUT for CLB

In Fig. 4, the block diagram of a 64-bit LUT has been given which utilizes a two 3 × 8 decoders for decoding the row and column address. The unit memory cell (UMC) holds a single-bit data constructed using a D latch. The 64-bit RAM used in designing a memory-based LUT represents a parallel pipelined architecture. The data input and read/write inputs are given to UMC through programmable interconnects.

The UMC is selected by the decoders by using a AND gate operation to select the UMC. The output from the UMC is multiplexed through the OR gate, and the

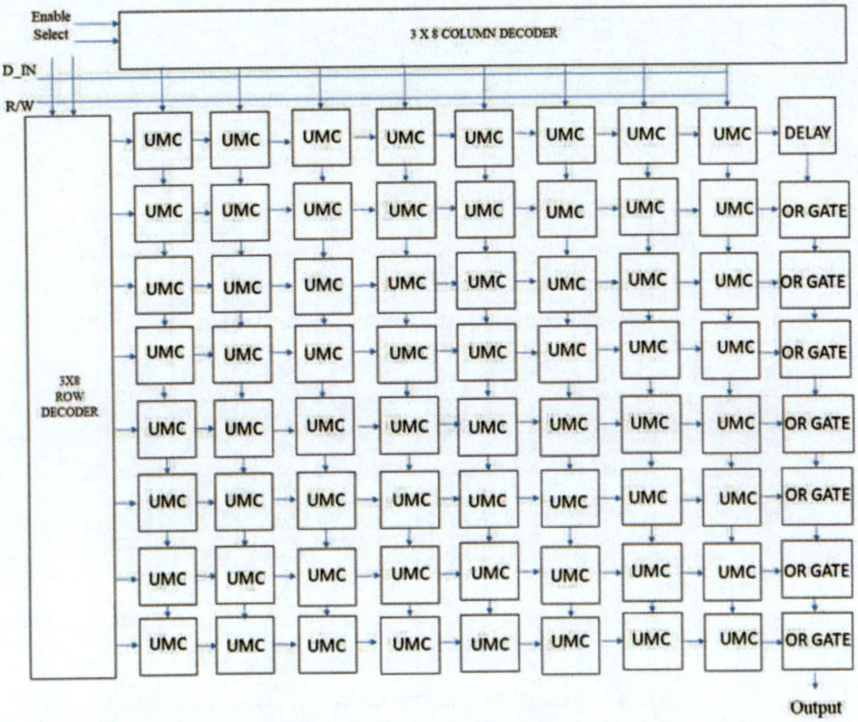

Fig. 4 Block diagram of 64-bit LUT for RAM

output function is obtained. Figure 4 represents a single LUT for CLB configuration. Multiple CLB connected together can be used for implementing FPGA based on the alignment. Figure 5 represents QCA layout implanted using QCA designer tool in which the row and column decoders are fixed, and the flexibility is given for the programmable interconnects. The zone-based clocking is provided for decoder and UMC, whereas the clocking for the programmable interconnects is provided as the information flows from the decoder outputs. Outputs are ORed between the cells, and the final output is obtained as shown in Figure 4. The design has utilized a cell count of 25,721, with an area of 56.19 μm^2 with a total clock phases 158 which includes clock phases of decoder, UMC and programmable interconnects.

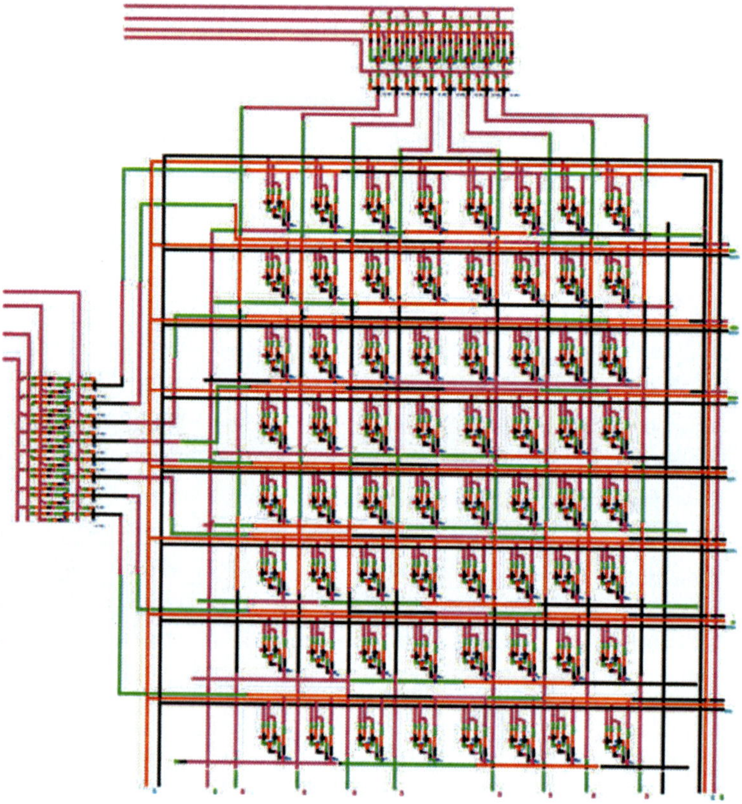

Fig. 5 QCA layout of 64-bit LUT for RAM

The proposed work utilizes effective design compared to the multiplexer-based LUT proposed in [7] and loop-based memory in [9], in terms of cell count, area and delay for the memory cell.

Comparisons	Architecture	Cell count	Area (μm^2)	Clock cycles
Kianpour [9]	Loop-based parallel memory	140	0.25	2
Pandiammal [7]	Multiplexer-based parallel memory	104	0.17	1.5
Proposed work	Loop-based parallel memory	88	0.16	1.5

5 Conclusion

In this work, a 64-bit parallel memory architecture using high-input decoder along with UMC utilizing the memory loop configuration has been proposed. The latency has been reduced by effective utilization of programmable interconnects and using fixed memory addressing decoders. The speed has comparatively increased by effectively using the decoder-based memory loop. The pipelined architecture is effective in controlling and maintaining the transmission of signals. The coplanar wire crossing gives the feasibility of implementations and fabrication. The future scope of the work proposed is extended for defect analysis and temperature stability.

References

1. Lent C, Tougaw P, Porod W (1994) Quantum cellular automata: the physics of computing with arrays of quantum dot molecules. In: Proceedings workshop physics and computation. Dallas, TX, 17–20, pp 5–13
2. International Technology Roadmap for Semiconductor (IRTS) (2007). https://www.irts.net.
3. Balijepalli H, Niamat M (2012) Design of a novel quantum-dot cellular automata field programmable gate array. In: 22nd international conference on field programmable logic and applications (FPL). Oslo, pp 611–614. https://doi.org/10.1109/FPL.2012.6339219
4. Amiri MA, Mahdavi M, Mirzakuchaki S (2008) QCA implementation of a MUX-based FPGA CLB. In: International conference on nanoscience and nanotechnology. Melbourne, Vic., pp 141–144.https://doi.org/10.1109/ICONN.2008.4639266
5. Crocker M, Sharon Hu X, Niemier M, Yan M, Bernstein G (2008) PLAs in quantum-dot cellular automata. IEEE Trans Nanotech 7(3):376–386
6. Crocker M, Niemier M, Sharon Hu X (2012) A Reconfigurable PLA architecture for nanomagnet logic. ACM J Emer Technol Comput Syst 8(1), Article No. 1. https://doi.org/10.1145/2093145.2093146
7. Pandiammal K, Meganathan D (2016) Design of multiplexer based 64-bit SRAM using QCA. Aust J Basic Appl Sci 10(1):207–213
8. Niamat M, Panuganti S, Raviraj T (2010) QCA design and implementation of SRAM based FPGA configurable logic block. In: 2010 53rd IEEE international midwest symposium on circuits and systems. Seattle, WA, pp 837–840. https://doi.org/10.1109/MWSCAS.2010.5548679
9. Kianpour M, Sabbaghi-Nadooshan R (2014) J Comput Electron 13:709. https://doi.org/10.1007/s10825-014-0590-z
10. Amlani I, Orlov A, Toth G, Bernstein G, Lent C, Snider G (1999) Digital logic gate using quantum-dot cellular automata. Science 284:289–291
11. Timler J, Lent CS (2002) Power gain and dissipation in quantum-dot cellular automata. J Appl Phys 91(2):823–831
12. Walus K, Vetteth A, Julien G, Dimitrov V (2003) Ram design using quantum-dot cellular automata. Proc Nanotech Conf Trade Show 2:160–163
13. Ottavi M, Vankamamidi V, Lombardi F, Pontarelli S, Salsano A (2005) Design of a QCA memory with parallel read/serial write. In: Proceedings of the IEEE symposium on VLSI, pp 292–294
14. Jayalakshmi R, Amutha R (2018) An approach towards optimisation of 3 to 8 decoder using 5 input majority gate with coplanar crossing in quantum dot cellular automata. In: AIP conference proceedings, 1966, p 020039. https://doi.org/10.1063/1.5038718
15. Walus K, Dysart TJ, Jullien GA, Budiman AR (2004) QCA designer: a rapid design and simulation tool for quantum-dot cellular automata. IEEE Trans Nanotechnol 3:26–31.https://doi.org/10.1109/TNANO.2003.820815

Application of CuO Nanoparticles as Biodegradable Lubricant Additive for Domestic Refrigerator Compressors—An Experimental Investigation

P. Vithya, G. Sriram, S. Arumugam, R. Sundararajan, and A. Manikanta

Abstract The application of nano-sized particles is increasing in refrigeration lubricants due to their considerable rise in thermal and heat transfer characteristics for improvement in coefficient of performance (COP) of the refrigeration system. The present study investigates the tribological behaviour of nano-CuO-based biodegradable refrigeration oil using four-ball tribometer. First, the pentaerythritol ester biodegradable refrigeration oil (PE) is blended with synthetic refrigeration oil (ISO 68 grade-POE oil) in the fraction of 50% by volume. Then the pentaerythritol ester-based biodegradable refrigeration oil blend (PE50) is mixed with CuO nanoparticles in the concentration of 0.1% to 0.3% (w:w) to prepare CuO nano-biodegradable refrigeration oil. The friction and wear experiments were conducted using the ASTM D 4172. The wear scar of the ball surfaces was analysed using a scanning electron microscope (SEM). The results show that CuO nano-based biodegradable refrigeration oil has good lubricity property. The tribological behaviour of nano-CuO 0.3% biodegradable refrigeration oil shows superior friction and wear reduction behaviour when compared other than nano-blends with PE50 and (ISO 68 grade-POE oil) synthetic refrigeration oil.

Keywords Refrigeration compressors · Biodegradable lubricant · Nanoparticle · Tribology

1 Introduction

In recent years, nanotechnology plays a major role in refrigeration systems endow with enhanced thermo-physical properties, heat transfer and tribological characteristics which in turn improving the performance of the system. Energy demand and environmental impact of the refrigerants are the main issues of the refrigeration industry. Several techniques have been carried out for increasing the coefficient of performance of the vapour compression refrigeration system [1, 2]. Alican et al. [3]

P. Vithya (✉) · G. Sriram · S. Arumugam · R. Sundararajan · A. Manikanta
Department of Mechanical Engineering, Sri Chandrasekharendra Saraswathi Viswa Mahavidyalaya, Kanchipuram, Tamilnadu 631561, India
e-mail: vithyagautham@gmail.com

T. Rajmohan et al. (eds.), *Advances in Materials and Manufacturing Engineering*, Springer Proceedings in Materials 7, https://doi.org/10.1007/978-981-15-6267-9_70

investigated the CuO nanoparticle incorporation in refrigeration lubricant and found the enhancement of COP by 11% due to excellent tribological properties in terms of friction coefficient reduction in the piston-liner mechanism of the compressor. Many researchers experimentally investigated the effect of nano-CuO concentrations ranged from 0.05 to 1% (w/w) on the evaporating heat transfer coefficient. The results showed that the increase in nano-concentration increases the evaporating heat transfer coefficient [4, 5]. Bi et al. [6] replaced the polyol ester (POE) oil with mineral oil with TiO_2 nanoparticles and investigated the energy consumption and freezing capacity of the refrigerator. The refrigerator performance was improved with 26.1% less energy consumption using 0.1% mass fraction TiO_2 nanoparticles compared to the HFC134a and POE oil. Sabareesh et al. [7] investigated the dispersing effect of low concentration of nano-TiO_2 in the mineral oil lubricant and concluded that coefficient of friction reaches a minimum value of 0.032 from the original value 0.047 lead to increase of 17% in the coefficient of performance owing to the addition of 0.01% (v/v) of TiO_2 in the lubricating oil. Xing et al. [8] reported the inclusion of fullerene C_{60} nanoparticles with concentration of 1, 2 and 3 g/L in the mineral oil reduces the friction coefficient by 12.9, 16.1 and 19.6% evaluated with pure mineral oil.

Earlier studies on bio-lubricant derived from various vegetable oils showed that bio-lubricant acquires promising lubrication characteristics [9]. Zulkifli et al. [10] studied the tribological behaviour of from palm oil-based TMP (trimethylolpropane) ester added with TiO_2 nanoparticles, concluded that the TiO_2 nanoparticles inclusion in the TMP ester at 160 kg exhibits good friction reduction by 15%. The application of TiO_2, WS_2 and CuO nanoparticles as anti-wear additives in synthetic lubricant (SAE20W40) and chemically modified rapeseed oil (CMRO) bio-lubricant showed that addition of CuO nanoparticles in the bio-lubricant decreases the friction coefficient by 27% and wear by about 47% as evaluated with a synthetic lubricant [11]. More attention in the last decades has been given to lubricant and additive chemistries to enhance heat transfer and compressor efficiency in the refrigeration system. The nanotechnology provides a better perspective for improving the performance of the refrigeration system and numerous research works are carried using bio-lubricant as an alternative to synthetic lubricant for engine oil applications. On the other hand, less work has been reported on using bio-lubricant as refrigeration oil. The work reported here presents an experimental investigation to understand the tribological behaviour of the addition of CuO nanoparticles (0.1% to 0.3% (w:w)) in the pentaerythritol ester-based biodegradable refrigeration oil blend (PE50) using four-ball tribometer using ASTM D4172. The wear scar of the ball surfaces was analysed using a scanning electron microscope (SEM).

2 Experimental

2.1 *Synthesis of Pentaerythritol Ester-Based Biodegradable Refrigeration Oil*

The synthesis of pentaerythritol ester-based biodegradable refrigeration oil by successive transesterification is clearly reported by the earlier studies of the authors [12]. The transesterification process of pentaerythritol ester was carried in two stages. The first stage is the process of rapeseed oil methyl ester using methanol in the presence of sodium hydroxide to produce formulation of rapeseed oil methyl ester. With the 1 L of rapeseed oil, 440 ml of methanol and 9.23 g of sodium hydroxide were added and mixed well using the mechanical stirrer. The reaction mixture was heated up to 55 °C in stirring condition of 300 rpm for 1 h. The reaction process is examined by thin layer chromatography until the complete disappearance of triglyceride. The second stage is the reaction of the methyl ester with pentaerythritol. The rapeseed oil methyl ester 500 ml was then mixed with xylene, 47.54 g of pentaerythritol and 27.37 g of para-toluene sulphonic acid. The mixed contents were heated in the atmosphere containing nitrogen up to 160 °C for 6 h. The yield of 85% pentaerythritol ester was obtained as the final product.

2.2 *Preparation of Biodegradable Refrigeration Oil Blend*

The pentaerythritol ester biodegradable refrigeration oil (PE) is blended with synthetic refrigeration oil (ISO 68 grade-POE oil) in the fraction of 50% by volume. The blending process of 750 ml of synthetic refrigeration oil and biodegradable refrigeration oil of 750 ml was carried out in three necked round bottom flask with mechanical stirrer of 2 L capacity. The contents were well mixed, and the mixture was heated to 55–60 °C at a stirring speed of 300 rpm for one hour. Finally, pentaerythritol ester biodegradable refrigeration oil (PE) blend was obtained after 60 min.

2.3 *Preparation of Nano-CuO-based Biodegradable Refrigeration Oil*

The CuO nanoparticle was used as the lubricant additive. The CuO nanoparticles average size of 30 nm was supplied by Nano Research, USA. Electronic analytical balance was employed for weighing the quantity of nanoparticles. The nano-CuO biodegradable refrigeration oil with the concentration of 0.1, 0.2 and 0.3 (w/w) was prepared using ultrasonic agitation for achieving good dispersion of the particles in the bio-lubricant. The uniform dispersion of nanoparticles in the PE50 was achieved by means of a standard ultrasonic agitator [Anna University, India], by sonicating for a period of 300 min The prepared nano-CuO biodegradable refrigeration oil is stable without any sedimentation for a substantially long time period of several days after its preparation, as a result of that agglomeration does not occur [9].

2.4 Tribological Investigation

The four-ball tribometer [supplied by M/s DUCOM Instruments Pvt. Ltd, (TR 500), India] were used to evaluate the friction coefficient and wear rate of the lubricant oils. The conducting tests are intended to study the anti-wear behaviour of lubricant under sliding contact by four-ball test. For testing the tribological properties of the lubricant oils, a constant load of 40 kg was applied with the rotational speed rate of 1200 rpm at 75 °C temperature for 1 h was employed for each lubricant oil samples. The wear occurred on the balls is subjected to the microscope to measure the wear scare diameter (WSD). For each sample, triplicate tests were done to get accurate results.

3 Results and Discussion

3.1 Properties of Nano-CuO-based Biodegradable Refrigeration Oil

The kinematic viscosity of synthetic refrigeration oil, PE, PE50 and nano-CuO-based PE50 was measured. Table 1 depicts the properties of CuO nano-based biodegradable refrigeration oil. It is evident from Table 1 that the viscosity @ 100 °C of the PE50 CuO 0.3% is more than that of other refrigeration oil blend samples. The results also indicate that the increase in viscosity index due to the addition of CuO nanoparticles is up to 0.3% (w/w) in PBRO50. The value of the viscosity index for PBRO50 CuO 0.3% is higher compared to the viscosity index of POE68 oil. The flash point is approximately the same for all oils. The results also indicate that the increase in viscosity due to the addition of nanoparticles is more prominent in the lower-temperature range. Earlier researchers stated and agreed the trend of viscosity variation obtained is in accordance with the results concluded by Kedzierski [13].

Table 1 Properties of CuO nano-based biodegradable refrigeration oil

Test parameters	Standard	POE 68	PE	PE50	PE50 + 0.1CuO	PE50 + 0.2CuO	PE50 + 0.3CuO
Kinematic viscosity @100 °C (cSt)	ASTM D445	9.3	11	10.54	11.4	11.9	12.5
Viscosity index	ASTM D2270	114	248	244	249	252	257
Flash point (°C)	ASTM D92	270	242	220	224	226	229
Pour point (°C)	ASTM D97	−39	−25	−28	−29	−29	−29
Total acid number mg KOH/g	ASTM D974	0.3	1.2	0.79	0.76	0.83	0.70

3.2 *Tribological Investigation of Biodegradable Refrigeration Oil*

3.2.1 Friction Behaviour

Figure 1 shows the variation in friction torque for various concentrations (POE68, PE, PE50, PE50 0.1CuO, PE50 0.2CuO and PE50 0.3CuO) of refrigeration oil. Figure 1 exhibits all the blends come under boundary lubrication (0.08–0.2 Nm). The addition of 0.3% CuO nanoparticles with PE50 refrigeration oil blend reduces

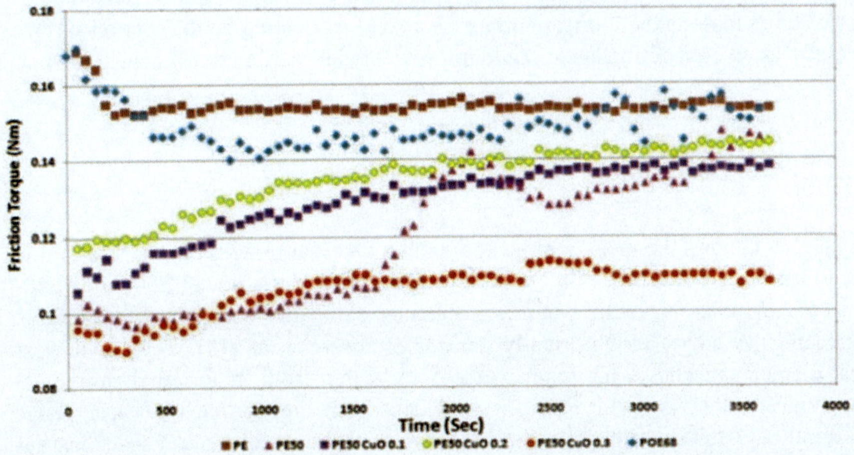

Fig. 1 Effect of CuO nano-biodegradable refrigeration oil on the friction torque

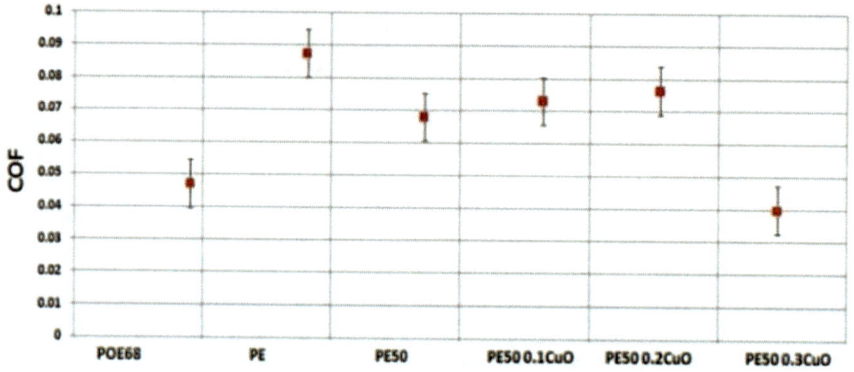

Fig. 2 Effect of CuO nano-biodegradable refrigeration oil on the friction coefficient

the friction torque by 33% and 28% compared with PE50 and POE68, respectively. It is also clear from Fig. 1; PE displayed the highest friction torque compared to other refrigeration oil samples. It can be seen that the friction torque increased with increasing friction time. The results from the tribological study can be observed that the refrigeration oil especially PE50 with 0.3% CuO exerts produces the lowest frictional torque of 0.09 Nm [8, 14].

Figure 2 shows the typical evolution of the coefficient of friction (COF) for different refrigeration oil. The friction coefficient of refrigeration oil significantly decreases with increasing the concentration of nanoparticles. The addition of 0.3% CuO nanoparticles with PE50 refrigeration oil blend reduces the friction coefficient by 39% and 24% compared with PE50 and POE68, respectively. This is due to the fact of CuO nanoparticles are spherical in shape which disperses in the base refrigerant oil lead to the improvement of lubrication between the friction surfaces [3, 12]. PE50 with 0.3% CuO exerts lowest frictional coefficient value 0.39. This is attributed to viscosity increase in the lubricant oil forms an oil film coating on the friction surface and acts as an antifriction agent could improve the lubricating performance [15].

3.3 *Wear Behaviour*

Figure 3a–f shows the mean wear scar diameter and scar images of refrigeration oil blend samples POE68, PE, PE50, and 0.1, 0.2, 0.3 nano-CuO-based PE50. The PE50-0.3 CuO exhibits minimum wear scar diameter compared with other refrigeration oil blends. A large WSD normally designates severe wear [14]. The mean wear scar diameter reaches a minimum value of 0.40 mm from the original value 0.73 for a 0.3% CuO nano with PE50. It is evident from Fig. 3f that wear scar image of PE50-0.3 CuO smoother, clearer and circular in shape, other wear scar image is unclear and rough. Wear debris showed in all the refrigeration oil blend samples. The surface morphology shows the layer of wear surfaces was excluded out in the sliding

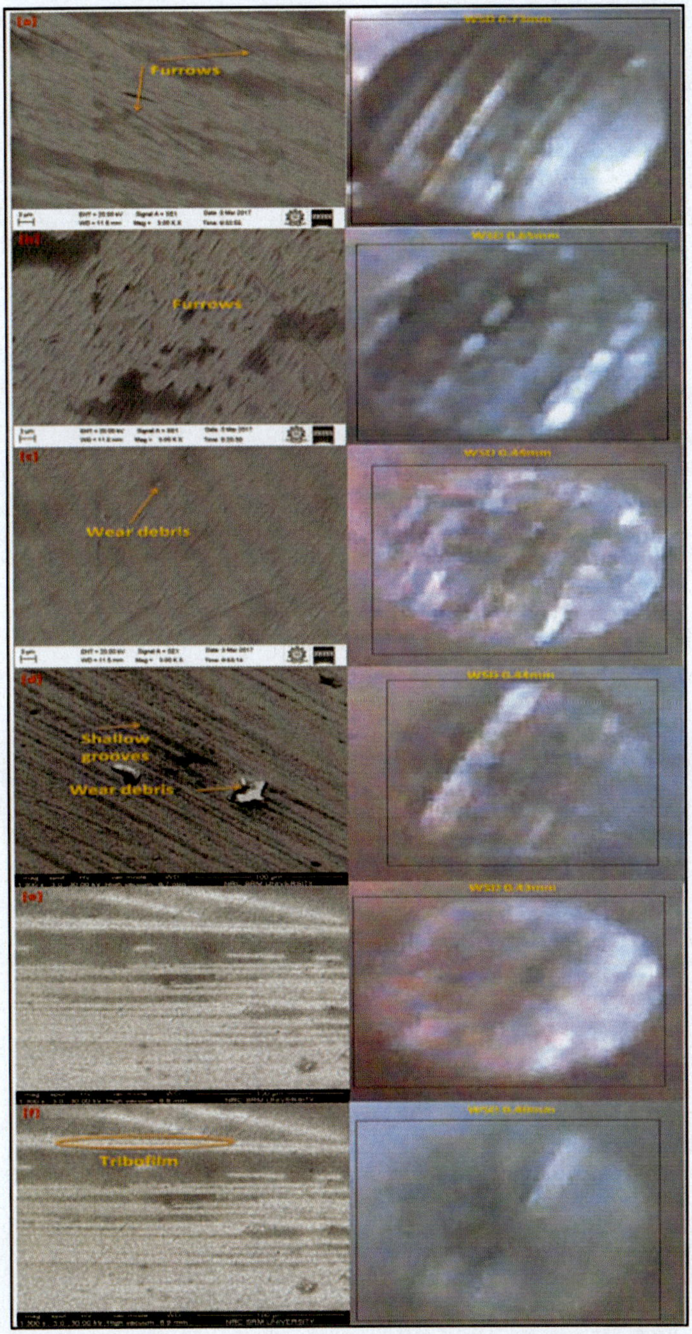

Fig. 3 SEM micrographs (magnification 3kX) and wear scar image on the ball surface **a** POE68, **b** PE, **c** PE50, **d** PE50 0.1CuO, **e** PE50 0.2CuO and **f** PE50 0.3CuO

direction of the rotating steel balls. In POE68 and PE, the sharp furrows and shallow grooves were present on the wear track, thus it indicates that it has lower friction reducing and anti-wear properties. In pure synthetic refrigeration lubricant, it has been severely scratched, whereas in the nano-based refrigeration blend lubricants, the scratches are less and indicate it has less wear scar as shown in Fig. 3d–f. The surface damage for all lubricants is reflecting adhesive wear. The tribochemical film could be formed on the surfaces lubricated with PE50-0.3 CuO which could be a reason of an increase in viscosity avoids direct metal contact and acts as anti-wear additive [16–18].

4 Conclusion

This study offers detailed insight into the effect of using nano-CuO as an anti-wear additive in biodegradable refrigeration oil on the improvement of its tribological behaviour in the refrigeration system. The outcome of this investigation is detailed below.

- The properties of nano-CuO-based biodegradable oil refrigeration oil satisfy the requirement of synthetic biodegradable refrigeration oil. The viscosity of CuO nano-based biodegradable refrigeration oil is higher than the synthetic refrigeration oil.
- In the case of frictional behaviour, the nano-PE50-CuO 0.3% has a lower coefficient of friction than other blends and synthetic refrigeration oil. The coefficient of friction for PBRO50 CuO 0.3% is reduced by 33% than synthetic refrigeration oil. The wear scar diameter for PE50-CuO 0.3% was least (0.40 mm, while that was maximum 0.73 mm for synthetic refrigeration oil).

References

1. Saidur R, Leong KY, Mohammad HA (2011) A review on applications and challenges of nanofluids. Renew Sust Energy Rev 15(3):1646–1668
2. Jia T, Wang R, Xu R (2014) Performance of $MoFe_2O_4$-$NiFe_2O_4$/fullerene-added nano-oil applied in the domestic refrigerator compressors. Int J Refrig 45:120–127
3. Yilmaz A, Erdem R, Erdem O (2018) Dispersion of Cuo nano particles in the lubricant of a vapor compression refrigeration system: an investigation on enhancing coefficient of performance.Int Acad Conf Rome 75–80. https://doi.org/10.20472/IAC.2018.042.048
4. Hao P, Guoliang D, Weiting J, Haitao H, Yifeng G (2009) Heat transfer characteristics of refrigerant-based nanofluid flow boiling inside a horizontal smooth tube. Int J Refrig 32:1259–1270
5. Fu L, Wang R, Cong W, Li Q, Wu Y (2008) Experiment study on performance of refrigerator using nano particle additive. J Xian Jiaotong Univ 42:852–854
6. Bi S, Shi L, Zhang L-L (2008) Application of nanoparticles in domestic refrigerators. Appl Therm Eng 28(14–15):1834–1843

7. Krishna Sabareesh R, Gobinath N, Sajith V, Das S, Sobhan CB (2012) Application of TiO_2 nanoparticles as a lubricant-additive for vapor compression refrigeration systems—an experimental investigation. Int J Refrig 35(7):1989–1996
8. Xing M, Wang R, Yu J (2014) Application of fullerene C_{60} nano-oil for performance enhancement of domestic refrigerator compressors. Int J Refrig 40:398–403
9. Adhvaryu A, Liub Z, Erhan SZ (2005) Synthesis of novel alkoxylated triacylglycerols and their lubricant base oil properties. Ind Crops Prod 21:113–119
10. Zulkiflia NWM, Kalam MA, Masjukia HH, Yunus R (2013) Experimental analysis of tribological properties of biolubricant with nanoparticle additive. Proc Eng 68:152–157
11. Baskar S, Sriram G, Arumugam S (2016) Tribological analysis of a hydrodynamic journal bearing under the influence of synthetic and biolubricants. Tribol Trans 60(3):1–36
12. Vithya P, Sriram G, Arumugam S (2018) Tribological analysis of biodegradable refrigeration oil for vapour compression refrigeration system using four ball tribometer. IOP Conf Ser Mater Sci Eng 390:012070. https://doi.org/10.1088/1757-899X/390/1/012070
13. Kedzierski MA (2018) Effect of CuO nanoparticle concentration on R134a/lubricant pool-boiling heat transfer. J Heat Transf 131(4):1–8
14. Shahabuddin M, Masjuki HH, Kalam MA (2013) Experimental investigation into tribological characteristics of biolubricant formulated from Jatropha oil. Proc Eng 56:597–606
15. Ruggiero A, Amato D, Merola R, Valasek M, Muller PM (2016) On the tribological performance of vegetal lubricants: experimental investigation on Jatropha Curcas L. oil. Proc Eng. 149:431–437
16. Padgurskas J, Rukuiza R, Prosyčevas I, Kreivaitis R (2013) Tribological properties of lubricant additives of Fe Cu and Co nanoparticles. Tribol Int 60:224–232
17. Yu HL, Xu Y, Shi PJ, Xu BS, Wang XL, Liu Q et al (2008) Characterization and nanomechanical properties of tribofilms using Cu nanoparticles as additives. Surf Coat Technol 203:28–34
18. Kato H, Komai K (2007) Tribofilm formation and mild wear by tribo-sintering of nanometer-sized oxide particles on rubbing steel surfaces. Wear 262:36–41

Study on Emission Analysis of Tamarind-Based Ground-Level Pyrotechnics

Manikandan Rajendran, Rajesh Shanmugavel, and Rajajeyaganthan Ramanathan

Abstract In this chapter, an attempt is made to change the chemical composition of the ground-level pyrotechnics (firecracker) to reduce the environmental pollution during bursting, without affecting the performance of the firecracker. The small percentage of tamarind seed powder (TSP) is used as a substitute for sulfur powder. The performance of tamarind-based firecracker (TBFC) is measured in terms of noise level produced as per the standards, and it is compared with a normal firecracker. The emission rate of normal and TSP-based firecracker is also compared. The experiment result reveals that the addition of the TSP along with normal firecracker (NFC) reduces the 20% of the sulfur content and 16% of nitrate content in the air. A sound level of NFC is 98.3, and 120.8 dBC is obtained with the addition of TSP along with normal firecracker composition. The noise level produced during the bursting of TBFC is 98 and 119.5 dBC. The TBFC achieved the standard noise level prescribed in the Government of India Regulation and also produced less atmospheric pollutants like CO, SO_2, and NO_3 by complete combustion toward making a green environment.

Keywords Pyrotechnics · Noise-level test · Emission rate of a firecracker

1 Introduction

Firework industries are one of the major revenue sources for the Indian government, and it gives direct and indirect employment opportunities for more than one lakh seventy thousand people [1]. The demand for firecrackers is increasing every day. To cater to the needs, 40% of the firecrackers are also imported from China. The usage of firecrackers during Diwali and other major festival is inevitable in Indian culture. Firecrackers are available in a different variety, and ground-level pyrotechnics is

M. Rajendran · R. Shanmugavel (✉)
Department of Mechanical Engineering, Kalasalingam Academy of Research & Education, Krishnankoil, Tamil Nadu, India
e-mail: s.rajesh@klu.ac.in

R. Ramanathan
Department of Chemistry, Kalasalingam Academy of Research & Education, Krishnankoil, Tamil Nadu, India

T. Rajmohan et al. (eds.), *Advances in Materials and Manufacturing Engineering*, Springer Proceedings in Materials 7, https://doi.org/10.1007/978-981-15-6267-9_71

commercially used by all people. The chemical mixture of the firecracker is 60% of potassium nitrate (KNO_3), 20% of sulfur (S) and 20% of aluminum (Al). Among the above, aluminum acts as fuel, and sulfur and KNO_3 act as igniter and oxidizer. During the bursting of various crackers, atmospheric pollutants are generated and pollute the environment. The metal concentration in atmospheric air was very high as compared during Diwali festival season than the normal days [2]. The existence of the particulate matter in the air was 753.3 μg/m3. The SO_2 and NO_x were in the range of 139.1 and 107.3 $\mu g/m^3$ during the Diwali season in Lucknow city. The observed ranges were higher than the normal days and exceeded the prescribed limit of National Ambient Air Quality Standards [3].

During manufacturing of crackers, workers are exposed to the aforementioned chemicals by inhaling the chemical powders which affect the health of the workers and lead to wheezing, asthma and lung maladies. Another problem is during the bursting of firecrackers which emit polluted gases which adversely affect the environment. The presence of NO_2 and SO_2 creates allergic inflammation effects on the airways of the young children [4]. The ecosystem is affected by the SO_2 [5]. SO_2 reacts with air and forms H_2SO_4, which leads to acid rain, and it affects the human and aquatic life [6]. The aforementioned problem is to be addressed without affecting the cracking performance of the firecracker. The performance of the firecracker is measured based on noise level and emission rate. It is desirable to have the proper chemical composition and combination to obtain the required performance. The reduction of the particle size from micron to nano-level increases the performance of cracker [7]. So, when using nanopowders the performance is improved but the hazardousness during handling is very high. The thermal behavior of the nanopowders studied using DTA and TGA curves; it was observed that nano-sized powders have high thermal energy content [8]. Scanning through the literature, it is found that there are no papers discussing the effect of TSP and the performance of TBFC.

In this work, an attempt is made to study the effect of noise level and emission rate by adding TSP along with NFC composition. The tamarind seed was obtained from the local garden and crushed to fine powder particle to make firecracker. It is observed from the ultimate analysis of TSP that it can be added into the cracker as a fuel because of the high combustion property and it consists of 47.76% of carbon content and 42.39% of oxygen content [9]. In order to understand the effect of TSP, the samples are prepared by using 5 wt% TSP and also tests are conducted as per the standards. The effect of a conventional type of fire is compared with industrial firecrackers.

2 Materials and Methods

The chemical mixture of the firecracker like KNO_3, Al and sulfur is locally available in Kaliswari Metal Mart, Sivakasi, Tamil Nadu, India. In this work, Al with 96% purity is used as one of the chemicals. KNO_3 and sulfur are purchased with the purity of 97.6 and 99.9%. The TSP was powdered in a ball mill to the desired particle size. The particle size of all the chemical powders is in the range of 45 nm. Weighing of chemical mixtures, sieving to the preferred size, mixing the powder to acquire a homogenous mixture of chemicals and filling them in the chapter container are the processes followed for producing the firecrackers. In this work, two different firecrackers are fabricated using different chemical compositions. NFC is fabricated as per the standard chemical composition (normal flash powder) followed in the firework industries (60% KNO_3, 20% Al and 20% sulfur). The TBFC is made by modifying the chemical composition of the existing chemical composition. The modification is made by reducing the sulfur content by 15% and by adding 5% of the tamarind seed powder (60% KNO_3, 20% Al, 15% sulfur and 5% TSP).

2.1 *Methods for Testing of Pollutants*

2.1.1 Estimation of Aluminum and Potassium

The amount of aluminum and potassium present in each of the flash powder samples is found by the Indian standard procedures (EPA 3050 B) by atomic absorption spectroscopy (AAS).

2.1.2 Estimation of Nitrate

The quantity of nitrate present in each of the flash powder samples is obtained as per Indian standard procedure (IS 14,684:1999 (RA-2014)).

2.1.3 Estimation of Sulfur

The amount of sulfur content present in each of the flash powder samples is found by adopting the Indian standard procedure (IS 6655-1972).

2.1.4 Auto-exhaust Gas Analyzer

The gases discharged during the bursting of firecracker are analyzed by auto-exhaust/multi-gas analyzer.

2.2 Noise-Level Test

As per the notification of Petroleum and Explosives Safety Organisation, Government of India, noise-level test is carried out. Noise is measured by noise-level meter (Model No. 824L), data provided here was measured in the east direction, and the details of measurement procedures are described before [10].

3 Results and Discussion

It is observed from the emission test that the addition of TSP on firecracker reduces the environmental pollutants. The amount of aluminum present in each of the flash powder samples is found to be 18.89% in normal flash powder, while TBFC had 18.66% of aluminum. Similarly, the quantity of potassium present in each of the flash powders is found to be 20.54% in NFC while 20.19% of potassium present in TBFC which is obtained as per standard procedures. This confirmed that all the chemicals are uniformly distributed and form a homogeneous chemical mixture as the compound was present in equal percentage in both NFC and TBFC compositions. The emission rate of firecrackers is given in Table 2.

The amount of sulfur content present in NFC is found to be 8.66%, while TBFC sample contains 6.93% of sulfur which is obtained as per Indian standard procedure.

3.1 Theoretical Analysis of Ground-Level Pyrotechnics

Aluminum combustion reaction

$$2Al + 3O_2 \rightarrow Al_2O_3 \tag{1}$$

Sulfur combustion reaction

Table 2 Comparison of emission rate of a firecracker

Test parameter	Test method	Emission rate (%)	
		Normal firecracker	Tamarind-based firecracker
Al (Al)	EPA 3050 B (by AAS)	18.89	18.66
Nitrate (NO_3)	IS 14,684:1999 (RA-2014)	34.76	29.13
Potassium (K)	IS3025-Part 45-1987 (RA 2014)	20.54	20.19
Sulfur (S)	IS 6655-1972	8.66	6.93

$$S + O_2 \rightarrow SO_2 \tag{2}$$

Carbon combustion reaction

$$C + O_2 \rightarrow CO_2 \tag{3}$$

$$C + O \rightarrow CO \tag{4}$$

Based on the above equations, the required amount of oxygen and lack of oxygen are estimated using theoretical combustion calculation as given in Table 3.

From the above combustion calculation, to complete the combustion of NFC the required level of O_2 is 1.667×10^{-4} kg. The available amount of O_2 is 8.471×10^{-5}. So, the lack of O_2 is 8.202×10^{-5} kg. The amount of oxygen required for complete combustion in TBFC (7.977×10^{-5}) is lesser than the normal cracker. So, TSP-based ground-level pyrotechnics helps to enhance the chemical reaction, make complete combustion and produce less pollutant. It is observed from bursting of an NFC the emission range of CO is 0.032% PPM, CO_2 is 0.020% PPM, and hydrocarbon (HC) is 15% PPM. Propane equivalent factor (PEF) is 0.546% PPM.

The carbon and oxygen derived from TSP are to stimulate the chemical reaction and make complete combustion. So, the CO range is reduced and the value is 0.018% PPM. The hydrocarbon content is 6% PPM. So, the TBFC increases the reaction rate and helped for the complete combustion, thereby reducing the emission rate.

Table 3 Calculation of amount of O_2 using combustion reaction

Firecracker	Cracker quantity (300 mg)	Quantity kg	Available kg of O_2	Required kg O_2		Lack of O_2 kg
Normal	60% KNO_3	1.8×10^{-4}	8.471×10^{-5}	–	–	8.202×10^{-5}
	20% Al	6×10^{-5}	–	Al as Al_2O_3	1.067×10^{-4}	
	20% S	6×10^{-5}	–	S as SO_2	5.999×10^{-5}	
TSP	60% KNO_3	1.8×10^{-4}	8.471×10^{-5}	–	–	
	20% Al	6×10^{-5}	–	Al as Al_2O_3	1.067×10^{-4}	
	15% S	4.5×10^{-5}	–	S as SO_2	4.499×10^{-5}	
	5% TSP	1.35×10^{-5}	9.106×10^{-5}	C as CO_2	1.708×10^{-4}	7.977×10^{-5}
				C as CO	1.612×10^{-4}	7.022×10^{-5}

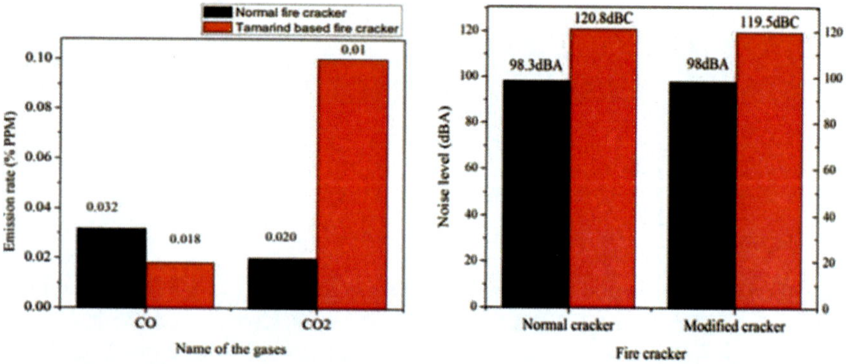

(a) Exhaust gas analyzer result (b) Noise level test result

Fig. 1 Exhaust gas analysis and noise-level test result of normal and TBFC. **a** Exhaust gas analyzer result. **b** Noise-level test result

Experimental result also agreed with theoretical data calculation. The exhaust gas analyzer result is given in Fig. 1a.

It is observed from the noise level test result the bursting of NFC releases the noise level of 98.3 dBA and 120.8 dBC. The noise level of TBFC is 98 dBA and 119.5 dBC. The dBA is impulse noise level, and dBC is peak noise level. The addition of TSP to the chemical mixture does not alter the noise performance of the TBFC. The noise-level result is given in Fig. 1b. In addition, the additions of TSP increase the combustion property, reduce the amount of sulfur content and attain complete combustion without any compromise.

4 Conclusion

Addition of 5% of TSP along with NFC reduces the sulfur content in the air which is 20% lesser than the NFC. The reduction of sulfur reduces the atmospheric pollutants during the bursting of firecrackers and makes the environment pollution free of SO_2 gas. The bursting of TBFC increases the chemical reaction rate, supplies more amount of oxygen to make complete combustion and releases less environmental pollutants. Addition of 5% of TSP is not affecting the sound level of the firecracker, and the sound level of TBFC is 98 dBA and 119.5 dBC and NFC; it is 98.3 dBA and 120.8 dBC. As per the standard, the recommended maximum sound level of firecracker is 125 dBA and 145 dBC. It was concluded that TSP can be mixed with the firecracker chemical composition to reduce SO_2 emissions without affecting the quality of NFC.

References

1. Ravi A, Gandhinathan R (2013) Analysis of safety climate in fireworks industries in Tamilnadu. Int J Sci Eng Res 4:760–764
2. Kulshrestha UC, Rao TN, Azhaguvel S, Kulshrestha MJ (2004) Emissions and accumulation of metals in the atmosphere due to crackers and sparkles during Diwali festival in India. Atmos Environ 38(27):4421–4425
3. Barman SC, Singh R, Negi MPS, Bhargava SK (2008) Ambient air quality of Lucknow City (India) during use of fireworks on Diwali festival. Environ Monit Assess 137(1–3):495–504
4. Greenberg N, Carel RS, Derazne E, Bibi H, Shpriz M, Tzur D, Portnov BA (2016) Different effects of long-term exposures to SO_2 and NO_2 air pollutants on asthma severity in young adults. J Toxicol Environ Health Part A 79(8):342–351.3
5. Gawande P, Kaware J (2015) Health and environmental effects of sulphur oxides—a review
6. Rahila R, Siddiqui M (2014) Review on effects of particulates; sulfur dioxide and nitrogen dioxide on human health. Int Res J Environ Sci 3(4):70–73
7. Azhagurajan A, Selvakumar N, Thanulingam TL (2011) Thermal and sensitivity analysis of nano aluminium powder for firework application. J Therm Anal Calorim 105(1):259–267
8. Selvakumar N, Azhagurajan A, Suresh A (2013) Experimental analysis on nano scale flash powder composition in fireworks manufacturing. J Therm Anal Calorim 113:615–621
9. Parveen M, Islam MR, Haniu H (2011) Thermal decomposition behavior study of two agricultural solid wastes for production of bio-fuels by pyrolysis technology. J Therm Sci Technol 6(1):132–139
10. Sharma M et al (2017) Status of pollution abatement measures: FIRE-CRACKERS (patakas), control of urban pollution series CUPS/88/2017–2018, U.P.C. Division, Editor. Central Pollution Control Board (CPCB), Ministry of Environment, Forests & Climate Change. Delhi, India, pp 1–83

Tribological Characteristics of Aluminum with Volumetric Fractions of Nano-Boron Nitride Composites

M. Ekambaram, M. Vetrivel, and D. Balaji

Abstract The tribological characteristics of aluminum metal matrix with different volumetric fraction of boron nitride nano-powder reinforcement were studied. The various volumetric fractions are 0.6 vol.% BN, 1.2 vol.% BN, and 1.8 vol.% BN which are added with pure aluminum. In each composition, specimens are prepared and tested on pin on disc wear testing machine for three different load conditions and three different sliding velocity conditions with sliding distance of 1000 m. The study suggests that 0.6 vol.% of BN with aluminum provides better wear resistance when compared with other compositions and pure aluminum.

Keywords Nano-boron nitride · Specific wear rate · Coefficient of friction

1 Introduction

In recent years, the demand for the new materials has been increased rapidly in industrial sectors. In aeronautical sectors, the need of metal matrix composites was raised due to its high specific strength and stiffness, lesser in weight, low density, and low thermal expansion [1–3]. The sliding characteristics of materials are notable in tribological systems. Hard particles reinforced aluminum-based composites possess good strength and tribological properties when compared to unreinforced aluminum and its alloys [4–9].The strength and tribological properties of composites are dependent on various parameters such as reinforcement particle size (micro or nano), reinforcement content, and reinforcement phase in the base material and type of fabrication process [10]. The reasons for damage in machine parts are due to more friction

M. Ekambaram (✉) · M. Vetrivel · D. Balaji
Department of Mechanical Engineering, Sri Chandrasekharendra Saraswathi Viswa Mahavidyalaya University, Enathur, Kanchipuram, Tamil Nadu 631561, India
e-mail: e4ekam@gmail.com

M. Vetrivel
e-mail: km_vetrivel@yahoo.com

D. Balaji
e-mail: balajibala81@gmail.com

T. Rajmohan et al. (eds.), *Advances in Materials and Manufacturing Engineering*, Springer Proceedings in Materials 7, https://doi.org/10.1007/978-981-15-6267-9_72

Table 1 Description of the materials used

Description	Aluminum (matrix)	Boron nitride (reinforcement)
Symbol	Al	BN
Particle size	130–180 μm	70–80 nm
Density (g/cm^3)	2.69	2.29
Purity (%)	98	99.8

and wear. An appropriate solid lubricant should be added to reduce the friction and wear. The addition of solid lubricants to the matrix results in better wear performance and reduction in coefficient of friction [11–16]. The Gr particles are used as solid lubricant in the aluminum-based matrix due to its low cost and yield better tribological properties with SiC, Al_2O_3, and B_4C particles [17–19]. Boron nitride is used as solid lubricant in the aluminum-based matrix due to its low density, high thermal conductivity, and high resistivity and also exhibits optimum reinforcement in Al-based matrices [20–28]. In present study, the boron nitride nano-powder is reinforced with pure aluminum matrix, and its tribological and mechanical properties were studied.

2 Materials and Experiment

2.1 Materials

The description of the materials used in the research work is mentioned in Table 1.

2.2 Specimen Preparation

Initially, the Al and BN nano-powder is taken in wt % and then compacted to a standard size (10 mm diameter and 30 mm height), and then, vol.% is measured using the formula given in Eq. (1). The volume of each specimen is also calculated by mass and density (using Archimedes principle) values, and the micro-hardness values of each composition are listed as given in Table 2.

Table 2 Micro-hardness value of specimens

Vol.% in specimen	Micro-hardness value (H.V)
0.6%—specimen 1	12.5
1.2%—specimen 2	8.1
1.8%—specimen 3	9.5

$$\text{Volume fraction of reinforcements } V_{r1} = \frac{V_{r1}}{V_m + V_{r1} + V_{r2}} \quad (1)$$

where in Eq. (1),

V_{r1} Volume fraction of reinforcement 1,
V_{r2} Volume fraction of reinforcement 2,
V_m Volume fraction of matrix.

3 Experimental Procedure

3.1 Wear Testing Procedure

The test specimens for wear testing were prepared using powder metallurgy route. ASTM G99 standard is used for wear testing. The specimens are tested in the pin on disc wear testing machine. There are three different load conditions, and three different sliding velocities are used for testing. The three different load conditions are 5, 10, and 15 N, and the three different sliding velocities are 1.5, 2, and 2,5 m/s. For the different load conditions and different sliding velocities, the specific wear rate (SWR) and coefficient of friction (COF) were calculated and tabulated in Tables 3 and 4.

Table 3 Specific wear rate of all specimens

S. No.	Sliding distance (m)	Sliding velocity (m/s)	Load (N)	Specific wear rate × E^{-9} m^3/Nm			
				Pure Al	0.6 vol.% BN	1.2 vol.% BN	1.8 vol.% BN
1	1000	1.5	5	1.5464	0.0010	0.0613	0.6591
2	1000	2.0	5	2.0286	0.0027	0.0621	0.5176
3	1000	2.5	5	2.8266	0.0031	0.0771	0.5151
4	1000	1.5	10	2.1024	0.0022	0.0831	0.5119
5	1000	2.0	10	3.3367	0.0026	0.0772	0.5294
6	1000	2.5	10	4.0359	0.0024	0.0981	0.6260
7	1000	1.5	15	2.3591	0.0035	0.0852	0.7546
8	1000	2.0	15	4.3288	0.0015	0.0218	0.9952
9	1000	2.5	15	5.6959	0.0014	0.0423	1.7700

Table 4 Coefficient of friction of all specimens

S. No.	Sliding distance (m)	Sliding velocity (m/s)	Load (N)	Coefficient of friction			
				Pure Al	0.6 vol.% BN	1.2 vol.% BN	1.8 vol.% BN
1	1000	1.5	5	2.75	0.18	0.52	1.68
2	1000	2.0	5	2.79	0.22	0.56	1.87
3	1000	2.5	5	3.02	0.38	0.77	2.41
4	1000	1.5	10	4.12	0.4	1.12	3.72
5	1000	2.0	10	4.58	0.6	1.46	4.17
6	1000	2.5	10	5.2	0.7	2.32	4.72
7	1000	1.5	15	6.6	0.5	2.85	5.6
8	1000	2.0	15	6.9	0.5	3.1	6.1
9	1000	2.5	15	7.1	0.9	3.6	6.6

4 Results and Discussions

4.1 *Specific Wear Rate (SWR)*

Table 3 and Fig. 1 represents the specific wear rate results of all specimens. The specific wear rate can be calculated from the below formula,

$$\text{Specific Wear Rate} = \frac{\text{Weight Loss}}{(\text{Density} * \text{Sliding distance} * \text{Load})} \frac{\text{m}^3}{\text{Nm}} \tag{2}$$

4.2 *Coefficient of Friction*

Table 4 and Fig. 2 represents the coefficient of friction results of all specimens. The coefficient of friction can be calculated from the below formula.

$$\text{Coefficient of firction} = \frac{\text{Firctional force}}{\text{Normal load}} \text{(No Unit)} \tag{3}$$

5 Conclusion

The results show that 0.6 vol.% of BN added with pure aluminum gives good wear resistance. Further increasing vol.% of the BN content increases the wear, which

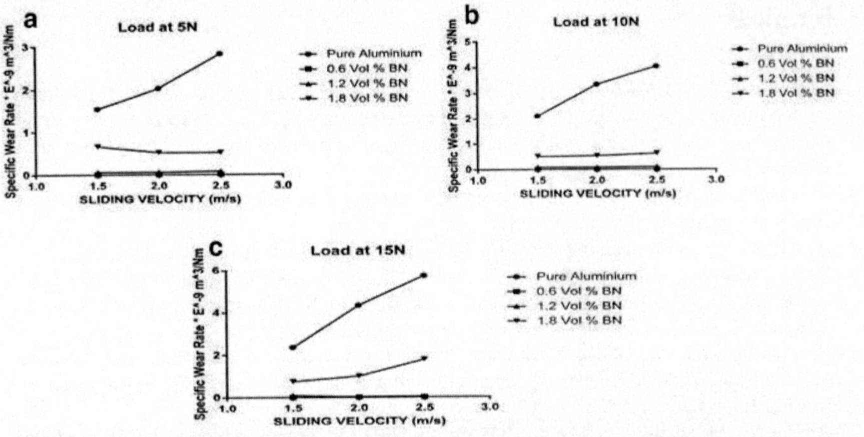

Fig. 1 **a** Specific wear rate at load 5 N, **b** specific wear rate at load 10 N, **c** specific wear rate at load 15 N

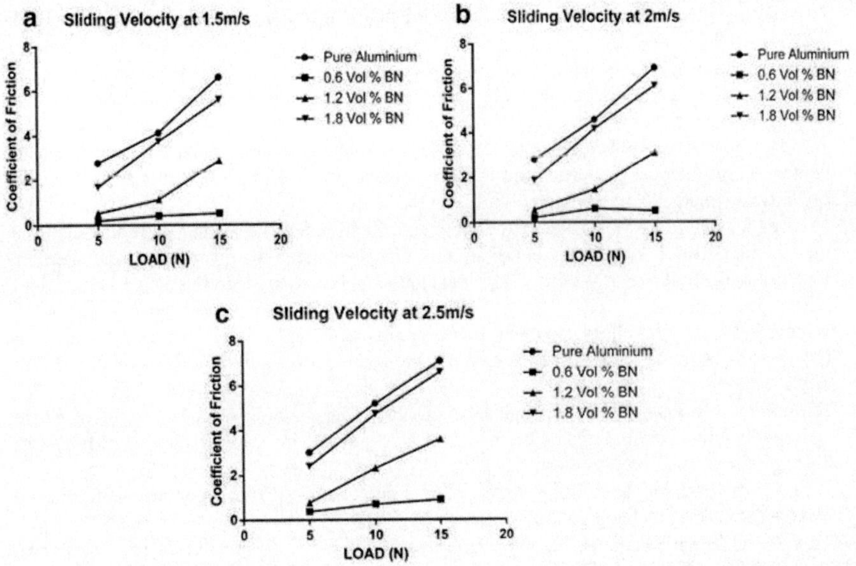

Fig. 2 **a** Coefficient of friction at 1.5 m/s, **b** coefficient of friction at 2 m/s, **c** coefficient of friction at 2.5 m/s

causes more weight losses. Increasing the sliding velocity also increases wear. The above study suggests that 0.6 vol.% of BN will provide good wear resistance.

References

1. Macke A, Schultz BF, Rohatgi P (2012) Metal matrix composites offer the automotive industry an opportunity to reduce vehicle weight, improve performance. Mater Process 170:19–23
2. Miracle DB (2005) Metal matrix composites-from science to technological significance. Compos Sci Technol 65:526–540
3. Simon SC, McMillan ML (2004) Automotive tribology overview of current advances and challenges for future. Tribol Int 37:517–536
4. Iredale R Manufacturing composites for automotive applications. Univ Bristol 10–11
5. Reddy AP, Krishna PV, Rao RN (2017) Al/SiC NP and Al/SiC NP/X nanocomposites fabrication and properties: a review. Proc Inst Mech Eng Part N J Nanomater Nanoeng Nanosyst 231:155–172
6. Soorya Prakash K, Gopal PM, Kavimani V (2017) Effect of rock dust, cenosphere and E-waste glass addition on mechanical, wear and machinability behaviour of Al 6061 hybrid composites. Indian J Eng Mater Sci 24:270–282
7. Al-Qutub AM, Khalil A, Saheb N, Hakeem AS (2013) Wear and friction behavior of Al6061 alloy reinforced with carbon nanotubes. Wear 297:752–761
8. Fallahdoost H, Nouri A (2016) Azimi A : Dual functions of TiC nanoparticles on tribological performance of Al/graphite composites. J Phys Chem Solids 93:137–144
9. Gupta P, Kumar D, Parkash O, Jha AK, Sadasivuni KK (2018) Dependence of wear behavior on sintering mechanism for Ironalumina metal matrix nanocomposites. Mater Chem Phys 220:441–448
10. Soorya Prakash K, Balasundar P, Nagaraja S et al (2017) Effect of reinforcement, compact pressure and hard ceramic coating on aluminium rock dust composite performance. Int J Refract Met Hard Mater 54:223–229
11. Akbari MK, Baharvandi HR, Mirzaee O (2014) Investigation of particle size and reinforcement content on mechanical properties and fracture behavior of A356-Al2O3 composite fabricated by vortex method. J Compos Mater 48(27):3315–3330
12. Kumar M, Murugan AM, Baskaran V, Hanumanth Ramji KS (2014) Effect of sliding distance on dry sliding tribological behaviour of aluminium hybrid metal matrix composite an alternate for automobile brake rotor—a grey relational approach. Proc Inst Mech Eng Part J J Eng Tribol 228(3):332–338
13. Hekner B, Myalski J, Valle N, Botor-Probierz A, Sopicka-Lizer M, Wieczorek J (2017) Friction and wear behavior of Al-SiC(n) hybrid composites with carbon addition. Compos Part B Eng 108:291–300
14. Hariharasakthisudhan P, Jose S, Manisekar K (2018) Dry sliding wear behaviour of single and dual ceramic reinforcements premixed with Al powder in AA6061 matrix. J Mater Res Technol: 1–9
15. Leng J, Wu G, ZhouQ et al (2008) Mechanical properties of SiC/gr/Al composites fabricated by squeeze casting technology. Scr Mater 59: 619–622
16. Carvalho O, Buciumeanu M, Madeira S, Soares D, Silva FS, Miranda G (2015) Dry sliding wear behaviour of AlSi-CNTs-SiCp hybrid composites. Tribol Int 90:148–156
17. Aruri D, Adepu K, Adepu K, Bazavada K (2013) Wear and mechanical properties of 6061–T6 aluminumalloy surface hybrid composites [(SiC + gr) and (SiC + Al2O3)] fabricated by friction stir processing. J Mater Res Technol 2:362–369
18. Ravindran P, Manisekar K, Rathika P, Narayanasamy P (2013) Tribological properties of powder metallurgy - processed aluminium self lubricating hybrid composites with SiC additions. MaterDes 45:561–570
19. Kavimani V, Prakash KS, Gunashri R et al (2017) Tribological behaviour predictions of r-GO reinforced mg composite using ANN coupled Taguchi approach. J Appl Res Technol 110:430–441

20. Basavarajappa S, Chandramohan G, Arjun M, Mukundan T, Subramanian R, Gopalakrishnan P (2007) Influence of sliding speed on the dry sliding wear behaviour and the subsurface deformation on hybrid metal matrix wear behaviour and the subsurface deformation on hybrid metal matrix composite. Wear 262:1007–1012
21. Ekambaram M, Vetrivel M, Balaji D, Afrid AS, Naveenkumar B, Manikanta DR, Amruthraj D, Krishna KJ (2018) Tribological characteristics of aluminium metal matrix with nano BN powder metallurgy composite. Mater Sci Eng 390:012085
22. Xue XM, Wang JT, Quan MX (1991) Mater Sci Eng A 132: 277
23. Shen L, Tan BJ, Willis WS, Galasso FS, Suib SL (1994) J Am Ceram Soc 77: 1011
24. Fujii H, Nakae H, Okada K (1993) Acta Metall Mater 41: 2963
25. Xue XM, Wang JT, Quan MX (1991) J Mater Sci 26: 6391
26. Xue XM, Wang JT, Zhao FM (1992) Mater Sci Lett 11: 199
27. Kobashi M, Choh T (1997) J Mater Sci 32: 6283
28. Nicholas MG, Mortimer DA, Jones LM, Crispin RM (1990) J Mater Sci 25: 2679

Powder Metallurgy Fabrication, Characterization and Wear Assessment of Al-BN-TiO_2 Composites

M. Ekambaram, M. Vetrivel, D. Balaji, Mandala Praveen Kumar, K. Venkatesh, S. Sudarsan Reddy, and Shaik Ahammad Basha

Abstract The tribological characteristics of aluminium metal matrix with different weight fraction of titanium dioxide nanopowder reinforcements along with 0.5 wt% of boron nitride nanopowder reinforcement were studied. The various weight fractions of reinforcements are 0.3 wt% TiO_2 + 0.5 wt% BN, 0.5 wt% TiO_2 + 0.5 wt% BN, and 0.7 wt% TiO_2 + 0.5 wt% BN are added with pure aluminium. In each composition, specimens are prepared and tested on pin-on-disc tribometer for three different load and sliding velocity conditions with constant distance. The study suggests that 0.3 wt% TiO_2 + 0.5 wt% of BN with aluminium provides better specific wear rate and coefficient of friction.

Keywords Hybrid composites · Specific wear rate · Coefficient of friction

M. Ekambaram (✉) · M. Vetrivel · D. Balaji · M. Praveen Kumar · K. Venkatesh · S. Sudarsan Reddy · S. A. Basha
Department of Mechanical Engineering, Sri Chandrasekharendra Saraswathi Viswa Mahavidyalaya University, Enathur, Kanchipuram 631561, Tamil Nadu, India
e-mail: e4ekam@gmail.com

M. Vetrivel
e-mail: km_vetrivel@yahoo.com

D. Balaji
e-mail: balajibala81@gmail.com

M. Praveen Kumar
e-mail: praveenkumarm935@gmail.com

K. Venkatesh
e-mail: venkykushal777@gmail.com

S. Sudarsan Reddy
e-mail: sudharsansuda515@gmail.com

S. A. Basha
e-mail: ahammadbasha555@gmail.com

T. Rajmohan et al. (eds.), *Advances in Materials and Manufacturing Engineering*, Springer Proceedings in Materials 7, https://doi.org/10.1007/978-981-15-6267-9_73

Table 1 Description of the materials used

Description	Aluminium (matrix)	Boron Nitride (reinforcement 1)	Titanium di Oxide (reinforcement 2)
Symbol	Al	BN	TiO_2
Particle size	130–180 μm	70–80 nm	5 nm
Density (g/cm^3)	2.69	2.29	3.78
Purity (%)	98	99.8	99.8

1 Introduction

In the recent years, the metal matrix composites have emerged with various broader applications in the aerospace, structural, automobile and defense sectors [1–3]. It has been reported that Al composites exhibit remarkable performances by the inclusion of small amount of ceramic reinforcements [4, 5]. To increase the wear resistance of aluminium-based composites, researchers were added various ceramic reinforcement particles such as alumina (Al_2O_3), boron nitride (BN), silicon nitride (Si_3N_4), titanium dioxide (TiO_2), tungsten carbide (WC), tungsten trioxide (WO_3), titanium carbide (TiC) and titanium diboride (TiB_2) in the aluminium matrix [6–10]. In hybrid aluminium matrix composite materials, high strength and wear capabilities of ceramic particles are combined with lightweight, low cost and lubricating properties of the secondary reinforcements. It has been revealed that the presence of lubricating particles considerably reduces the wear rate [11–15]. Boron nitride used as solid lubricant in the aluminium-based matrix due to its low density, high thermal conductivity and high resistivity and also exhibits optimum reinforcement in Al-based matrices [16–28].

2 Materials and Experiment

2.1 *Materials*

The description of the materials used in the research work is mentioned in Table 1.

2.2 *Specimen Preparation*

Each pellets of size 10 mm diameter and 30 mm height was made using powder metallurgy route, for all the three different compositions of titanium dioxide nanomaterials. The specimen composition and its micro-hardness value mentioned in Table 2.

Table 2 Micro-hardness values of specimens

Specimen composition	Micro-hardness value (H.V)
Specimen 1—Al + 0.5 wt% BN + 0.3 wt% TiO_2	30.5
Specimen 2—Al + 0.5 wt% BN + 0.5 wt% TiO_2	12.9
Specimen 3—Al + 0.5 wt% BN + 0.7 wt% TiO_2	14.1

Table 3 Specific wear rate of all specimens

S. No.	Sliding distance (m)	Sliding velocity (m/s)	Load (N)	Specific wear rate × E^{-9} m^3/Nm			
				Pure Al	0.3 wt% TiO_2	0.5 wt% TiO_2	0.7 wt% TiO_2
1	1000	1.5	5	1.5464	0.0012	0.0025	0.0020
2	1000	2.0	5	2.0286	0.0013	0.0021	0.0019
3	1000	2.5	5	2.8266	0.0007	0.0050	0.0028
4	1000	1.5	10	2.1024	0.0005	0.0007	0.0006
5	1000	2.0	10	3.3367	0.0007	0.0009	0.0008
6	1000	2.5	10	4.0359	0.0004	0.0017	0.0012
7	1000	1.5	15	2.3591	0.0004	0.0015	0.0008
8	1000	2.0	15	4.3288	0.0010	0.0021	0.0018
9	1000	2.5	15	5.6959	0.0002	0.0069	0.0019

3 Experimental Procedure

3.1 Wear Testing Procedure

ASTM G99 is the standard used and the experiments conducted on pin-on-disc wear testing machine. The three different load conditions are 5, 10, 15 N and the three different sliding velocities are 1.5, 2, 2.5 m/s were used. The specific wear rate (SWR) and Co efficient of friction (CoF) were calculated and tabulated in Tables 3 and 4 for above mentioned conditions.

4 Results and Discussions

4.1 Specific Wear Rate (SWR)

Table 3 represents the specific wear rate results of all specimens. The specific wear rate can be calculated from the below formula,

Table 4 Coefficient of friction of all specimens

S. No.	Sliding distance (m)	Sliding velocity (m/s)	Load (N)	Coefficient of friction			
				Pure Al	0.3 wt% TiO_2	0.5 wt% TiO_2	0.7 wt% TiO_2
1	1000	1.5	5	2.75	0.76	0.98	0.9
2	1000	2.0	5	2.79	0.8	0.99	0.94
3	1000	2.5	5	3.02	0.82	0.98	0.92
4	1000	1.5	10	4.12	0.66	0.82	0.8
5	1000	2.0	10	4.58	0.65	0.88	0.81
6	1000	2.5	10	5.2	0.70	0.95	0.78
7	1000	1.5	15	6.6	0.67	0.94	0.72
8	1000	2.0	15	6.9	0.69	0.93	0.76
9	1000	2.5	15	7.1	0.74	0.91	0.78

$$\text{Specific Wear Rate} = \frac{\text{Weight Loss}}{(\text{Density} * \text{Sliding distance} * \text{Load})}\frac{\text{m}^3}{\text{Nm}} \quad (1)$$

4.2 *Coefficient of Friction (CoF)*

Table 4 represents the coefficient of friction results of all specimens. The coefficient of friction can be calculated from the below formula,

$$\text{Coefficient of firction} = \frac{\text{Firctional force}}{\text{Normal load}}\,(\text{No Unit}) \quad (2)$$

4.3 *Comparison of SWR and CoF with Different Load and Velocities.*

As aluminium possesses higher value when compared with hybrid compositions, the graphs were plotted for hybrid compositions only to study its characteristics (Figs. 1 and 2).

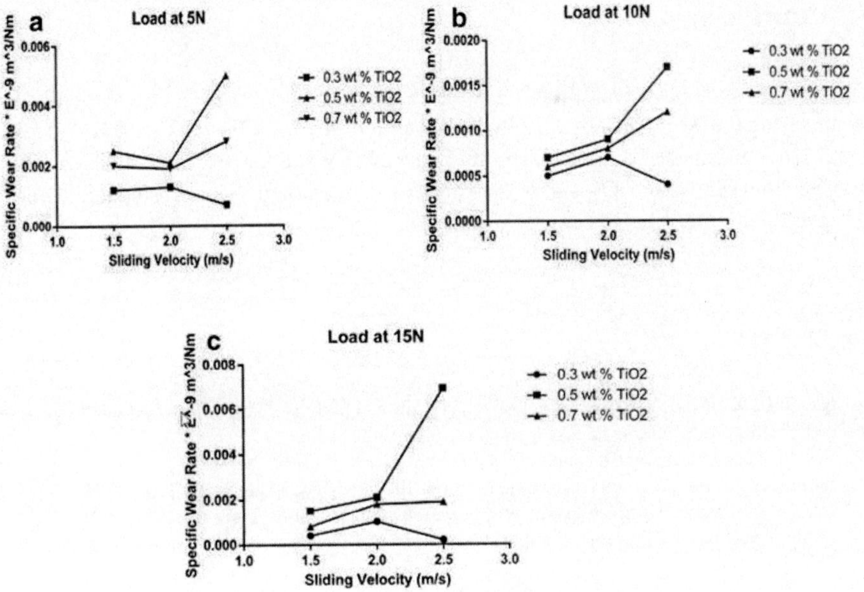

Fig. 1 **a** SWR at load 5 N, **b** SWR at load 10 N, **c** SWR at load 15 N

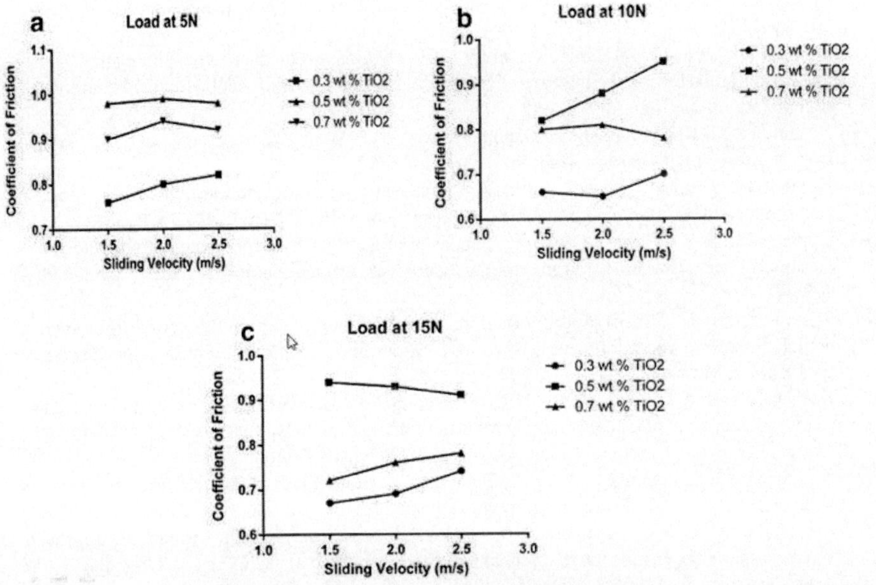

Fig. 2 **a** CoF at load 5 N, **b** CoF at load 10 N, **c** CoF at load 15 N

5 Conclusion

The result shows that an addition of reinforcements improves the wear resistance of the material. 0.3 wt% TiO_2 + 0.5% wt% BN and 0.7 wt% TiO_2 + 0.5 wt% BN gives better wear resistance which can be obtained from SWR and CoF graphs. The study suggests that 0.3 wt% TiO_2 + 0.5% wt% BN provides better wear resistance when compared with other hybrid compositions.

References

1. Kaushik NC, Rao RN (2016) Effect of grit size on two body abrasive wear of Al 6082 hybrid composites produced by stir casting method. Tribol Int 102:52–60
2. Prasad Reddy A, Vamsi Krishna P, Narasimha Rao R, Murthy NV (2017) Silicon carbide reinforced Aluminium metal matrix Nano composites-a review. Mater Today Proc 4:3959–3971
3. Iredale R Manufacturing composites for automotive applications. Univ Bristol 10–11.
4. Surappa MK (2003) Aluminum matrix composites: challenges and opportunities. Sadhana 28: 319–334
5. Rohatgi P, Schultz B (2007) Light weight metal matrix composites- stretching the boundaries of metals. Mater Matters 2:16–19
6. Reddy AP, Krishna PV, Rao RN (2017) Al/SiC NP and Al/SiC NP/X nanocomposites fabrication and properties: a review. Proc Inst Mech Eng Part N J Nanomater Nanoeng Nanosyst 231:155–172
7. Soorya Prakash K, Gopal PM, Kavimani V (2017) Effect of rock dust, cenosphere and E-waste glass addition on mechanical, wear and machinability behaviour of Al 6061 hybrid composites. Indian J Eng Mater Sci 24:270–282
8. Al-Qutub AM, Khalil A, Saheb N, Hakeem AS (2013) Wear and friction behavior of Al6061 alloy reinforced with carbon nanotubes. Wear 297:752–761
9. Fallahdoost H, Nouri A, Azimi A (2016) Dual functions of TiC nanoparticles on tribological performance of Al/graphite composites. J Phys Chem Solids 93:137–144
10. Gupta P, Kumar D, Parkash O, Jha AK, Sadasivuni KK (2018) Dependence of wear behavior on sintering mechanism for Ironalumina metal matrix nanocomposites. Mater Chem Phys 220:441–448
11. Soorya Prakash K, Sathiya Moorthy R, Gopal PM, Kavimani V (2016) Effect of reinforcement, compact pressure and hard ceramic coating on aluminium rock dust composite performance. Int J Refract Metal Hard Mater 54:223–229
12. Karbalaei Akbari M, Baharvandi HR, Mirzaee O (2013) Investigation of particle size and reinforcement content on mechanical properties and fracture behavior of A356-Al O composite fabricated by vortex method. J Compos Mater 48(27):3315–3330
13. Kumar M, Megalingam Murugan A, Baskaran V, Hanumanth Ramji KS (2015) Effect of sliding distance on dry sliding tribological behaviour of Aluminium Hybrid Metal Matrix Composite (AlHMMC): An alternate for automobile brake rotor – A Grey Relational approach. Proceedings of the Institution of Mechanical Engineers, Part J: J Eng Tribol 230 (4):402–415
14. Aruri D, Adepu K, Adepu K, Bazavada K (2013) Wear and mechanical properties of 6061–T6 aluminum alloy surface hybrid composites [(SiC+Gr) and (SiC+Al2O3)] fabricated by friction stir processing. J Mater Res Tech 2(4):362–369
15. Ravindran P, Manisekar K, Rathika P, Narayanasamy P (2013) Tribological properties of powder metallurgy – Processed aluminium self lubricating hybrid composites with SiC additions. Mater Des 45:561–570

16. Hekner B, Myalski J, Valle N, Botor-Probierz A, Sopicka-Lizer M, Wieczorek J (2017) Friction and wear behavior of Al-SiC(n) hybrid composites with carbon addition. Compos Part B Eng 108:291–300
17. Hariharasakthisudhan P, Jose S, Manisekar K (2018) Dry sliding wear behaviour of single and dual ceramic reinforcements premixed with Al powder in AA6061 matrix. J Mater Res Technol: 1–9
18. Leng J, Wu G, ZhouQ et al (2008) Mechanical properties of SiC/gr/Al composites fabricated by squeeze casting technology. Scr Mater 59: 619–622
19. Carvalho O, Buciumeanu M, Madeira S, Soares D, Silva FS, Miranda G (2015) Dry sliding wear behaviour of AlSi-CNTs-SiCp hybrid composites. Tribol Int 90:148–156
20. Basavarajappa S, Chandramohan G, Arjun M, Mukundan T, Subramanian R, Gopalakrishnan P (2007) Influence of sliding speed on the dry sliding wear behaviour and the subsurface deformation on hybrid metal matrix composite. Wear 262:1007–1012
21. Ekambaram M, Vetrivel M, Balaji D, Afrid AS, Naveenkumar B, Manikanta DR, Amruthraj D, Krishna KJ (2018) Tribological characteristics of aluminium metal matrix with nano BN powder metallurgy composite. Mater Sci Eng 390:012085
22. Xue XM, Wang JT, Quan MX (1991) Mater Sci Eng A 132: 277
23. Shen L, Tan BJ, Willis WS, Galasso FS, Suib SL (1994) J Am Ceram Soc 77: 1011
24. Fujii H, Nakae H, Okada K (1993) Acta Metall Mater 41: 2963
25. Xue XM, Wang JT, Quan MX (1991) Mater Sci 26: 6391
26. Xue XM, Wang JT, Zhao FM (1992) Mater Sci Lett 11: 199
27. Kobashi M, Choh T (1997) J Mater Sci 32: 6283
28. Nicholas MG, Mortimer DA, Jones LM, Crispin RM (1990) J Mater Sci 25: 2679

Optimization in Tribological Behaviour of Al-BN-TiO_2 Composites Using D-Optimal Method

M. Ekambaram, M. Vetrivel, and D. Balaji

Abstract The aluminium hybrid composite processed powder metallurgy and wear behaviour was experimentally investigated on aluminium with 0.5 wt% nano-BN reinforcement combined with nano-TiO_2 weight per cent reinforcement of 0.3, 0.5 and 0.7% at different sliding velocities ranging from 1.5 to 2.5 m/s and different normal loads ranging from 5 to 15 N. The effect of specific wear rate and coefficient of friction was optimized using D-optimal method. The results predicted by the developed model are concurring well within the measured values. The correlation coefficient of model, regarding the specific wear rate and coefficient of friction, was about 0.9891 and 0.9805, respectively, with conform the degree of accuracy of the mathematical model. The optimal condition attained the minimum specific wear rate and coefficient of friction at sliding velocity 1.5 m/s at normal load 14.36 N for Al + 0.5 wt% BN + TiO_2 0.3 wt% composite. The optimal predicted value of specific wear rate was 0.00043e-9 mm^3/Nm and coefficient of friction 0.66, respectively.

Keywords Optimization · Specific wear rate · Coefficient of friction

1 Introduction

In the recent years, to increase the wear resistance of aluminium-based composites researchers added various ceramic reinforcement particles such as alumina (Al_2O_3), boron nitride (BN), silicon nitride (Si_3N_4), titanium dioxide (TiO_2), boron carbide (B_4C), tungsten carbide (WC), tungsten trioxide (WO_3), titanium carbide (TiC) and titanium diboride (TiB_2) in the aluminium matrix [1–5]. In hybrid aluminium matrix

M. Ekambaram (✉) · M. Vetrivel · D. Balaji
Department of Mechanical Engineering, Sri Chandrasekharendra Saraswathi Viswa Mahavidyalaya University, Enathur, Kanchipuram, Tamil Nadu 631561, India
e-mail: e4ekam@gmail.com

M. Vetrivel
e-mail: km_vetrivel@yahoo.com

D. Balaji
e-mail: balajibala81@gmail.com

T. Rajmohan et al. (eds.), *Advances in Materials and Manufacturing Engineering*, Springer Proceedings in Materials 7, https://doi.org/10.1007/978-981-15-6267-9_74

composite materials, high strength and wear capabilities were combined with light weight, low cost and lubricating properties of the secondary reinforcements. It has been revealed that the presence of lubricating particles considerably reduces the wear rate [6–10]. Boron nitride is used as solid lubricant in the aluminium-based matrix due to its low density, high thermal conductivity and also exhibits optimum reinforcement in Al-based matrices [11–19].

1.1 D-Optimal Method—Response Surface Methodology

The mathematical model has been developed based on the experimental results, with an objective to determine specific wear rate and coefficient of friction and also to study interrelationship between the test variables of Al-BN-TiO_2 hybrid composite. A response surface equation is formulated as

$$Y = \beta_0 + \sum_{i=1}^{m} \beta_i x_i + \sum_{i=1}^{m} \beta_{ii} x_i^2 + \sum_{i=1}^{m} \sum_{i<j}^{m} \beta_{ij} x_i x_j + \varepsilon \tag{1}$$

where β_0, β_i, β_{ii} and β_{ij} are regression coefficients, x_i ($i = 1, 2, 3, \ldots m$) are design variables, ε is the random error, Y is the response and m is total number of design variables. To analyses statistically the model, the analysis of the variance (ANOVA) was used. Fisher's F-test (overall model significance), its probability $p(F)$, determination coefficient R^2 (used to measure the fitness of regression model) were also studied by ANOVA.

2 Materials and Experiment

2.1 Materials

The description of the materials used in the research work is mentioned in Table 1.

Table 1 Description of the materials used

Description	Aluminium (matrix)	Boron nitride (reinforcement 1)	Titanium dioxide (reinforcement 2)
Symbol	Al	BN	TiO_2
Particle size	130–180 μm	70–80 nm	5 nm
Density (g/cm^3)	2.69	2.29	3.78
Purity (%)	98	99.8	99.8

Table 2 Specimen representation and its composition

Specimen representation (based on eight percentage of TiO_2)	Composition
Sample 1	Al + 0.5 wt% BN + 0.3 wt% TiO_2
Sample 2	Al + 0.5 wt% BN + 0.5 wt% TiO_2
Sample 3	Al + 0.5 wt% BN + 0.7 wt% TiO_2

Table 3 Experimental factors and their levels

Symbol	Parameter	Unit	Parameters		
			Level 1	Level 2	Level 3
A	Load	N	5 N	10 N	15 N
B	Sliding velocity	m/s	1.5	2	2.5
C	Weight fractions of TiO_2	–	0.3	0.5	0.7

2.2 Specimen Preparation

Each pellets of size 10 mm diameter and 30 mm height were made using powder metallurgy route, for all the three different compositions of titanium dioxide nanomaterials.

3 Experimental Procedure

3.1 Wear Testing Methodology

Wear test was conducted on pin on disc wear testing machine with all the hybrid specimens as per ASTM G99 standard and recorded output values of wear rate, frictional force, coefficient of friction based on the design variable mention Table 2.

Specific wear rate (SWR) is calculated by wear volume per unit distance and load and coefficient of friction with a sliding distance of 1000 m (Table 3).

4 Results and Discussions

4.1 RSM Modelling

The model for Specific wear rate (**SWR**) and coefficient of friction (**CoF**) was developed from the regression coefficient and represented in Eqs. 2–7 as follows,

$$\textbf{Specific Wear rate at 0.3 wt\% TiO}_2 = 0.011730 - 1.37928E^{-003} * \text{Load} - 2.40033E^{-003} * \text{Sliding Velocity} + 2.67288E^{-004} * \text{Load} * \text{Sliding Velocity} + 3.45686E^{-005} * \text{Load}^2 - 3.43137E^{-004} * \text{Sliding Velocity}^2 \quad (2)$$

$$\textbf{Specific Wear rate at 0.5wt \% TiO}_2 = 1.09529E^{-003} - 1.17722E^{-003} * \text{Load} + 3.11238E^{-003} * \text{Sliding Velocity} + 2.67288E^{-004} * \text{Load} * \text{Sliding Velocity} + 3.45686E^{-005} * \text{Load}^2 - 3.43137E^{-004} * \text{Sliding Velocity}^2 \quad (3)$$

$$\textbf{Specific Wear rate at 0.7 wt\% TiO}_2 = 0.011872 - 1.45046E^{-003} * \text{Load} - 1.57915E^{003} * \text{Sliding Velocity} + 2.67288E^{-004} * \text{Load} * \text{Sliding Velocity} + 3.45686E^{-005} * \text{Load}^2 - 3.43137E^{-004} * \text{Sliding Velocity}^2 \quad (4)$$

$$\textbf{Coefficient of Friction at 0.3 wt\% TiO}_2 = 0.30343 - 8.98504E^{-003} * \text{Load} + 0.45194 * \text{Sliding Velocity} - 6.38983E^{-003} * \text{Load} * \text{Sliding Velocity} + 6.21569E^{-004} * \text{Load}^2 - 0.077843 * \text{Sliding Velocity}^2 \quad (5)$$

$$\textbf{Coefficient of Friction at 0.5 wt\% TiO}_2 = 0.81271 - 0.035491 * \text{Load} + 0.36366 * \text{Sliding Velocity} - 6.38983E^{-003} * \text{Load} * \text{Sliding Velocity} + 6.21569E^{-004} * \text{Load}^2 - 0.07784 * \text{Sliding Velocity}^2 \quad (6)$$

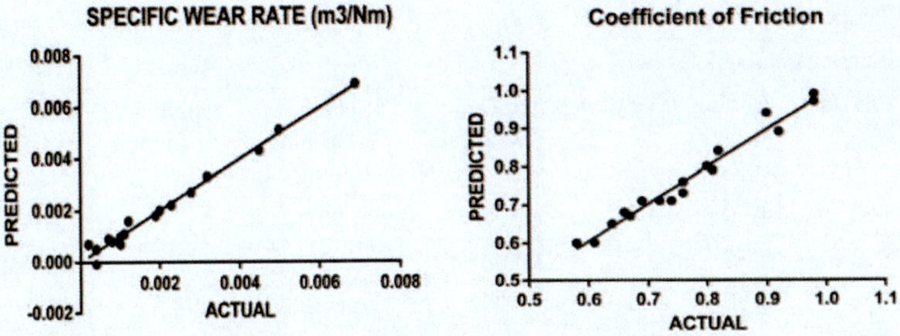

Fig. 1 Experimental (actual) and predicted values of SWR and CoF

$$\textbf{Coefficient of Friction at 0.7 wt\% TiO}_2 = 0.82581 - 0.025926 * \text{Load} + 0.29925 * \text{Sliding Velocity} - 6.38983E^{-003} * \text{Load} * \text{Sliding Velocity} + 6.21569E^{-004} * \text{Load}^2 - 0.077843 * \text{Sliding Velocity}^2 \quad (7)$$

4.2 *Predicted Versus Actual*

From Eqs. (2–7), the predicted value determined and compared with the experimental results (Fig. 1; Tables 4 and 5).

From the Table 6, the developed models were used for multiple response optimizations by desirability function approach to obtain maximum strength coefficient and minimum strain hardening exponent. The desirability is found to be 0.868 as acceptable and excellent significant based on Harrington's rating system [20]. The optimized specific wear rate 0.0004 E^{-9} m^3/Nm and coefficient of friction 0.6676 for load 14.36 N and sliding velocity 1.5 m/s, when the TiO_2 weight percentage 0.3 (wt%), are selected from Table 6. Based on the optimization result 0.3 wt% TiO_2 at sliding velocity 1.5 m/s with 15 N load was tested, and the result represents specific wear rate is 0.00042431 E^{-9} m^3/Nm and coefficient of friction 0.649423; hence, the model is significant and optimized results are verified.

Table 4 Predicted and actual values of SWR and CoF

Exp. No.	Process parameters levels			Actual	Predicted	Actual	Predicted
	Load (N)	Sliding velocity (m/s)	TiO_2 wt%	Specific wear rate E^{-9} m^3/Nm	Specific wear rate E^{-9} m^3/Nm	Coefficient of friction	Coefficient of friction
1	15.00	1.50	0.7	0.0008	0.0008	0.72	0.71
2	15.00	2.50	0.7	0.0019	0.0018	0.58	0.6
3	5.00	1.50	0.5	0.002	0.0020	0.98	0.97
4	15.00	1.50	0.3	0.0004	0.0005	0.67	0.67
5	15.00	2.50	0.3	0.0002	0.0007	0.74	0.71
6	5.00	2.50	0.3	0.0007	0.0009	0.82	0.84
7	5.00	2.50	0.5	0.005	0.0051	0.98	0.99
8	5.00	2.00	0.3	0.0023	0.0022	0.8	0.8
9	5.00	1.50	0.7	0.0045	0.0043	0.9	0.94
10	5.00	1.50	0.3	0.0032	0.0033	0.76	0.73
11	15.00	2.50	0.5	0.0069	0.0069	0.61	0.6
12	5.00	1.50	0.5	0.002	0.0020	0.98	0.97
13	10.00	1.50	0.3	0.001	0.0010	0.66	0.68
14	10.00	2.50	0.3	0.0004	-0.0001	0.76	0.76
15	15.00	2.00	0.3	0.001	0.0007	0.69	0.71
16	15.00	2.50	0.7	0.0019	0.0018	0.58	0.6
17	15.00	2.50	0.5	0.0069	0.0069	0.61	0.6
18	10.00	2.00	0.7	0.0012	0.0016	0.81	0.79
19	15.00	1.50	0.7	0.0008	0.0008	0.72	0.71
20	15.00	1.50	0.5	0.0011	0.0011	0.64	0.65
21	15.00	1.50	0.5	0.0011	0.0011	0.64	0.65
22	5.00	2.50	0.7	0.0028	0.0027	0.92	0.89

5 Conclusion

According to the analysis of variance (ANOVA), the accuracy of developed model conformed by R^2 and p value. The predicted and measured values have excellent agreement. Further, optimum combination of process parameters to attain the minimizing the specific wear rate and minimizing coefficient of friction can be predicted by the statistically developed models.

Table 5 ANOVA based RSM Models

Source	Specific wear rate (SWR)				
	SS	DF	MS	F-Value	$p > F$
Model	8.100E−005	11	7.363E−006	82.75	<0.0001
Residual	8.898E−007	10	8.898E−008		
Total	8.189E−005	21			
R^2	Specific wear rate (SWR) **0.9891**				
Source	Coefficient of friction (CoF)				
	SS	DF	MS	F-Value	$p > F$
Model	0.36	11	0.032	45.76	<.0001
Residual	7.057E−003	10	7.057E−004		
Total	0.36	21			
R^2	**0.9805**				

Table 6 Best solution for optimization

Number	Load (N)	Sliding velocity (m/s)	TiO_2 wt%	Specific wear rate (E^{-9} m^3/Nm)	Coefficient of friction	Desirability
1	14.36	1.5	0.3	0.00043631	0.667699	0.868
2	14.31	1.5	0.3	0.000435676	0.667737	0.868
3	14.25	1.5	0.3	0.000435136	0.667786	0.868

References

1. Reddy AP, Krishna PV, Rao RN (2017) Al/SiC NP and Al/SiC NP/X nanocomposites fabrication and properties: a review. Proc Inst Mech Eng Part N J Nanomater Nanoeng Nanosyst 231:155–172
2. Soorya Prakash K, Gopal PM, Kavimani V (2017) Effect of rock dust, cenosphere and E-waste glass addition on mechanical, wear and machinability behaviour of Al 6061 hybrid composites. Indian J Eng Mater Sci 24:270–282
3. Al-Qutub AM, Khalil A, Saheb N, Hakeem AS (2013) Wear and friction behavior of Al6061 alloy reinforced with carbon nanotubes. Wear 297:752–761
4. Fallahdoost H, Nouri A (2016) Azimi A : Dual functions of TiC nanoparticles on tribological performance of Al/graphite composites. J Phys Chem Solids 93:137–144
5. Gupta P, Kumar D, Parkash O, Jha AK, Sadasivuni KK (2018) Dependence of wear behavior on sintering mechanism for Ironalumina metal matrix nanocomposites. Mater Chem Phys 220:441–448
6. Ravindran P, Manisekar K, Rathika P, Narayanasamy P (2013) Tribological properties of powder metallurgy – Processed aluminium self lubricating hybrid composites with SiC additions. Mater Des 45:561–570
7. Aruri D, Adepu K, Adepu K, Bazavada K (2013) Wear and mechanical properties of 6061–T6 aluminum alloy surface hybrid composites [(SiC+Gr) and (SiC+Al2O3)] fabricated by friction stir processing. J Mater Res Technol 2(4):362–369

8. Kumar M, Megalingam Murugan A, Baskaran V, Hanumanth Ramji KS (2015) Effect of sliding distance on dry sliding tribological behaviour of Aluminium Hybrid Metal Matrix Composite (AlHMMC): An alternate for automobile brake rotor – A Grey Relational approach. Proceedings of the Institution of Mechanical Engineers, Part J: J of Eng Tribol 230 (4):402–415
9. Karbalaei Akbari M, Baharvandi HR, Mirzaee O (2013) Investigation of particle size and reinforcement content on mechanical properties and fracture behavior of A356-Al O composite fabricated by vortex method. J Compos Mater 48(27):3315–3330
10. Soorya Prakash K, Sathiya Moorthy R, Gopal PM, Kavimani V (2016) Effect of reinforcement, compact pressure and hard ceramic coating on aluminium rock dust composite performance. Int J Refract Metals and Hard Mater 54:223–229
11. Basavarajappa S, Chandramohan G, Arjun M, Mukundan T, Subramanian R, Gopalakrishnan P (2007) Influence of sliding speed on the dry sliding wear behaviour and the subsurface deformation on hybrid metal matrix composite. Wear 262:1007–1012
12. Ekambaram M, Vetrivel M, Balaji D, Afrid AS, Naveenkumar B, Manikanta DR, Amruthraj D, Krishna KJ (2018) Tribological characteristics of aluminium metal matrix with nano BN powder metallurgy composite. Mater Sci Eng 390:012085
13. Xue XM, Wang JT, Quan MX (1991) Mater Sci Eng A 132: 277
14. Shen L, Tan BJ, Willis WS, Galasso FS, Suib SL (1994) J Am Ceram Soc 77: 1011
15. Fujii H, Nakae H, Okada K (1993) Acta Metall Mater 41: 2963
16. Xue XM, Wang JT, Quan MX (1991) J Mater Sci 26: 6391
17. Xue XM, Wang JT, Zhao FM (1992) Mater Sci Lett 11: 199
18. Kobashi M, Choh T (1997) J Mater Sci 32: 6283
19. Nicholas MG, Mortimer DA, Jones LM, Crispin RM (1990) J Mater Sci 25: 2679
20. Harington J (1965) Ind Quality Control 21: 494–498

Influence of Aroma Skin Reinforcement on Erosive Behavior of Polyester Composites

Jayamani Manivannan, Shanmugavel Rajesh, Kalimuthu Mayandi, and Nadir Ayrilmis

Abstract This research work is aiming to provide a solution to the people who are living in the comfortless or in desert areas. In the desert region, the construction of building with conventional material will be difficult. In this proposed work, the composite panels are fabricated from aroma coffee skin reinforcement, reinforced with polyester resin. The hand layup technique is used to fabricate the composites for different wt% of composites (2.5, 5, 7.5 and 10%). In desert region, the panels are subjected to heavy wind blows; thus, it affects the life of the panel. To estimate, the erosion wear rate of the fabricated composite, the composite is subjected to erosion studies. The erosion studies is conducted as per ASTM standard G76. The stand-off distance and flow rate of the abrasive particles were kept constant. Alumina oxide is used as erodent material with the size of 50 μm. The behavior of fabricated composites is measured at regular interval. It is observed that 2.5% of reinforced composite is having less erosion rate than the 5, 7.5 and 10%. The reason for failure of the composites is analyzed and discussed.

Keywords Polyester · Aroma coffee skin · Hand layup · Time · Erosive wear

1 Introduction

Nowadays, polymers and its composites are playing a vital role on replacing the conventional materials in many engineering applications. In addition, composites are focused toward to operate in dusty environment where it should possess better erosion resistance. As on date, performance of the polymer composites is improved by adding hard particulate fillers such as fiber fillers, metal and ceramic particles are being used [1]. Investigation carried out by Jung et al. reveals that the addition of organic,

J. Manivannan · S. Rajesh (✉) · K. Mayandi
Department of Mechanical Engineering, School of Automotive and Mechanical Engineering, Kalasalingam Academy of Research and Education, Krishnankoil, Tamil Nadu 626126, India
e-mail: s.rajesh@klu.ac.in

N. Ayrilmis
Department of Wood Mechanics and Technology, Faculty of Forestry, Istanbul University, Cerrahpasa, Bahcekoy, Sariyer, 34473 Istanbul, Turkey

T. Rajmohan et al. (eds.), *Advances in Materials and Manufacturing Engineering*, Springer Proceedings in Materials 7, https://doi.org/10.1007/978-981-15-6267-9_75

inorganic and metal particles with varying aspect ratio is capable of producing a good composites, and it can be effectively used for industrial applications such as electrodes and heaters [2]. The addition of low-cost particulate in making of polymer composites is most beneficial, and thus, the overall cost of components is reduced [3–5]. Srivastava et al. studied the angle of impingement and erodent velocity on erosion behavior of epoxy composite made with E-glass fiber and fly ash as reinforcement and particulate filler [6]. Wang et al. investigated the performance on wear resistance of SiC and ZrO_2 particulate reinforced polyetheretherketone composites with varying wt% of reinforcement [7–9]. The erosion wear behavior of aroma skin-reinforced polyester composites is not reported in the literature.

The aim of the present work is to fabricate a new composite of aroma skin-reinforced polyester composites with different wt% (2.5, 5, 7.5 and 10%) and to study the performance of erosive wear resistance of the newly fabricated composites.

2 Experimental Details

2.1 Materials

In this work, polyester resin and aroma coffee skin were chosen as matrix and reinforcement material. The polyester resin, catalyst and accelerator are brought from VB Ltd, Madras, Tamil Nadu, India. Unsaturated polyester resin, methyl-ethyl-ketone-peroxide (MEKP) and cobalt naphthalene (CN) are used as matrix, catalyst and accelerator materials. The collected aroma coffee skin contains dust particles; it is removed by water wash and dried at room temperature. Dried aroma skin powder is processed in ball mill to reduce the size of the aroma skin flakes. The processed aroma skin is sieved, and particles are selected in the range 100–150 μm for fabrication of composites.

2.2 Specimen Preparation

Aroma coffee skin is reinforced with unsaturated polyester resin to fabricate the composite. The composite plates are prepared by hand layup method. Two percent MEKP and 2% CN as catalyst and accelerator are mixed thoroughly in unsaturated polyester resin followed by reinforcing material (ACS). Composites of five different wt% (2.5, 5, 7.5 and 10% as ACS filling) are prepared by following the same procedure. The castings are put under load of 24 h at room temperature for proper curing. The fabricated plate is cut into required size as per ASTM G76 standards. The prepared specimen for erosion wear test is shown in Fig. 1. The detailed wt% of reinforcement for the erosion study is listed in Table 1.

Fig. 1 Specimens prepared for erosion test

Table 1 Detailed composition of the specimen

Composite	Composition
CP1	Polyester + 2.5 wt% aroma coffee skin
CP2	Polyester + 5 wt% aroma coffee skin
CP3	Polyester + 7.5 wt% aroma coffee skin
CP4	Polyester + 10 wt% aroma coffee skin

2.3 *Erosion Test Apparatus*

Figure 2 shows the photographical view of air jet erosion tester. The experiments is conducted as per ASTM G76, on all the selected specimen. Alumina oxide with particle size of 50 μm is used as erodent material. The initial mass of the specimen is precisionly measured in electronic weigh balance with an accuracy of 0.0001 g. The specimen placed on holder, the size of the holder and specimen is 25 × 25 × 3 mm. The experiment is conducted by keeping standoff distance 10 mm and flow rate of abrasion particles 3.3 g/min as constant. The weight loss of the composite material after 2 min is noted for further analysis. The same procedure is repeated for varying wt% of composites. Table 2 shows the erosion parameters used to conduct the studies.

3 Results and Discussion

The erosion test is carried out as per ASTM standards for 2.5, 5, 7.5 and 10% of aroma coffee skin-reinforced polymer composites. A graph is plotted to understand the erosive behavior of composites with respect to time for varying wt% of composites. From the experimental results, it is studied that the rate of weight loss of the eroded samples is varying in accordance with the parameter of different wt% with regular time interval. The reason for increase in erosive wear rate of the composites is due to brittle nature of the materials. Addition of particles influences the strength of the materials, and increase in hardness increases the strength. Therefore, the materials

Fig. 2 Air jet erosion tester

Table 2 Erosion test parameters

Fixed parameters		Variable parameters	
Erodent material	Alumina oxide	Time (mins)	2, 4, 6 … 60
Particle size of erodent (μm)	50	Reinforcement (wt%)	2.5, 5, 7.5 and 10
Angle of impingement (α, °)	90		
Velocity of impact (m/s)	100		
Erodent discharge rate (g/min)	3.3		
Testing temperature	RT		
Standoff distance (mm)	10		
Diameter of nozzle (mm)	3		

in ductile nature become semi-ductile or brittle nature. It is well known that some of the researchers studies shown that an angle of impingement is the most important factors in the erosion studies. For ductile materials, the maximum erosion rate occurs at 15–30° angle, while for brittle materials the erosion rate is maximum at 90° angle. The maximum erosion rate occurred at an impingement angle of 60° which clearly indicates that the composites response to the erosion studies as neither ductile nor brittle in nature. This behavior of material is called semi-ductile manner [10]. In this work, the variation of erosive wear rate of polymer composites with regular time interval is plotted as shown in Figs. 3 and 4 for 2.5%, 5%, 7.5% and 10% of ACS reinforcement, respectively, and keeping other parameters as constant. From Figs. 3 and 4, it is understood that the addition of particulate reinforcement shows that the significant reduces in the rate of material loss and simultaneously increases in the erosive wear rate of the composites. However, the composite is made with 2.5% of ACS reinforcement shows better erosion resistance than the other wt% (5, 7.5 and 10%) of ACS reinforcement. Figure 3 indicates that the erosion wear rate takes place in test conditions at 90° of impingement angle, and after a certain interval of time, the erosion wear rate for all the composites should achieve a constant state.

Fig. 3 Variation of erosion rate with time interval for different wt%

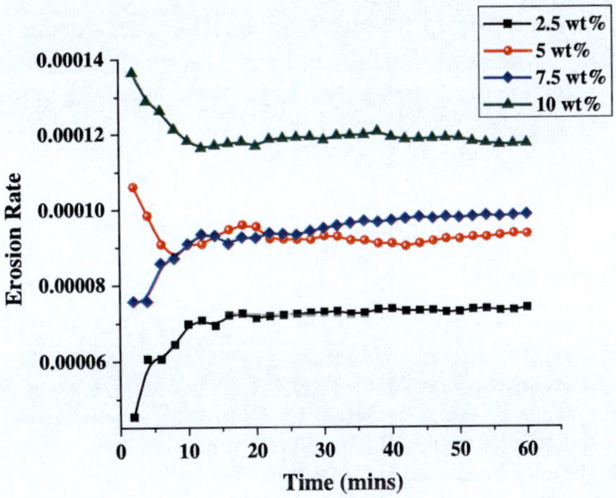

Fig. 4 Variation of average erosion rate with time interval for different wt%

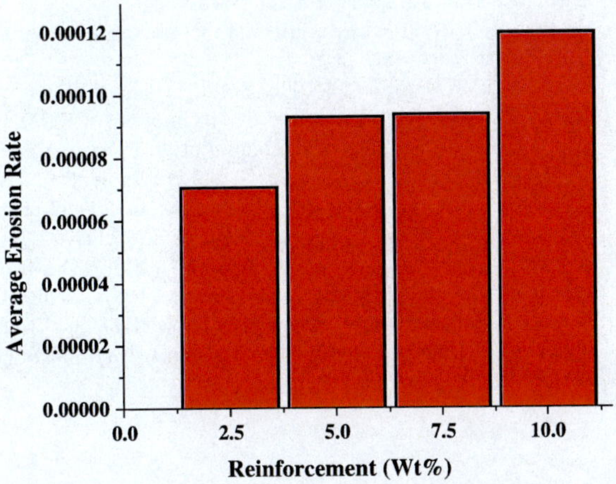

4 Conclusions

The following observation are made from the erosive wear studies of aroma skin-reinforced polyester composites.

(i) The composites successfully fabricated using aroma skin reinforcement for varying wt% with simple hand layup technique, without defects.
(ii) Addition of reinforcement significantly alter the erosive wear behavior of the materials.
(iii) The erosion wear resistance of aroma coffee skin-reinforced polyester composite improves erosion resistance property.
(iv) Among the four, 2.5% of ACS reinforcement gives better erosion resistance of the composite. The erosion wear rate of 5, 7.5 and 10% of ACS reinforced composites is relatively higher than 2.5% under similar test conditions.
(v) These composites may be recommended because of its nature as biodegradable and eco-friendly for the applications of lightweight vehicles and partition boards.

References

1. Gregory Sawyer W, Freudenberg Kevin D, Praveen B, Schadler Linda S (2003) A study on the friction and wear behavior of PTFE filled with alumina nanoparticles. Wear 254:573–580
2. Jung-il K, Kang PH, Nho YC (2004) Positive temperature coefficient behavior of polymer composites having a high melting temperature. J Appl Polym Sci 92:394–401
3. Kranthi G, Satapathy A (2010) Evaluation and prediction of wear response of pine wood dust filled epoxy composites using neural computation. Comput Mater Sci 49:609–614
4. Unal H, Findik F, Mimaroglu A (2003) Mechanical behavior of nylon composites containing talc and kaolin. J Appl Polym Sci 88:1694–1697
5. Rothon RN (1999) Mineral fillers in thermoplastics: filler manufacture and characterization. Adv Polym Sci 139:67–107
6. Srivastava VK, Pawar AG (2006) Solid particle erosion of glass fibre reinforced flyash filled epoxy resin composites. Compos Sci Technol 66(15):3021–3028
7. Wang QH, Xue QJ, Shen WC, Zhang J (1998) The friction and wear properties of nanometer ZrO_2-filled polyetheretherketone. J Appl Polym Sci 69(1):135–141
8. Wang QH, Xue QJ, Shen WC (1998) The tribological properties of SiC whisker-reinforced polyetheretherketone. J Appl Polym Sci 69(12):2341–2347
9. Wang QH, Xue QJ, Liu WM, Chen J-M (2000) The friction and wear characteristics of nanometer SiC and polytetrafluoroethylene filled polyetheretherketone. Wear 243(1):140–146
10. Patnaik A, Satapathy A, Mahapatra SS, Dash RR (2009) Tribo-performance of polyester hybrid composites: Damage assessment and parameter optimization using Taguchi design. Mater Des 30:57–67

Optimization of Wear Reduction on Al–TiO_2–Gr Powder Metallurgy Composites Using D-Optimal Method

D. Balaji, M. Vetrivel, and M. Ekambaram

Abstract The aluminum hybrid composite is prepared in powder metallurgy route aluminum with equal wt% nanoTiO_2 and nanoGr reinforcements (5 wt% TiO_2 + 5 wt% Gr). The wear behavior was experimentally investigated on three different specimens weight of same volume such as 4, 5, 6 g at different sliding velocities of 1.5, 2.0, 2.5 m/s and normal loads of 5, 10, 15 N were considered. The effect of specific wear rate and co-efficient of friction was optimized by D-Optimal method. The results predicted by the developed mathematical model are compatible well within the measured values. The correlation co-efficient of model, regarding the specific wear rate and co-efficient of friction, were about 0.9895 and 0.8468, respectively. The optimal condition of specific wear rate and co-efficient of friction is attained at sliding velocity 1.62 m/s and normal load 10 N with specimen weight of 5 g. The predicted value of specific wear rate is 5.5548e-6 mm^3/Nm and co-efficient of friction is 0.23, respectively.

Keywords Powder metallurgy · D-optimal · Wear rate · Coefficient of friction

1 Introduction

Metal matrix composites have a prospect for resistance in wear over the unreinforced materials and also have increased mechanical properties [1]. The aluminum alloys are quite attractive due to their low cost, low density, its capability to strengthen by precipitation, high thermal and electrical conductivity, good corrosion resistance, and improved tribological properties [2, 3]. This paper is concentrated on the special

D. Balaji (✉) · M. Vetrivel · M. Ekambaram
Department of Mechanical Engineering, Sri Chandrasekharendra Saraswathi Viswa Mahavidyalaya University, Enathur, Kanchipuram, Tamil Nadu 631561, India
e-mail: balajibala81@gmail.com

M. Vetrivel
e-mail: km_vetrivel@yahoo.com

M. Ekambaram
e-mail: e4ekam@gmail.com

T. Rajmohan et al. (eds.), *Advances in Materials and Manufacturing Engineering*, Springer Proceedings in Materials 7, https://doi.org/10.1007/978-981-15-6267-9_76

metal matrix composite composed of aluminum, titanium di-oxide in nanoscale, and graphite [4–8]. Titanium di-oxide (TiO_2) possesses more hardness with superior wear resistance and corrosion resistance. This hybrid aluminum matrix composite improves the tribological and self-lubricating properties of the composites [9–12]. RSM can be an excellent approach to study a process response and to figure out the best correlation among the parameters of a process. This is achieved via developed models based on the statistical methods to investigate the relation between the inputs and outputs of any process [13]. The optimal input parameter is investigated to minimize specific wear rate and co-efficient of friction. A desirability function-based simultaneous optimization technique is used in this work [14].

1.1 D-Optimal Method-Response Surface Methodology

The mathematical model has been developed based on the experimental results, with an objective to determine specific wear rate and coefficient of friction, and also to study inter-relationship between the test variables of Al–TiO_2–Gr hybrid composite. A response surface equation is formulated as

$$Y = \beta_0 + \sum_{i=1}^{m} \beta_i x_i + \sum_{i=1}^{m} \beta_{ii} x_i^2 + \sum_{i=1}^{m} \sum_{i<j}^{m} \beta_{ij} x_i x_j + \varepsilon \quad (1)$$

where β_0, β_i, β_{ii}, and β_{ij} are regression co-efficients, x_i ($i = 1, 2, 3, \ldots m$) are design variables, ε is the random error, Y is the response, and m is total number of design variables. To analyze the model statistically, the analysis of the variance (ANOVA) is used.

2 Materials and Experiment

2.1 Materials

The description of the materials used in the research work is mentioned in Table 1.

2.2 Specimen Preparation

Each pellets of size 10 mm diameter and 30 mm height was made using powder metallurgy route for all the three different weights of the compositions. The compositions of the materials are 90 wt% Al + 5 wt% TiO_2 + 5 wt% Gr.

Table 1 Description of the materials used

Description	Aluminum (matrix)	Titanium di oxide (reinforcement 1)	Graphite (reinforcement 2)
Symbol	Al	TiO_2	Gr
Particle size	130–180 μm	5 nm	400 nm
Density (g/cm^3)	2.69	3.78	2.66
Purity (%)	98	99.8	98.0

3 Experimental Procedure

3.1 Wear Testing Methodology

Wear test was conducted on Pin on Disc wear testing machine with all the hybrid specimens as per ASTM G99 standard and recorded the output values of wear rate, frictional force, and coefficient of friction based on the design variable mentioned in Table 2.

Specific wear rate (SWR) is calculated by wear volume per unit distance and load and co-efficient of friction with a sliding distance of 1000 m (Table 3).

Table 2 Specimen representation and its composition

Specimen representation	Weight in grams
Sample 1	4
Sample 2	5
Sample 3	6

Table 3 Experimental factors and their levels

Symbol	Parameter	Unit	Parameters		
			Level 1	Level 2	Level 3
A	Load	N	5 N	10 N	15 N
B	Sliding velocity	m/s	1.5	2	2.5
C	Weight in grams	–	4	5	6

4 Results and Discussions

4.1 RSM Modeling

The model for specific wear rate (**SWR**) and co-efficient of friction (**CoF**) was developed from the regression coefficient and represented in Eqs. 2–7 as follows

$$\begin{aligned}\textbf{Specific Wear rate at 4 gram} = {} & 30.96999 - 2.03471\text{E}^{-003} * \text{Load} \\ & - 39.87146\text{E}^{-003} * \text{Sliding Velocity} \\ & + 3.19703\text{E}^{-004} * \text{Load} * \text{Sliding Velocity} \\ & + 0.25724\text{E}^{-005} * \text{Load}^2 + 6.77976\text{E}^{-004} \\ & * \text{Sliding Velocity}^2\end{aligned} \quad (2)$$

$$\begin{aligned}\textbf{Specific Wear rate at 5 gram} = {} & 66.08371\text{E}^{-003} - 8.63364\text{E}^{-003} \\ & * \text{Load} - 42.88157\text{E}^{-003} * \text{Sliding Velocity} \\ & + 3.19703\text{E}^{-004} * \text{Load} * \text{Sliding Velocity} \\ & + 0.25724\text{E}^{-005} * \text{Load}^2 + 6.77976\text{E}^{-004} \\ & * \text{Sliding Velocity}^2\end{aligned} \quad (3)$$

$$\begin{aligned}\textbf{Specific Wear rate at 6 gram} = {} & 28.75369 - 4.83166\text{E}^{-003} \\ & * \text{Load} - 33.06627\text{E}^{-003} * \text{Sliding Velocity} \\ & + 3.19703\text{E}^{-004} * \text{Load} * \text{Sliding Velocity} \\ & + 0.25724\text{E}^{-005} * \text{Load}^2 + 6.77976\text{E}^{-004} \\ & * \text{Sliding Velocity}^2\end{aligned} \quad (4)$$

$$\begin{aligned}\textbf{Coefficient of Friction at 0.3 wt\% TiO}_\mathbf{2} = {} & 2.74614 - 1.76861\text{E}^{-003} \\ & * \text{Load} - 2.35067 * \text{Sliding Velocity} \\ & + 8.14865\text{E}^{-003} * \text{Load} * \text{Sliding Velocity} \\ & + 4.30785\text{E}^{-005} * \text{Load}^2 + 0.53167 \\ & * \text{Sliding Velocity}^2\end{aligned} \quad (5)$$

$$\textbf{Coefficient of Friction at 0.5 wt\% TiO}_2 = 3.40871 - 0.13463 * \text{Load} - 2.07710 * \text{Sliding Velocity} + 8.14865\text{E}^{-003} * \text{Load} * \text{Sliding Velocity} + 4.30785\text{E}^{-005} * \text{Load}^2 + 0.53167 * \text{Sliding Velocity}^2 \quad (6)$$

$$\textbf{Coefficient of Friction at 0.7 wt\% TiO}_2 = 2.59231 - 8.59678\text{E}^{-003} * \text{Load} - 2.20192 * \text{Sliding Velocity} + 8.14865\text{E}^{-003} * \text{Load} * \text{Sliding Velocity} + 4.30785\text{E}^{-005} * \text{Load}^2 + 0.53167 * \text{Sliding Velocity}^2 \quad (7)$$

4.2 *Predicted Versus Actual*

From Eqs. (2)–(7), the predicted values are determined and compared with the experimental results as shown in Tables 4 and 5 (Fig. 1).

From Table 6, the developed models were used for multiple response optimizations by desirability function approach to obtain maximum strength coefficient and minimum strain hardening exponent. The Fisher's *F*-test (overall model significance), its probability *p*(*F*), and determination coefficient R^2 (used to measure the fitness of regression model) were also studied by ANOVA [15]. The desirability is found to be 0.968 as acceptable and quite significant based on Harrington's rating system [14]. The optimized specific wear rate 5.55482 E^{-6} mm^3/Nm and co-efficient of friction 0.23 for load 10.0 N and sliding velocity 1.62 m/s, when the TiO_2 weight percentage 5 (wt%) are selected from Table 6. Based on the optimization result, 0.3 wt% TiO_2 at sliding velocity 1.5 m/s with 15 N load was tested and the result represents specific wear rate as 0.00042431 E^{-9} m^3/Nm and co-efficient of friction as 0.649423; hence, the model is significant and optimized results are verified.

5 Conclusion

According to the ANOVA, the accuracy of developed model is confirmed by R^2 and *p*-value. The predicted and measured values are in complete agreement. Further, the optimum combination of process parameters to attain the minimizing specific wear rate and minimizing co-efficient of friction can be predicted by the statistically developed models.

Table 4 Predicted and actual values of SWR and CoF

Exp. No.	Process parameters levels			Actual	Predicted	Actual	Predicted
	Load (N)	Sliding velocity (m/s)	TiO_2 wt%	Specific wear rate E^{-6} mm^3/Nm	Specific wear rate E^{-6} mm^3/Nm	Coefficient of friction	Coefficient of friction
1	5	1.5	4	5.16	6.65	0.5	0.47
2	5	1.5	4	5.16	6.65	0.5	0.47
3	10	1.5	4	41.25	39.75	0.5	0.53
4	10	1.5	4	41.25	39.75	0.5	0.53
5	5	2.5	4	12.88	9.89	0.23	0.29
6	10	2.5	4	57.47	58.97	0.41	0.38
7	10	2.5	4	57.47	58.97	0.41	0.38
8	5	1.5	5	2.05	4.26	0.99	0.88
9	7	1.5	5	5.15	2.75	0.53	0.63
10	10	1.5	5	3.72	4.36	0.33	0.27
11	5	2.0	5	2.78	2.67	0.59	0.79
12	10	2.0	5	11.56	10.77	0.27	0.20
13	5	2.5	5	6.04	4.48	1.14	0.97
14	7	2.5	5	7.88	9.37	0.70	0.74
15	10	2.5	5	20.05	20.57	0.34	0.40
16	5	1.5	6	4.64	0.66	0.45	0.50
17	10	1.5	6	18.01	19.77	0.52	0.53
18	10	1.5	6	18.01	19.77	0.52	0.53
19	7	2.0	6	12.38	13.28	0.51	0.37
20	5	2.5	6	7.26	10.70	0.44	0.47
21	10	2.5	6	47.74	45.79	0.51	0.53
22	10	2.5	6	47.74	45.79	0.51	0.53

Table 5 ANOVA-based RSM models

Source	Specific wear rate (SWR)				
	SS	DF	MS	F-value	$p > F$
Model	7640.97	11	694.63	85.46	<0.0001
Residual	81.28	10	8.13		
Total	7722.25	21			
R^2	**0.9895**				
Source	Coefficient of friction (CoF)				
	SS	DF	MS	F-value	$p > F$
Model	0.77	11	0.070	5.03	<0.0083
Residual	0.14	10	0.014		
Total	0.90	21			
R^2	**0.8468**				

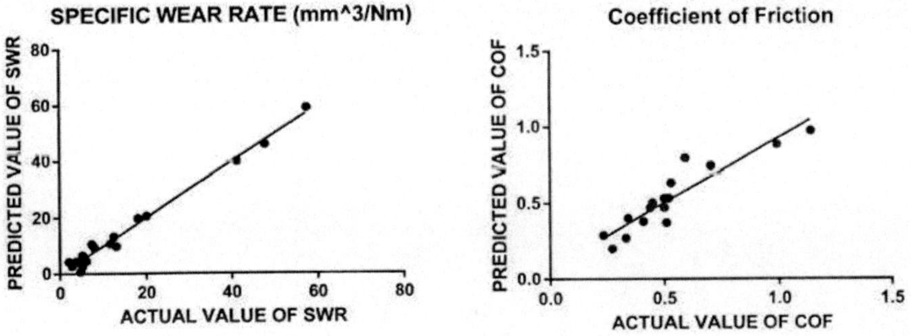

Fig. 1 Experimental (actual) and predicted values of SWR and CoF

Table 6 Best solution for optimization

Number	Load (N)	Sliding velocity (m/s)	TiO_2 wt%	Specific wear rate (E^{-6} mm^3/Nm)	Coefficient of friction	Desirability
1	10.00	1.62	5	5.55482	0.23	0.968
2	5.00	2.10	4	6.94407	0.232564	0.953
3	5.00	1.91	6	3.16099	0.361761	0.915

References

1. Gurcan AB, Bakar TN (1995) Wear behavior of AA 6061 aluminium alloy and its composites. Wear 188:185–191
2. Dandekar CR, Shin YC (2011) Effect of porosity on the interface behavior of an Al_2O_3 aluminium composite a molecular dynamics study 71: 350–356

3. Mahadavi S, Akhalghi F (2011) Effect of the graphite content on the tribological behavior of Al/Gr and Al/30 SiC/Gr composites processed by In Situ Powder Metallurgy (IPM) method. Tribol Lett 44(1): 1–12
4. Suresha, Sridhara PK (2010) Mater Des 31: 1804–1812
5. Padmavathi KR, Ramakrishnan R Int J Eng Technol 04: 3368–3373
6. Rousses Papagiannakis G (2014) Rev TecInguniv Zulia 2: 33–36
7. Naher S, Brabazon D, Looney L (2003) Simulation of the stir casting process. J Mater Process Technol 143–144:567–571
8. Singh M, Goyal K, Goyal DK (2015) Univ J Mech Eng 3(4): 142–146
9. Moslahshirazi S, Akhlaghiand F, Li DY (2016) Tribol Int 103: 620–628
10. Mina Bastwros MH, AmalEsavia MK, Wifi A (2013) Wear 307: 164–173
11. Zhang XQ, Peng YH, Li MQ, Wu SC, Ruan XY (2000) Study of workability limits of porous materials under different upsetting conditions by compressible rigid plastic finite element method. J Mater Eng Perform 9(2): 164–169
12. ChKaushik N, Rao RN (2016) Tribol Int 96: 184–190
13. Vetrivel M, Senthilvelan S (2012) J Arch Des Sci 6512
14. Harington J (1965) Ind Qual Control 21: 494–498
15. Fisher RA (1951) Design of experiments. Oliver & Boyd, Edinburgh, UK

Effect of Welding Speed on Advanced CMT-Welded AA 6061 Grade Aluminum Alloy Joints

S. T. Selvamani, S. Velmurugan, S. J. Hariharan, and K. Palanikumar

Abstract Joining of aluminum alloys by fusion welding is a challenging process due to its low melting temperature and the formation of solidification defects such as pores and warm holes in the weldment. The CMT welding process is an emerging fusion welding process to join the low melting temperature non-ferrous materials such as aluminum, copper, and silver thin sheets. So far, in this research work, lot of efforts has been taken to increase the AA 6061 grade aluminum alloy joint strength with help of advanced CMT welding process. The results showed that, the ultimate tensile strength of CMT-welded joint has been increased 30% than the other welding process and the microstructure analyses were confirmed that there is no defects in the joints.

Keywords CMT welding · AA 6061 alloy · Tensile strength

1 Introduction

Joining of non-ferrous metals by the fusion welding process is a challenging process due to its low melting temperature and the formation of solidification defects. The high heat generation is the phenomena to generate defects in the welded joints. Normally, the existing fusion welding such as TIG, MIG, and laser beam welding process will generate continuous heat during the welding and it causes the pores, cracks, and warm holes during solidification. This will lead to reduce the joints efficiency and strength. AA 6061 grade aluminum alloy finds broad use in welded structural members such as truck and marine frames, railroad cars, and pipelines [1, 2]. Mayur [3] investigated the mechanical properties of aluminum alloy AA-5083 by

S. T. Selvamani (✉) · S. Velmurugan
Vel Tech Multitech Engineering College, Chennai, Tamil Nadu, India
e-mail: selva_balu@rediffmail.com

S. J. Hariharan
Technical Module Slip Control System, Brakes India Pvt. Ltd, Padi, Chennai, Tamil Nadu, India

K. Palanikumar
Sri Sairam Institute of Technology, Chennai, Tamil Nadu, India

T. Rajmohan et al. (eds.), *Advances in Materials and Manufacturing Engineering*, Springer Proceedings in Materials 7, https://doi.org/10.1007/978-981-15-6267-9_77

tungsten inert gas welding using the filler wire AA 5356 grade and showed improved tensile strength with optimum hardness. Hussain [4] studied about TIG welded AA 6351 grade aluminum alloy and found that the effect of welding speed has more influence on the tensile strength of the joint. Also, the results showed that the depth of penetration decreases with increasing bevel height. Finally, at the welding speed of 0.6 cm/s, the maximum tensile strength was found but, the strength of joint is weaker than the base metal and the heat affected zone, strength increased with decreasing heat input rate. Sivashanmugam [5] has studied GTAW- and GMAW-welded AA-7075 grade aluminum alloy to enhance the mechanical and metallurgical properties and achieved the more impact strength of 6 J for GTAW and 4 J for GMAW, respectively. Lakshman [6] has studied the TIG welded AA-5083 grade aluminum alloy to understand the effect of welding parameters on tensile strength. Susil [7] has compared TIG- and FSW-welded AA-6106-T6 grade aluminum alloy to study the mechanical properties. Ghazvinloo et al. [8] analyzed the effect of arc welding current, voltage, and welding speed on mechanical properties of the joint of AA 6061 aluminum alloy and the results revealed that increase in heat input decreases the fatigue life whereas impact energy initially increased and then decreased. Several researchers compared TIG-welded joints with FSW-welded joints for aluminum alloys to choose the right process and it was found that FSW process is effective than TIG-welded joints for most of the aluminum alloys [9–12]. From the above literature survey, it is observed that a study of fusion welding process to increase the joints efficiency on aluminum alloy is scarce. Hence, an attempt was made in the work to find the effect of the welding speed on the tensile strength of the weld pool by CMT welding process to improve the mechanical and metallurgical properties. The macrostructures and microstructures were examined and the results are presented.

2 Experimental Work

High strength aluminium alloy AA 6061-T6 of size 150 mm × 70 mm × 3 mm sheets were cut into required size and the butt joints were fabricated with help of CMT welding process by different welding speeds. The specimen surfaces of the sheets were cleaned from the dirt, grease, and the other foreign material by using the dirt removers, cleaning agents. The edges of the plates were prepared by the grinding process to make them smooth before the fabrication of joints. The process is performed using the ER 4043 filler wire of diameter 1.2 mm and using 99.9% argon as shielding gas. The selected input variables for CMT welding process such as wire feed rate, welding current, voltage, and inert gas flow rate were kept constant with the help of trials test. The photograph of the fabricated CMT-welded joints are shown in Fig. 1b. ASTM—E8 standard were followed for conducting the tensile test [12]. Tensile test was carried out in 100 KN, electro-mechanical controlled universal testing machine (Make: FIE-BLUE STAR, India; Model: UNITEK-94100). The microstructure examinations were carried out in the cross sections of the welded joints. The specimen for examinations was initially cut by EDM, and progressively

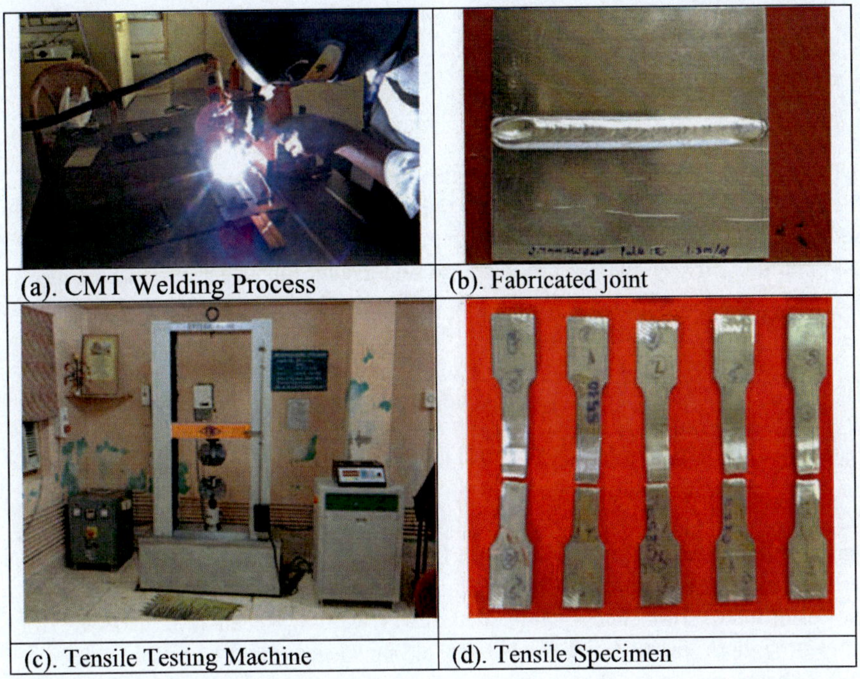

Fig. 1 Experimental work

grind using different grades of emery paper, then polished using diamond paste, and finally etched with modified keller reagent. Figure 1 shows the experimental work of CMT welding process and Table 1 shows the CMT welding process parameter used to fabricate the joints.

Table 1 CMT welding process parameter to fabricate the joints

Trial test	Welding speed (m/min)	Current/voltage/wire feed	Gas flow rate	Arc length correction	Pulse dynamic correction
Trial 1	0.25	(I)100 Amps (V)14 V (WF) 6.2 m/min	15L/Min	1.4	1.0
Trial 2	0.30				
Trial 3	0.35				
Trial 4	0.40				
Trial 5	0.45				

3 Results and Discussion

3.1 Tensile Test

The aluminum alloy AA 6061 grade joints were fabricated using advanced CMT welding process by varying its speed at the interval of 0.25–0.45 m/min and other predominating process parameters are kept constant with the help of trial examination. Totally, five joints are fabricated to understand the effect of welding speed on tensile strength of the joints. As per ASTM—E8 standard, the specimens are prepared and tensile tests are carried out on fabricated joints.

The result showed that the tensile strength reaches a maximum value of 241 MPa and then decreases to 180 MPa for the welding speed ranges from 0.25 to 0.45 m/min. From the result, it is understood that the reduction in tensile strength is caused by insufficient metal penetration at high welding speed. The optimum range in welding speed improves the metal penetration during the advanced CMT welding process, thereby improves tensile properties without metallurgical defects. The lower and higher welding speeds lead to poor bonding due to low heat generation and high heat generation, respectively. The optimum heat generation creates a quality weld. The result also shows that a moderate level of welding speed was necessary for better strength of joints. A lower level of welding speed cannot be recommended to produce better response. The maximum ultimate strength of the joint was fabricated by advanced CMT welding 241 MPa and the joint efficiency is obtained as 70% which is better than other fusion welding. The Fig. 2 shows the tensile strength of joints for various heat input and a sample of load versus displacement curve. It is observed that the advanced CMT welding process can produce a better joint than any other welding process with AA 6061 grade aluminum alloys without solidification defects.

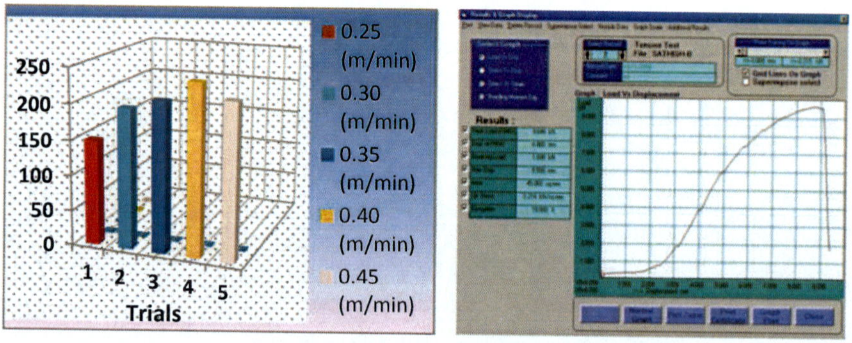

Fig. 2 Tensile strength of joints and load versus displacement curve

3.2 *Macrostructures and Microstructures*

The optical macrograph and microstructure of the advanced CMT-welded AA 6061 grade aluminium alloy joints obtained with an optimal setting of parameters is shown in Fig. 3, respectively. Defects such as voids, cracks, and unbonded zones are not observed in and around the weld area for weld fabricated by welding speed of 0.40 m/min. Compared to the parent metal, grain growth was visualized at the weld area (fusion zone) and the grain growth is due to the elevated temperature during CMT welding. Dendritic structure appeared at fusion zone was due to the faster heating and cooling of the weld metal. Figure 3b shows a clear visualization of dendritic spacing at weld site. At heat hazard region (HAZ), the dendritic structure gets reduced and relatively coarse grain structure was observed at HAZ (Fig. 3c). The results observed from the microstructure defines that the structure consists of interdendritic network of aluminum silicon eutectic in the matrix of aluminum solid solution.

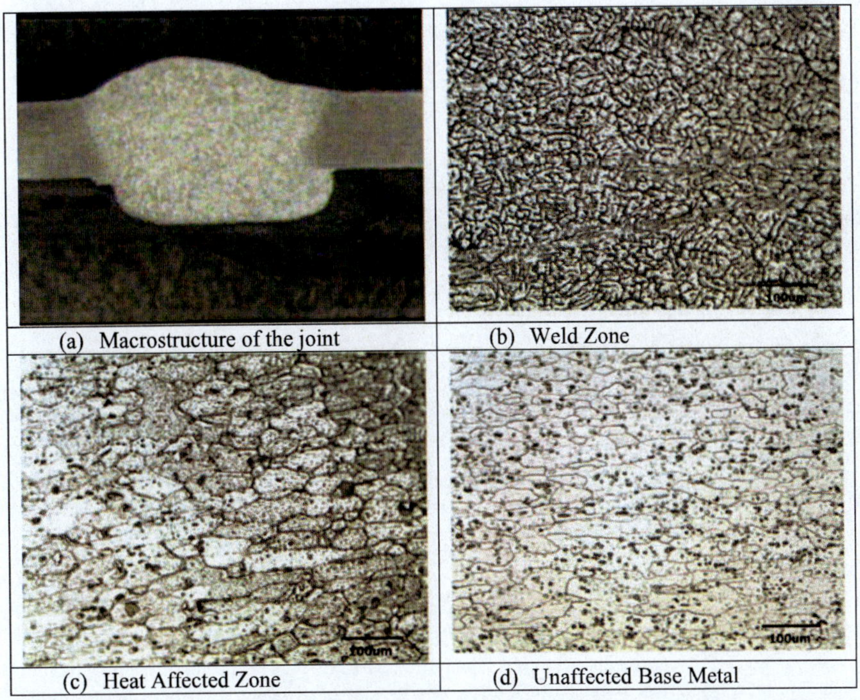

Fig. 3 Macrostructure and microstructure analyses of CMT-welded joints

4 Conclusions

The present work made a study on the effect of welding speed on tensile strength of the joints obtained by the CMT welding processes with AA 6061 grade aluminum alloy and the following conclusions can be drawn:

- The joint fabricated by the CMT welding process was found to possess better metallographic and tensile strength.
- The joint fabricated by a welding current of 100 A, inert gas flow rate of 15 lpm, welding speed of 0.40 m/min, and an arc voltage of 14 V exhibited superior tensile properties compared to other joints.
- The microstructure of the weld samples was observed that the structure consists of interdendritic network of aluminum silicon eutectic in the matrix of aluminum solid solution.
- Defects such as voids, cracks, and unbonded zones are not observed in and around the weld area for weld fabricated by welding speed of 0.40 m/min

Acknowledgements The authors are grateful for the DST-SERB for providing funding under EEQ (F. No. EEQ/2016/000114) and also wish to thank, The Chairman, Col. R. Rangarajan, Vel Tech Institutions for providing support to carry out the research work.

Reference

1. MujiburRahman ABM, Kumar S, Gerson AR (2010) The role of silicon in the corrosion of AA6061 aluminium alloy laser weldments. Corr Sci 52: 1969–1975
2. Mandal NR (2005) Aluminium welding, 2nd edn,Narosa Publishing House, Kharagpur, pp 15
3. Mayur S, Pavan KM, Sachin LS, Chandrashekar A, Kumar BSA (2013) Effect of welding current on the mechanical and structural properties of TIG welded aluminium alloy AA-5083 3: 431–438
4. Hussain AK, Lateef A, Javed M, Pramesh T (2010) Influence of welding speed on tensile strength of welded joint in TIG welding process 1(3): 518–527
5. Sivashanmugam (2010) Investigation of microstructure and mechanical properties of GTAW and GMAW joints on AA7075 aluminum alloy. Front Autom Mech Eng FAME
6. Singh L, Shah V, Singh NK (2013) Study the influence of TIG welding parameters on weld characteristics of 5083 aluminum alloy. Eng Sci. Innov Technol 2(5)
7. Kamat SK, Kumar A (2013) An experimental investigation of mechanical properties of Al 6106 T6 alloy joined by Friction Stir Welding and TIG welding. Innov Eng Technol 3(1):246–253
8. Ghazvinloo HR, Honarbakhsh-Raouf A, Shadfar N (2010) Effect of arc voltage, welding current and welding speed on fatigue life, impact energy and bead penetration of AA6061 joints produced by robotic MIG welding 3(2)
9. Singh G, Kang AS, Singh K, Singh J (2017) Experimental comparison of friction stir welding process and TIG welding process for 6082–T6 Aluminium alloy. Mater Today Proc 4:3590–3600
10. Baijusasidharan K, Narayanan P, Prakash RS (2014) Tensile and microstructural characterestics of dcsptig welded and friction stir welded AA 2219 aluminium alloy. Int J Des Manuf Technol 5:121–129

11. He Z, Peng Y, Yin Z, Lei X (2011) Comparison of FSW and TIG welded joints in Al-Mg-Mn-Sc-Zr alloy plates. Trans Nonferrous Metals Soc China 21:1685–1691
12. Cabello Munoz A, Ruckert G, Huneau B, Sauvage X, Marya S (2008) Comparison of TIG welded and friction stir welded Al–4.5Mg–0.26Sc alloy. J Mater Process Technol 197:337–343

Heat Transfer Analysis on Advanced CMT Welded Low Carbon Steel Joints

S. T. Selvamani, S. Velmurugan, and K. Palanikumar

Abstract The zinc-coated low carbon steels are widely used in the automotive sectors due to its strength and low cost. Welding of zinc-coated steel is commonly delaying the solidification process due to growth of the zinc vapor bubbles and high vapor pressure in the weld pool during the welding. To overcome the above problem, the advanced CMT welding process can be used successfully to join the zinc-coated steels without defects because of its controlled heat input. Therefore, in this research work the zinc-coated steel butt joints are fabricated by altering the CMT welding process parameters and the temperature distribution is measured using the K-type infrared thermometer during the process. Finally, with the help of ANSYS Workbench the simulation is carried out to predict the temperature distribution and compared with the experimental work to understand the thermal history. As the results show that the predicted maximum temperature is 2200 °C and the experimentally measured maximum temperature is 1850 °C, respectively.

Keywords Zinc-coated steel · Heat transfer · Simulation

1 Introduction

The applications of steel sheets in daily life are constantly expanding. The galvanized steel sheets have strong growth in fields where corrosion resistance and cost reduction by process omission are important. It offers superior economy in construction, automotive applications, electric appliances and electronic equipment. Also, the treated zinc-iron alloys are used for its outstanding corrosion resistance, excellent surface, and weldability. The advanced CMT welding is an emerging fusion welding process which is suitable to weld ferrous and nonferrous materials. Paulo Roberto discussed that the thickness increases, the bead width and the penetrated

S. T. Selvamani (✉) · S. Velmurugan
Vel Tech Multitech Engineering College, Chennai, Tamil Nadu, India
e-mail: selva_balu@rediffmail.com

K. Palanikumar
Sri Sairam Institute of Technology, Chennai, Tamil Nadu, India

T. Rajmohan et al. (eds.), *Advances in Materials and Manufacturing Engineering*, Springer Proceedings in Materials 7, https://doi.org/10.1007/978-981-15-6267-9_78

depth decrease, because of the increase in the heat dissipation. This phenomenon is also influenced by the behavior of the heat flow (2D or 3D), depending on the welding conditions [1]. Ahsan resulted that the maximum temperature of CMT-GMAW was measured within 1650° to 2700 °C with different heat input conditions [2]. Amin developed heat source model for better understanding of the weld pool behavior and design in CMT welding [3]. Zhang resulted that in the unsolidified weld pool, the solidification rate of the mushy zone increases with the time as a result of drop in latent heat and temperature gradient [4]. Dutta explained that for joining of high carbon steel by GTA welding resulted that the experimental temperature cycle is extreme near to the HAZ. But, owing to rapid heat loss through the movement of gas tungsten arc welding torch the attained temperature in other locations is less for the same time interval [5]. Attarha has reached the temperature about 550 °C in the point at 3 mm distance from the fusion line and found that the temperature was reduced in a nonlinear trend because of the variation of the material's thermal properties and the local heating of the welding torch temperature [6]. From the above literature survey, it can be understood that the heat transfer analysis for advanced CMT welding is very scanty. Therefore, in this research work a Gaussian heat source has been developed in the finite element analysis (FEA) to predict the temperature distribution for CMT butt welded of zinc-coated low carbon steel using ANSYS tool. To model the three-dimensional heat transfer process, the Solid 90 elements and a moving coordinate have been introduced. As a result, the temperature distributions of the weld at a various process parameter was obtained and compared with experimental values.

2 Experimental Work

The zinc-coated low carbon steel (150 × 75 × 3) in mm was taken as a parent metal, and the chemical compositions are shown in Table 1. The ER 70 S-6 filler wire with 1.2 mm in diameter was used to fabricate the butt joints. The advanced CMT welding machine 400TPSi which is used to fabricate the joint is shown in Fig. 1b. The process parameters were used to perform the experiment which is given in Table 2. By altering the process parameters, the induced temperature were measured with the help of the K-type high-temperature infrared thermometer at various distances from the weld and shown in Fig. 1c. The fabricated joints were machined by electrical discharge machining (EDM) to prepare the specimens for various destructive and nondestructive testing. A finite element model were designed and simulated with

Table 1 Chemical composition of parent metal

Grade/element	C	Mn	P	S	Si	Cu
AISI 1005	0.048	0.341	0.006	0.007	0.022	Nil
ER 70 S—6	0.08	1.53	0.009	0.01	0.88	0.18

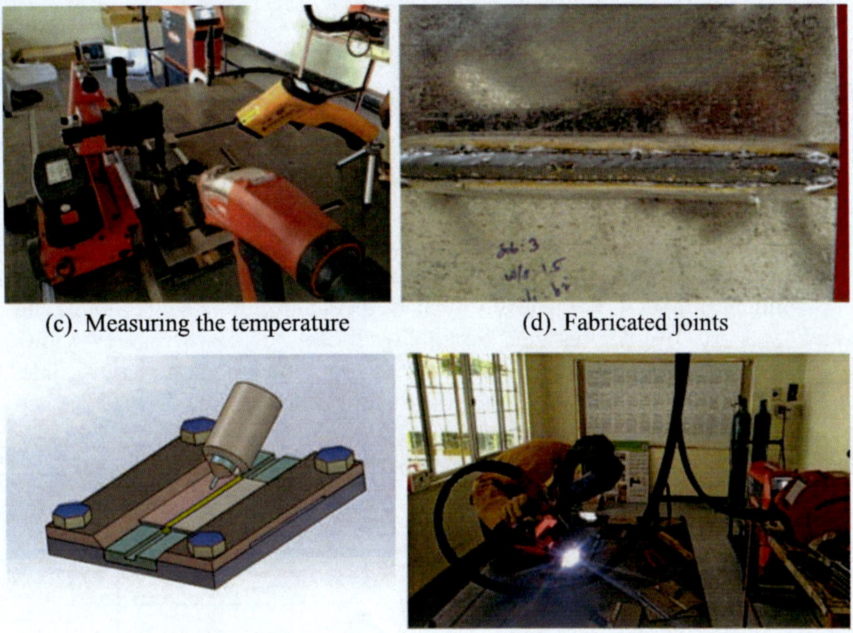

(c). Measuring the temperature

(d). Fabricated joints

(a). Schematic diagram of process

(b). Fabricating the Joints

Fig. 1 Experimental work

Table 2 CMT parameters used for fabricating the joints

S. No.	Current (A)	Voltage (V)	Wire feed rate (m/min)	Gas flow (L/Min)	Welding speed (mm/s)	Heat input (J/mm)
1	210	15.9	6.9	15	6	445
2	235	16.4	7.8	15	6	513
3	265	17.1	8.1	15	6	604

the ANSYS Workbench using Solid 90 elements. The Fig. 1 shows the details of experimental work.

3 Results and Discussion

In the present study, the heat distribution induced in zinc-coated sheet 3 mm in thickness during CMT welding process is experimentally measured and further validated using FEA. The volumetric and Gaussian heat source models are simulated in transient heat transfer analysis which is performed with three different heat inputs. From Fig. 4, the temperature distributions are found in transient condition for various heat input showed that, the temperature values are varied with heat concentration, but the radius of weld pool remains constant. The joints fabricated with different heat inputs by varying the process parameters, its microhardness are examined carefully to realize the effect of heat input. The obtained microhardness of the weld zone is 80% higher than that of base metal hardness at a maximum heat input of 604 J/mm, 45% higher in minimum heat input of 445 J/mm, and 70% higher in optimum heat input of 513 J/mm. Figure 2 shows the microhardness for optimum heat input joint. On the other hand, the joint fabricated by maximum and minimum heat inputs are showed pores, blowholes, cracks, and voids. But, the defects are free in the joint fabricated by optimum heat input. Comparatively, it showed better result than other joints and its micro- and macrostructures are shown in Fig. 3 for the evidence. Figure 4 shows the 3D temperature distribution contours for optimum heat input at various time. The experimentally measured maximum distribution of temperature is 1850 °C, and the corresponding simulated temperature is 2200 °C. The results predicted using ANSYS Workbench is in good agreement with measured values as shown in Fig. 5.

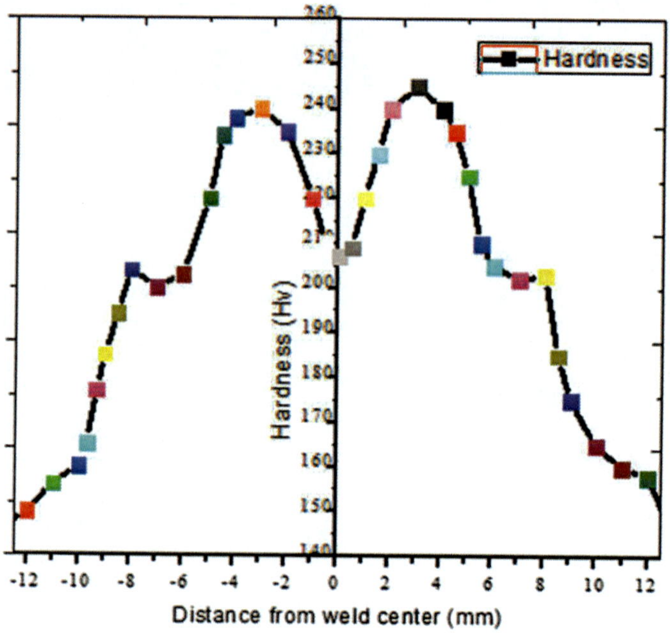

Fig. 2 The microhardness of optimum heat input joint

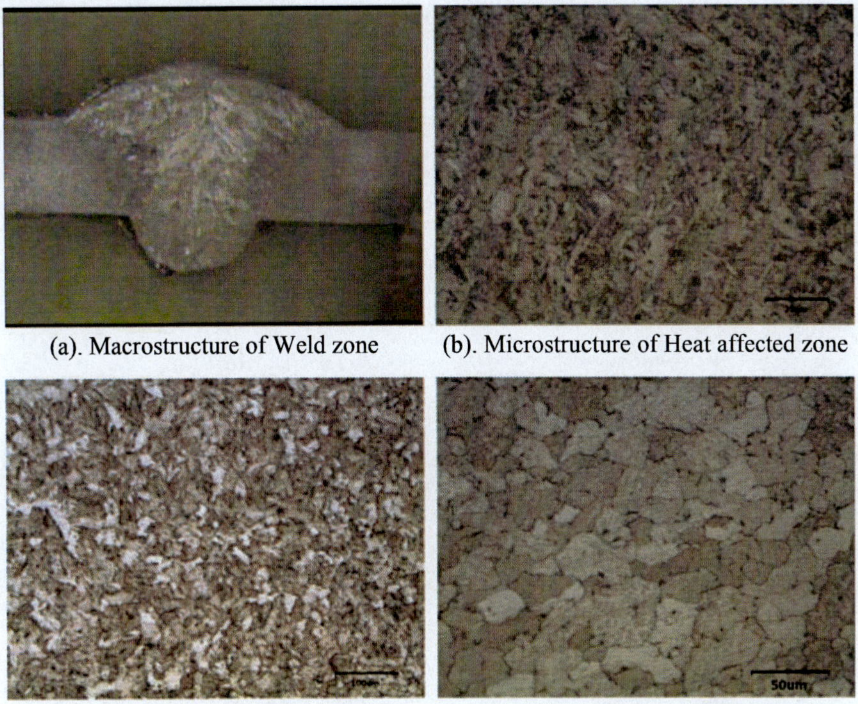

(a). Macrostructure of Weld zone (b). Microstructure of Heat affected zone

(c). Microstructure of Weld zone (d). Microstructure of Base metal

Fig. 3 Microstructure and macrostructure of fabricated joint at optimum heat input

4 Conclusions

The joints are successfully fabricated with different heat input, and its temperature distributions are measured.

- The results predicted using ANSYS Workbench is in good agreement with measured values
- The joint fabricated by optimum heat input showed 70% higher in weld zone and 66% higher hardness in heat affected zone than that of the base metal
- The joint fabricated by maximum and minimum heat input showed pores, blow holes, cracks, and voids. But, the defects are free in the joint fabricated by optimum heat input

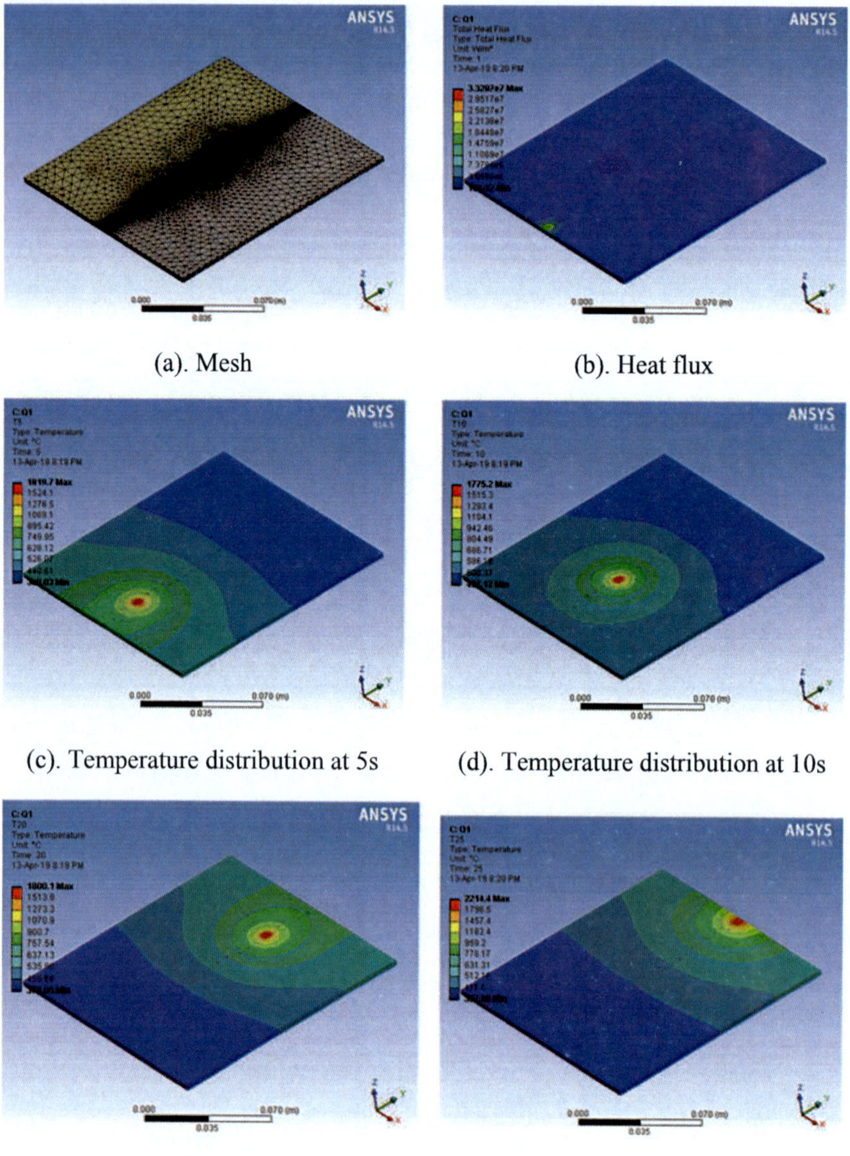

(a). Mesh

(b). Heat flux

(c). Temperature distribution at 5s

(d). Temperature distribution at 10s

(e). Temperature distribution at 20s

(f). Temperature distribution at 25s

Fig. 4 Temperature distribution contours

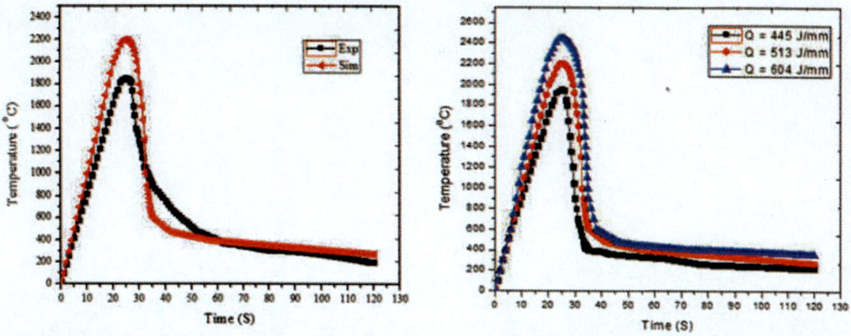

(a) Comparison of Experimental and Simulation

(b) Heat Transfer at various Heat input

Fig. 5 Graph of temperature distribution

Acknowledgements The authors are grateful for the DST-SERB for providing funding under EEQ (F. No. EEQ/2016/000114) and also wish to thank, The Chairman, Col. R. Rangarajan, Vel Tech Institutions, for providing support to carry out the research work.

References

1. Teixeira PRF, Araújo DB, Cunha LAB (2014) Study of the Gaussian distribution heat source model applied to numerical thermal simulations of TIG welding processes. Sci Eng J 23(1):115–122
2. Ahsan MDRU, Kim YR, Ashiri R, Cho YJ, Jeong C, Park YD (2016) Cold metal transfer (CMT) GMAW of zinc coated steel. Weld J 95:120–132
3. Azar AS (2015) A heat source model for cold metal transfer (CMT) welding. J Thermal Anal Calorimetry
4. Zhang W, Roy GG, Elmer JW, DebRoy T (2003) Modeling of heat transfer and fluid flow during gas tungsten arc spot welding of low carbon steel. J Appl Phys 93:5
5. Dutta J, Narendranath S (2014) Experimental and analytical investigation of thermal parameters developed in high carbon steel joints formed by GTA welding. J Mech Eng 44: 2
6. Attarha MJ, Sattari-Far I (2011) Study on welding temperature distribution in thin welded plates through experimental measurements and finite element simulation. J Mater Process Technol 211:688–694